Ultra-Large Aircraft,
1940–1970

To: Fred 11/02/19

Also by William Patrick Dean

The ATL-98 Carvair: A Comprehensive History of the Aircraft and All 21 Airframes
(McFarland, 2008; softcover 2015)

Ultra-Large Aircraft, 1940–1970

The Development of Guppy and Expanded Fuselage Transports

William Patrick Dean

McFarland & Company, Inc., Publishers
Jefferson, North Carolina

Library of Congress Cataloguing-in-Publication Data

Names: Dean, William Patrick, 1944– author.
Title: Ultra-large aircraft, 1940–1970 : the development of guppy and expanded fuselage transports / William Patrick Dean.
Description: Jefferson, North Carolina : McFarland & Company, Inc., Publishers, 2018. | Includes bibliographical references and index.
Identifiers: LCCN 2017057401 | ISBN 9781476665030 (softcover : acid free paper) ♾
Subjects: LCSH: Transport planes—United States—History—20th century.
Classification: LCC TL685.7 .D37 2018 | DDC 629.133/340423—dc23
LC record available at https://lccn.loc.gov/2017057401

British Library cataloguing data are available

ISBN (print) 978-1-4766-6503-0
ISBN (ebook) 978-1-4766-3015-1

Cover image of the Aero Spacelines B-377PG first engine startup, August 1962, Van Nuys, California courtesy Lloyd S. Jones

Printed in the United States of America

McFarland & Company, Inc., Publishers
Box 611, Jefferson, North Carolina 28640
www.mcfarlandpub.com

Table of Contents

Acknowledgments

Extreme outsize cargo aircraft slowly evolved from the logistical needs of World War II. Ultra-large postwar transports were not identified as a separate class of aircraft before 1962; however, after the development of the Aero Spacelines "guppy" the term has been retroactively applied. The "guppy" class of aircraft is the creation of John M. "Jack" Conroy, a bigger than life character with an incredible imagination. In 1961 Conroy set the standard for extreme high-volume heavy lift aircraft and wide-body airliners when he built the Aero Spacelines B-377PG Pregnant Guppy. I am privileged to have met members of his family and many fascinating individuals who knew and worked for Aero Spacelines.

I am especially grateful to Jack Conroy's youngest daughter, Angelee, and his widow, Milbrey, for their assistance. Angelee recalled stories of her Dad and very generously gave hours of her time reviewing documents, letters, press releases and photos from her personal collection obtained from Dottie Furman, who managed Aero Spacelines public relations. Ms. Conroy provided updates of new material throughout this project. We are most indebted to her for an introduction to legendary pilot Clay Lacy. This work would not have been complete without Captain Lacy's account and stories of his adventures with Jack Conroy, which he summed up by stating, "He was my best friend."

I am indebted to Tom Smothermon of Goleta, California, who has worked tirelessly reviewing original Aero Spacelines files, documents and photographs which were placed in his charge. He has provided a wealth of information to make this volume possible and has continually been available to provide photos and documents. The direction of this project changed after meeting Tom and reviewing the volume of data and photographs he placed at our disposal. He has contacted many sources on our behalf and introduced us to former Aero Spacelines staff, who related their stories.

We offer a special thanks to British aviation enthusiast Michael Zoeller for his artwork and support. French aviation historian Pierre Cogneville (custodian, F-BPPA) generously provided information and photos from his personal files. NASA Guppy flight crews were very helpful with technical data, recounting stories and providing their personal logs, files and photographs. Their professionalism in handling these incredible aircraft to achieve the NASA mission is most impressive.

We are indebted to the American Aviation Historical Society for access and use of photographs from the society's vast archives. Many of the photos provided by Tom Smothermon could have been taken by Aero Spacelines photographer Clyde Bourgeois, however, the photographer was not credited on any of them. The balance of photos and data reproduced came from the author's collection of 50 years. Every effort has been made to credit the original photographer or source. When we began collecting material many years ago on heavy lift aircraft this volume was not foreseen. Therefore many names and sources were not recorded. We appreciate the material provided by the following individuals. Apologies are extended to anyone who has been inadvertently not recognized or whose name does not appear in the following list:

Don Ahrens (USAF), Jim Babcock (Aero Union), Margy Bloom, Edward J. Bodenmiller (USAF), Andrew Boehly (Pima Air and Space Museum), Shy and Clyde Bourgeois (ASI), John Breitenbach (NASA), Jim Bugbee (FAA), Bob Burns (NASA), Bobby Cameron (USAF), Joel Carlisle (Conroy Aviation), Frank Clark (Clark Aviation, ASI, Hughes Corp.), Steve Clark, Pierre Cogneville, Chloe and Kayla Dean, Linda and Patrick Dennis, Keith Evans, Wayne Fagan (DHC), John Ferguson (ASI), Dottie Furman (Furman Associates), Charles "Chuck" Gillespie (NASA), Larry Glenn (NASA), Cole Goldberg, Hayden Hamilton (AAHS), Seth Hamond (ASI), Robert Herridge (ASI), Dan Hill (NASA), Paul Hooper, Ted Hunt (Air Systems), Glenn Iverson (World Airways), Lloyd Jones (ASI), Bob Kirby (Conroy Aviation), Tony Landis (Edwards Flight Test), Tim Lawton (SBA Airport Authority), Scott Marchand (Pima Air and Space Museum), James Meer, Olie Okpysh (Avionics West), Anvar Orgakov (USSR Space Program), Harry Parish (Ogden-Allied), Stu Prince (Conroy Aircraft), Robert Rivers (NASA), Tom Roberts (ASI-Tracor), Cal Taylor (USAF), Ken "Vern" Sanders (NASA), Don Stratman (ASI), Bruce Stratton (ASI), Richard Vandervord (Customs Southend), Nichlolas Veronico (NASA), Jerry Walterreit (USAF ret.), George Warner (Conroy Aviation), John Wegg (Airways), Fred Weir (ASI), Andrew Wiegert (Clay Lacy Jets), Nicholas Williams, and Stuart J. Williams (NASA).

Preface

This project began with the intent of presenting a comprehensive chronological review of the development and construction of expanded fuselage purpose built transport aircraft from 1940 to date. It soon became obvious each event was connected by unique characters and fascinating side stories which led to the revolutionary design change in cargo aircraft in 1961. Futurists in the 1930s predicted the development of flying giants but it was considered science fiction because the technology did not exist. Transport aircraft development floundered for decades with little sense of urgency. Large aircraft development prior to World War II was confined to flying boats. Landing gear which could support heavy airframes was not yet developed and there were few paved airfields capable of handling large aircraft. Early military planners viewed air transports primarily in terms of passenger travel. The needs of World War II suddenly pressed commercial aircraft on the drawing boards into military service. Most manufacturers did not have any commercial designs in progress when the war ended. They were left to exploit designs paid for by military contracts to develop new commercial aircraft.

We begin by relating the slow progress of large transport aircraft development sprinkled with both well-intended and dubious characters seeking financial gain. Conventional aircraft engineering theory limited fuselage diameter relative to wingspan and power plants. Aircraft were designed with a small fuselage and large wings. The focus was to transport multiple small cargo items instead of a single outsize item. The trend continued with postwar aircraft designs but there were never enough cargo items between two points to make them cost effective.

Boeing, Convair and Douglas attempted to interest airlines in passenger conversions of lumbering, fuel guzzling military designs. The problems encountered resulted in disastrous consequences, which have been related in detail. There were only a few large military aircraft in the prewar era. In 1948 Douglas Aircraft attempted to meet the need for larger military capacity by expanding the existing C-74 Globemaster airframe to produce the C-124 Globemaster. The conventional theory was maximum fuselage diameter had been reached.

You will be introduced to John M. "Jack" Conroy, former actor, pilot and swimming pool salesman who turned established aircraft design upside down in 1961. He defied accepted aerodynamic theory by believing there were no limits to the size an aircraft could be built. Conroy is not widely recognized for his achievements in American aviation. Yet he created a class of aircraft which revolutionized air transport. He brought a new era to aviation by building a giant aircraft with an interior the size of a gymnasium. The idea of transporting NASA rocket boosters inside an airplane actually sprang forth in the mind of aircraft broker Lee Mansdorf, who had acquired the existing fleet of unwanted Boeing B-377 Stratocruiser airliners. In an effort to recover his investment he challenged Jack Conroy with the idea of modifying the aircraft for NASA contract work.

The adventures and efforts of Conroy, an amazing bigger-than-life character who lived a lifetime before he was thirty years old, are presented in detail. He learned to fly as a young man while appearing in a dozen movies. He left Hollywood for work in Hawaii and was at Pearl Harbor on 07 December 1941. Within days he joined the Army Air Corps, becoming a B-17 captain in Europe. He survived being shot down over Germany on his 19th combat mission, was captured and became a POW. After the war he became a commercial pilot and joined the Air National Guard, where he set a coast to coast speed record in an F-86 jet. He was an accomplished pilot but a most unlikely choice to build aircraft. With no experience in aircraft manufacturing, he enthusiastically accepted Mansdorf's challenge. Respected aerodynamicists scoffed at him but he was determined to do the impossible. His Frankenstein approach of grafting parts of obsolete airframes together resulted in a new aircraft type.

The building of an ultra-large diameter aircraft did not suddenly occur in 1961. Our account begins with the slow development beginning in 1940. The record is spotted with well-intended individuals who opposed large aircraft development as well as those who attempted to skirt rigid aviation regulations.

Early aircraft development has been researched in detail. Much of the post–1961 data comes directly from Aero Spacelines company files. I was privileged to have been granted access to these records, which had been stored away for decades. They were rescued from a hangar with a leaking roof just before being lost forever. We relate stories of working for the most unpredictable Jack Conroy as told to us by Aero Spacelines staff.

The Pima Air Museum granted us access to a C-124, C-97 and the original Aero Spacelines B-377 Super Guppy. Angelee Conroy provided us access to her Dad's personal files, letters and photographs. Reviewing them with Ms. Conroy provided a perspective and insight into his personality, charm and incredible drive to succeed. Aero Spacelines and NASA flight crews provided photographs and personal accounts of flying these giants. To add clarity the author relies on his experience with Douglas C-124 and Boeing C-97 aircraft while serving in the United States Air Force. The combination of these sources and research of non-traditional aircraft development has been brought together with stories, photographs and personal accounts not previously presented.

These aviators, businessmen and engineers are occasionally recognized for their achievements but seldom as legends in aviation. They are the dreamers, tinkerers and risk takers. Their endeavors are connected by building on the successes of each other. The achievements of Jack Conroy, an American hero, are the primary focus of this work, but this is not a biography of one man. He is one of a few dedicated men who set new standards of aircraft development and made history.

This is more than a technological history. It is an evolving accounting of how obstacles were overcome to build revolutionary aircraft and the eventual need to transport NASA components, which put man on the moon. The original B-377 airframes were subject to catastrophic failures and reliability issues which could have made them a bad choice for modification. The author's approach is to relate aviation history as told by people who were there before it is lost in the recesses of time. The intent is to weave the back stories with the facts and technical data to connect the many personalities and produce an appreciation of the sacrifice of advancing an idea to fill a need. Years of research revealed actual incidents which read like a fictional thriller with heroes and villains. The technical breakthroughs are laced with political events, deal making and greed with a touch of mystery and intrigue. Some of the characters surface multiple times over decades.

It is not possible to relate every event in the development and service of multiple aircraft types created by different manufacturers and modified by even more. Selected incidents and side stories along with statistics present an overall view. This is not intended to be an all encompassing single reference source for ultra-large guppy aircraft. The amount of information reviewed for this work is too large to be included in a single volume. This is an inside look taken from actual documents. There are other printed works which present minute technical details specific to aircraft types. Every effort has been made to assure accuracy and completeness. Corrections and additional information are always welcome.

The narrative is divided into four parts with chapters focusing on individual aircraft types. We begin with ideas of the 1930s and follow the development to the revolutionary change in 1961 when traditional aircraft design theory was thrown out. We focus on the evolution of ultra-large heavy lift transports built between 1940 and 1970. Larger "Conroy Class" guppy aircraft have been built since 1970; however, they are only refined versions of the original 1961 Aero Spacelines design. The text is generally organized in chronological order. Our objective is to be informative and entertaining. There are overlaps and flashbacks because multiple events were taking place at the same time. This accounting is complemented with many unpublished photographs, appendices, fleet data, glossaries, notes, bibliography and index.

Introduction

The evolution of extremely large transport aircraft began in the 1930s with the idea of producing multi-engine long-range airliners. The outsize volumetric aircraft of today are the end result of a few forward thinking individuals who predicted a future need for large transports. Prior to World War II, designers and engineers struggled to create four-engine aircraft capable of lifting a payload of 6,000 pounds. Today two-engine aircraft can uplift over 100,000 pounds of cargo and four-engine transports can handle over 300,000 pounds. The few early commercial and military fuselage designs which preceded the introduction of the Aero Spacelines guppy class of transports of the 1960s could carry more weight but none could accommodate the volume.

The military in ancient times was not capable of moving men and materials on short notice. Soldiers marched to the battlefield with their weapons while others followed carrying equipment and struggling with carts and carriages. Ships were utilized to move heavy equipment over greater distance and remained the primary mode of transporting war materials for centuries. At the dawn of aviation, some consideration was given to large lighter than air flying machines but not for cargo. They were considered more as tenders for small fighter aircraft or bombing platforms. Commercial aviation was primarily passenger transportation and not considered for volume cargo transport. The only mode for outsized commercial cargo was by sea. Rail was practical up to a point but restricted by the size of tunnels or overpasses.

The need for a large four-engine commercial or military transport was not even considered urgent prior to World War II except for one visionary. William Patterson was at the helm of United Air Lines in 1934 when he proposed long-range passenger transports. The volumetric transports of today can be traced back to the development of the four-engine aircraft. Patterson, who was a banker, became interested in the future of air travel while putting together the merger of Pacific Air Transport and Boeing Air Transport to form United Airlines.

Patterson proposed a four-engine airliner with a 2,200-mile range to Donald Douglas. He was able to convince the powers at American, Eastern, Pan Am and TWA to each put up $100,000 toward the $3 million development cost of a new long-range transport. As a result, the tri-finned DC-4E was developed, and it first flew 07 June 1938. Eventually Pan Am and TWA withdrew and opted for the Boeing B-307 Stratoliner, which was essentially a B-17 with an expanded fuselage.

The single Douglas DC-4E was flown on United's routes for evaluation and promotion. It was large and generated public attention but proved to be too complex for the times and plagued with maintenance problems. Patterson was not defeated by its poor performance. United, American and Eastern went back to Douglas in 1939 for a smaller and less complex design, which became the aircraft known as the DC-4. The outbreak of World War II put new airliner production in jeopardy. The Douglas DC-4 was placed on low priority while combat aircraft were being developed. In short order the war effort required a long-range transport to keep supply lines open. The DC-4 design was already beginning commercial service with 61 on order. It was quickly commandeered and transitioned to the military C-54 transport. The aircraft Patterson envisioned served well but the war brought into focus the need for truly large transports. A high-volume ballooned out fuselage had not been imagined until the Army needed larger capacity to move war machines worldwide. As a result, Douglas proposed a scaled-up successor to the DC-4 transport designated the C-74 Globemaster (originally designated DC-7).

World War II bomber development produced the revolutionary Boeing B-29, which was long and sleek, becoming the most well-known American military bomber. However, the program was almost cancelled because of the nearly impossible engineering challenges. Postwar necessity and cost factors of developing newer aircraft to meet the Cold War mission prompted Boeing to upgrade the B-29 to the B-50 with larger engines. Wartime technology was also being transitioned into commercial aviation. To capitalize on government funded wartime research and development, the B-29/B-50 wings, tail and fuselage were utilized to produce the double deck XC-97 military transport. A commercial B-377 version was developed for the airlines; however, it was expensive to operate and plagued with problems from the beginning. After a relatively short life of ten years in commercial service, airlines were anxious to unload the B-377 Stratocruiser, which had been outclassed by jets.

Prior to the wartime advance in reciprocating (piston) power plants, large aircraft required so many small engines they were not practical or efficient. The question then became, did the development of larger engines provide the incentive to build larger aircraft or did the need for larger aircraft prompt the development of larger engines? Whatever the case, development of the reciprocating engine peaked by 1950, limiting the size of airframe design. Until the jet age, with exception of the DC-4, all large heavy lift military aircraft (C-74, C-124, B-36, XC-99, B-50, C-97, B-377) were powered by the reciprocating Pratt & Whitney R-4360 Wasp Major engine. All cargo was transported internally and limited by the size of the cargo doors. The proposals of the 1960s to transport extremely large cargo externally in a piggyback configuration were not practical. The aircraft in service were already at the maximum end of their operating range because of airframe stresses and engine limitations. Some were augmented with jets, and even the smaller DC-4s had been fitted with rocket assist takeoff (RATO) in high payload situations.

The World War II military need initiated the process of developing larger cargo aircraft. When America was challenged to explore outer space in the 1960s, NASA was faced with the task of transporting extremely large hardware over long distances. It was soon evident that putting a man on the moon required logistics which had never been done before. There was an immediate need for high-volume aircraft which could accommodate rocket boosters too large to be transported overland.

Transport aircraft are seldom profiled in great detail. Many volumes have been written about frontline military combat aircraft. Classic bombers like the B-29 are well documented, and hot fighter aircraft thrill the public at air shows and museums. The development and operation of transport aircraft is tucked away in the recesses of time quietly making history. Early transports were not sophisticated and developed from existing technology on very limited research and development budgets.

The majority of surplus Boeing B-377 commercial fleet was acquired by aircraft broker Lee Mansdorf at bargain prices with no specific plans for their use. The defining moment in modern ultra-large transport aircraft design came in 1961 when Mansdorf conceived the idea of a greatly expanding the fuselage of the Boeing 377 Stratocruiser for the express purpose of transporting rocket boosters for NASA. He discussed his ideas of a blimp like transport aircraft with John M. "Jack" Conroy. The design was so revolutionary it was considered impossible but Conroy refined the idea into a workable design. Much of the engineering data was eventually borrowed by major aircraft manufacturers to create today's wide-body commercial airliners. Conroy predicted a commercial need for giant transports; however, it was decades later before the market came to the conclusion heavy lift high-volume transports were necessary.

Initially this revolutionary class of extreme transports, which do not fit the normal definition of cargo aircraft, did not have a name. Today they are commonly known as "guppies." The name came from a rather condescending government official in Washington during a 1961 presentation by Jack Conroy to acquire funding to build the first expanded fuselage B-377 for NASA.

This somewhat ugly group of aircraft was created from surplus airframes built on rather tired World War II technology. Both Convair and Douglas Aircraft produced large transports prior to 1961. However, Jack Conroy is credited with taking aircraft building beyond all known limits of the time. Only four guppy types were built under his direction, three expanded B-377s by Aero Spacelines and one CL-44 by Conroy Aircraft, but his influence is seen in every outsize transport and wide-body aircraft built since 1962. The idea of an extremely large cargo aircraft began in the 1940s when Consolidated-Vultee designed the expanded fuselage XC-99, which was derived from the components of the cigar shaped Convair B-36 bomber. Douglas Aircraft produced the competing C-74, which technically is a guppy when considering Douglas openly stated it was a scaled-up derivative off the DC-4. Whether the C-74 is a guppy can be debated; however, it provided the base for the double deck C-124 Globemaster, which was a guppy created by doubling the size of the C-74 fuselage.

Heavy Lift Hybrid and Guppy Class aircraft are truly remarkable flying machines. They are a special class of transports with a cavernous cargo hold capable of hosting sporting events. They can accommodate just about anything which can be delivered to an airfield. These aircraft usually perform behind the scenes in the shadows of remote cargo ramps. They are like the stage hands in a great theater going about their work unnoticed. The public glorifies the star but the show could not go on without those doing the heavy lifting in the dark. With the exception of the few single examples produced by Aero Spacelines, special purpose aircraft capable of uplifting huge bulk were still considered science fiction in 1962. Only the most radical thinking aerospace engineers and aerodynamicists believed it would ever be practical to build aircraft with extreme diameter fuselages and operate them regularly.

The space program became a priority during the Cold War. The political need to beat the Soviet Union in space development was paramount. Werhner von Braun suggested it could be done by landing on the moon. The military had been transporting ICBMs by air with the Douglas C-124 and C-133. Initially these aircraft were employed by NASA but were soon outgrown as space components increased in size. NASA actually issued contracts to study the possibility of modifying existing aircraft to gigantic proportions. In 1960 the Marshall Space Flight Center at Huntsville, Alabama, let a contract to Douglas Aircraft to study the feasibility of external transport of rocket components on the back of existing aircraft. A plan was submitted to attach large pods on top of the turboprop Douglas C-133. The plan was reviewed on 26 September 1961 by Wernher von Braun and his team and promptly rejected. A practical external transport application did not become operational until 1976 with the Boeing 747 Shuttle Carrier Aircraft (SCA).

Cold War transports served their missions but the space program required lift capacity greater than any previous military operation. To meet the challenge aircraft engineers evaluated existing designs which could be modified or had components applicable to building larger aircraft. By using existing technology and off the shelf parts, tooling costs could be kept in line and the production time greatly reduced. NASA needed an alternate transport from the time-consuming seagoing barges in order to speed up delivery of the Apollo boosters and Saturn S-IV rocket motors from California for testing. Major aircraft manufacturers were reviewing the problem, however, it takes years to develop and certify a new aircraft type. The NASA logistics budget did not have funds for developing transport aircraft. Jack Conroy circumvented the system and leapfrogged the major aircraft manufacturers by developing a volumetric transport from surplus aircraft using private funds on speculation. He believed he could finance a commercial aircraft venture from the profits of a NASA contract. The guppy design set the standard and is the most pronounced member of the heavy lift volumetric family of aircraft.

Today only the largest aircraft companies are capable of creating outsize aircraft. The expense is prohibitive because they are not needed in numbers which could offset the research and development cost. In the past military contracts paid for the development of new technology and the building of production facilities. Manufacturers attempt to capitalize on military aircraft development by transferring designs and technology to commercial transports, which can be produced in numbers. Aero Spacelines was able to produce a few special purpose aircraft by using existing airframe parts and engines. A similar example exists today with the Airbus Beluga being produced from existing A300 engineered components or the Boeing 747LCF built by expanding the fuselage of existing B-747 aircraft.

Whether military or commercial, it is all about logistics. The review of these unique aircraft stirs the imagination of what is to come. The most recent Boeing 747 LCF is just the next step in the development of volumetric lift. The number of individuals who set the pace in modern cargo aircraft development is quite small. Multi-engine long-range aircraft evolved over decades before the Aero Spacelines B-377PG and subsequent models forever changed outsize cargo transport. The development of the Convair XC-99, Douglas C-124, and Aviation Traders ATL-98 are all relative in creating the atmosphere which generated ultra-large aircraft conversions. The cast of characters in giant aircraft development covers the spectrum from bankers, engineers, aviation legends and entrepreneurs to mobsters and nefarious deal makers.

The evolution of outsize aircraft from 1930 to 1970 is considered "The Early Years." Although the last NASA operated Aero Spacelines Super Guppy Turbine was built in 1983, it was not a new design but an upgrade of the original 1965 Aero Spacelines Super Guppy. Jack Conroy left Aero Spacelines in 1967 and founded a new company to develop other aircraft and pursue internal and external transport designs. Each milestone in aviation became the inspiration for the next. Conroy's creative spirit is found throughout the evolution of guppy aircraft. After 55 years NASA continues to operate a Conroy designed Super Guppy aircraft in 2017.

1

Convair B-36 Peacemaker

Building a Giant

The Consolidated Aircraft Corporation was formed on 29 May 1923 in East Greenwich, Rhode Island, by Major Reuben H. Fleet from the combined assets of Gallaudet Engineering and the Dayton Wright Airplane Company. The company moved to Buffalo, New York, in 1925 and finally to San Diego in 1935. Consolidated Aircraft Corporation merged with Vultee Aircraft Incorporated on 17 March 1943. Although it was referred to as Convair, the name was not official until 29 April 1954.[1] For the purpose of this review Convair is used in all references to Consolidated Aircraft or Consolidated-Vultee after March 1943.

The idea of enlarged fuselage aircraft existed in the 1930s but was limited by technology. The evolution of an expanded fuselage transport slowly evolved from the early 1940s while America was in the depths of World War II. Military planners believed Europe could fall if strike aircraft could not be positioned a great distance from hostilities. The War Department issued a request to aircraft manufacturers for proposals on long-range bombers capable of striking Europe from bases in America. Designers at Boeing and Douglas were skeptical of it being possible. Conversely, engineers at Consolidated Aircraft believed it could be done and accepted the challenge.

The development of a large bomber produced the conditions for an outsized cargo transport as a support aircraft. The resulting Consolidated XC-99 transport in all probability was never identified as a guppy type aircraft during its operational life. It has a volumetric fuselage blended with the wing, engines, landing gear and empennage components from an existing aircraft design. By using multiple components from the existing B-36 design, the XC-99 is technically the first American guppy cargo aircraft. Building an outsize fuselage aircraft using existing airframe components was a new concept. Aircraft manufacturers prior to World War II did incorporate wings or subassemblies from other models to reduce design cost. A good example is Boeing adapting the wings from the XB-15 bomber to create the commercial B-314 Clipper seaplane or B-17 wings on the B-307 airliner.

There are some interesting size comparisons between the XC-99 of the 1940s and the Mini-Guppy of 1967. The XC-99 fuselage exterior was 19 feet, 3 inches high with two decks and 13 feet, 5 inches wide. The Aero Spacelines (ASI) B-377MG Mini Guppy measured 15.5 feet high from the 13-foot wide cargo floor and 18 feet, 2 inches wide at the midpoint height of the fuselage. The Mini was a smaller aircraft but able to accept much larger cargo.

The "guppy" identity for outsize transport aircraft came sometime in 1961. During Jack Conroy's presentation to NASA in Washington for an expanded fuselage B-377 transport, a government bureaucrat exclaimed this thing looks like a "pregnant guppy." Despite the cool reception Conroy's design of a modified Boeing B-377 aircraft became the standard for all expanded volumetric airframes from the 1960s forward. All of the recognized examples of purpose built volumetric aircraft were built for the military, NASA or the aerospace industry except the commercial ATL-98 Carvair. The practice of expanding an existing fuselage to a transport was initiated by Boeing by adding an upper deck to the B-29 creating the C-97. Convair took it a step further by expanding the B-36 to the giant XC-99.

Development of the XC-99 is a result of several factors. The urgency for rapid military response increased the need to move men and materials over great distance on short notice. Secondly, Consolidated was attempting to reduce the cost of designing a separate transport for B-36 support. The aircraft manufacturer was on the verge of being granted a major bomber contract. Designers could utilize basic components already engineered to meet the military transport request. The most lucrative factor was capitalizing on technology already developed for the military. Developing a commercial airliner from a military design could produce revenue long after the government bomber contract ended. The giant cargo version went through a series of design changes from the original artist's concept. The numerous changes and upgrades to the XC-99 coincided with the B-36 development and were directed at obtaining an order from the military for multiple cargo aircraft. The design changes were also incorporated into a commercial version in an effort to meet the requirements of Pan American World Airways.

The skeleton of the XB-36 clearly shows the wing spar with single tire main gear and empennage, which are the same as those used on the XC-99. The bomber fuselage with high wing configuration is only 12.5 feet in diameter. This is small when compared to the double deck midwing XC-99 fuselage, which has an outside height of 20.5 feet (Bill Norton collection of AAHS archives).

When the Army Air Force reluctantly ordered a single example of the XC-99 in 1942, it was obvious it would be a "one off" aircraft. This did not stop Consolidated from continuing to work on a larger commercial model, which more closely resembled a very fat B-36. An advance design with turboprop power was studied at the request of Pan Am. Finally, an all jet version was presented with the swept wings and tail utilizing the YB-60 bomber design. There is enough similarity in all the XC-99 designs to see the relation to the development of the B-36 Peacemaker.

The base for the XC-99 is unique in American military aviation. The design was conceived because of a presumed wartime need for a very long-range strategic bomber. In 1941 military planners believed England might fall to Nazi Germany and a massive U.S. based bombing campaign would be necessary. At the time, America did not have a long-range strategic bomber capable of flying 5,700 miles round trip and the concept of in-flight refueling was considered complete folly. In April 1941, the USAAC opened design competition for a super long-range strategic bomber. The technology for a bomber capable of carrying 10,000 pounds of bombs while flying at 45,000 feet with a 10,000-mile range did not exist and was generally considered impossible.

By August 1941 the requirements were reduced to an aircraft capable of flying at 40,000 feet with 10,000 pounds of bombs with a 4,000-mile range. Both Boeing and Douglas were not sure such an aircraft could be built with current technology. The management at Consolidated Aircraft was convinced its engineers could design such an aircraft as they pursued the project with optimism. Consolidated subsequently submitted a proposal for the Model 36 twin tail six-engine pusher type aircraft.

The U.S. government issued a contract on 15 November 1941 for two XB-36 aircraft. The war need for cargo aircraft produced a second contract, signed later in 1941, for a large transport aircraft designated the XC-99. It was based on using common components mated with an expanded fuselage to reduce cost. On 11 December 1942, Aircraft Production Contract number 4535 AC-34452 was awarded to Consolidated Aircraft Corporation "to construct a fleet of long-range bombers and cargo aircraft" for the Army Air Force.[2] Interestingly, the contract specifically noted "cargo aircraft." It also required the first two bombers to be completed before additional aircraft would be ordered.

The ability to transport bulk cargo and vehicles to support combat operations has always been a problem for military planners. As air operations became more sophisticated with the development of large bomber fleets, the need for spare engines and spare parts increases. The size and weight of replacement aircraft engines in the numbers needed to support a strategic bomber fleet dictated fuselage designs with greater capacity and huge cargo doors larger than anything previously built. The expense of developing a separate low production transport for military operational support was not considered cost effective. The solution was to develop a support aircraft for the B-36 operation with interchangeable parts utilizing the B-36 base airframe.

B-36 Peacemaker Contract

The XB-36 remained a low priority even after the military saw a need for a long-range strategic bomber and signed a production contract. The Consolidated B-24 and B-32 were the immediate need. Priorities changed in April 1943, when it became apparent China could not hold out much longer against the Japanese. The U.S. did not have any aircraft with the range required to bomb Japan. There were problems with the Boeing B-29 needed for the war in the Pacific. The requirement that the two XB-36 aircraft had to be completed before more were ordered was waived in May 1943. On 23 July, the Army issued a letter of intent to purchase 100 aircraft at a cost of $1,750,000 each. Ultimately 385 were produced.

A complete detailed accounting of B-36 development is quite an undertaking. There are several outstanding works with great detail. The objective here is to present a reasonable general description to become familiar with this mammoth Cold War bomber from which the XC-99 was derived. The B-36 by any standard was a giant aircraft. By comparison the XC-99 is even larger than the B-36 base airframe. The same wing and power plants are used on both aircraft sharing the same wingspan; however, because the XC-99 fuselage is wider with a mid-wing configuration it has slightly less wing area. The distance from inboard prop arc on the B-36 is 9 feet, 7 inches and the XC-99 is reduced to 8 feet, 8.5 inches.

Another difference is found in the tail section. The B-36 has more wing area but the reverse is true with the horizontal stabilizer. The XC-99 has 1048 square feet tail plane with a wider span of 76 feet, 9 inches, compared to the B-36 with 979 square feet at 73 feet, 5 inches wide. The same is true with the dimensions of the vertical fin of the XC-99, which has more surface area at 601 square feet compared to 481 square feet of the B-36. The top of the vertical fin of the XC-99 is over ten feet taller at 57 feet, 5 inches, compared to 46 feet, 9 inches, of the B-36. The comparisons demonstrate the XC-99 is an early volumetric giant or guppy built up from an existing airframe.

The B-36 was only active for approximately ten years and over half of the time was taken up with working out the problems and installing upgrades. It was a lumbering giant with only two things going for it: range and high altitude operation. With a max Allowable Takeoff Gross (ATOG) of 276,506 pounds and crew of 14 including a four man relief team, it could fly at 40,000 feet and carry 10,000 pounds of bombs over 10,000 miles. Even more impressive, it could carry a 72,000-pound bomb load 5,800 miles as opposed to the B-29 carrying 20,000 pounds 2,900 miles. The B-36 is 1.85 times heavier than the B-29, but can carry ten times the bomb load almost twice as far.[3]

Surface to air missiles were not yet developed to strike at high altitude and fighters could not climb to 40,000 feet. The B-36 pushed the structural technology envelope of the day because of its size and the many new electronic weapon systems incorporated into the design. The advancements were overshadowed by the Pratt & Whitney R-4360 engines, which were at the maximum limits of post–World War II technology. Operating them at high altitude presented an entire new set of problems never envisioned when they were designed. Despite the shortcomings the B-36 served the purpose of challenging Soviet aggression until the jet powered B-52s could become fully operational.

The cigar-shaped six-engine 162-foot XB-36 with conventional cockpit configuration dwarfs the C-87 chase plane. The military specified two bombers must be completed before more aircraft were ordered. The bomber diameter is small when compared to the double deck XC-99, which had nearly 50 percent more fuselage above the wing (NARA collection of AAHS archives).

In addition to the obvious mechanical and technical problems, the B-36 cost was a continual point of controversy. It was under constant scrutiny by Congress and survived six different moves to cancel the project during development. The Air Force and Navy were in a dispute on which branch of the military would control atomic weapon delivery systems and Congress was continually holding hearings on the B-36 being worth the expense.

When the first XB-36 was introduced Consolidated was not in the position to produce all the parts and systems required. The production cost were so great, Consolidated was unable to enlist subcontractors to tool up to build subassemblies for only two aircraft. Unless they could be assured of

New York designer Henry Dreyfuss was engaged to develop the raised greenhouse cockpit configuration for the YB-36 and all subsequent models including this B-36H. The cockpit of the B-36 was designed with a narrow pedestal and a large auxiliary engineer's panel to manage the six-engine (later 10-engine) aircraft. It was an excellent example of a heavy aircraft requiring more engines than is practical to maintain (Lockheed collection of AAHS).

additional contracts the cost could not be justified. This did not change until 1943 when the Army Air Force issued a letter of intent for 100 aircraft.[4]

The initial XB-36 was designed with conventional cockpit configuration with a smooth fuselage and streamlined nose. The design was revised in December 1941, with constant-diameter fuselage with a blunt nose more reminiscent of the Consolidated B-32 and a double tail of the B-24. The final XB-36 design was similar to the Boeing B-29 with greenhouse cockpit windows contoured with the aircraft skin to reduce drag. The pilot seating was also similar to the B-29 with no center pedestal and an open area through to the glass nose. The configuration was revised again to a raised greenhouse cockpit on the YB-36 and all subsequent production models. New York designer Henry Dreyfuss was engaged to revamp the design of the cockpit configuration. The final B-36 cockpit configuration had a very large separate engineer's panel. By comparison The XC-99 cockpit was quite different with a very wide pedestal with the engineer's seat directly behind. The majority of the engineer's gauges and controls were on the center console with a smaller panel to his right.

Many of the engineering concepts on the B-36 were revolutionary. It was understood the Pratt & Whitney R-4360 power plants available were at their maximum performance range. Drag had to be reduced to fly at the speeds and altitudes set forth in the specifications. The standard method of pop-riveted skin was not acceptable and even flush rivets created some drag. There was also the problem of reducing weight, which was addressed by using magnesium skin for a portion of the fuselage. Consolidated invested $3,000,000 in developing a new metal bonding technique for the magnesium skin. It proved very successful at more than thirteen times stronger than conventional riveting. The method was used on nearly a third of the B-36 skin.

The size of the airframe presented problems with the 1,500 psi hydraulic system, which was standard on all aircraft of the time. The B-36 required additional accumulators, pumps and servos. Weight reduction was paramount, prompting Consolidated to develop a 3,000 psi system eliminating the need for additional systems and components. The new system was so successful it set the industry standard, which is still in use today. It is also the first aircraft to use 3-phase, 400-cycle AC electrical current. AC motors weigh only 25 percent of DC motors, creating a tremendous weight reduction in electrical components.

The early B-36 was designed with eight fuel tanks. This is unique when compared to smaller aircraft with more tanks. Dividing the space into additional tanks would have required more walls, baffles and fuel pumps. The record is not clear in noting the fuel tank configuration was directly related to a weight saving measure. Out of the total production of 385 aircraft, 289 were built with eight fuel tanks. Sixty-two B-36B models were fitted with only six tanks, and the 33 B-36J models were fitted with 10 fuel tanks. The tanks were huge, with the eight tank total capacity of 21,116 gallons and a volume 17,724 cubic feet. This huge fuel capacity gave it the range required for a global bomber.

The original Model 36 double tail design, which was similar in appearance to the smaller Consolidated B-24, presented many challenges for the engineers. The horizontal stabilizer was 67 feet wide (68.4 feet in some drawings) with 24-foot-high rudder end plates. There was concern from the beginning that operational stresses could cause a structural failure resulting in one or both sides breaking off. A double tail of these dimensions did not become reality on any large operational transport until the Russian AN-225 was conceived in 1980, first flying in 1988.

Consolidated engineers proposed a design change from a twin tail to a stronger single 47-foot vertical stabilizer, which satisfied two concerns by reducing drag and increasing stability. Even more beneficial, it reduced airframe weight by 3,850 pounds. The contract with the Army was amended in October of 1943 to accept the design change of a single vertical stabilizer on the XB-36 and YB-36. This change was also applied to the XC-99 tail configuration.

B-36 and XC-99 Simultaneous Development

Military planners were acutely aware in 1940 that no war can move any faster than the speed which men and materials can be assembled at the front lines. The War Department indicated an increasing need for a transport plane large enough to move war materials in volume to Britain. Allied merchant shipping was taking heavy losses in the Atlantic from the Germans. The War Department issued requirements for large transport aircraft which could move large quantities of materials by air avoiding German U-boats.

The Army requested design proposals from major aircraft manufacturers for a super long-range strategic bomber in April 1941. Convair engineering believed the company could capitalize on the War Department request for a long-range transport by adapting the B-36 design. Other manufacturers also submitted proposals, including Howard Hughes with his giant Hercules HK-1 flying boat. In mid–1941, Convair presented the Model 37, a super transport design mating an enlarged fuselage to wings, tail and landing gear of the XB-36 bomber. The war department reacted favorably to the design by agreeing such an aircraft could fill the need if it could be built.

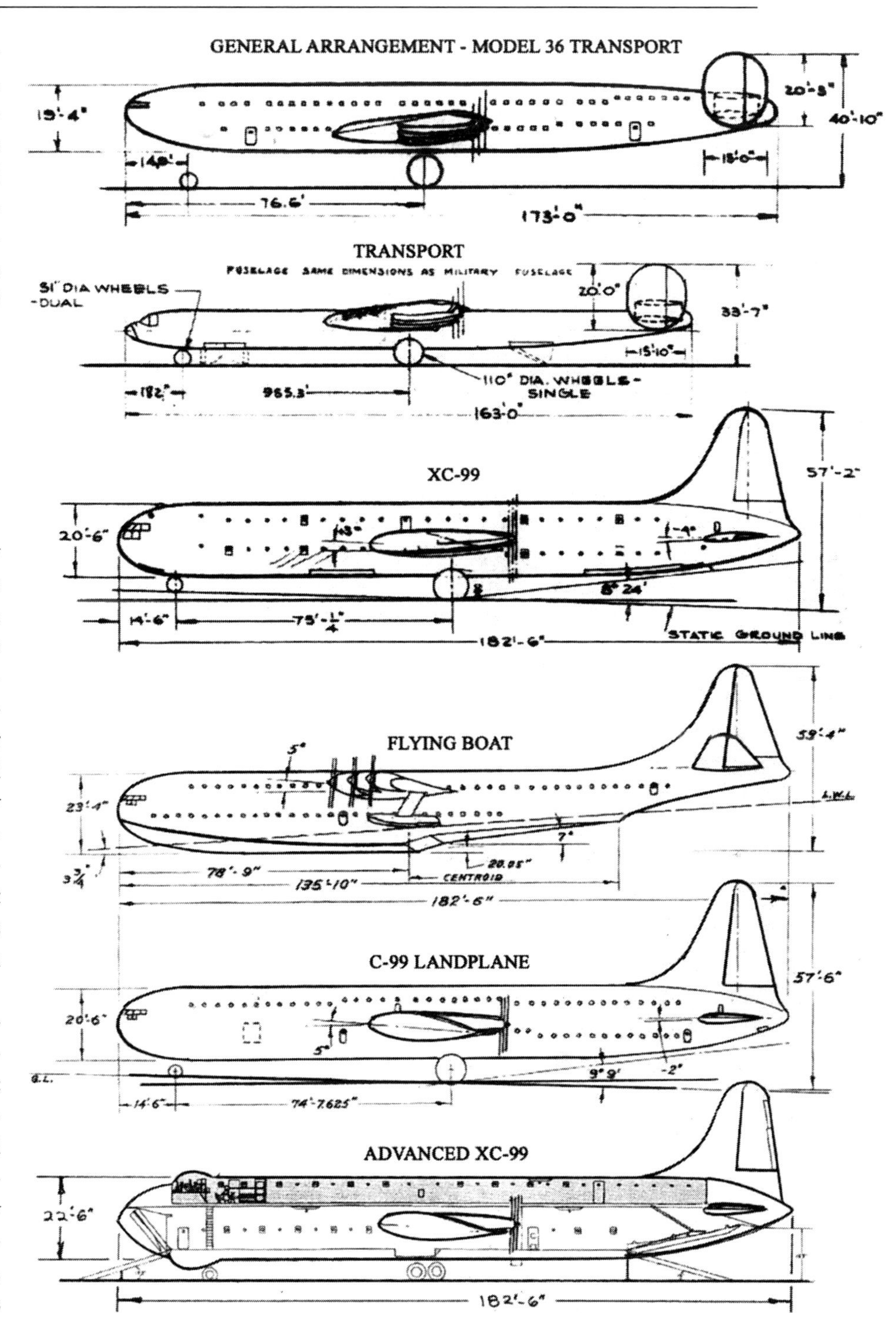

Progressive drawings show the development of the XC-99. In May 1943 the proposed Convair Model 36 transport had an enlarged fuselage with a twin tail. By August the fuselage size was similar to the XB-36 with the same tail. By 1944 the design was very close to the aircraft actually produced with a conventional tail. In 1945 Pan Am became interested in a commercial version and requested studies on a flying boat. By November 1949 Convair proposed a refined and expanded version with a B-36 cockpit arrangement for the military. It was offered to Pan Am as the advanced C-99 airliner (author's collection).

Consolidated had multiple marketing objectives. The military needed the aircraft and would purchase them in large numbers if they could be built. Consolidated engineers believed the B-36 program could be expedited by testing the engines, landing gear and general aircraft systems on a transport version. If the B-36 were successful and ordered in large numbers Convair could also expect orders for a transport to support the Air Force bomber operation. The cargo version could be built faster because it does not require sophisticated electronics, armament, defensive weapons and engineering of bomb bays. Aircraft manufacturers are acutely aware military aircraft production is cut back when war ends. If the military XC-99 was successful, Consolidated would have a transport aircraft which could easily be modified for commercial airline use with the research and development paid for by the war effort.

In the beginning the Army would only commit funds for two bomber aircraft, the XB-36 and YB-36. The military's priority was focused on production of the B-24 and B-32 bombers for the war. The development of the B-36 was placed on low priority because there was not yet an immediate need. The Army Air Force signed an agreement for a single XC-99 transport on 31 December 1941. In 1942, the contract for building a large cargo aircraft was let for $4,600,000 with the stipulation the XB or YB-36 would fly three months before the cargo version.

Initially the outsized transport was based on a modified version of the Model 36 (XB-36) design. It was a more streamline single deck design with a 12-foot, 6-inch diameter fuselage and twin tails. As the need was determined to transport volume, it was revised to a 173-foot pressurized double deck fuselage with an elliptical cross section of 13 feet, 6 inches wide by 19 feet, 4 inches high. The twin tail fins, 230-foot wing, engines and landing gear would be used directly from the XB-36. It was projected to carry a cargo load of 68,000 pounds with a range of 3,880 miles with a crew of five. The range was reduced to 3,250 miles for transport of 400 troops or 100,000 pounds of cargo.[5]

The engineering and materials of the time were not considered sufficient to withstand stress forces which would be exerted on the twin vertical tail planes mounted so far away from the centerline of the fuselage. The decision was made to change to a single vertical fin on the XB-36, which would be more structurally sound and also reduce weight. The Design of the Model 37 (CV-37) XC-99 evolved further with the decision to use the same single vertical fin design from the Model 36 (XB-36).

The payload requirements requested by the military prompted another change in August 1942. The XC-99 planned fuselage length was increased to 182.5 feet, which is 20 feet, 4 inches longer than the XB-36. The cross section was increased to 14 feet, 3 inches wide by 20 feet, 6 inches high. The crew complement was set at eight (six in the cockpit) consisting of pilot, copilot, two flight engineers, navigator, radio operator and two loadmaster—scanners. Two scanners were to be stationed in the cargo compartment rear lower deck. Their job during flight operations was to observe the engines and landing gear and report any malfunctions or out of the ordinary conditions.

The B-36 project moved to Fort Worth, Texas, in 1942 along with the XC-99. The XB-36 fuselage mockup was laid out adjacent to the planned XC-99 mockup. The San Diego plant was still very busy building the B-24 for the war. By June 1943, a round production nose had been finalized for the XC-99 with contoured cockpit windows. The range was set at 1,720 miles with 100,000 pounds of cargo and maximum range of 8,100 miles with 10,000 pounds of cargo. The number of crewmembers was increased to ten.[6] The normal operating gross aircraft weight was estimated at 265,000 pounds. However, under extreme circumstances and provided there were favorable atmospheric conditions, the MAX ATOG was set at 295,000 pounds. This would allow a cargo load of 117,000 pounds and still maintain a 500-foot per minute rate of climb.[7]

Aviation experts and the press began speculating on Consolidated's new cargo aircraft shortly after it was revealed to the public in 1942. Land planes of this size were considered pure folly into the 1940s because the technology had not previously existed. Any proposed aircraft design of these dimensions was only considered possible as a flying boat. The demands of war accelerated aircraft innovation and production; however, public imagination was still limited. Engineers proposed burying the engines in the wings, resembling German or Russian configurations. The XB-36 and XC-99 pusher configuration with six-engines on the trailing edge of the wing was a radical departure from standard military aircraft designs.

Historically, variations of military aircraft eventually find their way into commercial aviation after hostilities end. No commercial aircraft had ever been produced with the engines and propellers on the rear of the wings. The press predicted this would be the first pusher commercial airliner because of America's exceptional ability for innovation. However, the XC-99 never became a trendsetting airliner. The only other aircraft which compares in demonstrating how impressive the XC-99 was in the 1940s is the giant Russian Antonov AN-225 built in 1980.

There is an interesting configuration comparison to the six-engine Russian AN-225. The XC-99 was originally designed with six engines on the front of the wing and a 59-foot wide twin tail. The twin tail was revised before production because technology of the 1940s was not sufficient to produce a unit capable of withstanding high side stress forces. The six engines were moved to the rear of the wing to reduce aerodynamic drag. The XC-99 with a 230-foot wingspan in the 1940s is impressive when considering it took 41 more years before the 290-foot wingspan Ukraine built AN-225 first flew.

Convair press releases proclaimed the XC-99 six engines would produce a combined 30,000 horsepower, which exceeded any aircraft ever flown before. The press speculated it would possibly be equipped with a pair of inline engines in each nacelle with a common shaft driving the propeller because the thin light wing could not accommodate the giant experimental Allison 3420. It was further stated radial engines were out of the question because the Pratt & Whitney R-4360 of more than 3,000 horsepower was still in early development. It was assumed air cooled pusher mounted radial engines would overheat on the ground because they could not be properly ventilated. In reality the press was partially correct because the rear mounted engines did suffer from cooling problems which were never completely resolved.

The aft section of the XC-99 lower deck is 9.5 feet high. Two airmen demonstrate scanner positions for verification of landing gear and engine checks. The ladder leads to the upper cargo deck, which is slightly wider with a 7.5-foot ceiling. The rear belly sliding cargo loading door is open while the aft clamshell doors are closed (Consolidated, author's collection).

Because it was being presented as a high tech giant, speculators concluded it could only be powered by a secret gas turbine of 5,000 horsepower, which the public was not aware, existed. In reality, there were no plans for inline or turboprop power because neither had yet been developed which could produce 5,000 horsepower. The idea was to produce a cargo aircraft to test the B-36 engine configuration. The XC-99 would have the same Pratt & Whitney R-4360 power plants already planned for the XB-36.

A contract to produce the XB-36 and YB-36 was signed shortly before an agreement for a single XC-99. The cargo version closely resembled the XB-36 in the original artist renderings. Once the design of the cargo version was finalized and the cockpit configuration of the YB-36 was changed, the XC-99 fuselage did not resemble the B-36. They still shared the same wings, landing gear and empennage. Contrary to what Convair had in mind the military intended the XC-99 to be a "one off" experimental aircraft for data collection for use in future aircraft development.

The completed XC-99 would be more than a double deck B-36. The fuselage was a variation of a figure eight configuration very similar to the Boeing Stratocruiser without the indentation at floor level. The comparison can be made because the B-377 Stratocruiser was a figure eight double deck Boeing B-29. The prototype XC-99 was also a figure eight with a volume of 16,000 cubic feet on two levels and a conventional cockpit window configuration. It could move the equivalent of ten railroad cars or 400 troops in a single lift. Unlike the first two Aero Spacelines B-377 guppy types which came later with an expanded fuselage built on top of an existing airframe, the XC-99 was a new fuselage sharing B-36 systems engineering, wings, engines, tail and landing gear.

The Model 37 twin deck fuselage was set at 20.5 feet high (9.5 feet lower and 7.5 feet upper deck) by 14.25 feet wide giving it the capability of handling some military vehicles. The flight deck was similar to the B-36 in layout, but wider and positioned more forward and fully contained within the fuselage. The standard windscreen design was similar to the Boeing C-97, which was the largest operational production transport fleet of the day.

When compared to the C-97 the proposed XC-99 could transport four times the cargo or troops. This was quite impressive when considering an empty weight of 135,232 pounds was only 40 percent heavier than the C-97 but it can lift four times the cargo. The operating dry weight was set at 265,000 pounds with a maximum gross wartime overload takeoff of 322,000 pounds.

The aircraft size was obvious by the number of windows. Military cargo aircraft generally have a window ratio of one to four compared to commercial airliners. The XC-99 has 82 windows (43 on left and 39 on right) with 12 emergency exits. A commercial version was proposed to Pan Am with 104 windows. Unlike airliners of today the windows were double-spaced. This kept it from having well over two hundred windows.

The XC-99 was equipped with two electric powered sliding cargo doors in the belly, one forward and one aft of the wing. In addition there was a tandem pair of clamshell doors in the extreme rear aft of rear sliding door. The series of doors in the aft belly could be opened to accommodate very long items. The sliding doors could be opened during flight to drop cargo. The upper and lower decks were fitted with a pair of 4,000 pound capacity (8,000 combined) overhead cranes on tracks running the length of the cargo hold. It is not clear which manufacture first designed the overhead crane. The Boeing C-97 was the first cargo aircraft to fly with a mobile crane system. Similar systems were installed on the Douglas C-74 and C-124. Cargo could be lifted from the ground to either deck then moved into position for transport. The floor height and ramp angle made it difficult but a pair of floor winches could be used on the lower deck to pull vehicles into the lower cargo compartment.

The idea of building very large aircraft as a secondary invasion craft was conceived in 1942. All previous military invasions relied on ships which were slow. The idea of aircraft which could be built to a size capable of moving large forces on short notice was viewed positively by the Army but with some skepticism. The war department advised Convair the project was low priority because the current need was to produce bombers for the war. In an effort to generate a large military contract Convair's marketing team proposed to the war department a plan to build a fleet of forty giant C-99 transports, which could uplift an entire army division in a single flight movement.

The decision was made in 1944 to move the XC-99 project back to California. The original twin tail Model 36 mockup remained in Texas. The revised single tail design was re-designated Model 37. Partial mockups of the fuselage and certain sections were built at San Diego. The XC-99 received a set of early B-36 wings with six fuel tanks similar to those produced for the XB and YB-36. The total fuel capacity of those wings was 21,262 gallons. There are three tanks in each wing with the gallon capacity of 2,269 outboard, 4,212 inboard and 4,150 in the center. The left and right wing systems were independent of each other and did not interconnect. The fuel tank supply to engines can cross feed within each wing. This is the same configuration of the XB-36 and YB-36. The majority of subsequent models had eight fuel tanks except the B-36B with six tanks and the B-36J with 10 tanks. The early six tank wings were shipped by railcar from the Convair plant in Fort Worth to San Diego.

Once the inner wings were installed and construction was fairly well along, the enormous airframe had to be moved outside because there was no building with a high enough

The XC-99 was so large it had to be moved outside to attach the wings and vertical fin, which was 10 feet taller than the B-36. Installation of the rudder had been completed and the engines were hung by 11 March 1947. The wings and tail could not be installed inside because there was not a building at San Diego large enough to house the aircraft. It would be six more months before the first flight (USAF, NARA collection of AAHS archives).

ceiling to complete the construction. The 57.5-foot vertical fin and rudder was 10 feet higher than the B-36 partly because of the larger diameter fuselage. The wingspan was the same as the B-36 at 230 feet; however, the XC-99 fuselage was 20 feet longer at 182.5 feet.

The XC-99 shared systems and assemblies from the B-36 to reduce engineering cost. This included the same wing, engines and single wheel main gear. The horizontal stabilizer was the same but had a greater span because of the wider XC-99 tail cone. The vertical stabilizer fin and rudder are the same; however, the XC-99 has an extended bottom, making it 10 feet, 4 inches taller. A shared system was created in order for the XC-99 to use off the shelf B-36 engines. It is called the "power egg" concept where the P & W R-4360-25 engines are packaged in a quick-change unit. Both aircraft use the 19-foot three-blade Curtis Electric propellers in a pusher configuration. The B-36 and XC-99 were originally fitted with round-tip propeller blades. These props remained on the XC-99 until 1951 when they were replaced with square-tip units. The B-36 continued with the round-tip propellers, which were slowly phased out. The XC-99 performance tables indicate a higher cruising speed of 250 miles per hour; however, the actual speed was limited to 220 because of propeller limitations.

Pan American Airways Super Clipper

As allied forces began to make progress in the war, Convair officials understood orders for all aircraft would be drastically reduced when hostilities ceased. The end of the war could impact the B-36 program. The possibility of a Cold War need was not fully understood, leaving speculation the B-36 could be cancelled. Even in peacetime the military requires large transports to supply peacekeeping forces around the world. A margin of preparedness must be maintained in the event hostilities were reignited. In addition there is always a need to maintain large transports for disaster relief.

The Army had not expressed any serious interest in ordering more than one XC-99 for evaluation. Because of the size it was considered a cargo handling test aircraft to determine reliability and operating cost. Military planners were not projecting a need for giant transports for possibly 20 years. War had made the world seem smaller and there was anticipation of postwar long-range commercial travel requiring new aircraft. Production of the commercial Boeing B-314 Clipper had ended in 1941, but Pan Am and other airlines were still thinking in terms of the lumbering flying boats. Considerable technology had been developed during the war. Convair planners believed postwar long-range commercial aircraft would be land based as opposed to the prewar flying boats. In an effort to capitalize on the XC-99 and benefit from technology gained in the B-36 program, Convair had been quietly working on designs for a postwar commercial airliner. During the war in 1943, a commercial version was pitched to several airlines as the next generation of luxury airliners with the style, comfort and prestige of the Clippers. Juan Trippe of Pan American Airways was the only airline executive to express serious interest. New York designer Henry Dreyfuss was engaged to create preliminary interior layout designs.

As the war continued Convair actively pursued the postwar airliner market. By 1944, the Boeing B-29 bomber was able to operate from the islands in the Pacific putting it in striking range of Japan. On 19 August a contract was signed with Convair for 100 B-36 aircraft at a cost of $160,000,000. This was the incentive Convair needed to press for a contract to build military transports while promoting a commercial version C-99. Pan American's annual report for 1944 featured a centerfold of a giant airliner and outlined it merits in the postwar era of air travel. The company's president envisioned a 400-passenger coast-to-coast airliner with ticket prices equal to bus fares.

Convair's chief engineer at San Diego, Ralph Bayless, along with Robert Hoover held extensive talks with Trippe during January 1945. Trippe believed the aircraft would allow Pan Am to offer $90 one-way tickets from San Francisco to Hawaii. Bayless and Hoover were overoptimistic in predicting the aircraft could be built to operate with low fares. Press releases touted the commercial C-99 as a flying double deck ship with upper deck lounges and private lower compartments. One side of the aisle would have large comfortable seats with compartments like a train car on the other. "The plane will have virtually all luxuries afforded by finest of ocean-going accommodations."[8]

In reality the XC-99 was always a one-of-a-kind military aircraft because of the operating expense. The Pratt and Whitney R-4360 engines consumed huge amounts of fuel. The fuel cost alone far exceeded what commercial airlines could afford. The engines were not practical for commercial aircraft. The four-engine Boeing B-377 would eventually prove this and the C-99 had six of the P & W R-4360s. However, the early proposals of a commercial C-99 version convinced Pan Am to sign a contract for the land version. Convair officials were hopeful this would shore up the costly development programs and produce multiple commercial C-99 aircraft.

Pan Am immediately began promoting the new super airliner. The airline placed an order in February 1945 for fifteen 204 passenger commercial versions, twelve for Atlantic service and three for the California to Hawaii route. It was dubbed Super Clipper with a proposed gross weight of 320,000 pounds. The C-99 was projected to have a 4,500-mile range with a maximum speed of 370 miles per hour at 20,000 feet with upgraded engines. Initially it would be

Pan American placed an order for 15 giant 204-passenger airliners in February 1945. The airline began an extensive advertising program with artist drawings of a giant aircraft and plush interiors with staterooms and lounges (Pan Am/author's collection).

powered by the P&W R-4360, but it was projected to be upgraded to more efficient turboprop engines once the technology became available.

Henry Dreyfuss designed a complete passenger interior for the commercial version with nine staterooms and 12 berths in addition to the standard cabin seating. The Super Clipper would cruise between 319 and 343 miles per hour flying between New York and London in just over nine hours. The C-99 was projected to carry 67 percent of the total postwar North Atlantic traffic of all ships by all nations.[9] Pan Am planners were still not thinking in terms of land-based transports. They continued to think in terms of Clipper flying boats and ocean liners of the prewar era, seeing huge growth in a market which did not exist.

While the military version was being built, Pan Am press releases gave the public the impression that mass commercial air transportation would be commonplace by carrying the average man at rates he could afford.[10] Operational test were going well for the XC-99 when Pan Am officials began to question the viability of the civilian land version as originally presented. During the same period Howard Hughes was building the giant Hercules HK-1 seaplane for the military. It was powered by eight of the Pratt & Whitney R-4360 engines. Rumors were circulated predicting if Hughes was successful there could be commercial applications after the war.

The Convair XC-99 was designed and built using prewar technology. The size alone made it slow and lumbering. The extreme weight was concentrated on the main landing gear limiting the aircraft to only a few airfields in the country. Pan Am marketing still believed the 1930s Clipper glamour would continue after the war. As negotiations progressed the airline requested a preliminary design study of a Convair C-99 flying boat version which closely resembled the Hughes HK-1 and postwar British built Saunders Roe SR.45 Princes of the 1950s.

The size dictated a different fuselage design than the prewar Boeing 314 flying boats. They were fitted with sponsons for lateral stability in the water. The XC-99 was so large that it would not be practical with the additional weight of sponsons. The proposed flying boat version only had wing floats for stability. The wide beam of 13 feet, 7 inches works in favor of a flying boat of this size because it makes it easy to get up on to the aquaplaning step. However, the large hull has the opposite effect on getting into the air. It takes an extreme amount of power to overcome the static tension of the water to become airborne. The design was the same length as the land based XC-99 transport.

In typical flying boat fashion, the wing was moved from mid-fuselage to a high wing configuration. It was positioned 10 feet forward with the engines moved to the front of the wing in tractor configuration. The upsweep of the boat bottom eliminated almost all of the lower deck behind the wing. Convair engineers considered this to be one more example of how the original Model 36 and 37 could be modified to become another XB-36 variant. However the amount of changes made it practically a brand new design using the XC-99 cockpit and wings.

Production of a C-99 flying boat presented multiple challenges with estimates it would not be put in service until the mid–1950s. If the project had gone forward it would have been competing with the British built Saunders-Roe SR.45, which proved to be a failure. The SR.45 was reviewed by the U.S. Navy as a test aircraft for nuclear power because it could support the reactor weight. Interestingly, the XB-36 was also converted to a NB-36H for development test of a nuclear powered aircraft.

Convair continued development on a new multi-wheel landing gear design. If successful the landing gear problem could be resolved for the XB-36 and XC-99 long before a flying boat could be built. Pan Am eventually came to the conclusion the Clipper days were long past. There was no longer a commercial need for giant flying boats. Convair continued

The completed XC-99 and XB-36 shared the single 110-inch main landing gear, which is visible with large gear doors. Because the aircraft weight was concentrated on the main gear, there were only a few airfields with runways thick enough for it to land. More airfields would have to be built or the landing gear redesigned. This was also a factor in Pan Am cancelling the order for 15 aircraft (AAHS archives).

to negotiate with the airline for a land based C-99 passenger model.

XB-36 Bomber Rollout

The first XB-36 (42-13570) bomber was rolled out on 08 September 1945, which was six years after the contract was signed and six days after the War in the Pacific ended. It would be another year before the XB-36 would fly and two more years before the XC-99 first flight. The original intent was for the XC-99 to fly first as a test platform for the B-36 engines, landing gear and general aircraft systems.

On 08 August 1946 Convair test pilots Beryl A. Erickson and Gus S. Green took the XB-36 up for the first time on a 37-minute flight. At the time the bomber was the largest and heaviest aircraft ever flown; however, it did not meet expectations. When designed it was estimated to have a cruising speed of 369 miles per hour. When completed the aircraft was 13,000 pounds heavier at 278,000 pounds. The increased weight reduced the service ceiling to 38,200 feet and the speed to 230 miles per hour.

During 1946–47, pilots at Convair logged 117 hours on 53 flights in the XB-36 and the Air Force flew an additional 160 hours of test flights. There were many problems with such a radical new heavy bomber design. The XB-36 test program was impacted when the aircraft was almost lost on a 26 March 1946 test flight.

Beryl Erickson and Gus Green along with flight engineer J.D. McEachern took off with six Convair test engineers, three AAF observes and two Curtiss-Wright techs on board. At 700 feet, just after takeoff, Erickson called for gear up and Gus Green moved the landing gear levers to the up position. As the gear was coming up, the hydraulic actuator on the extremely heavy 8,550 pound single wheel right main gear exploded. The gear was designed with the single 110-inch wheel. This was necessary in order to be thin enough to retract into the wing because there were no nacelle landing gear bays with the pusher engines. Additionally the technology had not progressed far enough for multiple wheel bogie type gear.

During the test flight, the sideways retracting gear swung back down with tremendous force, damaging the number four nacelle. Fuel lines were ruptured in addition to the already damaged hydraulic system. The right gear did not lock back down when it dropped and continued to swing in the airstream. Anticipating the gear would collapse on landing and rupture the fuel tanks on right side, Erickson decided to circle Fort Worth, Texas, to burn off fuel.

After six hours he ordered the other twelve men to bail out before he and Green attempted a landing. With no flaps, brakes or steering, they set the plane down expecting the worse. The gear did not collapse. Conversely to what was expected the aircraft drifted off the runway to the left coming to a stop in the dirt. The XB-36 survived this incident. There was no airframe structural failure. All damage was confined to the landing gear, sheet metal, and fuel lines. The aircraft was repaired and test flown but it was not until over a year later on 14 August 1947 when it was flown to 38,000 feet.

Flight test and static test were of the utmost importance to meet the promised performance. To test the limits of the airframe a production B-36A-1-CF 44-92004 was structurally tested to destruction at Wright-Patterson Air Force Base in August 1947. It was extremely important to determine the limitations of the airframe because nothing this size had ever been built and no data was available. Up until this time most airframe engineers did not believed it possible an aircraft of this size could be built strong enough to support its own weight and still get off the ground.

Landing Gear Development

The XB-36 experienced limitations and suffered malfunctions from the beginning because of the single wheel main gear. The XB-36, YB-36 and XC-99 were originally

A dual tandem configuration with eight 56-inch tires was developed and tested. This configuration gave excellent weight distribution, allowing the aircraft to land on any field. However, the gear could not be retracted without an extreme modification of the landing gear bays and the installation of large blister housings on top and bottom of the wing (NARA collection of AAHS archives).

fitted with a 110-inch single wheel main landing gear. The design concentrated so much weight on a single wheel that it cracked runways and taxiways. The tire alone weighed 1,475 pounds and carried a 225-pound inner tube. The huge brakes added another 735 pounds to the assembly.

The single wheel gear design was incorporated primarily because brakes had not been developed for multi-truck bogie type landing gear. Also, the pusher engine configuration did not provide nacelles large enough to house the retracted gear in the configuration found on more conventional aircraft. The main gear retracted sideways into the wing, leaving little room for multiple sets of wheels.

Major General Edward Powers realized as early as 1945 the small number of airfields which could handle the weight on a single wheel landing gear would limit the B-36 operation. The B-36 was a strategic bomber which required versatility in being able to operate from multiple locations. It was not practical from a defensive standpoint to operate a strategic bomber from only a few airfields. It left the nuclear bomber strength of the Air Force vulnerable. If the Soviets did attack they only had to strike a limited number of airfields to destroy the United States' ability to retaliate.

General Powers proposed rather than build special runways at a limited number of bases it would be more cost effective to redesign the landing gear for a better footprint. More evenly distributing the aircraft weight would allow the B-36 to operate from most military fields, giving versatility to the weapons system. The same reasoning applied to the XC-99 transport version, which was also built with the single wheel main gear, restricting it to only a small number of airfields. Restricting a military cargo transport to only a few fields defeated the purpose of building it. Military cargo aircraft are expected to support the mission wherever needed. A new multi-wheel landing gear configuration could insure worldwide operation.[11]

Convair engineers agreed with General Powers' assessment; however, the technology and strength of materials was just being developed, delaying the production of a new landing gear system. Eventually the brake problems were resolved, allowing the development of a bogie type landing gear with 56-inch tires. Engineers concerned with weight distribution and braking designed two different configurations. The first design with eight narrow 56-inch high tires mounted in dual tandem gave excellent weight distribution, but presented problems because the width of four tires was too wide to retract even into modified wheel wells. A blister was designed for the top of the wing to increase the wheel well and house the gear. It could only be so big or the drag would be unacceptable. The second design with four wider tires produced the equivalent amount of

Top: **The redesigned four-wheel single tandem configuration reduced the width by three feet with a blister modification added to the gear doors and the top of the wing for clearance when retracted. The new design was 2,600 pounds lighter than the original single wheel landing gear (AAHS archives).**

Bottom: **The Goodyear track landing gear system developed for heavy aircraft to land on unimproved fields was installed on the XB-36 and tested on 29 March 1950. The aircraft successfully took off and landed without damage, but the test proved the gear to be impractical and the project was abandoned (Lockheed collection of AAHS archives).**

tread on the ground and reduced the width of the first design by approximately three feet. The new landing gear was 2,600 pounds lighter than the original single wheel unit. The XB-36 was returned to the factory on 27 May 1948 for multiple modifications. The wings were modified to accept the four-wheel bogie type gear and a blister added to the top of the wing. A matching blister or bubble modification was also added to the landing gear doors to provide clearance. The blisters added a drag component; however it was considered tolerable because of the ability to operate to most any airfield. The same modification was made to the XC-99.

Convair experienced numerous engineering problems with the B-36 and sub-models from the beginning because the designs exceeded the technology of the day. In addition the XB-36 problems with the single wheel main gear, the aircraft was underpowered, prompting the several engine upgrades. To increase performance Convair proposed adding jet pods outboard of the piston engines in 1948. Eventually dual jet pods containing a pair of GE J47-11 engines were added to each wing beginning with the B-36D model.

The XB-36 was officially turned over to the Air Force for flight testing and ground training in 1948. Goodyear was developing a track type landing gear system for possible use on soft fields. It very closely resembled the tracks on a caterpillar tractor. The B-36 program was experiencing enough problems without adding more. Military planners decided to test the Goodyear system on the XB-36 because of associated problems with the large single wheel main gear. The track landing gear was being tested on multiple aircraft. Because the XB-36 was the heaviest aircraft of the time and had been restricted to operate from selected fields, the Army was considering all possibilities. The XB-36 was put into the test program for the track system along with a B-50 and C-82.

By testing the track landing gear on aircraft of varying weights a wide range of data could be compiled on the viability of the system. The V-belt system was designed to allow large heavy aircraft to land on unimproved runways. The gear looked like part of a caterpillar tractor and consisted of one-inch thick 16-inch-wide belts on a multi-wheel bogie gear. In an effort to reduce weight and maintain strength the system used magnesium wheels. However, it still weighed 5,000 pounds more than a conventional landing gear.

The system was first tested in 1950 by making taxi runs at Fort Worth. The moment of truth came on 29 March when test pilots Beryl Erickson and Doc Witchell took off on the track gear. The aircraft landed safely but the screeching sound was almost unbearable. It was concluded the system was not practical for an aircraft of the XB-36 weight and possibly any other aircraft.[12]

The XB-36 remained in the role of a test aircraft until retired on 30 January 1952. It was then modified to NB-36 configuration as part of the nuclear aircraft development program. The airframe was used for mockup purposes on the development of special parts and systems. Because of the numerous modifications and systems tested on it, the airframe had seen better days by 1957. Unfortunately, it was not preserved and met its end after being turned over to the fire department at Carswell Air Force Base at Fort Worth for training purposes.

2

XC-99 Behemoth

First Flight and Testing

Convair chief of flight test Russell R. Rogers (left) and project engineer Robert (Bob) Hoover review the XC-99 test program. The cockpit configuration is very wide with a conventional window configuration. The width of the pedestal console is quite different and over twice as wide as the XB-36 from which it was derived (USAF collection of AAHS archives).

While the B-36 was undergoing flight testing and upgrades the single XC-99 was being completed. The opinion at Convair was it would lead to the production of more aircraft. However, the army saw it as a test aircraft intended for evaluation of future transport development. On 24 November 1947, the XC-99 (43-52436) made the first flight from San Diego, commanded by Convair test pilot Russell R. Rogers with Beryl Erickson in the right seat. The flight engineers were Mel Clause and B.B. Gray. The huge aircraft with 12,000 pounds of test equipment onboard lumbered down the runway rotating at only 95 miles per hour. It climbed out and leveled of at a cruise of 195 miles per hour. The flight lasted for one hour before returning to Lindbergh Field in San Diego. The approach was long and slow landing at 120 miles per hour, which was just above the stall speed.

The XC-99 was restricted to only three airfields in the country because of the concentration of weight on the single wheel main landing gear. It had the same wing as the B-36, which was not thick enough for multi-wheel bogie type landing gear. The main landing gear with a single oversize 110-inch wheel had a combined total weight of 18,738 pounds. This huge single wheel system was created for the B-36 because the positioning of the pusher engines had been carried over to the XC-99. A new main landing gear had to be developed which would be thin enough to retract inside the wing. There had been no alternative when the landing gear was developed for the XB-36.

The XC-99 became more operationally functional in December 1948 when the new four wheel truck type main gear with improved brakes was installed, giving it the ability to land at any large airfield. The top of the wing and landing gear doors were modified to allow sufficient clearance for the gear to retract. The term thin wing is misleading because it was more than six feet thick at the wing root. However, it was thin relative to the size of the aircraft when considering all other conventional aircraft designs stow the gear in the nacelles. The installation of blisters on top of the wing resulted in additional drag which was calculated to have been offset by the 2,600 pound lighter gear.

The end result was a 300,000-plus pound transport which could land on the same runway as a 73,000-pound Douglas C-54 with no bigger footprint. It flew for the first time with the new landing gear on 24 January 1949, but not without incident. The landing gear doors malfunctioned on the right main, preventing it from retracting. After a safe landing and inspection, the problem was corrected with an adjustment of the gear door mechanism.

The XC-99 remained in a test program until 26 May 1949, when it was turned over to the Air Force 7th Bombardment Group at Carswell Air Force Base, Texas, for evaluation. The 7th was chosen because of operational experience with the B-36. On 09 June 1949, Captain Deane Curry of the 492nd Bomb Squadron made the first flight with an Air Force crew from Carswell. The giant transport was flown on a local circuit making six landings. During the remainder of the month Captain Curry made one night and four daylight test flights. On the final flight the aircraft suffered an engine failure resulting in an emergency landing at Kelly Air Force Base.

The XC-99 was known to be underpowered from the first test flight. The situation was compounded on hot summer Texas days when it struggled to get airborne. The decision was made to replace the R-4360-25 engines with more powerful 3,500 hp R-4360-41 units. Work began on 22 June 1949, on what was considered a simple engine swap. It was soon discovered the engine upgrade would require modifications to the wing structure. A new engine oil warning system was installed along with an upgraded fire extinguishing system. The fuel tanks were scheduled for a re-seal while it was in for heavy work.

Pilot Russell Rogers and XC-99 project engineer Robert Hoover discuss the aircraft problems on 16 March 1948, nine months before the upgrades and new landing gear were installed. The size of the aircraft is obvious. The enormous 20.5-foot double-deck fuselage is a foot larger than the upper section of the Aero Spacelines B-377 Pregnant Guppy, which was not built until 1962 (NARA collection of AAHS archives).

Commercial C-99 Turboprop

Pan Am continued to review a giant new Clipper Type to boost the prestige of the carriers' worldwide network but the piston powered C-99 was cost prohibitive. Once again the carrier inquired as to the possibility of Pratt & Whitney T-34 or Allison or T-56 turboprop power plants. After considerable interior design and advertising expense there was a need to justify the cost and save face. Pan Am had extensively promoted the passenger C-99 Clipper. The prevailing theory was turboprops consume less fuel and would possibly bring operational cost in line, making the six-engine giant cost effective. With turboprops being state of the art technology at the time, Pan Am marketing believed a turboprop airliner could put the carrier on the cutting edge of air travel. A review was conducted of the possibility of using the Allison turboprops but no serious design drawings were done.

Convair was dealing with fuel-thirsty underpowered reciprocating engines and design deficiencies on cargo loading. The single XC-99 had a floor line which was too high for straight in-ramp loading. A ramp system was proposed; however, it was too steep and considered impractical. Vehicle loading was restricted to the size of belly doors. The overhead crane system was an improvement over initial manual loading methods but still slow and labor intensive. The lack of drive on/off capability was similar to the problem the military experienced with the side loading Douglas C-74, which prompted expanding the fuselage design to the drive on frontloading Douglas C-124.

In an effort to save the XC-99 military and commercial program, engineers went back to the drawing board and created an advanced C-99 design. It had front and rear clamshell doors to improve loading capability. The advance version was submitted to the Air Force as an alternative to the Douglas and Boeing proposals. Convair marketing also believed a commercial version would appeal to the commercial airlines.

It was quite different from the original XC-99 with an even larger fuselage. The cross section was 15.5 feet wide and 27.5 feet high with 12 feet usable height on the lower deck and 6.5 feet on the top producing a volume of 21,715 cubic feet. It used the B-36 wings mated to a larger 182-foot long fuselage. Drawings of the advanced version appeared more like a very fat B-36 with the same raised greenhouse cockpit on top of the fuselage with an expanded lower half. The cockpit configuration was positioned rearward on the upper deck to accommodate the forward clamshell cargo doors. It featured a lower floor line with adjustable cargo ramps for both front and rear lower deck straight in loading. The upper deck and cockpit was pressurized for passenger comfort.

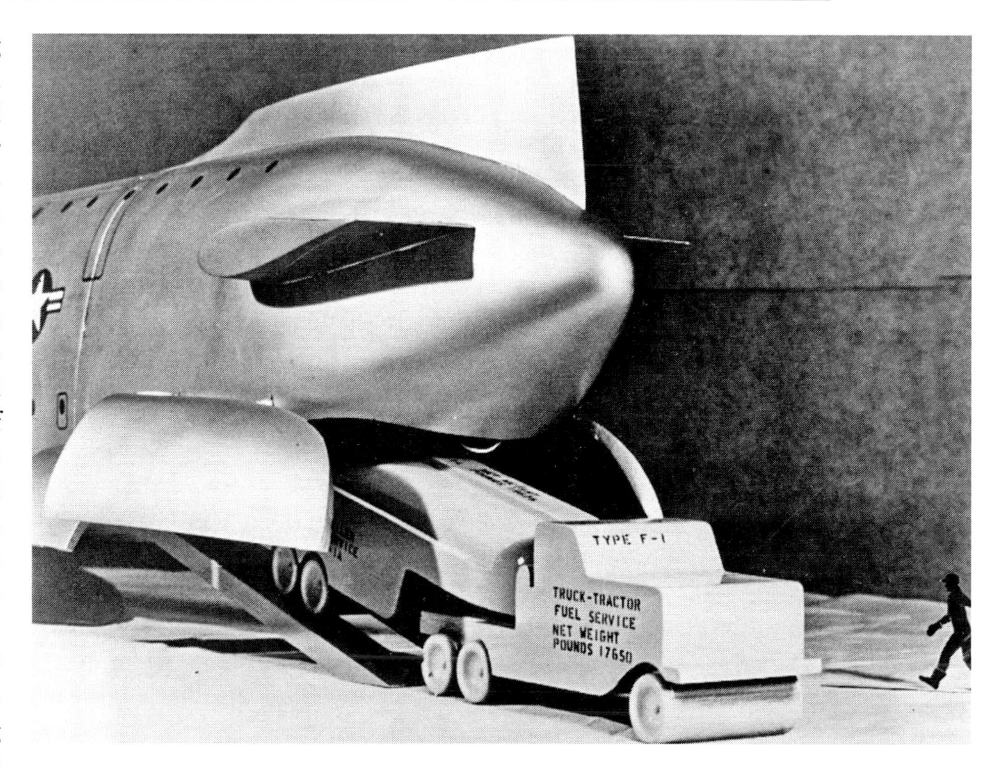

The 1949 advanced design C-99 featured a much lower floor line and rear clamshell doors, making drive on/off loading easier. The lower floor allowed for lower angle ramps for more practical loading of large vehicles. A ramp system was designed for the production XC-99; however, it proved to be too steep for heavy vehicles (Consolidated, author's collection).

The advanced C-99 was proposed with a B-36 raised greenhouse cockpit configuration, which was moved rearward for higher ceiling clearance in the forward cargo bay. Although closer to the ground with a lower floor line, the clamshell doors are similar to the Douglas C-124 Globemaster design with raised cockpit. The fuselage was set at 182 feet long and 27.5 feet high. The cargo capacity was an astounding 21,715 cubic feet, which exceeded any land-based plane ever produced (Consolidated, author's collection).

The nose loading doors were similar in design to the Douglas C-124. To provide a smooth continuous cargo floor the nose gear was moved outside and below the fuselage mounted in a blister pod similar to the configuration of the later ATL-98 Carvair conversion of the DC-4. In five years the already large 1942 Model 36 transport design with split tail had ballooned into a truly giant airframe. The landing gear design had progressed to the multi-wheel bogie type; however, Convair was still confronted with technical problems with strut strength and brakes.

Negotiations were renewed with Pan Am in November 1949. The design team headed by Fort Worth Chief Engineer J.W. Larson presented the completely redesigned advanced C-99. The presentation coincided with the heavy cargo version proposed to the military with drive on-off capability. The giant fuselage of the advance design further supports the premise of the XC-99 being a guppy. It used an expanded fuselage with B-36 greenhouse cockpit, wings power plants and tail. The proposed new C-99 production version was priced at $1,500,000, which was comparable to the cost of the Douglas C-124 at $1,600,000 and Boeing C-97 at $1,200,000, and yet they were much smaller four-engine aircraft.

XC-99 Operations

In March 1950 the XC-99 was scheduled for fuel tanks resealing. The XC-99 and the early B-36 models suffered from excessive fuel leaks. The aircraft were designed with a "wet-wing" to reduce weight. The fuel tanks were walled chambers between wing bulkheads vis-à-vis the rubber bladder tank system. The superior bladder system which was installed on the B-377 Stratocruiser was a series of large rubber fuel cells placed in the chambers of the wing between the bulkheads. Although heavier, the bladder tanks had an advantage of being self-sealing. The XC-99 has approximately 26,000 rivets which protruded into the fuel tanks. The tanks required regular resealing with a rubbery material which was dabbed on each rivet and allowed to dry. It is a slow labor-intensive dangerous process. On 20 March 1950 at approximately 10 a.m. maintenance crews were purging the left inboard fuel tank with carbon dioxide to clear fumes before entering for the resealing process. The 4,212-gallon tank exploded without warning, blowing out the leading edge of the wing between the fuselage and nacelle air intake of number three engine. Eight workers were injured when they were blown or jumped to the ground from maintenance stands. Seven of the airmen were treated and released but Sergeant J.R. Hendrick was admitted to the Lackland Air Force Base hospital suffering from burns.

The left wing of the XC-99 was severely damaged on 20 March 1950 during routine fuel tank resealing of number three inboard tank. A static spark from improperly grounded equipment ignited fumes blowing out the front portion of the tank and wing. Fortunately the inboard fuel cells were located 10 feet out from the fuselage, preventing damage to the interior. The repairs took 90 days to complete (USAF, author's collection).

A subsequent investigation determined the explosion was caused by static electricity in servicing equipment which was not properly grounded, igniting the highly volatile fumes of 115/145 avgas. Fortunately, the XC-99 was built with the intent of being produced as a postwar commercial airliner. For the safety of the passengers, the inboard wing tanks were located ten feet outboard from the fuselage. As an added precaution a supplemental bulkhead was installed between the fuel tank and the fuselage to block any possible fuel leaks from entering the passenger cabin. This feature reduced the fuel capacity from the standard B-36 wing.[1] This precautionary measure served its purpose well causing all the blowout of the March 1950 incident to go forward preventing structural damage to the airframe or blowing into the cargo compartment.

After considerable repairs taking

approximately 90 days the aircraft was returned to service on 25 June. A 2.5-hour test flight was scheduled for 03 July commanded by Kelly Air Force Base director of maintenance Colonel Fredrick Bell and Captain M. W. Neyland. After the test was deemed successful the Air Force proposed a test mission of considerable distance. The mission would test all the modifications and determine if they were properly carried out and insure operational readiness.[2]

The advanced C-99 had the nose gear mounted lower to achieve an unobstructed floor in the forward cargo bay. The gear retracted parallel to the fuselage with half of it outside. A blister pod nose gear door very similar to the type designed for the ATL-98 Carvair conversion of the Douglas DC-4 covered it, as seen in this artist rendering (Consolidated, author's collection).

On 12 July 1950, the aircraft ferried to San Diego to uplift a consignment of R-4360 spare engines and propellers needed for B-36 operations at San Antonio. The mission dubbed Operation Elephant transported 101,266 pounds of cargo consisting of ten R-4360-41 engines and 16 Hamilton Standard props. The XC-99 departed San Diego on 14 July at a takeoff gross of 303,334 pounds for the flight back to Kelly Air Force Base.[3] The flight of 1,150 miles was accomplished without serious incident. In the first three months of testing the XC-99 was flown 150 hours and conducted a demonstration landing with a full load on a 5,900-foot runway.

While Pan Am continued to evaluate the commercial potential of an advanced commercial C-99, the Air Force began operational testing on 06 September 1950 of the single XC-99 at Kelly Air Force Base. The testing was to be routine round trip flights to McClellan Air Force Base in California. Only four months into the Korean conflict, American C-54s were flying continual supply missions from McCord Air Force Base, Washington, across the Pacific.

Although Pratt & Whitney R-2000 engines were very reliable, the extreme hourly usage of the C-54s was taking a toll on power plants. The continual flying created a critical shortage, threatening the mission. The C-54 engine shortage became so critical the Air Force decided to change a XC-99 evaluation flight to an operational supply mission. On 10 October 1950, the test flight was conducted to supply the spare engines needed to maintain the operation. The XC-99 was dispatched to McCord Air Force Base, Washington, with 42 spare R-2000 engines on board, 27 on the main deck and 15 on the upper deck. After departing Kelly Air Force Base the XC-99s R-4360s engines began to overheat combined with excessive fuel consumption. It was determined the engine cooler doors were stuck in the open position creating excess drag increasing fuel burn.

After five hours, one of the spotters reported the prop on the number two engine appeared to wobble. It became progressively worse and was shut down. The aircraft diverted to Kirtland Air Force Base in New Mexico. Upon inspection the propeller was found to have a crack in the shaft. The mission was delayed for the replacement of the engine and propeller. After repairs the mission was completed successfully.

In November 1950, the U.S. Post Office evaluated the concept of transporting mail with the C-99 commercial version as a challenge to a 95 percent rate increase proposed by the railroads. The size of the aircraft brought ideas of actually sorting mail while in flight because it could carry the equivalent of four postal mail cars. The mail sorting system would be installed near the center-wing on both levels of the 130-foot usable cargo bay utilizing each end of the aircraft for simultaneous loading. Cost analysis of 100,000 pounds between Chicago and Los Angeles was estimated at 4.7 cents per ton mile. The positive numbers prompted the Post Office to consider asking the Air Force to place the XC-99 on loan for a test program. The Air Force refused the request and the plan did not continue past the planning stages.[4]

By the end of 1950, Pan Am still had not made a decision to purchase the redesigned larger C-99. Turboprop power was still not available and there was no firm date when it could be expected. Airline financial planners had major concerns the fuel and oil consumption for piston powered six-engine giant would be too costly for commercial operations, especially when fares were projected to be extremely low.

Fuel and oil costs eventually became a problem for Pan Am, United, Northwest and Overseas American even with the smaller Boeing 377 Stratocruiser and it was an aircraft with only four of the fuel hungry Pratt & Whitney R-4360 engines. The proposed six-engine commercial C-99 provided too much capacity for the amount of passenger or cargo traffic of the time. A study of the Hawaii route concluded the three aircraft would carry 150 percent of the traffic on the route making load factors impossible to be profitable. In addition airline officials were concerned it would be obsolete by the time the commercial version went into service.

3

YB-60 and Advanced XC-99

Swept Wing Prototype

The B-36, as first designed, did not meet the expected speed and altitude requirements of the mission. Convair experienced numerous design and engineering problems with the XB-36 from the beginning which continued with the submodels. The B-36 was underpowered, which led to the addition of paired jet pods on the later models. The plan to install the jets pods was first proposed on 05 October 1948. The units were the same as the twin J47-GE-11 inboard pods on the Boeing B-47 bomber. The engines were modified for the B-36 to burn standard 115/145 aviation gasoline. Standard B-36B model 44-92057 was the first to be fitted with the jets as a test program. With ten engines the aircraft performance was considerably improved but it was still a lumbering giant when compared to new jet bomber designs in progress. A rather desperate attempt was made to save the B-36 program in 1950 with an all jet swept wing version.

Convair submitted an unsolicited competing proposal to the Boeing B-52 on 25 August 1950. It consisted of an upgraded turboprop version of the B-36. The Material Air Command authorized Convair on 03 March 1951, under a change order to the contract AF-33(038)-2182, to upgrade two uncompleted B-36F aircraft to B-36G standards as a competing design with the B-52. The proposed B-36G upgrade consisted of a swept wing design with turboprop engines to be upgraded to turbojets when available. Convair considered it as an opportunity to compete with the B-52 being developed by Boeing. It was generally assumed the B-52 would replace the B-36. It had not been determined if the B-52 would be powered by turboprop or pure jet engines.

The YB-60 mockup was set up with the J-57 turbojets. Production problems at Pratt and Whitney caused a delay in receiving the engines. Engineers produced a second design reverting back to four turboprop engines with counter rotating props mounted on the jet pylons. It was estimated the YB-60 would be retrofitted with the turbojets by June 1953. The problems with the YB-60 project continued to compound because of the J-57 engine delays. The many engineering changes brought the possibility of the entire project being canceled. Convair attempted to save both bomber and transport projects in 1951 by proposing an expanded fuselage commercial transport version of the C-99 using the YB-60 swept wing and empennage configuration.

The swept wing design presented additional problems by shifting the center-of-gravity rearward creating a tendency to be tail heavy. The design was revised to all jet power in an effort to remain competitive with the Boeing B-52. The turboprop powered B-36 design was replaced with eight paired P&W XJ-57-P-3 turbojets configured similar to the B-52. It was expected the mounting of the engines on pylons ahead of the wing would compensate for the tail heavy imbalance. The extent of the upgrades and changes caused the designation to be changed from B-36G to YB-60. The designation change was necessary to continue with the program as a new aircraft. If it remained as a B-36 there would not be sufficient funding because it would be considered an upgrade to an existing model. By changing the designation to XB-60 it could be funded as a new design.

The military placed the planned YB-60 in the category of an alternative aircraft for the purpose of evaluation and comparison. The YB-60 was not really a competitive design because it was a B-36 fuselage with swept wings and tail. The loaded weight was 100,000 pounds less than the B-52 and the max takeoff weight over 180,000 lower. It was a swept wing B-36, which was not capable of filling the role of the B-52.

Work began on converting the two B-36F airframes, 49-2676/49-2684, to the redesigned tail and swept wings. The 35-degree sweep of the new wing reduced the B-36 wingspan by 24 feet. The new larger wing chord increased the wing area to 5,239 square feet. There were many problems with the building and testing the YB-60 design. The center-of-gravity problems existed because the swept wings shifted more weight toward the rear. During final assembly it became apparent the lack of fuel in the empty fuel tanks contributed to tail heavy tipping problems. Large weights were placed on each end of a shaft inserted through the nose wheel axle similar to a barbell to counter the tail heavy problem. This might work on the ground; however, considerable ballast would be required for flight, which reduced payload and presented ad-

When parked side by side it is obvious that the YB-60 is little more than a swept wing and tail version of the production B-36. The swept wings placed the engines farther back, changing the center of gravity and creating an extreme tail heavy problem. With no fuel in the tanks weights had to be placed on the nose gear to prevent tipping (USAF, courtesy Tony Landis, EFTC).

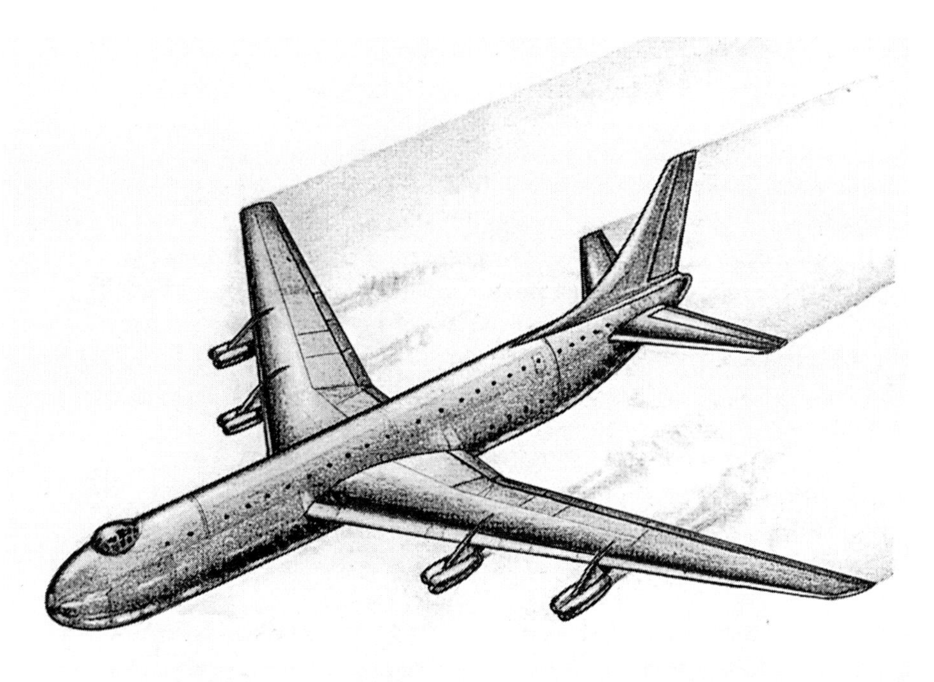

Convair attempted to revive the commercial C-99 in 1951 with Pan Am by proposing an all-jet version with eight engines. It was nothing more than the proposed XC-99 advanced version with B-36 cockpit and YB-60 wings and tail. The design was a hybrid combi with a passenger cabin on the upper deck and front and rear clamshell doors for the lower cargo deck (Convair, author's collection).

ditional balance problems as fuel was burned off.

The YB-60 program died a slow death but not before a XC-99 transport derivative of the swept wing giant was presented to Pan Am as a commercial jet transport.

Advanced C-99 Transport

The Convair engineering team suggested the YB-60 plans could be modified to produce an all jet-powered transport for both the military and commercial airline market. An attempt was made in 1951 to rescue the failing YB-60 program and at the same time regenerate commercial sales for the XC-99. Another effort was made to revive Pan Am interest in a commercial jet powered C-99 (Model 6) design. In reality, it was nothing more than the proposed advanced version of the C-99 with B-36 cockpit which had been previously presented to Pan Am with the YB-60 swept wings, empennage and eight jet engines.

It was inevitable, with the same senior engineering staff involved with both the XC-99 and YB-60 projects, a proposal would surface to save one of the programs. The YB-60 was in trouble because it was never a serious competitor with the superior Boeing B-52 bomber and it had serious design flaws. The advanced commercial YB-60 inspired swept wing design C-99 transport had an even larger fuselage cross section of 15 feet, 8 inches and a height of 22 feet, 7 inches with a volume of 22,800 cubic feet. It was even larger than the planned straight wing advanced C-99. The design also featured front and rear loading cargo doors. The cockpit configuration was moved back utilizing the B-36 type greenhouse canopy and cockpit configuration. It was a type of hybrid "combi" (combination) which would allow cargo to be transported on the lower deck with a pressurized upper deck for passenger comfort.

As the postwar commercial aircraft increased in size, airlines were confronted with the problems of longer turn times reducing utilization. It was reasoned the advance swept wing C-99 would satisfy airline needs to load and unload cargo quickly. Loading in and out of both ends while passengers were deplaning from the upper deck could achieve minimum turn times. The single production XC-99 utilized overhead cranes and belly doors to lift pallets of cargo which was time-consuming yet far superior to hand loading methods of previous aircraft. A front and rear loading drive on/off would have been superior. The concept did not become reality on a production transport until the Lockheed C-5A flew in June 1968 with a very similar configuration to the advance XC-99 with an upper passenger deck.

Pan Am was initially impressed with the YB-60 inspired transport version of the C-99 and the concept of a quick-turn large jet aircraft. Convair presented it as having both commercial and military applications. It could give the Air Force intercontinental jet cargo capability for the first time. In reality all jet cargo capability was not introduced until the Lockheed C-141A transport in 1963, which was delivered to the Air Force in 1965. When the C-141 came on line it was faster than the proposed jet version C-99. However, it could only carry 70,847 pounds compared to the projected 100,000 pounds of the Convair Advanced swept wing C-99.

The Convair design team was ahead of its time by proposing transports with both front and rear loading. The military did not acquire a double deck transport with front and rear loading until introduction of the Lockheed C-5A Galaxy in 1968. The C-5 had a payload of 180,000 pounds and later versions maxed out at 263,200 pounds. The C-99 jet powered airliner proposed to Pan Am more easily compares to the Boeing 747-100, which went into service in 1970. It could carry 374 passengers and 60,000 pounds of cargo below deck.

In retrospect, it appears Convair was struggling to save a contract rather than looking into the future of air transport. The jet powered C-99 Model 6 was based on pre–World War II technology married to swept wing jet innovation of the day. It could have been built, but like the B-36 it would have been a stop gap measure until aircraft design technology caught up with the need. As an airliner there was not enough commercial potential to justify the cost. The concept was probably too far ahead of the time. The airlines were not capable of filling high capacity passenger aircraft until 1969 with the introduction of the Boeing 747.

A single YB-60 was completed while the second was built up to a rolling unit without control surfaces. The military never opened competitive bids for a B-36 replacement. Boeing was selected from the outset to build the B-52. The YB-60 did have eight jet engines similar to the B-52, but it was basically adding swept wings to the existing B-36 fuselage. It never stood a chance against the all-new B-52 design but created enough interest as a possible alternate aircraft for data comparison. The Air Force ordered two YB-60s for evaluation. The military appeared to be sufficiently interested in the idea of using multiple components and systems from an existing aircraft to cut production cost of a heavy bomber.

The YB-60 was larger than the Boeing B-52 with a wing of 5,239 square feet compared to 4,000 square feet of the Boeing Stratofortress. The added area of the YB-60 increased drag and caused it to cruise at 467 miles per hour compared to 521 of the Boeing B-52 design. The flight controls were also designed from older technology. This resulted in slower control response times than the B-52. The YB-60 max ATOG was 410,000 pounds compared to 450,000 for the B-52.

The YB-60 (49-2676) first flew on 18 April 1952 commanded by Beryl Erickson. Convair conducted 20 test flights for a total of 66 hours of flight time. Flight testing proved the YB-60 to have stability and control problems with a stall speed of 140 which is lower than the stall speed of the B-52. During the eighth test flight on 24 June 1952 the aircraft experienced a flutter condition resulting in the severe wrinkles and disintegration of the rudder. On 12 June 1952, all four of the APU units failed to operate above 17,000 feet. Phase II of flight testing was cancelled on 20 January 1953, after only four flights totaling only 15 hours and 45 minutes.

The YB-60 was determined to have a number of design flaws, stability problems and speed deficiencies. These problems prevented it from competing with the Boeing YB-52 regardless of possible cost savings. The YB-60 program was cancelled and the two aircraft were stored outside at Ft. Worth, Texas. The Air Force officially accepted the pair on 24 June 1954, with no funding appropriated for their use or storage. Both aircraft were broken up by hand on 01 July 1954 with workmen using torches and axes.[1]

The Air Force made the correct choice in choosing the B-52, which has evolved over the years. Beginning service in 1955, it is still functional today and has the possibility of becoming the only military combat aircraft to remain in active inventory for 100 years. When developing outsize transport designs after building the Pregnant Guppy in 1962, Jack Conroy reviewed both the B-36 and B-52 as possible airframes to modify. He believed, because of their size, both would be good base airframes for a volumetric transport with a 40-foot diameter fuselage. He reviewed the possibility of utilizing surplus B-52 airframes again when proposing his twin fuselage Virtus shuttle transporter in the 1970s.

XC-99 Outlives Swept Wing Upgrade

At the same time Convair was proposing an advanced swept wing transport derived from the YB-60, the XC-99 was being upgraded. During the first six months of 1951, the XC-99 was scheduled for multiple upgrades and modifications at Kelly Air Force Base. The Pratt & Whitney R-4360 engines burned high performance 115/145 octane aviation gas with anti-detonation properties to allow higher takeoff power settings. The fuel had a purple dye added for identification. The aircraft continued to experience fuel leaks, which was evident by the purple-red stains along panel seams on the bottom of the wings. The XC-99 was grounded for major upgrades, the resealing of the fuel tanks and the replacement of the round-tip propellers.

Propeller Problems

Originally both B-36 and XC-99 were fitted with round-tip prop blades. The XC-99 experienced buffeting in certain flight configurations but the reason was not readily identified. The buffeting problem is not unusual for very large transport aircraft with a high fuselage sides. The Aero Spacelines B-377 Pregnant Guppy experienced a similar buffeting problem during flap retraction. It was determined the enlarged upper airframe combined with the indention in the fuselage interrupted air flow. It was corrected by adding air deflectors to the sides of the fuselage at the wing trailing edge. All subsequent ASI B-377 Guppy conversions were built with a very large wedge shaped fairing on the sides of the fuselage to reduce buffeting. A solution to the XC-99 buffeting was more difficult to identify because the XC-99 has a slab side fuselage and pusher props. It was eventually determined the buffeting could be reduced by 50 percent by replacing the round-tip propellers with square-tip units.

Multiple other upgrades were scheduled while the tank resealing process was being done. The aircraft was originally built with a flight deck crew quarters fitted with a toilet, drinking water supply and three canvas bunks and/or two crew seats. There were also two additional toilets on the upper deck and three on the lower. Because of the long duration flights the accommodations needed to be upgraded for extra and relief crew quarters. Soundproofing was added to the cockpit and crew area. A galley was added with two hot plates, electric oven, refrigerator and dining table. A total of eleven bunks were installed and a small room for aircraft commander quarters. A cargo elevator was also installed in the fuselage to decrease loading time. Finally the main bogie type landing gear was upgraded and strengthened increasing maximum gross from 312,000 to 320,000 pounds. The XC-99 was returned to service making the first flight after modifications on 06 July 1951.

When the post–World War II long-range bombers and transports were conceived technology was still a carryover from wartime aircraft development. A number of new systems came from the war effort but technology was slow to evolve. New generation jet aircraft were already being designed by other manufacturers. The B-36 was a deterrent during the Cold War but the fleet was continually being upgraded. In all probability the B-36 spent more time in modification programs than in operational service. The XC-99 proved the viability of giant transports but it was obsolete by the time it was operational. The XC-99 was built from 1940s design and engineering. Military planners were aware of the advances being projected in aircraft design and opted for a single XC-99 to evaluate the role of a giant transports.

During the time in service the XC-99 went through many upgrades and modifications. The Air Force was impressed with the ability to uplift an enormous amount of cargo. The four 10,000 pound capacity self-traversing overhead hoist were an asset. Even with the cranes, loading 60,000 to 100,000 pounds of cargo was extremely slow making a quick turn-around impossible.[2] The aircraft was equipped with two electrically operated sliding cargo doors in the belly, one forward and one aft. A set of clamshell cargo doors was added aft of the original rear door to expedite loading. By having two sets of cargo doors the upper and lower deck could be loaded simultaneously.

Aircraft loading was time-consuming and labor intensive. The Air Force was very concerned with turnaround time especially if the XC-99 would ever be used for combat support. Studies determined it took ten men an average of 54 minutes to load each 10,000 pounds of cargo and 30 minutes to offload the same amount.[3] An effort to improve the problem was addressed in April 1952. Special loading containers which could be loaded and pre-staged were developed for the upper deck. They could be lifted into position with the 4,000-pound capacity overhead crane eliminating upper deck bulk loading.

The enormous upper deck could handle 13 of these containers, which were lifted into place and locked down by a bolt type system in the floor. The containers were only for the upper deck with the lower deck reserved for larger items and rolling military vehicles. The upper deck loading time was reduced to 30-minutes with the new container system while the complete aircraft loading time remained at three hours.[4]

The XC-99 was fitted with radar in June 1953 with 200-mile range and special electronic equipment giving it all weather capability. Installation of the 250-pound radome unit was completed placing the aircraft back in service on 01 July. Convair reported on the same day in *Convairiety*, the company employee newsletter, the Air Force had re-designated the XC-99 to C-99. This was repeated again in the 26 August issue. No other confirmation has been found in either Air Force or independent documents of this change in designation.

The aircraft, sometimes referred to as Behemoth, made its longest flight of 12,000 miles a month later in August 1953. Sixty thousand pounds of cargo was carried in each direction to Rhein-Main, Germany with stops at Bermuda and the Azores. During the Cold War in May 1953, it was called on again for a special mission to transport cargo from Dover Air Force Base, Delaware, to Keflavik, Iceland, in support of the Distant Early Warning (Dew) Line Project.

In all, it made six round trips flying 30,000 miles and transporting 380,000 pounds of cargo. A total of 31 maintenance technicians were dispatched for the mission. Surprisingly it performed well requiring only routine repairs and service. Over the 30 day period the aircraft logged 201 hours of flight. In a rather unique arrangement it was chartered out commercially by MGM Studios to transport a consignment of vintage bi-planes from Florida to Los Angeles for the filming of the John Wayne movie *Wings of Eagles*. The aircraft arrived in Los Angeles as a large crowd gathered. Upon arrival at Los Angeles the aircraft was rushed by a large crowd of mostly women who mistakenly believed John Wayne was on board. It was suspected that the studio circulated rumors of Wayne being on board as a publicity stunt to promote the film.[5]

Almost ten years passed between the XC-99 first flight and it being grounded in 1957. It served the Air Force for seven years primarily shuttling parts and spare engines to and from B-36 bases and depots. It made twice weekly flights between Kelly Air Force Base, Texas, and McClellan Air Force Base, California, in support of the B-36 program. The flight crews had mixed opinions of the aircraft primarily because it was large and slow.

Captain James C. Pittard, who logged more than 3,827 hours in the cockpit, commented it flew more like an airliner than a military aircraft because it was big, comfortable and quiet. Conversely Major James Douglas described it as handling different than any other aircraft he had ever flown but was more like flying the much smaller B-24 bomber. He described it as responding slowly but you have to stay out in front of it at all times. You had to understand it because it never let you forget you had 50 tons of cargo behind you.[6]

XC-99 Retired

The XC-99 flew more hours than any other experimental aircraft before the Air Force came to the conclusion it was not practical. There is no question if it was fully loaded it could fly farther with more cargo at the lowest cost than any aircraft in history, but it was seldom possible to fill it up to a single destination. It was a problem of diminishing return. It could fly cargo for 16-cents per ton-mile when fully loaded compared to 26-cents per ton-mile of any other aircraft. Air Force priorities were changing by 1955, with the concentration on the next generation of jet transports and bombers. The B-36 was being replaced by the Boeing B-52. The phase out of the B-36 placed the many common parts with the XC-99 in short supply.

The XC-99 was ahead of its time relative to capacity providing important data to the military. In reality it seldom operated to maximum capacity to any destination, which could not justify the cost of operation no matter how efficient. It would have to be full in one direction and transport at least 60,000 pounds on the return flight to be practical. A high load factor for cargo is difficult in today's market with far more efficient jet aircraft. It was impossible in the 1940–50s.

The Air Force dropped all studies of additional XC-99 production or a jet powered swept wing version using YB-60 wings and empennage. By March 1957, the giant cargo plane was no longer needed for B-36 support causing the Air Force to suspend XC-99 operations. The Douglas C-124 and other large transports of less capacity were able to meet the heavy lift needs more efficiently. The XC-99 contributed considerable data to military planners regarding cargo loading and handling techniques. It had flown more than 7,400 hours transporting 60 million pounds of cargo more than 1,500,000 miles.[7]

The XC-99 was ideal for long-range volume resupply and support missions. With all-purpose built volumetric giant types whether then or now, there is only a limited amount of cargo to be transported in volume between two points. While it is true the military must maintain cargo fleets, the aircraft are required to have a practical if not universal application. One of a kind behemoths are difficult to maintain and have limited use.

Supporting the B-36 program during the Cold War was the primary function of the XC-99. It was operated occasionally supplying spares for Korean conflict air operations; however, Convair planners believed it was the prototype for a new generation of military transports. The military saw it as a logistics experiment. It was the first of a series of special outsize transports which were built in limited numbers for specific missions. It served in a similar fashion as the limited production Douglas C-74 for the Air Force or the Aero Spacelines Super Guppy (B-377SG) for NASA. These aircraft have a built in obsolescence. Once the mission ends, data is compiled and projects are completed these giants are no longer in demand or cost effective. There are exceptions but this is the general rule.

XC-99 Grounded

The XC-99 flew more hours than any other experimental aircraft but cost was a factor because there was seldom enough cargo to any destination to fill it up. The B-36 phase out reduced the XC-99 support mission consequently operating cost could not be justified. The giant transport ceased operational flying in March 1957 for a heavy maintenance review because many structural time limits had been reached. During operational service a number of reinforcement patch repairs had been accomplished to keep it flying. In June 1957, the main spar was determined to be cracked. San Antonio Air Materiel Area (SAAMA) estimated it would cost more than $1,000,000 and 144,734 man-hours for structural repair, replacement and modifications to remain airworthy. It was permanently grounded in August 1957.

The XC-99 was the largest transport in the world at the time it was built. A commercial version was proposed but the application of this aerial giant was limited to bulk cargo service. The possibility of transporting outsize items or complete airframes and subassemblies for aircraft manufacturers was not possible because of the double deck configuration and lack of straight in loading capability. However, transporting other aircraft or airframe subsections by air in a semi-external configuration was attempted with a B-36 long before super guppies or mounting of the space shuttle on top of a Boeing 747.

The B-36 fuselage is 12.5 feet in diameter. It is less than half the size of the XC-99, but a single aircraft was modified for special transport work in 1957. Convair engineers developed a system to move a Convair B-58 Hustler aircraft from Fort Worth, Texas, to Wright-Patterson Air Force Base, Ohio, by suspending it under the fuselage of a B-36F-1-CF (49-2677) with the upper half extending into the bomb bay. To clear the B-58 wings the props were removed from the inboard number three and four engines of the B-36. The B-58 wings blocked and even covered a portion of the wheel wells preventing the retraction of the B-36 main gear. The decision was made to fly it at low speed and altitude from Fort Worth to Dayton with the landing gear down. The flight was accomplished without incident, but considered a onetime event and not a viable way to transport other aircraft on a regular basis for the aerospace industry.

Both the military and Convair declared the volumes of data collected on air material transport and methods of cargo handling with the XC-99 provided an information base for future transports. It was agreed the technology advances of the B-36/XC-99 program demonstrated there are no practical limits to the size an airplane can be built and still fly relative to power plants and wingspan.[8] At the time the size of the fuselage dictated the length of the wing, which determined the number of engines required. The restriction was the size of reliable power plants. Convair engineers expressed the belief no other aircraft of this magnitude would ever be built again. It is now known that assessment was very short sighted.

Five years after the XC-99 was retired John M. "Jack" Conroy flew the B-377PG Pregnant Guppy with a 19-foot diameter cargo compartment and a wingspan considerably shorter than the XC-99. In an ironic twist in 1964 Fairchild Aircraft proposed to NASA the M-534 conversion. It was a plan to convert surplus B-36 aircraft into low wing volumetric guppy class transports by expanding the fuselage in a similar fashion to the Aero Spacelines B-377 guppy. One of the reasons it was rejected was the problem of being top-heavy. Aero Spacelines proposed an entirely different guppy design to NASA using the B-36 wings, tail and cockpit. The fuselage was used as the backbone and spacing for the ribs on a 40-foot diameter fuselage. Conroy's high wing B-36 design addressed the vertical center of gravity problems experienced by low wing guppy aircraft.

The Jack Conroy 40-foot diameter B-36 guppy was designed with a high wing for better stability vis-a-vis the Fairchild low wing, which would have been extremely top heavy with a critical vertical center of gravity. The problem would have been acquiring the B-36 airframes, which were being scrapped to prove to the Soviets we were reducing our nuclear capability (Doug Ettridge, courtesy Angelee M. Conroy).

The retirement of the XC-99 placed the disposition of the aircraft in limbo. The initial plans for it to be scrapped were met with protest from former crewmen, military historians and citizens of San Antonio. The Air Force considered flying it to Dayton, Ohio to the Air Force museum, but the expense of making it airworthy for one flight was considered too prohibitive. The Air Force attempted to pacify all concerns by announcing on 19 July 1957 it would be placed up for sale. A San Antonio department store offered to donate a parking lot for display. The offer was conditional on the chamber of commerce purchasing it and setting up offices in the fuselage. The city council pointed out a major portion of the city would have to be torn down to get it to the parking lot. The mayor proposed flying it ten miles to the commercial airport for a tourist attraction. The proposal was met with the problem of making it airworthy. If it were repaired then there was no reason not to fly it to Dayton to the Air Force museum.

The XC-99 was put up for sale in July 1957. Ownership passed through several groups that were never able to raise funds for display. After 47 years of open display at San Antonio it had deteriorated to deplorable condition. The decision was made to move it to the Air Force Museum at Wright-Patterson Air Force Base for restoration. It was disassembled and sections transported by a Lockheed C-5B (EFTC, courtesy Tony Landis).

On 06 November 1957, the Texas Department of Disabled American Veterans (DAV) received the title to the aircraft with the requirement it be removed from Kelly Air Force Base within 90 days. The city of San Antonio had agreed to provide funds to move the aircraft but was unable to raise the cash. After multiple extensions from the Air Force, DAV director Lewis Buckingham was eventually able to obtain funding to move it to a field across from Kelly Air Force Base.

Initially it was maintained but over many years of neglect it fell in disrepair. This author, while serving in the USAF, was privileged to have seen it on display across from Kelly Air Force Base in 1965. Even in deteriorating condition it was an impressive monster. It remained there until 1993 when it was sold to the Kelly Field Heritage Foundation for $65,000. All efforts to turn it into a museum failed because of the amount of funds required to restore and maintain a display of this proportion.

The XC-99 with at Max ATOG weight of 322,000 pounds was not outclassed for 21 years. On 30 June 1968, the Lockheed C-5A Galaxy first flew from Lockheed-Marietta Georgia using the runway at Dobbins Air Force Base at the Marietta facility. The C-5A is longer with twice the gross weight but has a wingspan seven feet less than the XC-99 at 230 feet. The XC-99 maintained its record of the greatest wingspan of any land plane until 1980, when it was surpassed by the Soviet AN-225 at 290 feet. The AN-225 fuselage is 48 feet longer with a payload five times greater. However, the XC-99 remains the largest piston powered land plane ever built.

After 47 years of display at San Antonio plans were made to move the derelict XC-99 monster to its original final destination at Dayton, Ohio. It was disassembled and transported to the Air Force Museum at Wright-Patterson aboard a Lockheed C-5B which has front and rear drive on/off loading. Convair had proposed drive on capability for an advanced XC-99 model. The XC-99 will be reassembled, restored and displayed. The military and Convair officials stated in 1957 there would never be a need for another aircraft the size of the XC-99. The prediction was obviously short sighted. Severe corrosion prevented restoration. It was moved to the 309th AMARG at Davis-Monthan AFB in Tucson for indefinite storage.

4

Fairchild Stratos M-534

Outsized Booster Carriers

The XC-99 and the proposed super advanced C-99 were not the final chapter in proposed outsize aircraft designs using a B-36 as the base. The success of the Aero Spacelines B-377 Pregnant Guppy, which first flew in 1962, prompted NASA in February 1964 to solicit proposals for the development of a larger volumetric transport aircraft for all boosters described as "Large Booster Carrier Aircraft."

As the space program progressed it was still not possible to transport all the stages of the Saturn rocket by air. NASA had to rely on sea-going barges to move the largest booster stages from California to Huntsville, Alabama, and Cape Canaveral via the Panama Canal. The Aircraft Missiles Division of Fairchild Stratos Corporation, a subsidiary of the company now known as Fairchild-Hiller, proposed to NASA at Marshall Space Flight Center a B-36 "Outsized Booster Carrier" named the Stratos M-534.

The B-377PG had been successfully transporting the S-IV second stage of the Saturn I, rocket engines and oversized NASA cargo for two years. The Saturn S-IVB booster was too large for even the proposed second generation Aero Spacelines B-377SG Super Guppy. NASA maintained the next generation volumetric aircraft would have to transport a 91-foot-long 121,000-pound payload, which is 38.5 feet in diameter. This translates to an all up takeoff weight of over 363,000 pounds. The takeoff gross narrowed the list of possible airframes to only a few possible choices.[1]

Multiple competing proposals were submitted for consideration. Aero Spacelines had the edge because the B-377 Pregnant Guppy was the only booster transport in operational service. Until it flew most aerodynamicists doubted anything of extreme diameters could fly efficiently. Jack Conroy had already proposed a high-wing B-36 conversion. The construction of his second turboprop B-377VPG (Very Pregnant Guppy) was already under way. His design team was far ahead of most other competitors with a proposed all jet 30-plus-foot diameter Boeing 707 guppy conversion.

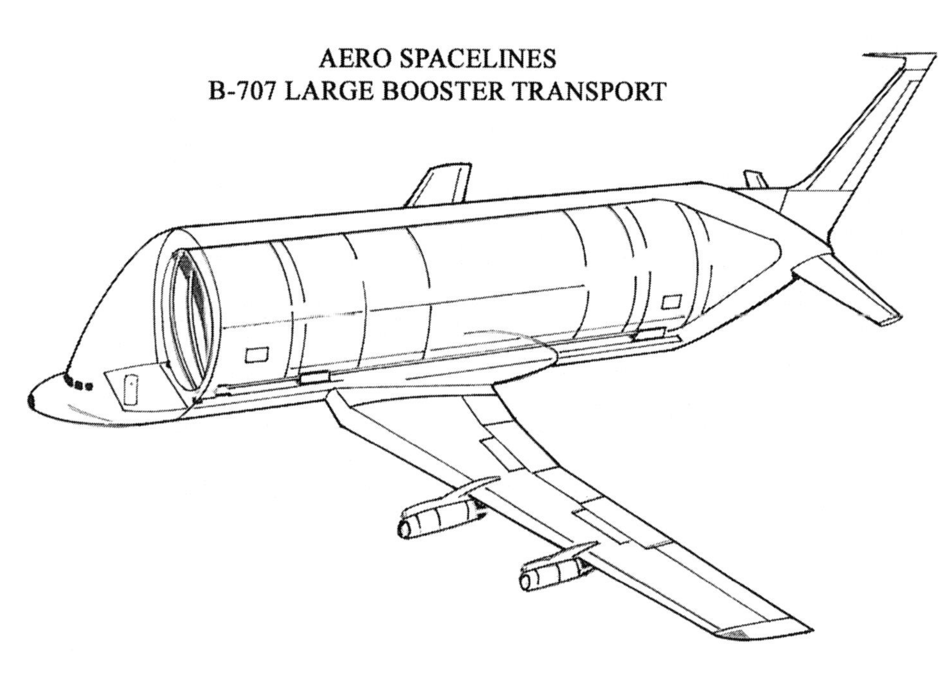

Jack Conroy proposed a 30-foot diameter guppy based on the Boeing 707 airframe. The biggest problem would have been acquiring a surplus airframe for conversion. The 707 was introduced to the commercial carriers in 1958. It had not been in service with the airlines long enough to be on the secondary market making the planned conversion cost prohibitive (Dottie Furman, courtesy Angelee M. Conroy).

Noted aviator Dee Howard submitted a 40-foot diameter bi-wing design powered by ten Curtiss-Wright R-3350 engines. It used part of the Douglas DC-7 fuselage and wing components. It is easy to understand why NASA rejected it because it was powered by Wright R-3350 engines, which were known to be problematic and routinely fouled sparkplugs. Aero Spacelines was already building the turboprop B-377SG and had proposed an all jet B-707 design.

On 16 March 1964, Fairchild

Stratos Corporation submitted summary report R-534-001 to NASA on the feasibility of a large transporter based on the B-36 which could transport the Saturn S-II stage booster rocket. In the process of selecting a base airframe Fairchild reviewed nine different existing large aircraft as possible base airframes. The study concluded there were only three candidates possibly suited for outsize volumetric modification. The Fairchild study noted all three existing airframes selected were deficient in some way but came close to meeting the criteria. The list included the Convair B-36, Douglas C-133 and the Saunders-Roe SR.45 flying boat.[2]

Dee Howard proposed a giant 40-foot diameter transporter. The 10-engine bi-wing design used Curtiss Wright R-3350 engines and DC-7 components. The upper wing with extended center had six engines. The lower wing was a standard DC-7 unit mounted at the bottom of the fuselage. The forward section of a DC-7 cockpit and fuselage was mounted on top. The multiple reciprocating engines were not practical, causing NASA to reject the design as too complex (Dee Howard Foundation).

NASA previously reviewed the C-133 as a "Heavy Vehicle Hauler" in 1960 with multiple modifications to transport outsize loads in a piggyback configuration. The C-133 had been flown at maximum weight of 300,000 pounds. The study concluded it would require a redesigned landing gear, modification and extension of the wings, and additional engines. After consideration it was rejected because of cost and time required for modifications.

The C-133 was already experiencing airframe issues which included excessive vibration and structural fatigue. In time, the C-133 was grounded because of a series of unexplained in-flight losses.[3] As a result of the study the rear cargo door was modified to a clamshell configuration to accept larger loads, which were primarily missiles. All subsequent C-133B models were built with the improved clamshell aft cargo doors. Even then, the payloads it could accommodate were small in comparison to the expanded fuselage or piggyback proposal.

It is surprising Fairchild would even consider the C-133 for modification in light of previous studies and known shortcomings. The Saunders-Roe SR.45 flying Boat was also a questionable choice because only three were ever built and only one flew. Jack Conroy reviewed the possibility of converting the flying boats to land planes before expanding the fuselage to a guppy configuration.

Fairchild determined the most viable choice to transport the booster stage was an expanded fuselage design utilizing the ten-engine B-36J airframe. The proposal was one of the more serious designs considered by NASA. The Convair B-36J aircraft has an all up weight of 410,000 pounds and the landing gear would not have to be reinforced. The aircraft has a span of 230 feet and a length of 168 feet which would require minimal fuselage extension. The high altitude bomber systems could all be eliminated which could help offset the weight of the expanded fuselage. In addition, it was already fitted with six Pratt &Whitney R-4360 engines and four J47-GE-19 jets.

The "M-534" was proposed in several engine configurations which were shown in artist renderings. One drawing has an additional single jet mounted on each side between the number three and four engines and the fuselage for a total of six reciprocating and six jets engines. Fairchild estimated it could be built in 18 months at a cost of $11.5 million. NASA solicited the proposal but no action was taken because there was no budget to fund aircraft purchases. Congress previously made it clear to NASA that no funds would be appropriated for aircraft manufacturing. The Aero Spacelines B-377PG under contract to NASA at the time was built on speculation with private funds for about $1.2 million. The larger B-377SG was being built at the time from private funds.

The Fairchild plan for construction appears similar to the process developed by Aero Spacelines to build the B-377SG Super Guppy. One of the problems encountered when building a large diameter fuselage is floor width. The B-36 fuselage is only 12.5 feet in diameter and is a high-wing aircraft. The new expanded fuselage would have been built above the wing spar converting it to a mid-wing configuration. In the case of the B-36 a floor resembling a platform would need to be built above the spar the length of the fuselage. The floor width would have possibly been more than 12.5 feet because the area from widest point on the fuselage is filled into create a slab side up to floor level. An expanded 35- to 40-foot diameter fuselage would then be built up from the widened floor. This in itself created an additional engineering problem. The Stratos cargo floor would have been approximately of 18.5 feet above ground in order to be above the wing. Because it loaded through rear clamshell doors under the empennage the bottom of the tail was over 58.5 feet above the ground. This would have placed the top of the vertical fin more than 80 feet above the ground.

A comparison gives perspective to the problems which would exist with the M-534 is the first two Aero Spacelines guppies. They were built up from the original 8-foot wide floor of the Boeing B-377 Stratocruiser. Large diameter cargo could only rest on the width of the narrow floor. Securing 18-foot wide cargo to an 8-foot floor posed a number of loading problems on the B-377PG. For this reason, subsequent Aero Spacelines guppies were built with a new 13-foot-wide floor for easier loading and better weight distribution. Securing cargo 40 feet in diameter to a 13-foot-wide surface presents balance problems. Loadmasters and flight crews were required to calculate not only horizontal center-of-gravity but vertical for fear of being top-heavy. The Fairchild Stratos M-534 design concentrated the load above the wing which could have drastically affected stability. In comparison the proposed Aero Spacelines 40-foot diameter B-36 guppy had the cockpit moved rearward with clamshell doors similar to the Douglas C-124 and a high wing concentrating the weight below for better stability. The proposed M-534 Stratos retained the B-36 cockpit configuration with the cargo weight above the wing.

The nearly 40-foot diameter Fairchild Stratos M-534 proposal was the end result of studying multiple airframes for possible conversion. The B-36J was selected as the base airframe; however, the aircraft were not available because they had been broken up for spares. The design appears to be extremely top heavy with the cargo load above the wing (NASA).

Concentrated weight high above the wing or horizontal centerline inhibits the angle of bank when turning. The B-36 Fairchild Strato M534 guppy with a 40-foot diameter load concentrated on a narrow floor even with the 230-foot wingspan is subject to being top-heavy and unstable. The design called for the complete B-36 empennage with a portion of the bombers aft backbone to be mounted on top of the 40-foot expanded fuselage. The high tail gives full clearance for the rear clamshell doors for straight in loading. However, the high mounted empennage adds additional weight far above floor level complicating the vertical center-of-gravity issue.

Although not specified it appears a new expanded center-wing similar to the Aero Spacelines guppies would be required to extend the wing root to the outside of the expanded fuselage. This would facilitate the fabrication of a wider floor for the 40-foot diameter fuselage. The B-36 main landing gear is 46 feet apart. This is quite narrow for an aircraft with a 40-foot high fuselage. A wider center-wing could position the landing gear farther away from the fuselage improving weight distribution and stability on the ground. A very high frontal forward section above the cockpit is grafted in above the B-36 cockpit nose deleting the rear half of the green house cockpit windows. The aft section of the lower belly fuselage forms a radical upward sweep to the rear of the new enlarged fuselage.

Vertical endplates were designed for the horizontal stabilizer to better control airflow from turbulence created along the expanded rear fuselage. The result is a huge aircraft with the tip of the vertical fin approximately 86 feet off the ground. This is taller than the Boeing 747, Antonov 225 or the Airbus 380, which is the only aircraft close with a vertical fin at 80 feet, 3 inches. Aircraft of this size are technological wonders today. The M-534 design would have no doubt tested the engineering and structural limitations of the 1960s.

In reality, the entire proposal was short sighted. Fairchild had no intention of funding the building of an aircraft on speculation NASA would approve and accept the design. There was no budget for such projects and would not be one. NASA was told by Congress it was not in the aircraft building business. Aero Spacelines was building aircraft on a budget of $1,000,000 with private funds because NASA could not provide financing. The agency certainly could not come up with $11,500,000 for Fairchild. It was a large corporation seeking profitable government contracts. There would be modifications and cost overruns which could be millions more.

The B-36 program was so costly the military defense budget committee looked at every opportunity to cut cost. The Air Force began planning the phase out of the B-36 in 1953 consequently orders for new spare parts were cut back. In 1957, update work was still being performed by Convair at San Diego to extend the B-36 life until the B-52 was completely phased in. The Strategic Air Command had retired the last B-36 Peacemaker to the Arizona desert in 1959. Fairchild planners miscalculated the availability by presuming there were many airframes and engines just sitting at Davis-Monthan Air Force Base waiting to be used. In reality they were not there because of parts shortages which were compounded by phasing out the aircraft. The Air Force began retiring the B-36 fleet to desert storage in February 1956. The Peacemakers were stripped of usable parts for use on bombers still in service and the airframes chopped up for salvage. The lack of new spare parts prompted the removal of many more used parts than are normally salvaged from retired aircraft. The B-36 phase out was slowed in 1958 because of defense budget cutbacks affecting B-52 delivery. The Air Material Command did not have the funds or manpower to remove all the usable spare parts and place them in storage for future use. Consequently many spare parts were not salvaged and destroyed. The Pratt & Whitney R-4360 engines were needed not only for the operational B-36 aircraft but for spares on other transports like the C-124 and C-97.

By the end of 1958, only 22 B-36J aircraft were still in Air Force inventory. In 1959, only five of the 385 B-36s airframes built were still in existence as all others had been scrapped for spares and melted down. Those five had been designated for static display purposes. Unless Fairchild could convince NASA to commandeer those five airframes, there were no B-36 aircraft available for volumetric conversion.

If NASA had been seriously interested in the M-534 conversion as an option it possibly could have pressured the Air Force to provide the B-36 airframes. A precedent had already been set when Marshall Space Flight Center convinced the Air Force to turn over the two YC-97J test aircraft to Jack Conroy to build the Aero Spacelines Super Guppy.[4]

As a result of multiple issues, the Fairchild M-534 volumetric transport did not progress past the initial proposal. NASA concluded it would take too long to develop such an aircraft and the cost would be too prohibitive. The reality of suitable airframes not being available appears to have never been considered. NASA budget conditions reduced the number of planned Saturn V launches and Aero Spacelines already had a plane in service to transport most of the components except the largest boosters. If the need had risen, Jack Conroy was already developing plans on larger capacity guppies with private investment which would not require funding from NASA. He had a proven design with the B-377SG making his proposal of a future outsize transport based on the B-707 and a 40-foot diameter B-52 more of a possible option.

The XC-99 is accepted theoretically as the first outsize volumetric transport. Building it by matching an expanded fuselage with multiple components from the B-36 began the process of blending components of existing airframes to produce giant special purpose transports. Combining parts to create a new aircraft was not new. Boeing used B-17 wings and tail to build the Boeing 307 Stratoliner, which had an expanded fuselage. After the loss of the B-307 prototype a completely new tail was designed and installed. Douglas used the DC-3 wings to build the B-23 Dragon bomber. However, these examples were not designed or intended as giant volumetric transports.

5

C-74 Globemaster I

Building a Global Transport

The Douglas C-74 Globemaster I was conceived in 1942 in response to the war needs for a practical long-range global transport. It was expected to be mass-produced; however, delays in production caused it to be an interim aircraft which flew after the war, laying the foundation for the Douglas C-124 Globemaster II. It is not recognized as a guppy type aircraft and does not physically compare with the rounded blimp image of the Aero Spacelines modified B-377 but the construction process was the same. The slab side fuselage and physical appearance of the C-124 has been more appropriately compared to a bumblebee. In comparison the first C-124 was built in a similar fashion to the Aero Spacelines B-377 guppy. Douglas removed the C-74 upper fuselage at floor level then built up an expanded top using the lower belly wings and empennage of the original Globemaster I to create the C-124.

The Douglas C-124 followed the Convair XC-99 and preceded the Aero Spacelines B-377 guppy in the progression of American outsize volumetric transports. It was a cost effective way to meet the military requirement for large drive on/off transport combined with the need to reduce research and development cost by using existing technology.

Only fourteen C-74s were built between 1945 and 1947. The design studies began in 1942 in the aftermath of the attack on Pearl Harbor. The military realized a need for a global transport because of the vastness of the Pacific. The C-74s served its mission well providing the Air Force with valuable data on a fleet of large heavy transport aircraft. Douglas promoted the C-74 as a scaled-up version of the wartime work horse Douglas C-54/DC-4 Skymaster.

After the C-74s were decommissioned from military service, options were taken to purchase the fleet by an assorted group of nefarious characters and entrepreneurs. Some of these individuals surfaced multiple times throughout this accounting of guppy aircraft development. This writing does not relate in detail the C-74 operational and unit history, missions and exercises. It is a review of the evolution of the C-74 airframe into the double deck C-124 Globemaster II with discussion of the many upgrades and running changes during production. Over the production life of the C-124 all systems were refined into a notable workhorse. Related here is a series of incidents and anecdotal accounts of the C-74 which occurred during and after military service. Operational and unit history of the C-74 and C-124 Globemasters can be found in a number of other outstanding works.

C-74 Design and Development

The first Globemaster C-74 design originated shortly after America was drawn into World War II. Douglas Aircraft had 61 civilian DC-4s on order or in production for commercial airlines when the Army realized the need for a global transport capable of lifting large loads over great distances. The transition of the DC-4 to C-54 military use was quite easy with a minimum of engineering changes. The need for a larger aircraft was immediate and more difficult to obtain.

The C-74 Project Group was formed at Santa Monica in early 1942 to create a large transport by using current technology to build a scaled-up version of the DC-4/C-54. The Army requested a full proposal from Douglas for one static and 50 production aircraft designated Model 415. The Army specified it must be capable of transporting two T-9 Aero tanks, two 105mm howitzers, jeeps and support vehicles, two caterpillar type bulldozers or 125 troops.

As the project moved forward a full scale wooden mockup of the fuselage was built. Load tests were conducted using 40 and 90 mm anti-aircraft guns, 105mm howitzers, T-9E1 tanks, and crated P-39, P-40, P-47 and P-51 aircraft.[1] Although it has a unique cockpit appearance, the design was a scaled-up high capacity version of the DC-4 (DC-6 and DC-7 series). In some respect the Douglas C-74 can be loosely considered a DC-4 guppy.

Initially it was designated as the Model 415. It was scheduled to be powered by four Wright R-3350 engines. Douglas senior engineers Arthur E. Raymond, Dr. W.B. Oswald and project engineer C.G. Brown met with Army Generals Hap Arnold and Oliver Echols to review the new trans-

The twin canopy Douglas C-74 Globemaster was first conceived in 1942 as a scaled-up version of the DC-4. The war effort was in desperate need of a large logistical transport. The simple self-sufficient design was built to compete with the giant Consolidated XC-99. There was serious doubt the six-engine giant was even possible. The four-engine C-74 was favored with 173-foot span with side and elevator loading. It was fitted with full span fowler flaps identified by the actuator housings to increase the wing surface for easier takeoff (USAF, author's collection).

The C-74 interior is equipped with overhead traveling cranes which can raise and lower the loading elevator visible in front of the jeeps. The cranes can also move heavy cargo within the cargo compartment for easy loading. Note the gun-port cabin windows, which were added requirements. Although capable of transporting heavy cargo it lacked drive on/off capability (USAF, author's collection).

port design. The meeting produced new military expectations and the aircraft requirements were soon changed. The gross weight was reduced from 145,000 to 125,000 pounds and the project was re-designated Model 415A. In March 1942 the Army issued a letter of intent officially designating the design as the C-74. On 25 June 1942 contract AC-27042 (cost plus fixed fee) was signed for $50,000,000 USD for 50 aircraft and one static test airframe.[2]

America was in the depths of World War II and the army needed a simple design basic cargo aircraft as soon as possible capable of moving men and materials worldwide. The new proposed transport potentially would be the largest land based transport in the world with a range of 7,000 miles. Consolidated-Vultee also proposed the XC-99 to the military in 1942. There was doubt the XC-99 could be completed because the technology for the systems designed into it had not been developed at this point. The C-74 had a wingspan of 173 feet, length of 124 feet and 6,800 cubic feet of cargo space in a 75-foot long cabin. The self-sufficient design ultimately did have a gross weight of 145,000 pounds. Wartime need placed the C-74 in a race to be completed before the Consolidated-Vultee XC-99. However, by January 1943 the C-74 as well as the Consolidated XC-99 was placed on low priority in favor of concentrating on other much-needed war projects.

The C-74 contract specified all design features were to be proven technology. This prompted engineers to borrow from the existing smaller DC-4 design. To speed up delivery there was no experimental model (XC-74), prototype or "A" model. Ultimately, only fourteen C-74-DL models were produced. The first C-74 was scheduled for delivery by Christmas day 1943.

Problems began to occur in short

order when the Wright R-3350 engines were delayed until April 1943. Even before the delay the performance data indicated the engine would not provide the rate of climb or altitudes needed for operations above bad weather. In March 1943 Douglas made the decision to switch from the Wright R-3350 to the heavier 3,000 horsepower Pratt & Whitney (P & W) R-4360-27 engines.[3]

The Pratt & Whitney R-4360 engines have oval cowling. The twin canopy configuration gave the C-74 a very strange appearance. The crew entry door in the belly is visible behind the very long nose gear. The side rail of the extending ladder is behind the left nose wheel. The twin scoops above the nose gear doors are air intakes for the heaters, which were mounted in the nose (USAF, author's collection).

Once the decision was made to switch to the Pratt & Whitney engines, everything forward of the firewalls had to be redesigned along with changes to the front spar wing structure and flight controls, which added more delays. The gross weight of the aircraft was increased back to 145,000 pounds with the heavier new engines to offset the increased Zero Fuel Weight (ZFW). The next setback occurred in October 1943 when Pratt & Whitney informed Douglas the R-4360 engines would also be delayed. The engines finally arrived on 15 December 1943. In addition to the delays caused by new engines, the Army contributed to the problem with request for 125 other design changes. They included dual nose wheels, gun ports in all cabin windows and a blister window for the navigator and engineer.[4]

The change from the Wright R-3350 to the larger Pratt & Whitney R-4360s placed the C-74 in position to become a test program to evaluate large volume heavy transports with the P & W R-4360-27 engine and study maintenance procedures. Programs were set up to gather data and review test missions. This change gave the USAAF an opportunity to obtain daily operational and long-range statistics which could be applied to other aircraft types being designed. It was not realized at the time but placing the C-74 in the category of a test aircraft ultimately limited the production and set up the circumstances for it to become a test article for the C-124 Globemaster II program.

In part the decision to switch to the Pratt & Whitney engines was for in-service testing of the R-4360 for the Northrop XB-35 flying wing and Consolidated-Vultee YB-36, which were slated to have the same 28-cylinder engines. Operational data was needed to evaluate cost of operation and reliability. A flight program was proposed to put 300 operational hours on the engines as soon as possible. Engineers were very anxious to receive the data to firm up performance expectations for the heavy bombers being produces by Consolidated and Northrop. As the evaluation progressed there were considerable problems with the R-4360-27 engines. Consideration was given to changing to the 3,250 horsepower R-4360-69 model engines.

A unique and strange design feature of the C-74 is the side-by-side twin canopy cockpit. It is the only transport ever put into production with this configuration. The twin bubble canopy design was tried on the Douglas XB-42 in 1942 but proven to be impractical for any aircraft. The double canopy C-74 configuration was created as an assumed safety feature on very large aircraft giving pilots almost 360 degrees of vision. Overall the C-74 was a very aerodynamically efficient design.

While, the double canopy gives the appearance of being more streamlined than previous cockpit configurations, it actually produced increased drag. Also, sitting in twin bubbles protruding from the line of the aircraft proved to be a communication nightmare for the pilots. As a result, in 1949 the remaining C-74s were modified to a standard greenhouse canopy cockpit configuration.

A variation of the greenhouse C-74 cockpit glass placement was incorporated in the design of the successor C-124

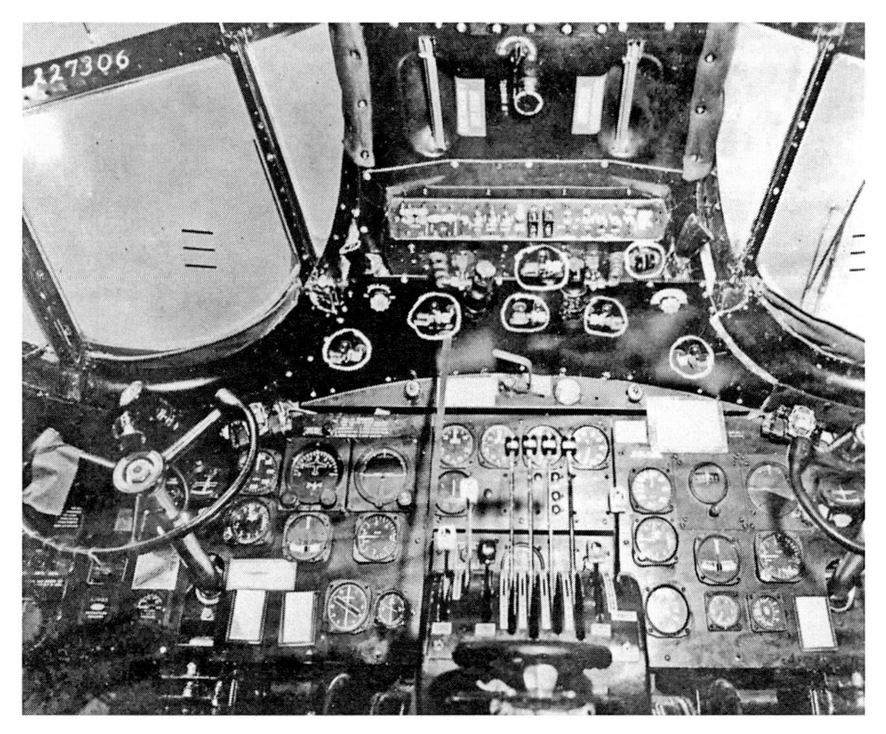

The original C-74 cockpit had bizarre twin canopy configuration. Each pilot had wide outside visibility with their heads in the canopy. Communication between them was very difficult because it was like they were sitting in two different cockpits. During evaluation the switches on the upper panel were marked for relocation (Douglas, courtesy Nick Williams).

The C-74 being completed with Pratt & Whitney R-4360 engines. The Fowler flaps extend the full length of the wing and the fuselage loading elevator doors are open. The twin canopies have not been installed. The open access door on top of the number four engine nacelle provides access to a large compartment big enough for an airman to stand up. The compartment can be accessed in flight via a crawlway in the wing (Douglas).

Globemaster II. Three C-74s were never upgraded to the standard cockpit windscreen configuration. They are number two (42-65403 c/n 13914) which crashed during flight testing in 1946, number four (42-65405 c/n13916) diverted to static test in 1946 then broken up in 1948 and number five (42-65406 c/n 13917) which was returned to Douglas in 1948 for conversion to the first C-124.

The C-74 weighed 85,000 pounds empty and 145,000 pounds loaded but has a wartime overload maximum ATOG of 165,000 pounds. At a span of 173 feet, 3 inches it was a giant of the day. Not even the Pregnant Guppy, Super Guppy with extended center-wing, CL-44 Guppy after conversion or the Airbus A300-600ST Beluga had a greater wingspan. The C-74 length at 124 feet, 1.5 inches is longer than the Boeing B-377/C-97 pre-guppy airframes.

These comparisons give some perspective on size of the C-74 fuselage relative to other airframes of the day. It was 43 feet, 9 inches high with a wing area 2,510 square feet. It cruised at 204 miles per hour with a max speed of 312 at 20,800 feet. The wing design was tested on wind tunnel models featuring electrically powered control surfaces. The most unique test came in development when a Douglas XTB2D-1 and later a Northrop A-17 was fitted with C-74 wing section sleeves, which slid over the existing wings like a glove. The test proved successful producing data which resulted in only minor changes in the wing design.

Douglas engineers decided to use the Fowler flap, designed by noted aerodynamic engineer Harlan D. Fowler. He designed an efficient retractable slotted airfoil which would provide increased lift of a large heavy aircraft on takeoff and landing with an acceptable lift/drag ratio. The full-span flap unit ran the entire length of the wing trailing edge. The retractable design provided increased lift at takeoff with minimum drag at cruising speeds. When the flaps were extended the outer aileron portion provided only slight action, however, when retracted the aileron operated in a conventional manner with greater control. The Fowler flap system, which was fitted to all 14 C-74s, was carried forward on 189 successor C-124s through aircraft s/n 51-182. The balance of

The flight crew observes loading procedures during testing. The C-74 was fitted with a 4,500-pound capacity swing out crane on the forward bulkhead for lifting cargo through the forward cargo door. This allowed both ends of the aircraft to be loaded at the same time. Even with the large forward door the aircraft was limited to cargo and vehicles under a certain size (DOD).

C-124A and C model aircraft production was fitted with the partial span flaps. The C-74 and C-124 were the only Douglas produced aircraft which used the full span Fowler flap design.

The C-74 featured a cargo-loading elevator in the aft fuselage floor behind the center-wing. The elevator allowed military vehicles to be lifted into the aircraft by a pair of 8,000-pound capacity overhead mobile winches. The cranes had a combined lift of 16,000 pounds. They could lift cargo and vehicles on the elevator platform or reposition cargo within the aircraft. One of the unique features of making the aircraft self-sufficient was the ability to perform engine changes at remote airfield locations. A specially designed hoist could be mounted on the nacelles. The cable from the overhead crane could be routed though a set of pulleys to lower the engine to the ground. A large cargo door on the forward left side of the fuselage was fitted with a 4,500-pound capacity swing arm hoist, which could be used to lift bulk cargo into the aircraft. The forward hoist and traveling cranes for the rear elevator expedited loading to both ends of the aircraft at the same time reducing turnaround time.[5] Even with much improved loading from previous transports the C-74 still lacked drive on/off capability.

Commercial Version

The expected postwar expansion in commercial airlines prompted Douglas to offer a C-74 passenger version designated DC-7 (not the later smaller production DC-7) to capitalize on the design and recover development cost. The new commercial passenger version was presented to Pan American's Juan Trippe in 1944 as a land based upgrade to the slow and boring long haul routes flown by the Clipper flying boats of the past.

Pan Am marketing still believed the public thought of air travel in terms of the large comfortable cabins of the Boeing 314 Clippers. The airline embraced the idea of a land based 108 passenger airliner because the design met the standard of the times of large aircraft with lounges and state rooms. These standards based on cruise ship travel did not change until the late 1950s, long after Pan Am ordered the civilian version of the Douglas C-74 (DC-7) and Convair XC-99.[6]

Trippe's strategy was to keep Pan Am on the cutting edge of worldwide passenger service. In 1944 the carrier placed an order for 26 C-74s at an estimated cost of $1,125,000 each with lounges and private staterooms for service on the newly proposed New York to Rio de Janeiro and Buenos Aires service.[7] Noted industrial designer Henry Dreyfuss was brought in to create the aircraft interior. By ordering both C-74s and XC-99 aircraft, it appears Juan Trippe was planning a major postwar fleet upgrade for Pan Am to maintain the carrier as the trendsetter in commercial aviation.

In August 1944, eight months after the original planned delivery date for first military C-74 aircraft, the first major assembly jig was completed. Six months later in February 1945 Pan Am ordered 15 giant Convair C-99 commercial airliners for the Trans-Atlantic and Hawaii routes. The Convair XC-99 was priced at $1,500,000 each. When considering the XC-99 was only $375,000 more and nearly twice as big, it appeared to be a bargain. Both the C-74 and C-99 were powered by the Pratt & Whitney R-4360 engines, which were costly to operate. The C-74 had been ordered by the military for the war effort. However, the first aircraft, 42-65402 was not rolled out until July 1945 less than two months before the war ended.

The civilian version of the Douglas C-74 was assigned the Douglas commercial designation of DC-7. It is not to be confused with the smaller production DC-7 model, which was the end model of the DC-4 airframe series. There is no reasoning for the designations. The DC-4 was the C-54 and the DC-6 became the C-118. The C-74 was an expanded and scaled-up DC-4 making it easy to understand why the DC-7 designation was used. Pan Am selected the name Clipper Type 9 to describe the newly proposed airliner. An industrial designer was employed to create elaborate renderings of luxury cabin interiors fitting for the affluent world traveler. All the appointments increased the aircraft weight to 162,000 pounds making the production cost eventually rise to a prohibitive $1,412,500 each in November 1944.

The postwar commercial air travel market was growing but not in the terms Pan Am expected. Market analysis showed the high passenger loads needed to justify the operational cost could not be guaranteed. Pan Am reconsidered the order and canceled in October 1945 opting for smaller equipment, which could show a profit with lower passenger loads. The cancellation ended any chance of a commercial version C-74 Globemaster I.[8]

Even though Pan Am cancelled orders for the C-74 commercial version (DC-7) in 1945, Douglas continued studies on upgraded turboprop models. In August 1946 a turbine model was proposed with Wright T-35-1 engines, which increased estimated gross weight to 218,000 pounds. The 4,000-shaft horsepower (shp) turboprop selected was being developed as the follow-up engine for large bombers fitted with the reciprocating Pratt & Whitney R-4360. The turboprop C-74 was projected to cruise at 352 miles per hour. A larger more advanced stretched version was proposed by inserting a 13-foot, 4 inch fuselage section behind the wing increasing the gross weight to 230,000 pounds. The cost was estimated to be a prohibitive $4,100,000.[9]

The biggest problem with the proposed Wright T-35-1 engine was the size. It was 59 inches in diameter, which required considerable engineering changes to the nacelles in order to fit it to the C-74 wing. Two years later Douglas again studied another engine upgrade for the C-74. Two proposals were considered. The first was the 4,000 shaft horsepower Allison XT-40 with 16-foot counter rotating props. The second was the Pratt & Whitney PT-2D turbine with four blade Hamilton Standard props.[10]

The PT-2D designation is the Pratt & Whitney internal company designation to the YT-34 turboprop engine, which was eventually fitted to the single Douglas YC-124B Globemaster II. It was originally proposed as the YKC-124B as a competing design with the Boeing KC-97 tanker. The tanker configuration was dropped before it rolled out as an YC-124B. The Pratt & Whitney YT-34 (PT-2D) engines were installed on two re-designated Boeing YC-97J and two Lockheed YC-121 aircraft as part of testing for use on the Douglas C-133. One of the YC-97J test aircraft and parts of the other were acquired by Jack Conroy with the assistance of Wernher von Braun. The engines, wings and cockpit nose from the YC-97J were used to build the Aero Spacelines B-377 Super Guppy. In a strange twist the engines did find their way to two guppy aircraft the YC-124B and the B-377SG. Douglas eventually suspended development studies of fitting turboprop engines to a C-74. The focus changed and they were fitted to the C-124B successor. After the USAF selected the Boeing KC-97 for aerial refueling in 1948, the program to produce a turboprop powered C-124 Globemaster was abandoned.

Pan Am ordered 26 of the 108-passenger commercial C-74s, which were designated as the DC-7 in 1944, at a cost of $1,125,000 each. At the time Douglas was still promoting the twin cockpit configuration although pilots were requesting a modification because of the flight deck communications problems. The changes did not come until May 1948 (Pan Am, courtesy Glenn Iverson).

Flight Testing and Military Evaluation

The C-74 was considered to be more complex than any large aircraft previously built. Crew selection was limited to only the most experienced pilots. In reality it turned out to be easier to fly than the DC-4 from which it was derived. Initially, pilot requirements were set at 2,000 hours with

Unlike the DC-4, the C-74 required a full flight engineer's panel situated directly behind and below the co-pilot. The pilots sat separate from the crew. A curtain could be drawn, allowing the lights to be turned up in the rear of the cockpit during night flight operations. The six manual fuel tank levers are visible on the far left of the engineer's table. This engineer's panel and cockpit configuration was carried over to the expanded fuselage C-124 (Douglas, courtesy Nick Williams).

250 hours as commander of B-29, C-54 or C-69. In addition pilots were required to have 500 hours of night flying and 75 hours of instrument time. Douglas Aircraft picked their best and most recognized test pilots to crew the first C-74 flight in August 1945.

Ben Howard was selected as pilot; John Martin, co-pilot; Jack I. Grant, flight engineer; and Christian B. Nielsen, flight inspector.[11] Howard who originally worked for Curtiss Aircraft came to Douglas in 1940. He set a name for himself racing aircraft. In 1937 he formed Howard Aircraft, which operated until 1944 where he built the now famous DGA cabin monoplanes. He was hired by Douglas in 1940 as a test pilot. Howard flew the first A-20 Havoc and A-26 Invader and is known as one of America's most outstanding engineer test pilots. He eventually became Assistant to the President of Douglas Aircraft.[12] (Ben Howard is not to be confused with Durrell "Dee" Howard who formed Howard Aero in 1947. Dee Howard proposed a 10 engine; 40-foot diameter DC-7 based Super Guppy to NASA in 1964.)

Co-pilot John Martin was a United Airlines pilot who also taught instrument flying for the U.S. Weather Bureau. He started as a crop duster and hunted coyotes from the air for sheep ranchers. Martin was also hired by Douglas in 1940. He was an engineer test pilot on the DB-7, B-23, DC-3, A-20, A-25, A-26 and C-54 aircraft.[13]

Nielsen served as a Navy pilot in 1938–39. Douglas placed him in charge of all engine testing for the Pratt & Whitney R-4360 program on the C-74. He also headed up the team conducting all wind tunnel testing of the C-74 at Wright Field. He was in charge of engine installation on the C-54s and was in the first flight crew. Neilsen had also flown as a crewmember on numerous test flights of the XB-19.[14]

Jack Grant began with Douglas Aircraft in 1930 as an experimental flight engineer. He was considered to be extremely qualified. He was also a member of the flight crew on the experimental Douglas XB-19, which was the world's largest bomber at the time. Grant was the engineer on the first C-54 flights and was in charge of all flight crews on the early A-26 flight test.[15] The C-74 was considered to be a "flight engineer's airplane." Because of the complexity and size, it required only the most qualified flight engineers.

The Douglas C-74 Globemaster I fleet was successful in demonstrating to the military the ability of a large aircraft to transport heavy cargo loads. The Air Force began to see a need for an airframe, which could not only lift tonnage but also transport outsize bulk items. The concept of volumetric lift was dawning as a need for future military operations. Convair and Boeing were also designing larger transports to meet military future needs and generate revenue after the postwar cancellation of combat aircraft orders.

The original wartime order of 50 C-74 aircraft and one static test model was reduced to only fourteen when the war ended. Ultimately only 12 entered active service. The second aircraft, 42-65403 crashed on 05 August 1946 during a demonstration flight with only 19.25 hours on the airframe. The number of operational aircraft was reduced again when the fourth aircraft 42-65405 was delivered to the Air Force 06 June 1946 and decommissioned on 06 August 1946 for static testing and eventually destroyed at Wright-Patterson Air Force Base.

The military need for volumetric outsize lift combined

with straight in drive on/off loading prompted Douglas engineers to design a new double deck configuration for the C-74 airframe by utilizing the same lower fuselage half, wings and empennage. The fifth C-74, 42-65406 which was the first delivered to an Air Force squadron on 19 September 1946, was selected as the base for the first C-124 Globemaster II project. It was returned to Douglas in June 1948 to be converted to an YC-124A. The aircraft was not decommissioned for the conversion but was considered an upgrade to an existing airframe retaining the C-74 Air Force identification 42-65406 after conversion to a C-124.

In March 1947 another C-74 was flown to Pope Field for evaluation by the Troop Carrier Command. The C-74s were not fitted with the waist passenger entry door, which became standard on the production C-124. The lack of waist doors is one of the reasons the military concluded the C-74 configuration was not suitable for paratrooper drops limiting the proposed universal military application. It was flown from Pope Field to Randolph Field, Texas, for further evaluation for casualty litter evacuation and as a troop transport. The rear fuselage elevator system contributed to a favorable review and acceptance for troop transport and casualty litter transport. The loading techniques, turnaround time and proper handling and securing of large vehicles such as tanks and trucks were evaluated.

The C-74 on display during evaluations in March 1947. Looking forward, the overhead traveling cranes with combined lift of 16,000 pounds are positioned over the elevator well. The troop seats along the sides are in the stowed position. It is obviously a large capacity aircraft; however, the relatively low ceiling prevented the transport of oversize cargo (USAAF).

Long-haul cargo delivery and logistical support flights were operated from Brookley Air Force Base in Alabama to evaluate the aircraft. The flights to Alaska, England, Hawaii, Iceland, Libya, Morocco, Panama, Puerto Rico and other destinations were reviewed to determine the response time in the event of international Emergency. World War II had tested the America's air logistics to put military materials in distance and remote locations. The growing threat of the Cold War presented an even greater need to reduce response time to anywhere on the globe.

Berlin Airlift

The Soviets closed all land routes to Berlin on 24 June 1948. The next day the Air Force was assigned the task of air lifting 25 tons of supplies to Berlin daily. Two days later on 26 June Mission "Vittles" (later named Berlin Airlift) began with the airlifting of 80 tons of cargo proving the task could be done. On 14 August a single C-74 Globemaster I (42-65414) was positioned at Rhein-Main Germany for airlift work. On 17 August it arrived at Berlin's RAF Gatow (EDBG) airfield in the British controlled sector near Havelsee Lake with 20 tons of flour. The British were transporting salt from Hamburg to Berlin by landing Short Sunderland seaplanes on the lake. The C-74 proved to be superior to British aircraft because it could be unloaded much faster with the onboard winch and elevator system. On 18 September the single C-74 made six roundtrips over 20 hours transporting 250,000 pounds of coal. A total of 24 C-74 Globemaster flights were operated to Gatow transporting 1,234,000 pounds of cargo.

When heavy equipment was needed for the construction of new runways at Tegel (EDDT) airport in the French sector of Berlin the C-74 was brought in because it could uplift very heavy loads; however, it was not front loading, which complicated the task. The equipment was too large to fit through the side loading door or the rear cargo elevator. The machinery was disassembled at Rhein-Main, Germany then loaded onboard the C-74 and flown to Tegel where it was re-assembled. The immediate need for a front-loading drive on transport was apparent.

Even with these limitations the value of C-74 operations did not go unnoticed. Soviet military officials claimed the cargo plane, which had rear opening cargo elevator belly doors, constituted an aircraft capable of being used as a bomber. The Soviet protest was a factor but the weight of the aircraft forced it to be the withdrawn from the Berlin operation. The runways were not stressed to accommodate a heavy outsize aircraft, which influenced the decision to return to Brookley Air Force Base at Mobile, Alabama. Although the C-74 proved too large for the "Vittles" operation, it continued to serve the mission by transporting spare engines for the lighter Douglas C-54 aircraft flying Berlin supply segments. The six weeks of outstanding performance in Germany reiterated to military planners the need for an even larger transport for future global operations.[16]

During the Berlin Airlift, the C-74 aircraft 42-65414 unloads sacks of flour at Berlin's Gatow airport while Berlin citizens and local children marvel at the giant aircraft. The aircraft also transported 250,000 pounds of coal in six round trips in just 20 hours (DOD, author's collection).

Phase I Modifications

The Douglas C-74 flew two years before the larger Convair XC-99, which was never intended by the military to be anything other than a test aircraft. Conversely, the original order for fifty C-74s indicated it would be built in large numbers. Ultimately only 14 were produced. The C-74 rush to production eliminated any XC or YC test models. It went into service as the largest operational transport produced to date with multiple revolutionary features, which were not proven at the time. As a result, many modifications and upgrades were progressively implemented. Phase I modifications began in the fall of 1947 when the first Globemaster I was returned to Douglas for a needed engine upgrade. The aircraft had been fitted with the problematic Pratt & Whitney R-4360-27 engines of which only 84 were produced. They were replaced with the improved 3,250 horsepower R-4360-49 engines in September 1947 retaining the Curtiss-Electric four-blade props.[17]

The C-74 twin canopy configuration was an untried feature that Douglas engineers believed would give the pilots greater visibility. It was the largest transport ever built at the time, which required considerable attention on the ground. In addition to the pilot communication problem the heat and glare off the fuselage prompted the painting of a flat black shield (Douglas).

Phase II Modifications

The twin canopy cockpit arrangement was another untried feature which was not well liked by the pilots. It was originally designed to give the pilots 360-

The Air Force and Douglas engineers met in May 1948 to review the cockpit modification. The result was a greenhouse design that improved pilot communication. The new canopy only partially addressed the glare and heat from the sun with sliding shades. The seventh C-74 42-65408 was the first returned to Douglas for installation of the new cockpit enclosure (Douglas, courtesy Nick Williams)

The C-74 42-65408 in service after installation of the improved cockpit windscreen. The aircraft was over utilized because it was the primary high-volume large transport in Air Force inventory. Even with the cockpit modifications other systems upgrades were needed. By 1952 the C-74 was experiencing maintenance problems and severe parts shortages (USAF).

degree visibility but it created the feeling the pilot and co-pilot were sitting in two different aircraft. Communication between the pilots was difficult and had to be conducted via the interphones. The sun reflecting off the shiny fuselage nose created such a glare it was often difficult for the pilots to see. A flat black coating was applied around the twin canopies as an attempt to deal with the glare.

On 03 May 1948 officials from the Air Force met with Douglas engineers at Long Beach to discuss cockpit modifications. The conference group agreed on a new conventional windscreen design to be installed during Douglas factory Phase III modifications. The new green house cockpit eliminated the twin canopies but afforded a wide range of visibility. The new design only partially addressed the glare and heat of the sun. Sliding shades were fitted to the overhead glass and a central hatch was installed in the rear portion, which could be opened on the ground. The seventh C-74, 42-65408 c/n 13919, was the first scheduled for the new "greenhouse" cockpit enclosure. It was returned to Douglas on 07 March 1949 for the upgrade. The last canopy modification of the 12 aircraft was completed in October 1949.

During the 1948 conference, multiple changes were reviewed including the placement of the autopilot; flight instruments, nose wheel steering and trim adjust. After consideration the Air Force decided it would not be cost effective to upgrade the remaining C-74s in service.[18] The military relegated the C-74 to a limited production short-term stop gap aircraft in favor expanding the fuselage to create the Globemaster II. It was determined the expense would be better served by incorporating the changes into the C-124 mockup and implement them on the production aircraft.

During the first two years of the Korean Conflict (1950–51), the C-74s were used primarily to move cargo and wounded personnel to and from Hickam Air Force Base in Hawaii and the continental United States. Because of the lack of spares, specifically system parts and spare engines, the operations over time became impractical. By 1952 the C-74 program was experiencing maintenance problems and severe parts shortages. The Air Force continuing need for the C-74 as a high-volume transport was taking a toll on the airframes. They remained in service through 1953 with increasing maintenance issues.

Phase Out Operations

In 1954 an additional program was implemented with needed upgrades. Three of the aircraft, 42-65411, 412 and 414 were fitted with nose radar units with a small radome. The Air Force utilization rate was reviewed and reduced after several in-flight engine failures. On 26 November 1954 one of the C-74s suffered a fire in number one engine. The fire was extinguished and it was landed safely. A similar incident occurred soon after emphasizing the question of aircraft reliability.

The 1703rd Air Transport Group (ATG) conducted a study in March 1954 titled "C-74 Modernization" which reviewed the maintenance problems and needed aircraft upgrades. It was presented to Continental Air Command (CAC) for evaluation. After reviewing the incidents of potentially catastrophic in-flight engine fires, the CAC recommended retirement and replacement of the fleet as rapidly as could be accomplished.

By 1955 C-74 crews began cross training on the C-124, which had entered service in May of 1950. The C-74 aircraft utilization was drastically reduced to two-hours per day in 1955 because of extreme high maintenance on the deteriorating systems. Shortly thereafter they were grounded for failures of fuel selector valves and deteriorating elevator pins. The C-74 had served well providing data for the military on long-range heavy lift multi-engine performance, which was used to develop standards for other aircraft and a data base for the C-124 replacement. The C-124 program was progressing so well the Air Force placed the C-74s in flyable storage at Brookley Field in Mobile, Alabama on 01 November 1955. Eleven of the fourteen C-74s built were subsequently retired from the Air Force on 31 March 1956 at Brookley.

Despite the low utilization over the previous six months the fleet made 45 cargo flights to Africa, Europe, Middle East and South America transporting six million pounds of cargo, one million pounds of mail and 1750 military personnel. In July 1956 the 6th ATS was deactivated and the crews transferred to the 1703rd Air Transport Group. Brigadier General George S. Cassady had accepted the first C-74 for the Air Force in 1945. After the decision was made to retire the C-74s, he requested and received permission to pilot the last C-74 from Mobile, Alabama, to Davis-Monthan Air Force Base at Tucson, Arizona, on 31 March 1956.[19] The aircraft were subsequently placed up for sale.

Globemaster I Civilian Operations

The Castro Connection

The following account is taken from testimony before the 86th Congress Senate Select Committee on improper activities in the labor field. It is relative to criminal activity within organized labor in the purchase of surplus military Douglas C-74 aircraft by Akros Dynamics Corporation. The witnesses were questioned by Senate members, which included Senators John F. Kennedy, Barry Goldwater and Sam Ervin. The hearings revealed the Teamsters union attempts to purchase the aircraft with pension fund money in order to profit from an aircraft sale to Fidel Castro and/or Fulgencio Batista.

This account is not directly related to the construction of the construction of the B-377 guppies. However, there is considerable cross involvement of certain individuals who resurfaced in relation to financial dealings of Unexcelled Chemical's acquisition of Aero Spacelines in the 1960s. The fate of the C-74 fleet and the part it played in the evolution of the ultra-large guppy transport is a unique. In addition the final owner of the last C-74 aircraft was ex–United Airlines pilot Orvis Nelson who founded Transocean Airlines. The carrier's B-377 Stratocruiser fleet was acquired by Aero Spacelines to be used as the base airframes used to build the guppy series of outsize transports.

Shortly after the C-74s were secured and processed at Davis-Monthan, the Air Force placed the lot of 11aircraft along with all spares and 26 Pratt & Whitney R-4360 engines up for bid. Major McCampbell, Chief Disposal Officer at Norton Air Force Base, San Bernardino, California, was responsible for compiling the catalog (NO. 04-607-S-57-21), which listed the aircraft and all parts up for bid. Major McCampbell reviewed all the aircraft stored at Davis-Monthan and parts inventory at Brookley and Norton Air Force Base. The C-74s were retired because of lack of parts and deteriorating condition. Why they were placed up for sale and any offers considered except from scrap dealers is a mystery.

In November 1956 a William Steiner of California began making inquiries about the aircraft and visiting Davis-Monthan, Air Force Base. In early 1957 Steiner who was later identified in senate investigation hearings as "the salesman or middleman" of an aircraft buying scheme perpetrated under the corporate name of Akros Dynamics. Steiner had a silent partner identified only as Mr. Patterson who was actually soliciting buyers on his behalf. Patterson's true identity was never revealed.

Steiner contacted an individual in Cleveland, Ohio, identified as Earl T. Benjamin offering him an opportunity to participate in a deal to purchase the surplus cargo aircraft for resale. Steiner, who is reported to have been a friend of Benjamin, stated the aircraft were in excellent condition and it would be easy to get them certified for civilian cargo service. Steiner told Benjamin he had investors who would come up with the majority of the purchase price. He stated he would introduce Benjamin to an individual identified as Fred Mingo of Lincoln Rebuilders in Cleveland who represented several other interested individuals.[20]

Benjamin was introduced to Fred Mingo but not told of the other investor Dominic Edward Bartone who was employed by Mingo and had questionable ties to individuals in Central America. Steiner explained to Benjamin the two of them could secure a loan for the difference between what Mingo and his investors could not provide and each could earn 25 percent on the profit from the resale of the aircraft. The deal almost immediately took on a nefarious nature when Mingo and Bartone went to California in an effort to purchase the planes themselves for resale and undercut Steiner and Benjamin but they were not successful.[21]

On 18 April 1957 Akros Dynamics was incorporated in the state of Ohio for the sole purpose of buying and reselling the eleven surplus C-74 aircraft. Earl T. Benjamin was named president, William Steiner vice president and Mike Zappone owner of Tony's Kamms Corners Restaurant in Cleveland was named Akros Proprietor. The other partners included a Farrell Pennsylvania Banker Guy Gully, Marvin L. Kast and a previously named partner of Steiner identified only as Mr. Patterson formerly with Allied Airlines. The airline is thought to be a postwar start-up or holding company; however, no records were located on the operation.

William Steiner called a meeting in 1957 just after Akros Dynamics was formed. Gordon Hamilton of Hamilton Aviation of Tucson, Arizona, had expressed an interest to the Air Force in obtaining the C-74s but was being overshadowed by the Akros group. He attended the meeting at the request of Steiner to consult on getting the C-74s airworthy. Two others in attendance were Earl T. Benjamin and Cleveland, Ohio restaurant owner Mike Zappone. They were introduced as the two principals of Akros. Also in attendance were Roman Catholic Priest Father Bernard Brady, Marvin L. Kast and Guy Gully who accompanied several unidentified individuals who seemed to be connected with Akros.[22] It is not clear why Father Bernard Brady was present except to possibly add legitimacy to the proposal.

Steiner had selected Earl T. Benjamin to participate because his credentials were needed as a legitimate business owner. He lived in Cleveland, Ohio, had previously participated in aircraft ventures, was a pilot and had reputable credentials as president of Welded Construction Company. In addition he was associated with the Port and Harbor Commission of Cleveland, Ohio. He also had a financial interest in his brothers company Aero-Ways Incorporated, an airplane charter and storage company at Cleveland Hopkins Airport.[23]

It should be noted the State of Ohio had a considerable Teamsters Union presence and was closely allied with Teamster senior officials. James "Jimmy" Hoffa was President of the International Brotherhood of Teamsters (IBT) at the time and maintained an office at IBT headquarters in Washington, D.C. It was reported in the hearings that Hoffa had some rather dubious syndicate friends who had a major presence in Cuba under the Batista regime.

Batista held a close relationship with American crime boss Meyer Lansky and Lucky Luciano for over ten years. Lansky was prominent in the Cuban gambling operation and had been given control of the casino and racetracks in exchange for kickbacks. It was in the best interest of the syndicate for Batista to remain in power. Conversely Castro became the darling of the American news media as the "Robin hood of the Caribbean" during the Cuban revolution. The syndicate believed that Castro could be influenced by cash the same as Batista and the gambling operations would continue.

From 1956 to 1958 Fidel Castro waged a guerrilla war against the Batista regime. In 1959 when Castro ousted Batista, the syndicate viewed Castro as an ally. Arms had been shipped to both Castro and Batista during the revolution in order to curry favor. It appears they were playing both sides believing they could buy Castro's cooperation in the event he was the victor. Edward Partin, who was a business agent for the Teamsters in New Orleans, later testified before the Senate Committee. He stated Hoffa and Teamster officials were in the business of ferrying arms from Florida to Cuba and were present on several occasions when boats were loaded under Hoffa's direction.[24]

The mob was working on multiple fronts to maintain its control of Cuban gambling. The purchase of the C-74 aircraft was in process before Castro defeated Batista. On 10 May 1957 the contracting officer at Norton Air Force Base offered the eleven C-74 aircraft and spares catalog for bid. The bids were to be opened on 19 July 1957. Akros submitted the only bid in the amount of $150,000 for seven of the aircraft in the catalog. The bid was for $20,089 for each aircraft and the balance for spares and parts.

Steiner either by design or simply being naïve inspected the aircraft and reviewed the photos of the interior provided by the Air Force. He convinced Earl Benjamin everything in the photos was included with the aircraft. However, the Air Force bid catalog had clearly stated the photos were for demonstration purposes only and did not include radio equipment, autopilot, navigation equipment, engine analyzers and numerous other electronics. Those items had since been removed and the bid was for the bare aircraft, mechanical spares and engines.

The Air Force refused the bid stating it was too low because the spare engines alone were worth more than $500,000. Akros submitted a second bid for $1,500,000 for all eleven aircraft and listed parts plus only two spare engines at $27,114 each for a total of $ 1,554,228. This alone demonstrates the group's unfamiliarity with large aircraft and the reliability of the Pratt & Whitney R-4360 engines. It would be impossible to operate 11 aircraft with only two spare engines. The contract was modified again to include a third spare engine for a total contract of $ 1,581,342. A legitimate operation would have known the aircraft would require a minimum of a spare engine for each aircraft.

The Air Force accepted the second bid. Pennsylvania banker Guy Gully put up $300,000 as a performance bond. At this point in time Gully and Benjamin jointly owned 65 percent of Akros stock. The other 35 percent of the stock was owned by Steiner and Patterson. Shortly after the contract was accepted Gully withdrew his funds maintaining they were committed elsewhere. Patterson, who had originally inspected the aircraft at Davis-Monthan in Tucson, appeared to be the self appointed salesman of the group. He had been seeking a buyer for the C-74s; however, after Gully withdrew his funds Patterson did not surface again as a partner.

Without Gully's investment the Akros group was left to find the other 19 percent financing elsewhere. The contract called for $500,000 down on 24 August 1958 and another $200,000 payment on 05 November 1958 with eight payments of $110,167.25 beginning 01 December 1958 at one half percent interest on the balance from 25 August 1958. The first four C-74s would be released upon receipt of the $500,000 payment. The four aircraft selected were 42-65404 N3182G, 42-65408 N8199H, 42-65409 N3181G, 42-65412 N3183G with others to be released as payments were received. The four aircraft had the following hours respectively 8,612; 8,496; 9,502; 5,825.

Steiner's original major investor in the aircraft buying scheme was Cleveland restaurant owner Mike Zappone. He had convinced Steiner that he commanded considerable assets. In reality Zappone had little cash and was unable to put together enough investors to finance even a portion of the 35 percent of the $1.5 million of his and Steiner's part. Steiner then brought in Cleveland businessman Alvin A. Naiman in an effort to re-form the group and secure enough capital to continue. Naiman was believed to have enough assets to secure a $300,000 loan. He owned Naiman Industrial Wrecking and Scrap Company and was also owner and President of Niagara Crushed Stone based in Ontario, Canada.

On 10 March 1958 Naiman and Zappone signed a contract with Earl Benjamin becoming part of the Akros investment group. Steiner told them he had several interested parties and if they could get the aircraft certified at a cost of $50,000 each they could sell them for $400,000 per aircraft. On 01 October 1958 Steiner resigned his position allowing Naiman to take over as vice president. Naiman then appointed his son Jack Naiman as Secretary-Treasurer of Akros.[25]

Earl Benjamin accompanied Alvin Naiman to Miami to apply for a loan from the Pan American Bank to purchase the aircraft. The bank vice president Maurice K. Lewis was immediately skeptical of the loan because Akros was set up solely to purchase the aircraft. The company had no assets and the only party verified to be even remotely interested in the aircraft at the time was Greek-Ethiopian Airlines. The bank required additional collateral for the loan and suggested a rather creative way to make the loan while at the same time getting bad debt off the banks books.[26]

The Pan-American bank held a note on Aircraft Instrument Company owned by Mr. Gus DeMeo in the amount of $419,000, which was in default. DeMeo's assets consisted of 39 surplus Consolidated PBY Catalina Flying Boats, one Lockheed Lodestar, another unnamed aircraft and a stock of aircraft parts which were estimated to be worth less than $100,000 in total. Naiman contacted Pennsylvania Banker Guy Gully who was one of the original Akros primaries. Naiman ask Gully if he could influence the deal as another banker. Gully contacted Mr. Lewis at the Pan-American bank in Miami suggesting "banker to banker" that Akros was actually an aircraft brokering company. As such it would take the mortgage on Mr. DeMeo's airplane assets to get it off Pan-American's books in exchange for the bank making the loan to Akros Dynamics.

It appears Gully was playing both ends of the scheme. DeMeo owed the bank $419,000 yet the loan would be for $500,000. Gully would pocket $81,000 for putting the deal together and ultimately saddle Akros with additional debt. Gully obtained a loan for Akros Dynamics in the amount of $840,000 of which $500,000 was to pay off DeMeo's loan to the bank leaving $340,000 for Akros. Another separate loan was arranged for $200,000 to the purchase the C-74s leaving Naiman (Akros) with $540,000 operating funds. Earl Benjamin as President of Akros used the $500,000 to pay the down payment to the Air Force. The $40,000 was used for insurance and to get one of the aircraft, 42-65408 (N8199H) transferred to Hamilton Aviation in Tucson and made airworthy. The mortgage was for sixty-days with a renewal on 17 October 1958 with an interest payment of $8,400. Steiner and Benjamin still maintained and possibly believed Greek-Ethiopian Airlines was interested in purchasing several of the aircraft for cargo work.[27]

It is also possible some of the players were told the aircraft were being acquired for Greek-Ethiopian Airlines to keep them from knowing about the intent to sell them in

Cuba. The less the investors knew of the real intent and the organized crime connection the better. The U.S. government was very concerned about the deteriorating situation in Cuba. If the real intent to purchase the aircraft was known investigators could have pressing questions for the investors. The first four aircraft were released by the Air Force and at least one moved to a storage facility leased by former Detroit crime boss Peter Licavoli who was living in Tucson.[28]

On 21 January 1959, with no sale of the aircraft on the horizon, Steiner brought in Dominick Bartone as an investor even though he and Mingo had previously tried to undercut the deal. Steiner told Naiman to introduce Bartone to Earl Benjamin as possibly a West Coast bail-out deal. Bartone, who would later be arrested for gun smuggling, brought in an individual named Jack LaRue, who co-owned International Trading Company Incorporated. Benjamin was not told Bartone had been negotiating with the Cuban dictator Fulgencio Batista to possibly purchase the aircraft. Bartone and LaRue claimed they could sell the C-74s for $400,000 each for a 10 percent commission. Bartone and LaRue took an option on two of the C-74s on 21 January 1959. Castro had taken over Cuba only 20 days earlier deposing Batista.[29]

U.S. Senate hearings on improper activity in the labor field determined Dominick Bartone was actually an arms dealer who was loosely associated with Jimmy Hoffa. It appears he intended to gain control of Akros Dynamics in order to secure the already in place financing with the Pan-American Bank to purchase the eleven C-74 Globemaster aircraft. His intent was to then sell them to Batista. Bartone and various partners tried to arrange a loan from the Teamsters to finance the deal but they had not been successful at the time of Batista's fall from power. The group supported Batista but had been playing both sides in the Cuban revolution. With the change in government the group attempted to sell the aircraft to Castro.[30]

After Castro overthrew the Batista regime and Akros was having problems obtaining a loan, James "Jimmy" Hoffa tried to obtain a $300,000 loan from the Teamsters Central States, Southeast and Southwest pension funds on behalf of his associates who had allegedly been running guns. During the Senate hearings, Edward Partin, business manager for the Teamsters in Baton Rouge, Louisiana stated, "The whole Akros Dynamics thing was purely and simply Hoffa's way of helping some of his mob buddies who were afraid of losing their business (casino and racetrack) in Cuba. They were trying to score points with Castro right after he moved in."[31]

Alvin Naiman by most accounts was a legitimate businessman. He was the major financial backer of the C-74 aircraft purchase and agreed to the loan with Pan American Bank. He demanded the other players give him complete control of Akros Dynamics in order to make the deal with Bartone's International Trading Company. Bartone made it clear he only wanted to negotiate with one person. Naiman was also becoming desperate because he had placed his business assets on the line and wanted to make a deal to get rid of the aircraft or have another investor buy him out.

The Senate investigation report determined on 11 Feb 1959 that Earl T. Benjamin signed over all Akros Dynamics stock along with his resignation and those of all Akros corporate officers. There was a provision stating if Bartone and LaRue could not make the "West Coast deal" all stock and the six resignations would be returned to Akros vice president William Steiner.

Naiman told Earl Benjamin he had taken care of the note with the Pan American Bank in Miami which was not true. The same day Naiman obtained the resignations and control of the company he met with attorney Herbert R. Burris and Louis Nuncio "Babe" Triscaro who was President of Teamsters Local 436 in Cleveland. Burris was introduced to Naiman as representing an interested banker and Triscaro said he could come up with the money if the deal looked good. Neither Herbert Burris nor Naiman were aware Triscaro had flown from Cleveland on 10 February 1959 and visited the Teamster Building in Washington, D.C., where Hoffa had his office. The same day an individual named Benjamin Dranow (aka Ben Davis) using the name Morris placed a call to New York from the Teamsters Building in Washington to Herbert Burris's father S. George Burris.[32]

The aircraft deal was collapsing because of financing. Payments to the Air Force for the aircraft were behind and the note with Pan American bank was also past due. Attorney Herbert Burris was dispatched by Benjamin Dranow to fly to Cleveland and meet Louis Triscaro. They discussed a deal to purchase the C-74s for a "Cuban" customer. Herbert Burris then called Dranow from Triscaro's office and told him the deal looked good. Naiman then told Earl Benjamin, former President of Akros Dynamics, that Dranow was a banker interested in financing the deal.

In reality Dranow was a front man for the Teamsters pension fund. In 1954 he had worked for John W. Williams Department stores. He was involved with the teamsters in some type of Teamster Jacket scam where he received $14.25 for every jacket sold in a money laundering operation. Herbert Burris had previously obtained a $1,400,000 loan from the Teamsters and his father George Burris obtained a $735,000 loan.[33] Dranow was later indicted for cashing a stolen $100,000 treasury bill and had been convicted of mail fraud in Minneapolis.[34]

It was agreed the trio of Naiman, Triscaro and Burris would fly to Miami to secure the loan. First they would fly to New York to present the deal to Burris's father, S. George Burris, to secure his approval. Then the next day, 12 February 1959, all four men flew to Miami to meet with Benjamin Dranow. The Senate investigators presented documentation indicating the plane tickets were paid for by the Teamsters Union.[35]

After spending the night in Miami, Alvin Naiman flew to Havana meeting Dominick Bartone who was already there. Although not verified it is suspected Naiman was flown to Havana by Chauncey Marvin Holt who was a friend of Detroit boss Peter Licavoli who lived in Tucson and Meyer Lansky who was attempting to maintain his Cuban gambling operation. Naiman and Bartone met with Cuban representatives of the supposedly potential buyer. Naiman then returned to Miami and met with Benjamin Dranow who said the deal looked good for obtaining the loan of $300,000 from the Teamsters. Dranow insisted Akros Dynamics, which was under the control of Bartone via Naiman, now be turned over to him.

Attorney Herbert Burris quickly drew up an agreement turning the corporation over to Benjamin Dranow for one dollar with the provision if the deal went through with Castro, Naiman would receive 15 percent of the profits. The agreement was open and did not contain Dranow's name. It would be added if the deal were completed. In the event Dranow could not close the deal with Castro the company would revert back to Naiman. From this time forward Naiman dealt only with Dranow's representative, attorney Abe Weinblatt.[36] Weinblatt was also involved in a separate Teamsters union land deal.[37]

Naiman was desperate for a loan because he was in default to the Air Force since August 1958 and the Pan American Bank since October 1958 and his industrial metal business was in jeopardy. Akros had taken possession of four aircraft and moved one of them. Any deal for the other seven aircraft would not be possible until the loans were satisfied. William Steiner originally contacted Hamilton Aviation in Tucson in August 1958 and asked Gordon Hamilton for a quotation to get C-74 N8199H airworthy and received a quotation of $6,000. Hamilton got the order to proceed in November 1958. He began work on the aircraft in December and completed it in February 1959.

Louis Triscaro, who was president of Teamsters Local 436, seemed very anxious to make all the arrangements to put the deal together. He called Gordon Hamilton in Tucson on several occasions to inquire about the progress. On 24 February Teamster local president Triscaro called Alvin Naiman in Cleveland to discuss the deal. On the 26th he called Koy Williams a Teamster official in Kansas City and a trustee for the pension fund. The next day he called Naiman again. Triscaro then made a phone call person-to-person to Mrs. Dranow in Las Vegas on 28 February. On 02 March he called Havana, Cuba and the next day he called George Burris in New York. The following day he called Teamster pension fund trustee Gene San Socci. All of this activity was followed by a phone call to Jimmy Hoffa at the Shorelands Hotel in Chicago on 11 March. Hamilton Aviation had completed the work on the aircraft and test flew it on 20 March. Naiman and Dranow each called Gordon Hamilton on separate occasions while the work was being done to inquire when the aircraft would be ready.[38]

Naiman had paid all monies up to this point. In February he told Gordon Hamilton that the operation would be led by Bartone. However, Dranow contacted Hamilton Aviation and made arrangements to have C-74 N8199H flown from Tucson to Havana on Friday 21 March 1959. He sent payment of $27,387.24 for expenses, gas-oil, radio equipment, and pay for crew etc. It included $10,637 to Hamilton Aviation and $4,750 for insurance. In addition he placed $3,500 in the Akros Dynamics account, sent $500 to Naiman, $1,800 to Bartone, $2,800 to Southern Pacific Railroad for shipping charges of parts from Mobile to Tucson and $1,000 to Central Truck Lines for delivery.[39]

Multiple flights were made to Havana by Bartone, Triscaro, and Naiman. Most of the flights were piloted by Chauncey Marvin Holt. Although he was not a crime family member, he became involved because of his close relationship with Meyer Lansky. Holt was a "jack of all trades" who became friends with Lansky's son and subsequently was offered work. He was an accomplished artist and art restorer, forger, pilot, mathematician and munitions expert. Considered to be fearless, he was a dubious character who freelanced for the CIA, FBI or anyone else who needed his services. He assisted in processing unregistered and stolen stock through Los Angeles broker Burt Kleiner. Holt claimed he became friends with Flying Tigers founder Robert Prescott because of his logistical assistance in CIA flight operations.[40] Holt really got around and was thought to be involved in some nefarious activity in Dealey Plaza the day President Kennedy was assassinated.

Alvin Naiman flew to Miami on 18 March 1959 to meet again with Louis Triscaro. Holt flew the two of them to Havana the next day to meet Bartone and Dranow and await arrival of the C-74 aircraft. All four of them remained in Havana until 22 March when all but Bartone returned to the United States.[41]

The March 21 C-74 flight from Tucson to Havana was under the command of Captain Ralph Johnson (formerly with United Airlines), Co-Pilot Captain Lee Evans, flight engineer George Bosley (Chief Pilot for Hamilton Aviation) and Gerald B. Juliani vice president of Hamilton Aviation who went along to coordinate and protect the interest of Hamilton Aviation.[42]

They were met by a large crowd at the Havana airport which included Naiman, Triscaro, Bartone and Dranow. Bartone was in charge of the group but no decision had been made as to when the aircraft would be flown again. With no scheduled return date the crew returned to Tucson. Dominick

Bartone was in the company of William Alexander Morgan at the airport. Morgan was an American fugitive who was sought by U.S. federal authorities. He was apparently a rebel leader who was given the rank of Major in the new Cuban army. He was also the head of the Cuban Provincial Police.

Pilot Marvin Holt recognized Morgan in the group. Holt originally met Morgan in 1950 but did not see him again until 1956 when he told Holt he was planning to join Castro. Holt stated in his book *Self-Portrait of a Scoundrel* that Morgan was, "the most devious, untrustworthy individual I have ever encountered." Despite his lack of character he was a devout Catholic. He was one of the most effective double and triple crossers of the day. Morgan was supposedly negotiating the deal for the C-74s for Castro.[43]

Prior to appearing in Cuba, Morgan was a driver for Dominick Bartone in Cleveland and Toledo, Ohio. He had been convicted for robbery in Ohio in 1948 and sentenced to five years. He escaped in 1949 and committed another robbery and was sentenced to three more years. He escaped a second time and committed another robbery and received five more years. He was arrested four times and escaped. Somehow he managed to find his way to Cuba where he became a leader in the rebel forces commanded by Che Guevara under Castro.[44]

On 28 March Dranow sent $3,800 and Naiman sent $2,200 to Maurice Lewis at the Pan American Bank in Miami for monies owed by Akros Dynamics. The next day Holt flew Alvin Naiman to Havana where the aircraft was still parked. Gordon Hamilton of Hamilton Aviation contacted Alvin Naiman and advised him the C-74 N8199H had to be returned to the United States. Hamilton had made the arrangements with the State Department to fly it out of the country and it was his responsibility to make sure it returned within six months.

On 29 March while in Havana, Triscaro told Naiman that Benjamin Dranow wanted out of the deal and was no longer interested. He said Akros Dynamic's corporate papers, resignations, etc., would be returned to him (Naiman) in Miami as previously agreed. This was very confusing because it appeared the Cubans were very interested in buying four to ten of the C-74s. In addition Maurice K. Lewis Jr., vice president of Pan American Bank, was also in Havana to review the deal and met with Dominick Bartone. Naiman (Akros Dynamics) was in default on a large loan and wanted the sale to go through with the Castro regime. Bartone met Lewis of the Pan-Am Bank and took him to the Havana Air Force base to inspect N8199H, which was guarded by the Cuban military. He met with Cuban officials and later stated at the U.S. Senate hearings he believed it was a bona fide transaction with the Cuban government subject to signing the purchase order.

In April 1959 attempts to complete the Cuban sale were quietly being negotiated because the Teamsters were secretly considering the loan. Marvin Holt stated he flew Dominic Bartone and Al Naiman to Havana at least four times in 1959 to negotiate the sale of the C-74s. He stated a loan with the Central Stated Pension Fund was in the works. Surprisingly on 01 April the Cuban government announced its intention to purchase four to ten C-74 Globemasters. Akros had title to four of the planes but a loan was needed to recondition the aircraft and get them to Miami for deliver to Castro's agents. Former Cuban president Carlos Prio Socarras, who had been deposed in a coup in 1952 by Fulgencio Batista, was supposed to underwrite the Castro purchase of the planes. Prio never produced any funds, causing the deal to sour.[45]

On 08 April 1959 Earl T. Benjamin received a package from (Teamster) attorney Abe W. Weinblatt of Miami containing the corporate kit of Akros Dynamics. It appeared the company and all the problems were now back in his lap; however Benjamin had resigned as President of Akros Dynamics prior to April 1959. Records indicate William Steiner was now President of Akros. Although Teamsters Local 436 president Louis "Babe" Triscaro set up meetings and attempted to facilitate the deal he denied he was involved in any way with Akros. However, when the company was in Dominick Bartone's control and later Benjamin Dranow's possession he made trips to Cleveland, New York, Washington and three flights to Havana with Naiman in a plane piloted by Holt. Triscaro was actively working with Naiman to put the deal together.[46]

The Teamster group had not been successful and the Castro deal appeared to be collapsing. Naiman was in desperate need of cash. He eventually put together a crew and the C-74 Globemaster N8199H was flown back to Miami. Louis Triscaro told Alvin Naiman the Teamsters had a loan agency, who may be able to help. On 13 April 1959 Bartone and Naiman went to Chicago and met with James C. Downs to apply for a loan from the Teamster Pension and Welfare Fund.

James Downs was the Chairman of the Board of Real Estate Research Corporation. He had been retained on 01 March by the Trustees of Teamsters Central, South East and South West Pension Funds in connection with a loan application. Bartone asked Downs if Mr. Hoffa had contacted him regarding a $300,000 loan. Downs stated he had not been contacted but told them to come back in the afternoon. Phone records indicated he called Hoffa in Washington. Hoffa told Downs it looked like a good deal and to approve the loan.[47]

Bartone and Naiman returned and Downs told them they would need additional collateral in the event the airplanes did not sell. Akros actually owned four of the C-74s but only one was flyable. Naiman agreed to put up Niagara

Crushed Stone which he owned, as collateral to secure the $300,000 loan. Dominick Bartone assured Naiman everything was fine because Dranow had worked with Hoffa and the loan would be approved.[48]

On 17 April 1959 James Downs learned a Canadian bank already held a $150,000 mortgage on Niagara Crushed Stone. Downs contacted the Teamsters on 24 April and told them this was a bad loan and he would not recommend it because there was no collateral. Alvin Naiman went to Hoffa's office in Washington on 01 May. While there Hoffa called Downs and asked what it would take to get the loan approved. Downs told Hoffa they needed to pay $156,000 to Dominion Bank of Crowland, Canada, to retire the loan on Niagara Crushed Stone and free up the stock as collateral for the $300,000 loan from the Teamsters Pension Fund. Otherwise there was no collateral. Naiman did not have the $156,000 so nothing was done. However, Naiman got a tip from an unnamed source on 07 May. He was told something was going on to approve the loan prompting him to call Downs. He learned Executive Secretary of the Teamsters Pension Fund Francis J. Murtha had asked the trustees to approve the loan, ignoring the recommendations of James Downs.[49]

Naiman, Triscaro and Gene San Soucie (original trustee of Central States Pension Fund) went to Hoffa's office. Hoffa told Naiman and Triscaro, "You have gotten yourself into a miserable deal down there with those airplanes." A few days later Stanford Clinton, who was the attorney for the Pension Fund, contacted Alvin Naiman and told him the loan was denied.[50]

The senate hearing proved Akros Dynamics was formed for the sole purpose to purchase the eleven C-74 Globemasters and at least four of them were acquired specifically for resale to Castro. Low level union leaders in Cleveland were involved from the earliest time. It is speculative and never determined if the Teamsters were interested in Akros Dynamics from the outset; however, it is conclusive at some point the Teamsters Union and James Hoffa were deeply involved in the scheme.[51]

Gun Running to Batista

After negotiations for the C-74s collapsed with the Castro regime and the syndicate could not gain favor with the new revolutionary government, it appears Dominick Bartone decided to support the opposition by supplying guns to pro–Batista forces in the Dominican Republic. It is suspected William Morgan, who was Bartone's former driver and inside man with the Castro regime, was expected to influence a new deal with Batista through Rafael Leonida Trujillo if not successful with Castro. This was never proven conclusively; however, a series of events including the collapse of the C-74 deal with Castro forces does leave the appearance of bad feelings and an attempt to return Batista to power.

Dominick Bartone contacted Hamilton Aviation in Tucson to inquire about an exhibition flight through South America. He had previously asked Gordon Hamilton for a price quote for gun turrets on the C-74 and other combat aircraft. Bartone then asked Alvin Naiman for permission to take the C-74 N8199H to Puerto Rico on a fourteen-day demonstration flight for review by some potential buyers. It was still in Miami after returning from Cuba. Later during the Senate Investigation hearings Alvin Naiman was asked why he would let a guy just take an aircraft valued at $400,000 off to Puerto Rico without more information about the interested parties. Naiman countered by stating he had only paid $75,000 for the aircraft.[52]

In April 1959 attempts to resurrect the Cuban sale were quietly being negotiated because the Teamsters were still secretly considering the loan. However, others in the group were attempting to sell the aircraft to Batista. Senate hearings determined an unidentified individual contacted the customs agent supervisor in Miami about possible need to move some guns and ammunition out of the country. The individual said he was not offering a bribe but someone else would be making payments if they could work something out. The operation mastermind was supposedly Dominican Republic strongman Rafael Trujillo.

The customs agent contacted the Deputy Commissioner of Customs, Chester A. Emerick and asked what he should do. Emerick discussed it with the FBI Bureau and the U.S. Attorney in Miami and was instructed to play along and report all meetings. The customs agents were told by the unidentified person they wanted to export $1,250,000 in arms in multiple shipments and would pay 20 percent of the value of the shipments or $100,000 total to the customs agents to look the other way. The customs agents were directed to agree to the deal by their supervisors. The unidentified contact instructed them to contact Augusto Ferrando, Counsel General of the Dominican Republic for instructions.[53]

Two customs agents went to the Dominican Consulate on 06 May 1959 where Ferrando gave them $400 as a "token of good faith" for meeting with him. Between 06 May and 20 May small shipment of arms began to arrive in Miami where they were stored in a warehouse. The customs agents met with Ferrando again on 19 May when he gave them $1,000 in one hundred dollar bills. He told them the arms were to be shipped to the Dominican Republic aboard a banana boat which was in Miami. The customs agents complained the $1,000 was not enough and they wanted $2,000. Ferrando agreed and said they would receive another thousand when the guns were on the boat. The customs agents were contacted again on 21 May and told the plans had been changed and the guns were now going by airplane. They were instructed to meet a man identified only as Dominick (Dominick

E. Bartone) in room 1103 of the Dupont Plaza Building that evening.[54]

Dominic Bartone had obtained a CAA ferry permit for a demonstration flight with the C-74 to San Juan, Puerto Rico, on behalf of Akros Dynamics. He stated there would be $65,000 in spare parts on board. For some reason one of the co-conspirators refused to let Bartone load the "parts." It is not possible to know what or if Bartone knew of the arms at this time. It appears the parts were the cover for the crates of arms. It is possible Bartone believed they were taking the aircraft covertly to Batista to complete the deal. The only qualified flight crewmember named was the pilot Samuel E. Poole. No other flight crewmembers listed were qualified on the C-74.

The C-74 is a complex aircraft which requires a minimum crew of three. It is known as a flight engineer's airplane because the four R-4360 engines are a handful to manage. The other flight crewmembers listed were New York attorney Sidney Neubauer, New Jersey arms dealer Charles Colle (aka Larry Lamarca), Miami Policeman Joseph Liquori, Leonard Trento, Dominick Bartone and Dominican Consulate Augusto Ferrando.[55] None of this group held a commercial multi-engine rating.

Dominick Bartone hatched a plan to fly to Puerto Rico to show the airplane to potential buyers. They would file a flight plan and as soon as the C-74 was in the air and leveled off out of Miami airspace the pilot would fake an engine failure and divert to Santo Domingo, Dominican Republic. There they would meet officials friendly to the Batista forces. Bartone had made arrangements to sell the C-74 for $400,000 to the pro Batista forces in the Dominican Republic.

The aircraft was parked at American Airmotive in Miami. About 10 a.m. on 22 May 1959 the group loaded the first shipment of arms consisting of 200,000 rounds of 45-caliber ammunition, 37 Garand M-1 rifles, and 21 machine guns. As soon as most of the conspirators were present customs and federal agents moved in and arrested them before they could depart. It appears the conspirators were not very knowledgeable of the law and could have gotten away with the plot. Deputy Commissioner of Customs Chester Emerick stated, what was most confusing was why they bothered to contact customs at all. They had received a CAA Ferry Permit to Puerto Rico, which did not require clearance from Customs, the State Department or anyone else. They were flying from the United States to Puerto Rico, a U.S. Territory.[56] In reality the ignorance of the law and the status of Puerto Rico caused the band of conspirators to make themselves known to authorities.

On 04 June 1959 a Federal grand jury in Miami indicted Leonard Trento, Augusto Ferrando, Joseph Liquori, Charles Colle (aka Larry Lamarca), Dominick E. Bartone, Samuel E. Poole Jr., and Sidney Neubauer. The group was indicted for conspiracy under section 371 of title 18, United States Code, to export arms and ammunition of war in violation of section 1934: of title 22, United States Code, and for bribing officers of the United States in violation of section 201 of title 18 of Unites States Code.[57]

The New Jersey arms dealer named Charles Colle (aka Larry Lamarca) owned Millville Ordnance Company in Union, New Jersey. It was a 7,000-square foot small business behind his home. In early 1959 Fulgencio Batista and two bodyguards visited Colle's warehouse and home in New Jersey. Batista negotiated a deal for weapons including 6,000 to 7,000 military carbines for his forces in the Dominican Republic.[58] Some of the arms negotiated during the meeting were on board the C-74 when the group was arrested at Miami. Two other co-conspirators, Virginia L. Bland and Julio Laurent were arrested at a location away from the airport.

Dominick Bartone was out on bail for the conspiracy charge when he was called to testify before the Senate hearings in 1959 on "improper activities in the labor field." He was excused because his testimony could prejudice the pending case against him in Miami for arms smuggling. Bartone had previously muscled himself into the position of vice president of both Alvin Naiman Corporation and Niagara Crushed Stone. He was still drawing $250 per week plus expenses from the companies.

On 18 October 1959 Bartone called a meeting with Alvin Naiman, Louis Triscaro and Mike Zappone in Naiman's Cleveland office. A heated argument ensued when Bartone stated he was taking over the company. He physically chased Naiman out of his own office. He stated the pending case against him in Florida for gun smuggling had been "fixed" and would never come to trial. He then produced two $50,000 cashier's checks drawn on a Latin American bank. He insinuated the checks came from Batista forces. He said there were still ten of the C-74s left to sell and he intended to delivery two of them to Nicaragua.[59]

Bartone literally took over Alvin Naiman's office in Cleveland and operated out of it as if it were his. It was reported Bartone, with $200,000 backing from a businessman in the Dominican Republic, gained control of Dinaloft Enterprises in Miami. The company supplied flight meals to Pan American World Airways and several other airlines.[60]

It appeared Bartone had gotten a taste for aviation and the deals and profits that could be made. He ultimately paid a fine for his participation in the C-74 gun running fiasco and was attempting to get into a more legitimate side of air operations. This was not the end of his interest in large aircraft. His name would surface again in the late 1960s as a Chauncey Marvin Holt associate. Holt was friends with Beverly Hills attorney and stock broker Burt Kleiner. Holt frequently ran unregistered or stolen stocks through Kleiner's brokerage firm. This would not be relative except Kleiner engineered the 1965 stock swap deal which allowed Unexcelled Chemical to acquire control of Aero Spacelines via Twin Fair department stores.

The C-74 N8199H seized in May 1959 was still impounded and stored at Miami where it would remain until September 1962. Earl Benjamin was no longer the president of Akros, however, he was still connected to the Akros operation and shopping the aircraft. Al Naiman was being sued for $500,000 by Coral Gables First National Bank, Boynton Beach State Bank and South Dade Farmers Bank.

In addition the Pan American Bank of Miami had filed suit against Naiman for $1,500,000. The claim was over $1,100,000 in monies advanced to Naiman for the Akros Dynamics purchase of the C-74 aircraft and the shuffle of the paper held on Aircraft Instrument Corporation. In December 1959 several representatives of Akros along with Naiman's accountant Burt C. Haddad went to Miami to try and work out a deal with the Pan American Bank for the sale of the aircraft.

In early February 1960 Haddad claimed he represented a middle-eastern group and may be able to sell the C-74s to the Lebanese government or an oil company in Lebanon. The Akros group signed a contract with Haddad giving him exclusive authority for 30 days to dispose of the aircraft. Haddad contacted Earl Benjamin on 24 March 1960 requesting the technical manuals for the C-74s. Benjamin called Naiman to determine where the manuals were located. Naiman said he would tell Haddad to get the manuals from Bartone. Haddad's efforts never materialized and the aircraft remained unsold.[61]

In the interim Earl Benjamin contacted Amos E. Heacock of Seattle who was a legitimate and respected businessman. Heacock was president of Air Transport Associates and Independent Air Carriers Conference. He was formerly a representative of Truman-Wasserman underwriters of New York before starting his own business. Heacock agreed to try and put together a deal to dispose of the four C-74s Akros owned and possibly all eleven aircraft. Truman-Wasserman expressed an interest in the aircraft prompting Heacock to go to Miami in March and April to meet with representatives of Akros Dynamics and the Pan American Bank. After the meetings Heacock contacted Earl Benjamin and said he would not negotiate with Haddad, Naiman or Bartone and termed them as "racketeers."[62]

All efforts to purchase or broker the sale of the four aircraft in possession of Akros failed. The situation became more complicated in April 1960. Pennsylvania banker Guy Gully who had greased the deal with the Pan American Bank was prosecuted in New York on charges brought on by the Securities and Exchange Commission for stock manipulation unrelated to Akros Dynamics.[63] FBI reports indicate Alvin Naiman was still trying to sell the aircraft outside the United States in 1960. On 06 July 1960 his phone records indicated he contacted an unnamed individual in Santiago, Chile who expressed an interest in the aircraft.

The Pan American Bank of Miami announced in September it had settled the $1,500,000 claim against Naiman for $832,000. The Pan American Bank filed suit against Pennsylvania attorney and Akros partner Guy Gully in 1961. The bank claimed Gully agreed to help dispose of a $419,000 debt of Aircraft Instrument Corporation owned by Gus DeMeo. As a result Gully caused Akros Dynamics to assume a debt of $500,000 of which he pocketed $81,000 difference.[64] He had also negotiated a second loan of $200,000 for Akros which was not satisfied.

The Pan American bank won a Judgment on 09 August 1961 in the Circuit Court of the Eleventh Judicial Circuit, Dade County, Florida in the amount of $847,033.95. The Circuit Court of Pennsylvania ruled on 01 August 1963, although Gully was served in Pennsylvania evidence existed he had a residence in Florida and the case should be moved to Miami for trial.[65] The single C-74 remained impounded in Miami and the other ten parked at Tucson.

Air Systems

Akros Dynamics had actually taken possession of four of the eleven C-74 Globemasters purchased in the original contract with the Air Force because of the first payment of $500,000. They are the only four C-74s, which received civilian registrations. Akros was only able to fly C-74 N8199H, which was ultimately impounded in Miami where it remained until September 1962. It was purchased along with two other C-74s with an option for the fourth by an investment group for the newly formed air operator Aeronaves de Panama operating as Air Services. The new cargo carrier headed by Orvis M. Nelson would operate under the name of Air Systems. There is no direct association of Nelson to the development of guppy aircraft. However, he was in the mix by association. He purchased the C-74s which were the inspiration and base for the C-124 Globemaster. Also, he was connected by the Transocean B-377s. Orvis Nelson was originally a captain for United Airlines, which flew Stratocruisers, and he was the former CEO of bankrupt B-377 operator Transocean Airlines. It was the largest contract air carrier in America. Nelson was operating and in the process of acquiring the balance of the existing B-377 fleet when Transocean ceased operations because of financial problems. These same Stratocruiser aircraft were acquired by Lee Mansdorf and subsequently Aero Spacelines for the production of B-377 guppies.

Nelson was able to get C-74 N8199H ferried from Miami to Oakland, where maintenance work was performed by Sierra Aviation Services. Two more of the Akros Dynamics

Globemasters N3181G and N3182G were made airworthy at Tucson and moved to Ontario, California. The fourth, N3183G, was flown to Long Beach and parked. The four aircraft had been released by the Air Force because Akros Dynamics had originally made a payment of $500,000 but never paid the balance for the eleven C-74s.

The original payment came from the loan Pan American Bank of Miami made to Alvin Naiman. Akros Dynamics made few if any mortgage payments for the other seven aircraft eventually defaulting. Prior to Nelson acquiring the four C-74s, the Air Force placed the other seven up for bid on 17 April 1962 but failed to receive any serious offers. There were a number of bids but all were less than the scrap value. None of the bids were accepted. The seven aircraft remained at Davis-Monthan Air Force Base until 1965 when they were broken up and sold for scrap.

Aeronaves de Panama

Three of the four Akros C-74s acquired by Nelson's Air Systems were to be operated under Panamanian registration as a country of convenience. Aeronaves de Panama took an option on the fourth aircraft pending cargo contracts in Europe. Nelson was known as one of the most colorful characters in American commercial aviation. He learned to fly in the Army Air Corps before resigning his commission to work for United Airlines. He flew for United from 1931 to 1943 reaching the position of senior pilot. In 1942 he became the first vice president of the Airline Pilots Association (ALPA) union. During World War II Nelson flew United Airlines contract missions for the Air Transport Command. He and several other pilots began formulating a plan to create their own airline when the war was over.

Nelson attempted to convince United's management of the potential of air operations in the Pacific without success. On 09 March 1946 United Airlines vice president of operations Jack Herlihy asked Nelson if he would be interested in sub-contracting the operations of the military airlift route between San Francisco and Hickam Air Force Base Hawaii. The problem was the operation had to begin in ten days. Nelson immediately left United and formed Overseas National Air Transport (ONAT) on 18 March. In June 1946 ONAT was incorporated into Transocean (TAL) with two surplus Army Air Force Douglas C-54s. Over time Transocean became one of the most significant and most unusual airlines of the post–World War II era.[66]

Nelson was confident the development of the Pacific was relative to commercial aviation. He had flown the Pacific extensively during the war and knew the market potential. Nelson lost a bid for the San Francisco—Hawaii contract to Robert Prescott's Flying Tiger Airlines in 1946 but took over TWA's contract to operate Philippine Air Lines East Asia operations. Under Nelson Transocean became the largest contract air carrier in the world with a fleet of 114 aircraft.

To increase market share he often skirted government regulations for non-schedule carriers. He developed a reputation of transporting anything anywhere. By 1957 he was in serious financial trouble prompting him to sell 40 percent of Transocean to Atlas Corporation. Atlas presented itself as a New York investment corporation but was more of a corporate raider. Atlas began selling off the subsidiaries and assets for cash leaving Transocean as a shell of what it once was.[67]

N3183G is one of three C-74s fitted with a radome. After the collapse of Akros Dynamics, three of the C-74s were acquired by Orvis Nelson's Air Services and registered with Aeronaves de Panama. An option was taken for the fourth Akros aircraft. It was ferried to Long Beach and parked with the intent of putting it in service when needed but the option was never taken up (author's collection).

The Atlas Corporation had acquired aircraft brokering firm The Babb Company in October 1952 after the sudden death of its founder Charles H. Babb. He became very wealthy in the 1930s acquiring and reselling surplus military aircraft. Babb's success came from converting surplus passenger aircraft to cargo-liners. Babb believed air cargo was the future. His most unique contribution to aviation came in 1939 when he designed and received a patent for his groundbreaking front-loading fuselage aircraft design, which he named the Air Truck. The all-metal high wing aircraft had a 100-foot span and a 60-foot fuselage with a swing nose. It was capable of uplifting an 11,500-pound payload. The nose swung past ninety-degrees to allow unobstructed straight in loading.[68] The front-loading door design is very similar to the configuration of Freddie Laker's ATL-98 Carvair. In December 1941 he received a second patent U.S. 2268009A for a front-loading flip-up nose design. It was similar today's B-747F and C-5. Babb's concept forever set the standard for front-loading cargo aircraft. The Babb Company made a fortune in 1951 when a major airline purchased a large number of surplus engines for $5,000 each. Babb had offered the engines to the same carrier in 1948 for $1,000 each but the airline declined.

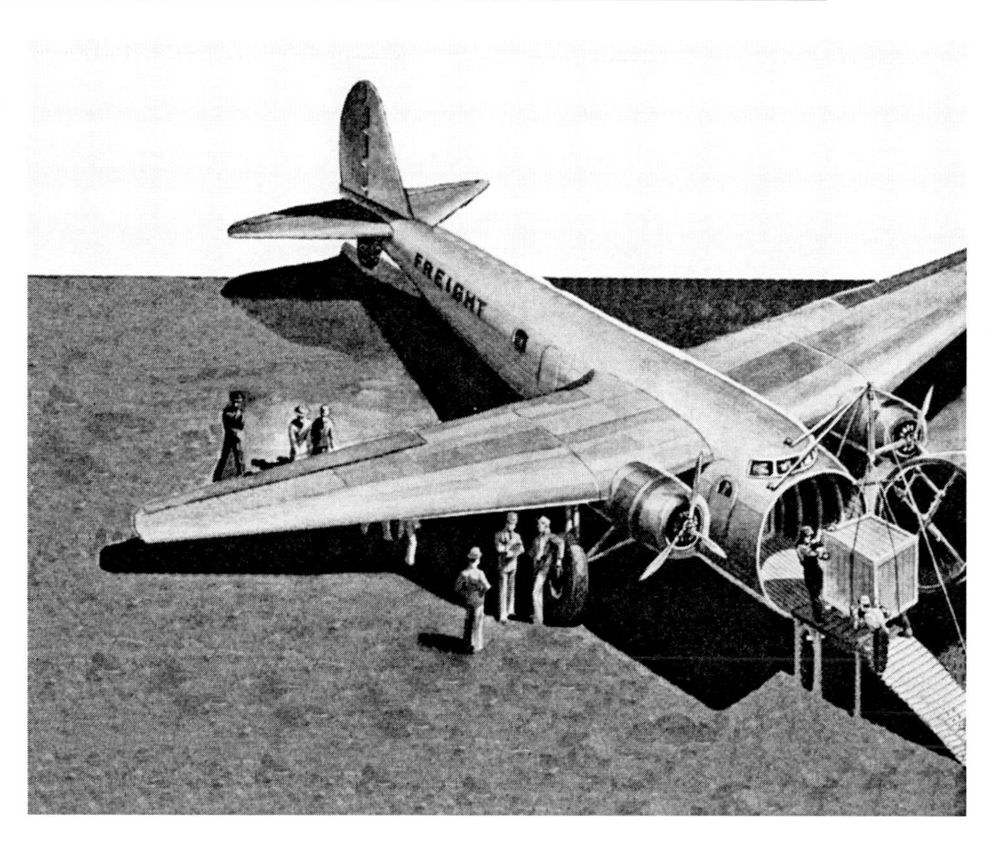

Charles Babb, a successful aircraft broker in the 1930s, believed there was a need for cargo aircraft. He designed and patented a new transport with a raised cockpit and front loading cargo nose door. This nose design is very similar to the ATL-98 Carvair configuration. The Aero Spacelines Super Guppy advanced the design by including the cockpit in the swing nose (Babb Company, author's collection).

Wall Street investment firm Atlas Corporation, founded by Floyd Bostwick Odlum and George Howard, saw the cash rich Babb Company as a corporate raiders dream. There are similarities in the takeover of several aircraft companies reviewed in this accounting of guppy aircraft development. The actions taken to acquire, control and extract profits from other companies like Consolidated-Vultee (Convair), The Babb Company, Transocean Airlines, Northeast Airlines and others changed events in certain incidents which affected the availability of B-377 aircraft in 1960. Atlas had enough control over XC-99 builder Consolidated-Vultee to have the ICBM missile developed for the military and space program named Atlas after the controlling company. The following statement from the 1954 Atlas Corporation Silver Anniversary Report gives some insight into how Atlas operated: a "special situation, as Atlas uses the term, means an investment in and involving not only primary financial sponsorship, but usually also responsibility for management of the enterprise." This typically involved reorganization, mergers or other measures to strengthen the company and a commitment "to stay with the investment until the essentials of the job have been done, and to move on then to another special situation."[69]

In reality Atlas Corporation made huge profits by acquiring and merging large companies then selling off divisions. The stock of the remaining shell of a company would be sold to investors and Atlas would move on to another deal.

The major air carriers including Northeast Airlines regularly complained to the CAB about Orvis Nelson's Transocean. The group filed protests expressing concern of such a large contract air carrier that was cutting into the regulated schedule airline market. In 1958 Orvis Nelson completed a deal to acquire through The Babb Company 14 of the surplus BOAC Boeing B-377 Stratocruisers and took delivery of eight.

The Atlas Corporation had purchased the Babb Company after the death of its founder. Many of its assets were sold off and the remaining merged into Transocean Airlines. Atlas had acquired 40 percent of Transocean before merging Babb and the Transocean family of companies into a single new corporation. Through stock swaps Transocean was given a 20 percent stake in the new company. However, Transocean was on a downward spiral because of the loss of cash flow from subsidiary companies that Atlas Corporation had been selling off.

While heading Transocean, Nelson created ten subsidiary companies including International Aircraft sales. Most of the sub-companies were sold off as soon as Atlas Corporation gained control of Transocean. The Transocean subsidiary companies had provided cash flow for the airline operation for years. Without the substantial cash flow of the sub-companies to offset the expense of operating the fuel thirsty Boeing B-377, the airline struggled to survive. The purchase of the B-377 Stratocruisers was a contributing factor in the demise of Transocean. The carriers B-377s were then acquired at bargain prices by Lee Mansdorf, eventually becoming the base airframes for Jack Conroy's Aero Spacelines Guppy conversions.

Nelson's Transocean had an excellent safety record as a non-schedule carrier. The heavy maintenance was contracted to Matson Air Transport. Matson had operated a small airline in the U.S. Hawaii market for a short period in attempt to acquire the routes.[70]

Several Matson and Transocean DC-4s were eventually converted to Aviation Traders ATL-98 Carvair guppies. Matson seriously considered purchasing Transocean when the financial difficulties began. However, the financial problems were compounded by the purchase of surplus BOAC Stratocruisers making the acquisition unattractive. The regulatory restrictions, corporate raiders, lack of cash and B-377 purchase led to the collapse of Transocean in 1960. Nelson moved to Europe but continued to try and resurrect the airline.

Even after the demise of Transocean, Nelson still maintained interest in sub-company International Aircraft Sales, which he created to purchase aircraft. He was anxious to return to the airline business. Transocean had been liquidated and the Boeing B-377 aircraft sold off leaving no chance of resurrection. He became aware of several aircraft brokers and underwriters who were attempting to interest potential cargo upstarts in the surplus Akros Dynamics C-74s. Only a few investors would even consider it but this was the perfect formula for Nelson. He was a risk taker, liked large long-range aircraft and was one of the few people who could start an airline with the ex-syndicate owned C-74s.

The Pan American Bank held the mortgage on four of the C-74 aircraft. After the experience with Akros and dealings with multiple suspicious characters, any serious offer to sell the aircraft getting them off the books was welcomed. Nelson had suffered serious financial setbacks, however, his reputation regardless of financial position was considered superior to Akros. He was a master at spinning deals and somehow was able to arrange financing to purchase three of the C-74s for Air Systems with an option for the fourth.

Nelson had learned through his worldwide contacts the milk cow producers in Denmark were seeking an air carrier to transport pregnant cows to several middle-eastern countries. Saudi Arabia and Iran were attempting to build up national herds to provide fresh milk for their citizens. He contacted a handful of former Transocean employees to put together three flight crews which included Don McAfee, Ed Landwehr, Jessie Morrison and Ted Hunt. Years later Hunt related his adventures with Air Systems in historical fiction titled *Flying Cowboys*.

Nelson's former airline Transocean had a policy of transporting anything anywhere which included goats, honey bees and even rabbits to South Korea. Transporting cows did not seem any different. Nelson saw it as the perfect opportunity to start another air carrier outside the constraints of the CAB and stringent U.S. regulations, used by other carriers to put pressure on Transocean.

The carrier had maintained a good maintenance and safety record. However, the condition of the C-74s was questionable from the outset. The military grounded them for lack of parts then corrosion and deterioration was a factor since they had been parked since 1956. It has been speculated Nelson may have known they needed considerable work and spares were in short supply but was so anxious to establish another air carrier the details may have been overlooked. He was the type of personality who never approached anything thinking it could not be done.

This was not new for Nelson In 1946 he had purchased surplus military C-54s to start Transocean. After military service they were converted to civilian DC-4 configuration by Matson Air Transport. Transocean maintained an excellent maintenance base to keep the carriers aircraft in good condition. When Nelson took delivery of multiple Boeing B-377 Stratocruisers for Transocean they had also been well maintained by first line commercial carriers United and BOAC. These C-74 Globemasters had been parked in the desert before being acquired by a collection of dubious characters for what was viewed by the U.S. Government as a convoluted ownership arrangement and possible criminal intent. It is doubtful any maintenance practices were followed on the C-74s once Akros acquired them. The players at Akros were only interested in selling the aircraft for a huge profit.

It would have been difficult if not impossible to get the C-74 Globemasters certified in the U.S. for several reasons. They had been involved in questionable money deals, disputes with the Air Force, organized crime connections, gun running and State Department embarrassment. A Senate investigation of organized labor had reviewed an effort by individuals connected to the Teamsters Union to sell the aircraft to both Castro and Batista. In addition CAB regulatory problems with Orvis Nelson at Transocean would prevent the FAA from ever certifying them.

In order to skirt all the potential registration problems Nelson opted to register the aircraft in Panama as a flag of convenience country. The aircraft were then leased to Air Systems. After N8199H was flown to Sierra Aviation service in Oakland in September 1962, it was painted with a red nose, vertical bar on the tail and cheat line along the fuselage.

All three Panamanian Globemasters displayed the name Heracles on the nose. Air Systems re-registered N8199H (42-65408) in Panama as HP-367. Shortly after N3181G (42-65409) and N3182G (42-65404) were ferried from Ontario to Oakland for overhaul where 42-65409 received Panamanian registration HP-379. The fourth C-74 N3183G (42-6541) had less than 6,000 hours on it when ferried from Tucson to Long Beach. It was expected Orvis Nelson would exercise the option as soon as Air Systems began to generate revenue. The owner of record remained the Pan American Bank of Miami. The aircraft remained at Oakland until 1964 when it was broken up.

HP-379 is the second of the C-74s Air Systems registered to Aeronaves de Panama. The aircraft were registered in Panama as a flag of convenience because it would have been impossible to get them certified in the United States. All three were trimmed in red and named Heracles (AAHS archives).

Fatal Flight of HP-385

The third C-74 acquired by Air Systems, 42-65404 (N3182G), was reregistered in 1962 to Aeronaves de Panama as HP-385. After the Globemasters Panamanian registration was final, they were ferried one by one to Copenhagen with the last arriving in March 1963. The maintenance was contracted to two Danish aircraft service companies, SAACO and Flying Enterprises. Because the Globemasters were limited production military aircraft neither of the contractors had experience in C-74 airworthiness and maintenance standards. Air Systems began operating charters all over Europe transporting anything from ships parts to livestock. The aircraft could transport up to 60 cows or 68 horses. The primary charters were cattle between Copenhagen and Tehran, Iran or Saudi Arabia. To insure proper handling of the animals Danish shippers required a veterinarian to travel on each livestock charter.

On 8–9 August 1963 veterinarian James Herriot was assigned to travel on HP-385 with a shipment of 40 cows from Gatwick (London) to Istanbul, Turkey. He noted the deplorable condition of the aircraft with bald tires, deferred instruments, jammed elevator loading hoist and landing gear which would not fully retract.[71] The C-74 suffered a fire in the number three engine en route to Istanbul. The aircraft made a three-engine ferry flight back to Copenhagen for repairs. With only three engines and excess drag from the partially retracted gear the C-74 barely cleared the Alps on the return flight.

Other accounts verify the Aeronaves de Panama C-74 aircraft were poorly maintained. However, some items identified as examples such as jammed loading hoist and bald tires may or may not affect airworthiness. The jammed hoist only prevents its use and bald tires are common on military aircraft because they are usually 28-ply rated. In the author's experience on C-124 Globemasters in the 1960s, they often flew on tires worn down through several plies with cord showing. This did not necessarily make them unsafe. However, the Curtiss Electric props on the P & W R-4360-49 engines were extremely corroded with the blades showing signs of cracks.

The crews had complained on numerous occasions of the poor condition. When compared to Orvis Nelson's Transocean aircraft, it was out of character for him to allow the warnings to go unanswered and to operate the C-74s in this condition. Transocean under his stewardship had an outstanding maintenance and safety record. Several crewmembers for fear of their safety resigned over the condition of the propellers.[72] Whether by choice or ignorance the disastrous decision was made to use C-124 weight and balance tables. This increased the maximum takeoff weight and maximized the amount of revenue cargo transported.[73] However, this allowed the aircraft to be overloaded beyond limits possibly in excess of 20,000 or more pounds.

On the evening of 09 October 1963 HP-385 departed Tirstrup, Denmark, with a crew of four, a veterinarian and an observer who was a personal friend of an official in the Danish government. On board was a consignment of 56-pedigree Jersey cows (in calf) and four bulls destined for Riyadh, Saudi Arabia. The aircraft was most likely over grossed leaving Denmark. The Globemaster was scheduled to make an intermediate stop for fuel at Marignane Airport at Marseilles, France (LFML), arriving at 01:15 a.m.

The C-74 original maximum takeoff weight when it was new was 165,000 pounds. It had later been increased to 175,000 pounds. After years of service combined with neglect it should have been restricted to a lower ATOG. The dry weight of the aircraft is 86,170 pounds. The consignment of cows has been estimated to have weighed at least 67,000 pounds and could have been as much as 74,000 pounds. The distance from Denmark to Marseilles is approximately 1,200 miles. The aircraft dry weight plus the cargo at a minimum of 67,000 pounds left 19,000 pounds (approx. 3,100 gallons) for fuel to be at Max ATOG. It is a safe assumption the aircraft was overweight leaving Denmark.

After refueling and oil service at Marseilles the aircraft left the blocks and was cleared by ATC for runway 31R to take off over the water. The crew misunderstood when they left the ramp and turned right on to the taxiway for runway 13L taking off toward the mountains. The Globemaster departed at 03:48 a.m. toward the southeast. It could have missed the mountains but overshot the turnoff point by two miles, barely clearing a 700-foot high ridge.

About seven-nautical miles south of the airport the aircraft made a sharp right turn and suddenly lost altitude slamming into an 800-foot high cliff, 20 feet below the summit. It erupted in a ball of fire destroying the aircraft and all on board. The airframe had a total of 8,901 hours. It is believed the aircraft was overweight because of previous incidents of using incorrect C-124 Globemaster II weight and balance tables and it was carrying a heavy fuel load for the 1,703 mile flight to Riyadh via Cairo.

One of the plane's engines was found approximately two miles from the crash site with one of the four propeller blades missing. Although not conclusive it indicates a catastrophic failure and loss of blade in-flight, which created an unbalanced condition causing the engine to separate from the wing. The investigation concluded the propeller failure was caused by severe corrosion.[74]

It has been speculated the sharp right turn was an attempt to return to the airfield. It is more likely the loss of an engine and propeller, which weighed 7,900 pounds, caused the left wing to rise placing the opposite wing in a downward attitude contributing to a slide. In a max weight configuration with an engine out it is not possible for the pilot to exert enough force on the rudder to keep the aircraft straight. The combination of an overweight aircraft, catastrophic propeller failure and the engine separating from the wing set up an unrecoverable situation. The loss of HP-385 prompted Danish authorities to suspend all Aeronaves de Panama/Air Systems flights out of Denmark.

The Douglas C-74 Globemaster as well as Nelson's Transocean commercial Boeing B-377s Stratocruiser was fitted with the Pratt & Whitney R-4360 engines. Depending on the airline some B-377s were fitted with the troublesome Hamilton Standard Hydromatic propellers and others with the Curtiss Electric units. Early in Boeing B-377 operation Pan American World Airways experienced multiple catastrophic failures of props separating resulting in crashes and loss of life.

The Aeronaves de Panama C-74s were fitted with the Curtiss Electric propellers, which Nelson had previously known to be relatively trouble free from Transocean experience. The ex–BOAC Stratocruisers acquired by Transocean under Nelson were fitted with the troublesome Hamilton Standard units. Transocean also acquired through the Babb Company, seven BOAC B-377s consisting of four ex–United and three ex–Sila aircraft, which had been changed from the Curtiss Electric propellers to the Hamilton's. Although those aircraft never entered service with Transocean, it is possible Nelson was thinking in terms of relative trouble free operational record of the Curtiss Electric units originally on these aircraft. This may explain why Nelson discounted the pilot complaints about the C-74 propellers. The C-74s did not have the Hamilton Standard units which previously lost blades. It can be argued Nelson should have realized prop corrosion could not be ignored. This oversight ultimately resulted in a disastrous failure and loss of life. Air Systems pilot Ted Hunt (Aeronaves de Panama) repeatedly reported the props were in deplorable condition. His continued complaints to Nelson regarding the condition of the propellers and request for replacements eventually forced his decision to resign.[75]

The Last C-74 Globemaster I

After Danish authorities prohibited Aeronaves de Panama from operating in Denmark, the company moved operations to Turin, Italy. It was reported one of the carriers other C-74s, HP-367, was involved in a gun-running situation and abandoned at Malpensa Airport, Milan. Within a few months the company ceased operations. The remaining employees abandon the other aircraft at Turin and escaped in a DC-7B to London just ahead of the authorities, who were attempting to recover monies owed. The gun running aircraft was eventually broken up at Milan in August 1972. The third Aeronaves de Panama C-74, HP-379, had been moved to Milan where it was left in deteriorating condition. Late in 1969 HP-379 was moved to Turin and painted in a false Communist Chinese livery for use in the movie *The Italian Job*. It was abandoned and damaged by fire on 11 June 1970. On 24 September it was being dismantled when it exploded. The ensuing fire killed two of the workers, leaving a tragic end to the last C-74 in existence.[76]

Ultimately Nelson's unwillingness to review the condition of the aircraft or lack of concern for safety ended in disaster which destroyed any chance of him being a major player in commercial aviation ever again. His demise in aviation began with the collapse of Transocean. The Aeronaves de Panama/Air Systems C-74 operation was merely a stopover on the way down. What he did after the collapse of Air Systems is sketchy. There were reports he was operating a DC-4 that either ran out of fuel or crashed in Nigeria or on an offshore island near Cameroon. It was rumored to be carrying a load of weapons, which was never verified.[77] The pilot without proper papers escaped, leaving the injured Nelson behind.

The C-74s after military retirement were involved in questionable and treacherous operations. Although, it was a landmark aircraft and for a period the largest production cargo aircraft in service it is not well remembered. History has explained it away as a limited production odd aircraft. However, it provided the base for the expanded fuselage C-124, which was one of the most significant transport aircraft to ever enter Air Force service. The C-74 suffered its share of problems but served the military for more than ten years flying 100,000 hours of accident-free operations which is an admirable record for any fleet of aircraft.

C-74 Globemaster I Disposition Summary

Globemaster I c/n 13913, AF 42-65402

The first C-74 flew on 05 September 1945 and was delivered to the USAAF in October. It was retired on 01 November 1955 and transferred from Brookley Air Force Base, Alabama, to Davis-Monthan with 9,168 total hours on the airframe and placed in storage. It never flew again and was scrapped in 1965 with the other six C-74s, which were acquired by Akros Dynamics but not delivered for nonpayment.[78]

Globemaster I c/n 13914, AF 42-65403

The second C-74 was delivered to the USAAF on 30 April 1946. It never saw any transport service. On 05 August 1946 it was being flown on the 23rd test flight to maximum speed test. The aircraft weighed 146,000 pounds. At 340 miles-per-hour the crew could not control the aircraft and it began to shutter violently. The outer wing had separated outboard of the engines. The four-man crew was able to bail out before the aircraft crashed in West Torrance, California, with only 19.25 hours on the airframe. There were no injuries on the ground.[79]

Globemaster I c/n 13915, AF 42-65404

The third C-74 was delivered to the USAF on 26 May 1948. It was the first C-74 to be certified with a gross weight of 165,000 pounds. It was retired with 8,612 hours on the airframe and stored at Davis-Monthan. Acquired by Akros Dynamics in 1959 it was brought out of storage and registered as N3182G. Subsequently acquired by Aeronaves de Panama in 1962 and registered as HP-385. With 8,901 hours it crashed with a consignment of cows on 09 October 1963 at Marseilles, France. All on board perished.[80]

Globemaster I c/n 13916, AF 42-65405

The fourth C-74 was delivered to the USAAF on 06 June 1946. The delivery was cancelled and the aircraft was flown to Wright-Patterson Air Force Base, Ohio, where it became the type test article. It was used in static stress test between August 1946 and November 1948 and destroyed.[81]

Globemaster I c/n 13917, AF 42-65406

The fifth C-74 first flew on 01 September 1946. It was the first aircraft delivered to the C-74 squadron. After two years of active service it was returned to Douglas Aircraft in June 1948 to expand the fuselage into the first YC-124 Globemaster II. It first flew as a C-124 on 27 November 1949 retaining the same tail number. It was used in testing until July 1954 when it was returned to the Air Force for active service. It was transferred to the Air Force Museum at Wright-Patterson in November 1957. Unfortunately it was not saved and eventually used for firefighting practice and destroyed in 1969.[82]

Globemaster I c/n 13918, AF 42-65407

The sixth C-74 was delivered to the U.S. Air Force and used for engine calibration test in December 1948. After testing the Air Force placed it in regular transport service. It was retired 01 November 1955 and transferred to Davis-Monthan Air Force Base, Tucson, for storage with 5,952 hours on the airframe. It never flew again and remained in storage until scrapped in 1965.[83]

Globemaster I c/n 13919, AF 42-65408

The seventh C-74 was delivered to the USAF in 1947. It was the first aircraft to have the cockpit canopy modification to standard configuration on 18 January 1949. It was retired from the Air Force on 01 November 1955 and transferred to Davis-Monthan for storage with 8,496 hours on the airframe. In 1959 it was purchased by Akros Dynamics and registered N8199H. It is the only C-74 that Akros took delivery and flew. The aircraft was flown to Cuba in an effort to interest Fidel Castro in purchasing the fleet. After the deal failed it was involved in suspicious activity and was confiscated on 22 May 1959 at Miami during a gun running operation. In 1962 it was purchased by Aeronaves de Panama and registered as HP-367. Subsequently impounded at Milan Italy on 11 March 1964 and allowed to deteriorate until broken up in August 1972.[84]

Globemaster I c/n 13920, AF 42-65409

The eight Globemaster was the third delivered to the C-74 squadron on 19 January 1947. It was dispatched to Manitoba, Canada, 12 May–01 June for cold weather operations testing. It was retired from the Air Force 01 November 1955 and stored at Davis-Monthan Air Force Base with 9,502 hours on the airframe. It was purchased in 1959 by Akros Dynamics and registered N3181G but never flown. It was acquired by Aeronaves de Panama in 1962 and registered HP-379. It was moved from Milan to Turin, Italy, in March 1964. It was then confiscated for non-payment of services. In 1969 it was repainted for use in the movie *The Italian Job*. It was damaged by fire on 11 June 1970. Two workers perished during dismantling operations on 24 September 1970 when the aircraft exploded.[85]

Globemaster I c/n 13921, AF 42-65410

The ninth C-74 was delivered to the Air Force in 1947. After twelve years of service it was retired from the Air Force on 01 November 1955 and transferred to Davis-Monthan for storage with 8,858 hours on the airframe. It never flew again and was scrapped in 1965.[86]

Globemaster I c/n 13922, AF 42-65411

The tenth Globemaster was the second delivered to the Air Force C-74 squadron in 1947. It was retired from the Air Force on 01 November 1955 and transferred to Davis-Monthan for storage with 9,033 hours on the airframe. It never flew again and was scrapped in 1965.[87]

Globemaster I c/n 13923, AF 42-65412

The eleventh Globemaster was delivered to the Air Force C-74 squadron in February 1947. It was evaluated by the Troop Carrier Command on 14 March. The aircraft was used for cruise control testing from 23 January to 01 February 1949. It was retired from the Air Force on 01 November 1955 and transferred to Davis-Monthan for storage with 5,825 hours on the airframe. In 1959 it was acquired by Akros Dynamics and registered as N3183G but remained in storage. The Pan American Bank of Miami took possession in 1963 after Akros defaulted. It was flown to Long Beach, California, for tentative purchase by Aeronaves de Panama. The option was never exercised and it was scrapped at Long Beach in 1964.[88]

Globemaster I c/n 13924, AF 42-65413

The twelfth Globemaster was delivered to the Air Force C-74 squadron in February 1947. It was the first C-74 authorized for regularly scheduled passenger (troop) flights on 21 March 1948. The C-74 set a round-trip speed record on 24 January 1949 from Brookley Air Force Base, Alabama, to Rhein Main, Germany. On 01 November 1955 it was retired from the Air Force and transferred to Davis-Monthan with 8,362 hours on the airframe. It never flew again and was scrapped in 1965.[89]

Globemaster I c/n 13925, AF 42-65414

The thirteenth Globemaster was delivered to the Air Force in March 1947. It was dispatched to Europe between 14 August and 26 September 1948 to become the only C-74 that participated in the Berlin Airlift. In April 1949 it set a monthly airtime flying record of 182 hours. On 18 November 1949 it set a trans–Atlantic record from Brookley Air Force Base, Alabama, to Marham, England, with 103 passengers on board. On 01 November 1955 it was retired and transferred to Davis-Monthan with 8,835 hours on the airframe. It never flew again and was scrapped in 1965.[90]

Globemaster I c/n 13926, AF 42-65415

The fourteenth Globemaster is the last C-74 built. It was delivered to the Air Force in April 1947. After twelve years of service it was retired on 01 November 1955 and transferred to Davis-Monthan with 9,231 hours on the airframe. It never flew again and was scrapped in 1965.[91]

Seven of the eleven retired C-74s purchased by Akros Dynamics remained at Davis-Monthan Air Force Base in storage because the Air Force never received payment. Akros made a down payment and took delivery of one aircraft. Eventually Akros received title to four of the C-74s with payments of cash and various loan scams. Aeronaves de Panama acquired three of the Akros aircraft and took an option on the fourth held by the Pan American Bank. The remaining seven C-74s remained in default at Davis-Monthan. They were never flown and remained in storage until all were scrapped in 1965.

6

C-124 Globemaster II

Development of an Expanded Airframe

The C-124 Globemaster was created by returning the fifth C-74 airframe c/n 13917 (AF 42-65406) to Douglas Aircraft at Long Beach where the top half of the fuselage was cut off at floor level. The lower front section was removed forward of the wing and replaced with a redesigned nose section with clamshell doors and self-contained loading ramps. The airframe was increased in size by building a double deck fuselage and grafting it on top of the lower half of the C-74. It was attached at floor level with the cockpit positioned above the clamshell doors producing the YC-124 front-loading aircraft.

The C-124 was unique in being the only mass-produced military guppy type aircraft. After the success of the hand built YC-124 using the C-74 airframe, 448 more Globemasters were produced from the ground up utilizing the same wings, engines, tail and systems developed for the Douglas C-74.

The Douglas C-124 model 1129A replaced the C-54 as the primary military long-range cargo transport beginning in the 1950s. Although the C-74 was much larger than the C-54 predecessor, it was not considered a primary long-range transport by the military. The C-124 served as the military's primary long-range oversize transport for 25 years. Because it was the only transport in Air Force inventory with the ability to move outsize items, it could not be retired until the Lockheed C-5A was in active service.

The C-74 from which the C-124 evolved was originally ordered in the 1940s to fill the need for a large global transport but it never saw service in World War II. The hostilities ended but the expanse of the Pacific inspired military planners to review the need for larger aircraft. It was not until after the Berlin Airlift when military planners were convinced a transport aircraft capable of uplifting bulk or outsize cargo over great distance was necessary. The C-74 Globemaster I proved to be a reliable heavy lift volumetric transport. It possibly would have been ordered in larger numbers if it had loading doors able to accommodate large wheeled military equipment or waist doors for troop drops. In 1947 Douglas began redesign studies to expand the fuselage size of the C-74. An upgraded design doubling the height of the fuselage with front-loading doors was put forward as the C-124 Globemaster II. It was the first military transport designed around front-loading cargo door. The cost was set at $1,646,000 per aircraft.

The forward airframe featured clamshell doors using an electric winch to deploy the loading ramps. The overhead traveling cranes brought forward from the C-74 were capable of moving cargo along the 77-foot cargo bay. The elevator platform behind the wing was also borrowed from the C-74. The first priority of the C-124 design was to handle oversize wheeled military equipment (trucks, jeeps, bulldozers, tanks, fire trucks, radar vans) and the second requirement was to handle extreme weight.

The C-124 large frontloading clamshell doors overcame the problems encountered with the forward side loading cargo door on the C-74. Very early artist concepts released in 1948 of the proposed C-124 have the same dual canopy design used on the C-74. This is an indication Douglas was reviewing a larger aircraft early in the C-74 program before they were retrofitted with a new cockpit canopy configuration. In 1949 the surviving C-74s were retrofitted with a traditional cockpit windscreen design.

This more traditional "greenhouse" cockpit configuration was necessary because of C-74 pilot communication problems. Subsequently a variation of the new C-74 cockpit windscreen configuration was designed for the production C-124. It was more streamline conforming to the lines of the airframe unlike the redesigned C-74 configuration, which was raised above the line of the fuselage. The fifth C-74 (42-65406) c/n 13917 never received the cockpit windscreen upgrade. It was still fitted with the twin bug-eye canopies when returned to Douglas in 1948 for conversion to the first C-124.

The visual impression of the C-124 is quite different from the base C-74 when viewed from the front. The C-74 has a tubular fuselage, which is a considerably high off the ground for clearance of the nearly 17-foot diameter propellers.

The high frontal fuselage of the C-124 with frontloading clamshell doors is quite massive. The fuselage forward of the wing is much closer to the ground to accommodate the grade of the ramp system. This allowed for a lower equipment bay forward of the spar which provides storage space for self-contained spare parts inventory and access to the wing crawlway.

The lower half of the aft fuselage, wings, large vertical fin and horizontal stabilizer are all taken from the C-74. Expanding the fuselage and adding clamshell doors to an existing design was a brilliant cost saving move for Douglas because most parts were already engineered. The new C-124 resulted in the lowest operational cost per ton-mile fleet of aircraft in Air Force inventory.

The retractable loading ramps weighed 942 pounds each. They are set at a steep 17 percent 30 degree slope because of the high cargo floor of the low wing airframe. The C-124 was designed to accommodate all motorized equipment used by the Army at the time. As an air drop platform, more than 180 paratroopers with equipment could be put on a drop zone. The self-contained elevator platform could be replaced with A-22 airdrop containers allowing equipment drops with the troops. In comparison the C-74 predecessor lacked waist doors making it impossible to deploy paratroops.

The basic airframe, wings, empennage and landing gear along with the twin overhead cranes, rear cargo elevator systems and cockpit design with forward facing engineer's panel had been proven in C-74 operational service. The YC-124 and all "A" series plus the first ten "C" models had the same forward facing flight engineer's panel as the C-74 behind the co-pilot. Most crews actually preferred this configuration to the revised side facing panel on "C" models after aircraft 51-5198.

The forward facing engineer's station divided the cockpit into separate compartments. A doorway curtain could close off the pilots. This was preferable at night allowing the lights to be turned up in the rear of the cockpit without affecting pilot visibility. The engineer, navigator, radio operator, loadmaster, crew chief or any extra crew using the three bunks, galley or lavatory could move around comfortably. It was not as easy on the later "C" models with the side-facing panel making the cockpit one large open area.

The "guppy" name for outsize aircraft expanded from previous types did not come until 1962 when Jack Conroy built the Boeing B-377PG by removing the existing fuselage at floor level to build an enormous new upper cargo hold to transport rockets for NASA. The same process transformed the C-74 into the C-124 making it an excellent example of specialized volumetric guppy transport.

Early C-124 artist drawings show the proposed double deck front-loading Globemaster II with the same twin bubble cockpit configuration of the C-74. This indicates how early the larger C-124 was being planned. The drawing has the four-blade props and no waist doors, the same as the C-74 (Douglas, author's collection).

Only one of the 14 C-74s was converted to a C-124. Twenty-one DC-4s became ATL-98s and multiple B-377s and C-97s were used to produce eight Aero Spacelines guppy types. By comparison, the success of only one C-74 conversion to C-124 was so impressive it resulted in Douglas building 448 Globemaster II aircraft from scratch which is the greatest number of any guppy type airframes produced.

The fifth C-74, 42-65406 only had 700 hours on the airframe when returned to Douglas at Long Beach in June 1948 for expanded fuselage conversion. The C-74 upper fuselage was cut off at floor level from the center-wing back and the section forward of the wing removed. A new double deck fuselage was built up in several sections, lowered into place and grafted to the lower half. The rear cargo elevator remained in the same position behind the wing with bomb-bay type outside doors. The empennage was raised to sit on top of the tail-cone of the new expanded double deck fuselage with an extended dorsal fin. The main landing gear was beefed up to handle heavier loads. The nose gear was shortened to fit behind the bulk-head at the lower end of the clamshell door below the loading ramp attach points.

The modifications resulted in the C-124 being 6 feet, 3.5 inches longer than the C-74 with a 77-foot cargo bay. The lower fuselage behind the wing remained the same as the C-74. A new set of lower compartments were added forward of the wing. The increased forward area allowed easier access to the wing crawlway because the C-74 had a very shallow forward compartment.

The C-124 was considered a self-contained aircraft which carried a supply of spare parts on board allowing it to operate into remote areas of the world. The crew chief or flight mechanic often assisted by the loadmaster had to be prepared to repair or replace a malfunctioning unit with parts from the spares carried in the WRM kit. In addition to spare hydraulic pumps, generators, etc., a considerable assortment of exhaust stacks and "W" clamps were stocked because the engine torque tended to break several on most landings. The late "C" models beginning with 52-932 were equipped with a turbine Auxiliary Power Unit (APU) also installed in the lower forward compartment with the exhaust on the right side of the fuselage. The forward bulk-head of "O" compartment is behind the nose gear wheel well.

The forward C-124 "P" compartment was large enough for an airman to stand up and al-

Top: **The double deck upper fuselage section of the first C-124 being lowered on the lower half of the fifth C-74 (42-65406). The C-74 elevator well is visible at lower center. Removing the old fuselage of the donor aircraft at floor level is the same method Jack Conroy used to create the Aero Spacelines B-377 guppies. Sections of the overhead crane tracks developed for the C-74 have been carried over and are visible in the top of the fuselage (Bob Williams collection of AAHS archives).**

Bottom: **Right door in lower "P" compartment forward of the center-wing allows access to the wings via a crawlway for inspection and access to engines three and four through the landing gear bay. Engines one and four have a compartment behind the engine large enough for an airman to stand up and service the engines in-flight (author's collection).**

lowed access to the wing crawlways. On long flights the crew chief periodically crawled out in the wings during flight to look for fuel leaks or mechanical failures. The compartment also contained an electronics bay with a radio rack. The adjacent "O" compartment contained the War Readiness Material (WRM) spares kit mounted on the forward bulkhead.

The first YC-124 conversion retained the C-74 wings intact with the P&W 3,250 horsepower R-4360-49 engines and four-blade Curtiss Electric props. All production C-124s were equipped with the Curtiss Electric three-blade electric propellers. The full-feathering constant speed unit with reversible pitch has an electric governor synchronization system. This allowed the control of constant engine speed either automatically or manually. The main gear is fitted with a limit switch which prevented reversing the props in-flight. At least one of the main gears has to be on the ground and the pilot's throttles must be pulled to the closed position before the props can be reversed for braking.[1]

Weights and Restrictions

The production "A" model was tested at a takeoff weight of 194,000 pounds with a 50,000-pound payload. The range was projected to be up to 2,500 miles. The Air Force began taking delivery of the first slick nose C-124A Model 1129A in May 1950. The later "C" or Douglas Model 1317 could transport a 70,000 payload on shorter distances. The original "A" model takeoff gross of 175,000 pounds was increased to 185,000 (max 194,500) on the "C" model and the performance manual allowed a max overload takeoff weight of 208,300 pounds.

The C-124 served the mission well until 1965 when the service life came into question. Aircraft 52-1075 crashed on 24 March near Cordova, Maryland. The right wing separated outboard of number four engine. The subsequent investigation found cracks in the wing spars prompting the grounding of the entire fleet. Up until this time the maximum ATOG had been pushed up to 194,500 with a max overload of 210,000 pounds. After the incident the C-124 was restricted to maximum takeoff of 187,500 pounds because of the aging wing spars.

Globemasters often took off at maximum weight and most likely overweight. This author can state from personal experience the C-124 is a very forgiving aircraft. While assigned to C-124 aircraft 0-15182, an "A" model converted to "C" standards, we departed from RAF Mildenhall knowing we were very heavy staging through Goose Bay, Labrador, en route to Dover Air Force Base, Delaware. The Globemaster always groaned and rattled on takeoff. On this morning the shaking felt excessive for what seemed like an eternity takeoff roll before slowly lifting off. The aircraft seemed to be hanging over the English countryside barely moving. The engines were maxed out as it continued to struggle for considerable time until enough fuel was burned off to finally begin to gain altitude. It was at this time the weights were rechecked. It became apparent we had taken off at well over 200,000 pounds. It was a very slow climb before reaching cruising altitude for the Atlantic crossing. Another Globemaster which left Mildenhall one half hour later passed us over the Atlantic and was already parked at Goose Bay when we arrived. The C-124 was a very forgiving aircraft but we were extremely fortunate there were no malfunctions.

The modified YC-124 first flew on 27 November 1949 carrying the same Air Force identification 42-65406 it wore as a C-74. It was considered a modified C-74 rather than a new type of aircraft. The fat guppy appearance of the expanded fuselage is obvious. The first C-124s did not have waist doors, which brought into question the viability as a troop transport (Douglas, author's collection).

The first C-124 retained the C-74 tail number 0-265406 when it flew on 27 November 1949 as a YC-124. Water tanks were mounted in the cargo bay to simulate load for stability and control test programs. The YC-124 test program lasted for four years flying over 350 hours. In 1954 0-265406 was refurbished with the latest C-124 modifications to date. It was eventually retrofitted with the APS-42 radome, wingtip heater pods and a tail heater identified by a scoop on the upper left empennage. The "C" models were receiving these upgrades on the production line. After being upgraded the converted C-74 was returned to the Air Force as an YC-124C model.

The YC-124, s/n 42-65406, was placed in an extensive test program. It is the only Globemaster to have the four blade Curtiss-Electric propellers, which were carried over from the C-74 from which it was constructed. It was eventually upgraded with a nose radome and wing heaters but retained the four blade props and removable engine cowling for the entire service life (USAF).

At a later date the YC-124 engines were upgraded to the improved 3,500 horsepower R-4360-35A series. The four-blade Curtiss-Electric props and smooth cowling without chin scoops were retained. In November 1957 it was turned over to the Air Force Museum at Patterson Field for display. In 1969 the museum was moved to Wright-Patterson Air Force Base but the first prototype Globemaster II could not be moved. It was eventually used for fire-fighting practice and unfortunately destroyed.

Engineering and Upgrades

The first Douglas YC-124 c/n 265406 was built up from the fifth C-74 and equipped with 3,250 horsepower R-4360-49 engines. The second aircraft, 48-795, designated YC-124A was the first Globemaster II built from the ground up. It was fitted with the 3,500 horsepower P&W R-4360-35 engines with 17-foot (16.6) Hamilton Standard three blade propellers as were the succeeding 11 production models (49-232 to 49-243) of the first block of 28 C-124A airframes. The engine cowlings or anti-drag rings on these aircraft were the same as the C-74. The engine cowling is not hinged and consists of six pieces. The upper and lower sections which contain the air induction system and oil cooler duct are rigidly supported by the nacelle and inter-lock with seams and latches at the 1–11; 3–9; and 5–7 o'clock positions. The four side sections are easily removable and have to be placed on the ground for engine servicing.[2]

The C-124A (Douglas Model 1129A) from ship 49-244 to the last "A" model 51-5187 was fitted with the P&W R-4360-20W engines rated at 3,500 horsepower with Anti-Detonation-Injection (ADI) system. The -20W engines as well as the -35 on the YC-124A and -49 on the YC-124 and C-74s were fitted with seven magnetos, one for each of the seven rows of four cylinders across the four banks. Servicing and changing magnetos was considerably time-consuming because the four-piece cowling must be removed. The magnetos were reduced to four, one for each bank of seven cylinders on the -63 engines. In addition engine access for service was greatly improved with the introduction of hinged orange-peel cowling with seams and latches at 2, 4, 8, and 10 o'clock positions. The cowl sections were hinged at the firewall allowing them to all open or just a single section to access a specific area. This was extremely convenient when the Basic Post Flight Inspection (BPO) was being pulled at a remote destination by a lone crew chief. He could inspect one side of the engine then close it up and move to the other side. The C-124 "A" models to aircraft 51-5187 had the 3,500 horsepower R-4360-20W engines. The C-124C models beginning with 51-5188 were equipped with the upgraded R-4360-63 engines, which were rated at 3,800 horsepower with ADI. Aircraft back to 51-073 were eventually retrofitted with the improved and more powerful -63 engines.

The C-124 was hurried into production with borrowed features from the C-74 because of the immediate need for an outsize transport. The first 14 aircraft including YC-124A (48-795) and the next 13 C-124A models (49-232 to 49-

244) were little more than double deck C-74s with clamshell nose doors. The fuselage was not fitted with waist doors for paratroop drops making them only suitable for cargo. They had an emergency exit on each side of the fuselage in the rear.

The first production batch of 13 C-124A models 49-232 to 49-244 at Long Beach. They were little more than double deck C-74s with clamshell nose doors. There were no waist doors or access for troops except through the nose loading ramps. The airframe on the center left is marked Ship-8 (Bob Williams collection of AAHS archives).

The brakes, autopilots and instrument visibility were recognized as problems with the early C-124 models in 1948. It was soon determined a number of airframe changes were required for operational reasons. As production progressed the C-124 airframe was continually updated as better systems were designed. There were so many upgrades, it is difficult to identify and describe all of them. As soon as engineering was complete on a particular modification it was applied to the next aircraft on the assembly line and retrofitted to earlier models in an effort to meet Air Force requirements and increase efficiency. This created a hodge-podge of configurations, which was challenging to maintenance technicians and flight crews.

Douglas Model 1129A C-124

The Anti-Detonation-Injection (ADI) system is an alcohol/water mixture injected into the engine to increase maximum power on takeoff. It suppresses detonation by cooling the cylinder head preventing engine damage. To prevent corrosion the mixture contains an oil emulsion additive. The 30 gallon ADI tank is mounted in outboard nacelles one and four adjacent to the GPU or putt-putt. Sixty gallons of ADI is enough to supply all four engines for about five minutes at max power.

The tanks are a chore for ground crews to service. The crew chief has to walk out on the wing to the outboard engines, open a hatch on top of the nacelle and lower his body into the compartment. The nacelle behind the engine is quite large making it possible for an average height airman to stand up behind the engine. The airman would open a hatch in the nacelle floor drop out a line to the service cart operator. The 20 pounds of nozzle and hose is pulled up through the hatch and the tank is filled.

All series of the P & W R-4360 engines consume a lot of oil. They either burn it or sling it out no matter how low the hours are on the engine. A freshly mounted low time engine can consume two gallons per hour easily. The C-124 does not have a central oil system; therefore, it is fitted with an 82.5-gallon oil tank behind each engine. The calculated oil burn is two gallons per hour per engine. This may be true with a new engine under optimal conditions on a run up stand but, in reality, most engines regardless of hours would use or throw out three to five gallons per hour.

After any flight of considerable distance the wings were slick with oil residue making it dangerous for airmen to walk out on the wings. The oil tanks would require servicing which requires throwing a line down from the wing and pulling a heavy hose up over the leading edge to each nacelle to top off the tanks. Even more challenging was wrestling the heavy fuel hose up over-the-wing to fuel the "A" models. It was always a bit unnerving pumping 115/145 octane gas into an open tank with the strong fumes rolling out from the spout. Any airman who has ever done it can tell you they are constantly evaluating the distance to the fire extinguisher or wondering if they will break a leg jumping off the wing in a fire.

Airframe

The first fifteen airframes, which include the YC-124 265406, YC-124A 48-795 and production C-124A aircraft

49-232 to 49-244, had ten windows and small troop doors on the upper level and 14 windows on the main cargo level with no rear waist door. Beginning with ship 49-245 the window configuration was changed on the upper level to eight windows and an emergency exit added at window five. The main cargo level was reduced to 12 windows and a full height waist doors added to the cabin before the last two windows.

The waist doors are approximately 10 inches thick and not hinged. The door is bevel cut around the edge with the outer side being slightly smaller than the inner side making it fit snuggly in the frame. It has a large actuating handle which moves a series of rods with pins into the jam locking the door in place. The lack of the waist door on the early model C-124 rendered them unsuitable for troops. The cargo compartment was hot and stuffy. After waist doors became a standard production features, in the summer months crews often flew with the doors open to reduce the heat in the cargo bay. The doors were removed and laid crossway on the floor over the opening. The reasoning was this would prevent anyone or cargo accidentally sliding out.

A pair of Globemasters over the Olympic Mountain Range on 03 July 1957. The cargo compartments were hot. It was typical even at altitude during the summer months to fly with the waist doors out and laid crossway across the opening. It was assumed this would prevent any troops or cargo from accidently falling out (USAF).

All C-124s have two windows forward in the main compartment along the slanted floor behind the loading ramps and two more in the lower forward "P" compartment. Aircraft with the forward facing engineer's panel have a side window on the flight deck. The window was eliminated beginning with production aircraft 51-5198 when the engineer's panel was mounted sideways. The first aircraft to have the engineer's window eliminated was turboprop YC-124B c/n 51-072. It is also the first aircraft to have a side facing flight engineer panel.

The C-124A and the C-74 share the same wing with the full span Fowler flaps. The C-124 early "A" model is easily identified because the flaps have eight hinge cover fairings over the full length of the each wing. This flap configuration is fitted to all "A" model C-124s through aircraft 51-182 of production block four. The last block of "A" models (5th block) from aircraft 51-5173 to 51-5187 came off the line with partial span flaps identified by only five flap actuator covers on the underside of each wing.

The control cables for the C-74 were routed under the cargo bay floor to the wing and not the walls to avoid the side opening forward cargo door. The rudder and elevator cables were routed to the rear in the overhead in the same manner as its little brother DC-4. The cable routing had to be redesigned for the C-124. The cables are under the floor of the raised cockpit above the nose doors. They exit the rear cockpit floor and are routed over to the side walls and then down under the outer edges of the cargo compartment floor to the center-wing then out to the wings.

The C-124 rudder and elevator cables exit the rear cockpit floor and were routed upward along the overhead and aft to the empennage. The cables for the engine controls were run from under the cockpit floor over to the walls and down then along the walls under the floor to the wings. Both routings were similar to the cable routing designed for the conversion of the DC-4 to the ATL-98 Carvair guppy.

The routing of the control cables from under the cockpit floor is both an engineering and safety problem with frontloading cargo aircraft. The cable routing along the sidewalls of the C-124 proved fatal on more than one occasion. An example of disastrous consequences from this design was brought into focus on 22 February 1957 when a C-124A model 51-0141 operated by Military Air Transport Service (MATS) suffered a catastrophic failure of number three engine departing Korea.

MATS flight 503 departed Seoul-Kimpo (SEL) at approximately 18:00 hours taking off on runway 32. The num-

C-124C looking forward into the nose with the ramps up in the stowed position. The right half of the upper deck can be seen on the right in the stowed position. The rudder and elevator control cables exit from under the floor of the cockpit to the left of the ladder beside the cockpit entrance hatch. They are directed to the ceiling in a cable tray then to the rear of the aircraft (Patrick Dennis).

C-124A-DL Globemaster 51-113 from production block four has the final window configuration with waist doors. The large diameter fuselage is obvious. This aircraft has been retrofitted with wing heater pods, the later "orange peel" cowling with chin scoop. However, the APS-42 radome has not been added (Douglas, author's collection).

ber three engine backfired and exploded as the landing gear was in partial retraction at about 900 feet. The reports are conflicting of both a separated propeller blade and shrapnel penetrating the fuselage in the main cargo and lower forward "P" compartment. The flying parts cut control cables in the wall of the main cargo cabin and severed hydraulic lines in the lower "P" compartment. Several passengers on the right side were fatally injured by the flying debris.

Before the out-of-balance condition could be brought under control the number four engine lost power because of damage from flying debris. With engine number one and two under full takeoff power the aircraft wanted to roll over. The aircraft commander reduced power to maintain control and both pilots stood on the rudder pedals in an effort to keep the aircraft straight. The Globemaster went down in the Han River. There were 10 crew and 149 army personnel passengers on board. Twenty-two perished including two pilots and another crewmember.

Heaters, Wings and Fuselage

The early "A" model C-124 was equipped with six gasoline fired wing heaters for leading edge anti-icing. The heaters are mounted in the wing crawlways and draw 115/145 octane fuel from the wing fuel tanks. Fresh air enters through a leading edge port outboard of number one and four engines and is ducted to the heaters. The heated air is then ducted to the leading edge of the wing. This system presented a fire danger in the event of a fuel leak and was not very effective in preventing ice formation. Until a more efficient system was engineered flight crews were advised to avoid operations in icy conditions.

Beginning with the production

of airframe 51-073 the "slick" wing tips were changed to a wing pod which contained a single gasoline fired combustion heater. The streamline pod added one foot to each wing tip. Ram air is forced into the front and heated then ducted to the wing leading edge. The addition of the heaters incurred a manageable drag penalty of ten horsepower per engine to maintain the same cruise as the slick wing models. Eventually all aircraft in service were retro-fitted with the wing pod heaters except the YC-124A 48-795 which was the first production airframe.

The first 13 C-124s had the same engineer's panel as the C-74. Aircraft 49-243 to 51-132 had the second design forward facing engineer's panel. The fuel tank levers were retained at the left end of the engineer's table. The placement of gauges on the panel was changed and a scroll checklist was installed adjacent to the fuel levers (USAF).

Fuel Tanks

The first eleven aircraft of the 28 built in block one, 49-232 to 49-243, had the same six tank 11,000-gallon system as the C-74 and the same usable amount. The balance of 11 aircraft from this group, ships 49-244 to 49-259 along with the 36 aircraft in block two, 14 in block three and 60 of the 110 in block four to aircraft 51-132 had an 11,000 gallon six tank system with different usable amounts. Notably, the six-tank aircraft 49-232 to 49-250 had a usable fuel of 10,931 gallons. Aircraft 49-251 to 51-132 which were also six-tank had a reduced usable fuel of 10,857 gallons.

All C-124 aircraft up to 51-173 were refueled manually over-the-wing by pumping fuel into an open tank. The crew chief would lower a strap with a hook on the end to the fuel truck driver then drag the hose up and fill the tanks one by one with the appropriate calculated gallons to complete the fuel load. As previously mentioned, over wing fuel servicing was never a job which anyone looked forward to doing.

The fuel was measured in by a gallon calculation from the meter on the fuel truck because there were no fill valves. The fuel valves to engines and cross-feed were manually operated with levers on the left side of the flight engineer's table.[3] Even more confusing, 58 of the 110 six tank "A" models in block four beginning with 51-073 were upgraded with a single-point refueling port on the bottom of the fuselage with a fill valve switch panel just forward of the wing. However, the six manual levers on the flight engineer's table were maintained.

Those who serviced the C-124s welcomed the installation of the single-point upgrade. It was a safety improvement in two respects. First, there was always the fear of fire when dispensing volatile 115/145 octane fuel over-the-wing from a nozzle into the open tank with all the strong vapors. Second, the wings were usually slippery from the oil blown back or leaking from the R-4360 engines, making it dangerous to wrestle the large fuel hose up from the ground over the leading edge of the wing.

The new single-point fueling port and electric panel on the fuselage was only for filing the tanks on aircraft 51-073 to 51-132. The airman in the cockpit at the panel during refueling was in charge while the assistant on the ground at the refueling panel followed his commands over the headset. The airman on the ground did not have gauges and could only open and close the fill valves. It was still a primitive six tank system requiring the flight engineer to watch the gauges on the panel and communicate to the airman on the ground when to close the valve.

Sixty aircraft from serial number 51-073 to 51-132 had this hybrid manual system. The electric control panel adjacent to the single point port had six switches to activate the tank fill valve. These 60 aircraft still had the manual fuel tank levers on the left side of the engineer's table. Before refueling the manual tank selector levers had to be in "Off" position with firewall shut off valves closed and mixture controls in "Normal" position.

The six manual levers on the panel were used during flight to transfer fuel within the aircraft to the engines and cross-feed. The manual fuel selector levers on the engineer's table were eliminated on aircraft 51-133 and all subsequent models. The forward facing panel was redesigned for the third time eliminating the levers. All fill, feed and transfer valves were upgraded to electric with a switch panel. The fuel system switch panel was mounted on the left side of the flight engineer's table. With the all electric 12 tank configuration, the capacity was increased to 11,216 gallons with 11,128 gallons useble. The usable fuel amount on all C-124s is based on level flight, three-degree nose up attitude.[4]

Flight engineers and crew chiefs welcomed the final version of the fuel panel because tank valves, boost pumps, manifold valves could be manipulated from the panel on the engineers table in the cockpit. The twelve tank configuration consisted of five in each wing and two tanks (number 6 and 7) in the center-wing fuselage. Tanks six and seven were rarely if ever used because they were lower than the engines and could not gravity feed in the event of boost pump failure which did occur on occasion.

In order to not use the center tanks it was necessary to distribute the fuel load in the wing tanks without overstressing the airframe. This required a review of the cargo load to evenly distribute the fuel required for the mission. The fuel weight distribution is critical to prevent too much deflection stress on the wing roots. Care is taken not to have too much fuel weight inboard relative to the cargo load. Fuel weight outboard in the wing can reduce deflection when transporting heavy cargo. However, as the fuel burns off the wing deflection increases which can increase stress on the wing root. When transporting heavy loads the idea is to burn the inboard fuel first to maintain enough weight outboard away from the centerline of the aircraft to reduce wing deflection. The twelve tank system made balancing the aircraft with heavy loads easier.

When fueling the aircraft it was customary to shut the valves off below the required load to allow the gauges to settle. Once comfortable with an accurate reading the valves were reopened to fine tune the amount in each tank and balance the tanks in each wing. During the entire refueling operation it was important to keep at least one tank valve open at all times to avoid over-pressure from the fuel truck pump which could cause damage.[5]

Fuel management has always been a challenge on heavy lift aircraft. Douglas continually made upgrades to the fuel system for safety, reliability and easier management. Beginning with "C" model aircraft 51-5198, which is the tenth airframe in block six, the engineer's panel was turned side-ways and the fuel system control panel moved from the left side of engineer's table to the right side of the engineer's instrument panel. The configuration was first tested on the side facing panel of the YC-124B where the fuel control panel was in the middle of the engineer's panel. It was determined to be more practical on the right hand side on the production "C" models.

The early C-124s experienced multiple fuel management problems and fuel tank leaks. The Globemaster had a "wet-wing" system where the fuel is stored in compartments between bulkheads. The tanks had to be periodically resealed to prevent leaks. One of the requirements on the walk-around check list was to look for those telltale red stains on the bottom of the wings. The leak problem was so severe on early models all Globemasters assigned to the 22nd Troop Carrier Squadron Heavy were grounded in December 1952. After fuel tanks were modified and resealed the transports were returned to service in February 1953.

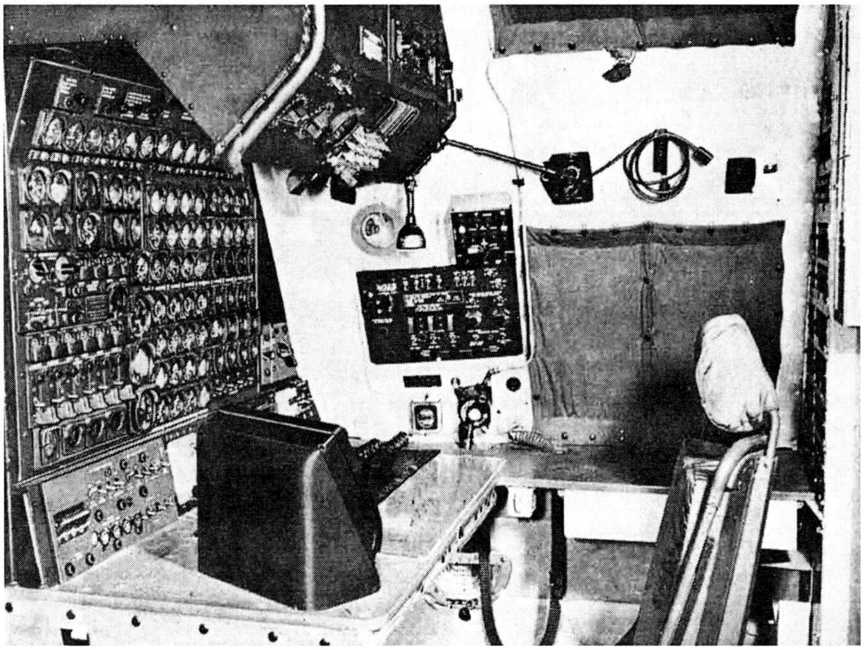

A third forward facing engineer's panel configuration was installed on aircraft 51-133 to 51-5197. The fuel system was changed to a 12 tank all electric fueling and transfer system. All tank levers were eliminated and an auxiliary electric fuel selector switch panel was mounted on the left side of the engineer's table (USAF).

Auxiliary Power, Putt-Putts and Turbines

All C-124 "A" and the majority of "C" models up to aircraft 51-7285 were equipped with a pair of small two-cylinder gasoline ground auxiliary power generators (APP). These units were originally installed on the C-74 in the nose. They were moved to the outboard nacelles and carried over to the

C-124. Known as "putt-putts," they are mounted in the number one and four nacelles on the outboard side. There were running changes in fuel supply to these units. Airframes 49-232 to 51-132 are supplied from fuel tanks one and four. Airframes 51-133 to 52-938 are supplied from tanks two and eleven. As technology progressed the final production block of "C" model C-124s, 52-939 to 52-1089 and 53-001 to 53-052 were equipped with a gasoline turbine APU unit. It was installed in "P" compartment deleting the putt-putts in the outboard nacelles.

The fuel for the APU units on airframes 52-939 to 52-944 is supplied from tanks nine and eleven and 52-945 to 52-1021 are from tanks one and twelve. The fuel supply was changed again on airframe 52-1022 to 53-052 to tank ten as a more practical way to manage fuel from a single tank.[6]

The engineer's panel was turned sideways beginning with the 10th C-124C, aircraft 51-5198, and all subsequent production. The electric fuel selector switch panel right of center on the engineer's panel was turned out at the far end for better visibility. This was the final cockpit configuration (USAF).

The APU turbine unit for the C-124 was introduced in late 1952 by the Solar Aircraft Company of San Diego and Des Moines. The unit named the "Mars" produced 50 horsepower and weighed less than 100 pounds contained in a compact two foot cabinet. It was mounted against the forward bulkhead in "P" compartment on the right side below the forward main cargo deck. The APU intake and exhaust is visible just below the leading edge of the wing on the right side of the fuselage adjacent to engine number three engine.

Photos and records show a portion of the next to the last block of aircraft had turbine APUs installed at the factory before delivery but records are not clear as to how many. In addition, some other "C" models were retro-fitted with the turbine units. Douglas continually upgraded the Globemaster II, for the Air Force making running assembly line changes as soon as they were available. Aircraft which were already in service were periodically retrofitted with newer systems.

Engines

The final C-124C (Model 1317) was vastly improved over the "A" model. The early aircraft were configured more closely to C-74. The continual incorporation of advancements on the C-124 assembly line made the final model far superior to the early aircraft. The YC-124 was fitted with the same 3,250 horsepower R-4360-49 engines which were on the C-74 which made it somewhat lacking in power. The production aircraft received the 3,500 horsepower R-4360-35 and -20W engines up to airframe 51-587. The engines were upgraded again to the more powerful 3,800 horsepower R-4360-63A engines with a two-stage supercharger which is manually operated. This became the final power package for all C-124C aircraft. The propellers were the same Curtiss Electric three blade units used on the "A" models at 17 feet (16.6) in diameter.

The engine changes over time improved the safety and reliability of the aircraft but not without catastrophic consequences. The C-124 was the primary transport in the supply chain for the Korean War. On 29 May 1953 a C-124A suffered an engine fire on number two engine because of an overheated generator. On 11 June a second aircraft suffered an engine fire because of a generator failure.

On 18 June a C-124A 51-037 departing Tachikawa Airbase Japan suffered a failure of number one engine. The aircraft commander with over 6,000 hours attempted to turn back to the field. The left wing stalled causing the aircraft to roll to the left in a slow spin. It crashed into a field 3.5 miles from the airfield. All 129 on board perished prompting an inspection and grounding of all C-124 aircraft.

The problem was traced to engine driven generators

overheating and bursting into flames. The generators were replaced with a different designed unit on some aircraft. However, only 17 aircraft were returned to service on 08 July 1953. Eight more in the Korean theater remained grounded at the end of the war until the new generators could be manufactured and installed.[7]

Operational History

Deliveries of the first C-124A to the Air Force began in May 1950. Both the YC-124 and YC-124A were extensively tested. The military was very anxious to determine the limits and capabilities of the Globemaster in every operational situation. The third C-124 s/n 49-232 which is the first production C-124 "A" model was placed in a test program on 04 May 1950 at Wright-Patterson Air Force Base. The aircraft logged 457 hours in multiple tests which included cross country transport cargo handling of bombs, F-86 jets, helicopters, missiles and spare engines. The tests were conducted on service altitude and 2 and 3 engine-out rate of descent.

On 23 May 1951 while conducting one of these test 49-232 crash-landed in a corn field near New Lisbon, Indiana. The aircraft caught fire and burned with only five of the twelve man crew surviving. The test program was immediately cancelled. All data accumulated during the testing was evaluated. Although there were 180 unsatisfactory reports from the test program, none were considered severe enough to ground the aircraft. The test results were considered sufficient to create the Flight Operations Handbook.[8]

Initially the C-124 was delivered to the Strategic Air Command (SAC) to begin operational service. It served as the primary transport for moving support equipment, men and nuclear weapons to operations throughout the world. The Military Air Transport Service (MATS) operated the C-124 Globemaster from 1950 to 1966 serving during the Korean and Vietnam War operations. They continued on with the Military Airlift Command (MAC) in reserve and guard units until 1974 when the C-5 became operational.

Possibly one of the most challenging missions for the C-124 came in 1952. Military planners had conceived a secret plan in 1949 for an extreme northern Cold War base directly across the pole from the Soviet Union. After Denmark joined NATO the United States embarked on a mission of building Thule Air Base (BGTL) 750 miles inside the Arctic Circle and only 900 miles from the North Pole in northern Greenland. There were no roads or supply routes overland. Virtually everything had to be brought in by air or sea.

The C-124 proved its worth as an outsize transport by uplifting a 29,000-pound snow removal machine. Globemasters became the supply line working in concert with the Navy to move men and materials. The navy dispatched 120 ships from Norfolk to move materials which the C-124s could not handle. The ships provided living quarters on board for the workers. It was a mammoth effort larger than any project of this type which had ever been attempted. The C-124 performance in the extreme cold exceeded all expectations.

As a result of the Globemasters exemplary performance in the subzero temperatures of Greenland, the C-124 was deployed for Task Force Forty-Three Operation Deep Freeze. The Globemaster proved invaluable for heavy lift supply operations in subzero temperatures during the 1955–56 building of a permanent research station in Antarctica. The Globemasters not only landed on the ice to support the Navy mission they also conducted aerial drops. The missions were not always without problems. As reliable as the C-124 proved to be in extreme cold engine failures did occur. Engines had to be changed on the ice before

Airmen of the 9th Troop Carrier Squadron change the number four engine in a record five hours at 25 below zero at Williams Field, Antarctica. The portable winch first designed for the C-74 mounts on top of the nacelle for such remote field operations. The firewall door is visible on the nacelle, which can be accessed via the wing tunnel for maintenance in the accessory section in flight. The APU exhaust is visible on the fuselage below the number three engine (USAF, A2C Edward C. Fohrer).

departing back to the staging base at Christ Church, New Zealand. On 21 November 1962 airmen of the 9th Troop Carrier Squadron changed the number four engine on a C-124 in a record five hours at 25 below zero at Williams Field, Antarctica. The aircraft was then returned to its air drop mission schedule. Two nine ton D-4 Caterpillar tractors were aerial dropped to a remote area of Antarctica. The successful drop was made by releasing the palletized dozers through the elevator well at an altitude of 1,500 feet using five 100-foot diameter parachutes. The outside temperature over the drop zone was 55 degrees below zero Fahrenheit. Only 30 minutes after arriving on the ice the tractor was operational.[9]

Jupiter IRBM being offloaded at Cape Canaveral Skid Strip. The C-124 was the primary transport for the main section of the rockets because of the size of the cargo hold. The Jupiter is a direct descendant of the German V2 being developed by von Braun at Redstone Arsenal in Huntsville, Alabama. The rear section was added at the cape where they were test fired (USAF).

The C-124 was determined to be universally adaptable to any military or government needs. When the rocket program was established at Redstone Arsenal in Huntsville, Alabama, in 1950 there was an immediate need for an outsize transport. In 1954 Wernher von Braun proposed using the Redstone rocket to launch artificial satellites. In 1956 the Army Ballistic Missile Agency (ABMA) was established naming von Braun as director of operations. Initially the space program relied on the military for rocket and hardware transport. Redstone and Jupiter rockets built at Redstone Arsenal for the early space program were moved to Cape Canaveral by C-124s. This was four years before Lee Mansdorf conceived the idea of a mission specific ultra-large transport for rocket boosters. When Mansdorf learned of the NASA logistical problems he envisioned an outsize transport based on the Boeing B-377. His vision eventually became the Pregnant Guppy. Up until this time the largest and only volumetric guppy type transport available was the Douglas C-124 Globemaster II.

The game changed drastically on 04 October 1957 with the Soviet launch of Sputnik. The space race was on and America was behind. President Eisenhower created the Defense Advanced Research Projects Agency (DARPA). The mission of DARPA was established to insure American technology was superior to any adversary and prevent another Sputnik surprise.

In 1957 the turboprop Douglas C-133 Cargomaster entered service as the new Air Force strategic transport. Because of the smaller diameter fuselage the C-124 remained the workhorse for military outsize cargo. The first U.S. attempt to launch a satellite was made with a Vanguard rocket which was transported by a C-124. On 31 January 1958 von Braun and his team successfully launched the Jupiter C missile, placing America's first satellite, Explorer 1, in orbit.

In October 1958, Project Mercury was authorized for the purpose of putting an American in orbit. A program did not exist to develop new rockets. For this reason the space program was forced to rely on the inventory of military Atlas ICBMs.[10] The C-124 Globemaster remained the primary aircraft to transport rocket boosters, hardware and nose cones from Huntsville to Cape Canaveral. The Globemaster was the only aircraft capable of transporting the massive Thor, Redstone, Atlas and Titan missiles.

By 1960 NASA was deep into the manned space flight program and still relying on the inventory of military missiles as launch vehicles. The C-124 Globemaster was slow and aging; however, no other aircraft in Air Force inventory had the ability to transport outsize cargo for the space program. The Mercury and Gemini capsules were being produce by McDonnell at St. Louis, Missouri. NASA had neglected to develop a serious logistics program relying on the military for missile transport.

The Mercury capsule was not too large to be transported overland. It weighed 2,422 pounds and measured 7 feet, 3 inches by 6 feet, 3 inches. However, because of the

Globemaster 52-1074 takes on an unidentified capsule for NASA at Cleveland Hopkins Airport. The C-124 was a primary transporter of military rockets and space hardware until the space items became too large for the aircraft (Cole Goldberg, opshots.net).

The Globemaster II in double deck configuration could transport 200 fully equipped troops. This view demonstrates how the upper deck can be lowered in multiple configurations and fitted with side troop seats. Decks on the right are retracted to stowed position. All decks were either retracted or removed to transport rocket boosters. The access hatch to the cockpit is visible in the far center with the ladder access from the main cargo deck (Patrick Dennis).

delicate nature of the instrumentation contained in the unit NASA specified it should be transported with minimal vibration and in the shortest time. It was mounted on a trailer at McDonnell for transport to Lambert Field in St. Louis. The trailer and capsule were loaded on to the C-124 as a single unit for transport to the Cape. The C-124 vibrated and shook so much it was often referred to as "big shaky" but it was big and reliable. The Saturn program was being developed; however, NASA logistics planners were not actively concerned with the lack of military aircraft capable of moving larger launch vehicles. At some point in time the size of the space hardware would exceed the capacity of the C-124.

Thor intermediate range ballistic missiles (IRBM) were deployed to military units in England beginning in 1958. The C-124 was called on to transport the missiles across the Atlantic. It continued to be the only U.S. military transport large enough for the task until the C-133 became operational. The Globemaster uplifted the first missile to England in September 1958 and continued to be the primary transport through 1961. Once the Douglas C-133 Cargomaster was produced in sufficient numbers and delivered to operational units it also transported the missiles.

The C-133 could lift considerably more weight than the C-124; however, the Thor missile was a very tight fit for the cigar-shaped fuselage. Until the C-5A became operational in June 1970, the C-124 remained the only Air Force transport with a sufficient fuselage cross-section and height to accommodate military oversize cargo. As the space program grew rocket boosters became too large for transport by air prompting NASA to utilize ocean going barges to move the outsize boosters from California to Alabama and Florida.

When the C-124A Globemaster II began service in 1950 it was used as a support aircraft with the Strategic Air Command (SAC). It also saw extensive action in the Korean conflict as part of the supply line. It was beginning to suffer age problems by the Vietnam War. However, there was no other transport which could take its place. It proved to be still quite capable of transporting up to 74,000 pounds of cargo including artillery pieces, light tanks, support vehicles, construction materials, radar vans and engineering equipment. During both military actions the C-124 transported as many as 200 troops or 127 casualty patients at a time.

C-124 s/n 53-0044 was purchased by a private buyer and flown to Las Vegas. The Globemaster was moved off airport for conversion to a bar-restaurant. The project was never completed, leaving the aircraft to deteriorate for years, with homeless people living in it from time to time. In 2001 it was broken up after the owners could not be located to move it to another location (author's collection).

Reserve and Air Guard C-124s

The last C-124C was delivered to the Air Force in 1955, ending a total production of 448 aircraft. Through the 1950s and 60s the C-124 was the principal heavy transport for the military and NASA. By late 1964 the "A" models were transferred from active military to Air Force Reserve units. The Military Air Transport Service (MATS) became the Military Airlift Command (MAC) in 1966. The C-124C models continued in active service as they were slowly transferred to Air Force Reserve units. When the Lockheed C-5A Galaxy was delivered to Air Force units in 1970 the military began the final transfer of the remaining regular Air Force C-124C models to Reserve and Air Guard units.

The Air Force Reserve 918th Troop Carrier Group under the 445th Military Airlift Wing Heavy based at Dobbins Air Force Base Georgia received the "A" models in late 1964 to replace the Fairchild C-123B Providers. The 116th ATG Air National Guard also at Dobbins received C-124s in 1966 operating on the adjacent ramp to the 918th Air Force Reserve unit. By 1970 all reserve units had received the last "C" models built.

The Georgia Air National Guard (ANG) is the first and last guard unit to operate the C-124. The 158th Air Transport Squadron under the 165th Air Group at Savannah, Georgia, received them in 1967. The last two, 52-1066 c/n 43975 and 53-0044 c/n 44339, were retired from the 158th ANG at Savannah in September 1974 and declared surplus on 15 November. The C-124 fleet had served admirably for 24 years as a both front line and secondary volumetric military transport. The Globemaster II was officially retired from Air Force inventory in 1974.

The two Georgia Air Guard Globemasters were transferred to Davis-Monthan Air Force Base for disposal along with all the stored C-124s. It was assumed these last two active examples would be displayed at one of the air base museums throughout the country. This was not the fate of 53-0044. It was purchased by a private owner and registered as N3153F.

Subsequently, it was flown to McCarran International Airport at Las Vegas for a commercial venture. It was parked in a field at Koval Lane and Reno Avenue outside the northwest corner of the airport for the purposes of becoming a restaurant. It was painted in a blue and white scheme and large windows were cut in the sides. It languished in various states of disrepair while homeless people lived in it from time to time. It remained in this condition until a major airport expansion and highway improvements were to be built on the land. Officials of Clark County, Nevada, notified the owner in 1993 the aircraft would have to be moved because of a street widening project. The aircraft remained on the property only a few feet from the edge of the road. Again in 2001 an attempt was made to get the owners to move it to a new location. After numerous lawsuits and court actions the owners could not be located and, regrettably, on 26 March another historical aircraft was destroyed.

The second of the last two active C-124s, 52-1066 received the honor it deserved by being transferred to the Wright-Patterson Air Force museum, where it was restored and is now on display indoors as 51-0135.

During active service 52-1066 was involved in an international incident during the Congo Airlift Operation SAFARI. Because of political unrest Belgium was forced to grant independence to the Congo on 30 June 1960. Eleven days later the premier of Katanga Province, Moise Tshombe, seceded from the republic followed by Kasai Province, throwing the Congo into civil war. The United Nations dispatched troops to quell the tension between Congo President Joseph Kasa-vubu and first democratically elected prime minister and rebel leader Patrice Lumumba. The rebel leader Lumumba was arrested by President Kasa-vubu and turned over to Tshombe and assassinated. Hostilities erupted, causing the United Nations (UN) to take action with an airlift only exceeded in scale by the Berlin Operation Vittles (Berlin Airlift). The UN Congo airlift was flown by an assortment of coalition military and civilian air carriers.

The American military action in the Congo airlift was flown by C-124s to the end of 1961. It was during one of these supply missions C-124 aircraft 52-1066 strayed off course, resulting in a diplomatic confrontation with a Communist nation. Three Globemasters squadrons from the 1607th Transport Wing (Dover Air Force Base) were staged at Chateauroux Air Base in France, which was a front line base for U.S. forces in Europe during the Cold War until 1966, when De Gaulle evicted all NATO forces.

The C-124 mission was to transport Pakistani troops from Karachi to Leopoldville (now Kinshasa). The fly route was established by diplomatic teams in advance but was a bit circuitous and backtracking in order to meet flight requirements of certain countries and not to step on any political toes en route. The round-trip course was complex and exhausting for the C-124 crews.

The first leg from Chateauroux France (LFLX) to Dhahran (OEDR) with a fuel stop at Athens took 14 hours. After a 15 hour layover for crew, the next leg to Karachi (OPMR) was nine hours. The Pakistani troops with gear and weapons boarded during a three hour layover. The C-124s flew nine hours back to Dhahran. The Globemasters were serviced at Dhahran while the crew took a 15 hour rest. The next leg to Khartoum (HSSS) was seven hours with a minimum ground time for servicing then back in the air for the final nine-hour leg to Leopoldville (FCAA).

The crews were secured in dorms at Lovanium University in Leopoldville for layover and much needed rest before a more direct return to France. The return flights from Leopoldville staged through Kano. Nigeria (DNKK). where they were refueled for the final 12 hour flight back to France. The next stop was Wheelus Air Force Base (HLLM) in Libya and a layover before the final leg to Chateauroux Air Base. This mission totaled six days covering over 13,000 miles in 67 hours in the air demonstrating the C-124 is indeed a globe master.

It was on one of these missions when 52-1066 was forced down. The fly route was across central France and northern Italy and down the Adriatic across the Straits of Otranto skirting Albania then across Greece to Athens. After crossing Italy and turning south down the Adriatic the C-124 got off course to the east and encountered Yugoslav Air Force fighters. After some tense moments the MiGs forced the Globemaster to land in Yugoslavia. The crew and aircraft were held until diplomatic negotiations secured their release. The aircraft and crew returned to Chateauroux, France, where the incident was reviewed.[11]

In a unique twist, two of the DC-4 ATL-98 Carvair guppy conversions were custom built for Luxembourg based Interocean Airways for the same ONUC United Nations operation in the Congo. The two Carvairs also participated in the airlift which supplemented the Italian Air Force C-119s (see Carvair chapter). (A detailed account of the Carvair mission in the Congo is covered chapters 10 and 11 in the book *The ATL-98 Carvair*.)

The C-124 Era

The C-124 initially came on line in 1950 as a support transport for the Strategic Air Command (SAC) to move nuclear weapons to air bases around the globe. It performed a secondary role of transporting personnel and support equipment for SAC exercises and deployments. Its size and fold up upper deck allowed it to move more cargo and vehicles farther and faster than any previous military transport. In comparison the Convair XC-99 could transport more cargo; however, it was a single experimental aircraft. It took longer to load and could not transport outsized vehicles because it had a stationary double deck. SAC operated fifty C-124s in four squadrons between 1950 and 1962.

By 1955 the Air Force achieved worldwide cargo reach to support any operation with the C-124 Globemaster II. It was the cheapest and most efficient air transport fleet per ton mile in Air Force service. It was a Cold War work horse with a commendable safety record for such a lumbering giant built of pre–World War II technology. It rattled, creaked and moaned but performed well despite demeaning references as "Old Shaky" and "Big Shaky" or "a million rivets flying in close formation."

The C-124 had the largest diameter fuselage of any production American military transport powered by piston engines. When the upper decks were folded up it gave the feeling of flying in a barn. Conversely the longer Convair XC-99 had the felling of flying in a very long tube. The C-124 Globemaster II survived well into the jet age because of the ability to airlift large bulky cargo over long distances and into primitive airfields. It was the only military outsize transport for 20 years before yielding in 1970 to the C-5A operated by the 437th Airlift Wing at Charleston Air Force Base.

The C-124 Globemaster cargo bay with the upper deck stowed gives the impression of flying in a barn. The decks could be lowered for a double deck troop transport or removed to accommodate missiles and high or wide cargo. It had the largest diameter fuselage of any piston engine American military transport (author's collection).

Last C-124 Flight

The last C-124 flight took place on 09 October 1986. McChord Air Force Base was a major operating center for C-124 throughout the aircraft's 20 year run. The Air Museum at the base near Tacoma, Washington, began searching in 1981 for a C-124 to display. Aircraft 52-0994 was eventually located at the Detroit Institute of Aeronautics at Willow Run, Michigan, where it was being used as a training aid. After considerable negotiation a deal was struck to trade the Globemaster for a more modern T-39 Sabreliner.

Over $40,000 was raised from private contributions to bring the C-124C up to airworthy standards for the flight to McChord. The maintenance team was headed by retired Master Sergeant Don Egan. The only set of airworthy P & W R-4360 engines was shipped from Travis Air Force Base to Willow Run and mounted on 52-0994 for the flight.

Lieutenant Colonel Gary Caldwell of the 4th Military Airlift Squadron at McChord was selected to command the flight with his hand-picked crew. After considerable work the aircraft was flown gear down the short distance from Willow Run to Selfridge Air Force Base, Michigan, where a retraction test was done to insure the gear was operable. More minor write-ups were cleared before it was released for the historic flight. On 09 October 1986, 34 years after it first flew the Globemaster lumbered into the air for the final three hour flight. After two hours of flight the number one engine began to trail light gray smoke. As they approached the field the engine began blowing thick white smoke and oil forcing the crew to shut it down. The last flight ended in true C-124 fashion as the crew made a perfect three engine landing. The engines were swapped out and returned to Travis Air Force Base, and the aircraft is now on permanent display at McChord Air Force Base.[12]

Turboprop YC-124B (YKC-124B) Model 1182E

As the C-124A was entering service with the Air Force in 1950, Douglas reviewed the possibility of a more powerful or an even larger transport. Engineers were aware it would not be possible to extract much more horsepower from the reciprocating Pratt & Whitney R-4360s. Studies were being conducted on the possibility of turboprop power. During operational testing to determine what changes and upgrades would be required to make the C-124A fulfill the Air Force mission, Douglas proposed an upgraded turbine powered model designated C-124B.

It would be powered by four Pratt & Whitney 5,500 shp YT34-P-1 turboprop engines fitted with three-bladed 18-foot Curtiss C-735S-B2 electric propellers. Douglas had previously conducted studies to adapt turbine engines to the C-74 Globemaster I. The low production numbers of the C-74 prevented the proposal from progressing very far. The idea was carried over to the C-124 with some optimism. The Air Force reviewed the turboprop proposal to increase performance and range. A government Letter of Contract AF33 (038)-14765 for the application of turboprops to the C-124 was submitted to Douglas on 29 July 1950. The turboprop YC-124B is a result of the contract to conduct a design study.

Douglas proposed the turboprop version to the Air Force as an air tanker designated YKC-124B as a possible alternative model to the Boeing KC-97 design. In 1950 the Air Research and Development Command (ARDC) placed an order for a single YC-124B.

The Strategic Air Command had considerable experience with the C-124 transporting nuclear devices and support equipment throughout the world. However, SAC won the bureaucratic battle for control of all USAF refueling assets. It had a close relationship with Boeing solidifying the SAC choice for the Boeing KC-97 tanker equipment. Consequently the YKC tanker configuration did not progress past the design stage and was changed to YC-124B before it ever flew.[13] Aircraft (51-0072) c/n 43406 was completed as an YC-124B high payload transport.

Surprisingly, with a cruising speed of 323 mph and a ceiling of 25,900 feet it was faster than the Boeing KC-97 tanker. The KC-97 cruised at 230 mph with a ceiling of 30,000 feet. The C-124 was capable of in-flight tanker service but it was large and cumbersome making it less desirable than the Boeing KC-97. In reality the Boeing KC-97 contract was never threatened by the Douglas proposal. Douglas was overly optimistic in believing it actually had a chance to win a tanker contract.

The only YC-124B was powered by Pratt & Whitney T-34 engines, which were the same as those used on the production Douglas C-133. They were tested on two Boeing YC-97J aircraft which were used to produce the Aero Spacelines B-377 Super Guppy. The YC-124B first flew on 02 February 1954 for one hour from Long Beach to Edwards Air Force Base, California (Bob Williams collection of AAHS archives).

In 1951 Douglas Aircraft Company celebrated the 30th anniversary with magazines ads showing cutaway photos and drawings of the C-124 describing its ability to lift 50,000 pounds of cargo. It was presented as the most versatile and flexible cargo transport in service with 65 on the production line. Douglas marketing was quite optimistic believing there was a very good chance because of the C-124 success the Air Force would place large orders for a turboprop version. The piston driven model was in continual upgrade to meet the Air Force needs. Douglas promotional ads stated, "A turboprop version would be a leap ahead."

The magazine ad copy stated, "Furthermore, propeller turbine engines can be readily installed in the current C-124 design, making possible more payloads, longer range and greater speed." This was more hype and overly optimistic press than reality. It had already been determined the C-124 airframe was not designed for the stress and increased speed achieved with turboprop power. It is ironic the problematic turboprop C-133 which was developed to augment or replace the C-124 was retired in 1971, three years before the last two reciprocating R-4360 powered C-124 Globemasters were retired in 1974.

Douglas planners entertained the idea of a tanker version C-124s from the beginning of the program. One of the special test missions conducted on the first production C-124A 49-232 at Wright-Patterson before it crashed was listed as "Simulated In-flight Refueling." The tests were conducted to generate and evaluate performance data as a proposed tanker. This was long before the YC-124B flew.

The same series of test were not conducted on the YC-124B because the proposal as an in-flight refueling tanker designation was dropped before it flew for the first time. The Air Force was not interested in a tanker version under any circumstances. The Air Force did express interest in the turboprop C-124 as a cargo carrying transport if it could be fully pressurized. Because aerial refueling tends to take place at higher altitudes Douglas had designed a pressurization system for the proposed YKC-124B in an effort to sell the USAF on the tanker version[14]

The turboprop "B" model had a projected speed of 400 plus mph because it was 60 percent more powerful than the piston powered production model. The performance was superior to the Boeing KC-97 with reciprocating P & W R-4360 engines. The YC-124B had an empty weight of 101,930 pounds with a cargo payload of 64,037 pounds at a design takeoff gross 200,000 pounds. The maximum overload takeoff is 235,000 pounds.

Overall the numbers are not an outstanding improvement over the later C-124C model. The service ceiling of the C-124B was set at 24,100 feet compared to 14,000 feet of the C-124C.[15] The weight penalty for the pressurization equipment on the C-124B model was not acceptable, therefore, only the cockpit was pressurized. Douglas proposed the YC-124B as an effort to extend the current production contract with the Air Force after the current orders for the C-124 cargo transport were completed.

Douglas flight engineer Duncan Hall was at the panel on the first YC-124B flight. The aircraft was the first C-124 to have a side facing engineer's panel. The 12 tank fuel management system was located in the middle of the panel. There are no levers on the table. The C-124B turboprop has propeller pitch controls on the panel, which is quite different from the production piston powered models (Douglas, author's collection).

The YC-124B prototype first flew on 02 February 1954 commanded by Frank Boyer and assisted by Frank Aitkin. The flight engineer was Duncan Hall. The aircraft departed Douglas at Long Beach for a 60 minute flight to Edwards Air Force Base for testing. This was only a year before the last production C-124C was delivered. In actuality, the YC-124B was always meant to be a one-of-a-kind test aircraft to test new systems for production models. Many improved new systems incorporated on the C-124B became standard on later "A" and "C" models. The features included single point refueling, 12 fuel tanks, electric fuel control valves, side facing engineer's panel with center mounted fuel control panel, NESA glass windshield, partial span flaps, APS-42 radome and wing tip heaters. The use of the Pratt & Whitney YT-34-P-6 turboprops was part of a program to test engines and systems for more efficient next generation transports. As a result the P & W T-34-P-1 engine was fitted to the Douglas C-133 which was produced between 1956 and 1961. The C-124 was a true workhorse for the military but there were restrictions because the cargo floor was 13 feet above the ground. The steep loading ramps up to the cargo floor made loading of ICBMs and rocket boosters a challenge.

In 1953 while the YC-124B was being built the Air Force issued specifications for a redesigned pressurized transport designated as Douglas Model 1324 or C-124X Globemaster III (not to be confused with the jet powered C-17 Globemaster III of today). The new design had a circular fuselage for ease in pressurizing the cockpit and cargo area. It was a radical departure from the production slab side C-124. The front clamshell doors were deleted for ease of pressurizing the cockpit and cargo area. The new design called for rear cargo doors similar to those on the C-133B. The C-124X wing was beefed up with an extended center-wing to increase the span and allow prop clearance from the new fuselage. Interestingly, this is the similar approach Jack Conroy took when building the Super Guppy. A new center-wing was built to move the engines away from the fuselage.

The planned C-124X retained the Pratt & Whitney YT-34-P-6 turboprop engines. The nacelles were expanded to house a new tandem bogie landing gear and the improved tail surfaces of the YC-124B were incorporated. The program was cancelled in February 1953 when the Air Force requested Douglas to proceed with the C-133 program. The YT-34 engines were carried forward to the C-133 and some of the design features were incorporated on the planned XC-132.[16]

The proposed XC-132 and production C-133 were designed as high wing aircraft with straight-in rear loading to alleviate the problems encountered with the C-124. The turboprop C-124B was critical in testing and developing systems for Douglas next generation transport designs. Engineers expressed concerns about airframe torque because the C-124B was so massive with slab sides.

The operational C-124 models were known as "Old Shaky" because they creaked and vibrated. Project engineers who worked on both the C-124B and C-133 indicated the bigger concern was the problem that torque created by turbine failure of an outboard engine which could result in extraordinary wind-milling and propeller drag. This alone creates severe stress forces on any fuselage. There was increased concern that an engine out could cause airframe failure on an aircraft this size. Considerable attention was given to testing and analyzing the problem.[17]

Visual first impressions of the C-124B convey the idea it was a standard C-124 with turboprop engines. The nacelles are obviously different with the T-34 turboprop engines however; closer inspection reveals multiple modifications to the airframe. The empennage is strengthened internally and visibly different from the production model. The horizontal stabilizer has a noticeable positive dihedral with squared ends as opposed to the standard production unit with zero dihedral and rounded ends. The area below the rudder extends straight down to the shorter tail cone. The tail cone does not have fairing from the trailing edge of the horizontal stabilizer. The vertical fin is 2 feet, 10.5 inches taller than the production model with a more squared top as opposed to the rounded top of the C-124A-C model. The dorsal fin is higher with a steeper angle from the top of the fuselage. The flight engineer's window was eliminated because it was the first C-124 fitted with the side facing engineer's panel. Also one of the lower forward compartment windows was deleted on each side. It is the only C-124 with a pressurized cockpit which required many changes which are not externally visible.

The YC-124B served as an engine test aircraft until October 1956. The aircraft was tested for two years and eight months providing valuable data which was incorporated into the development of the C-133 Cargomaster. The single YC-124B was retired to Redstone Arsenal in Huntsville, Alabama, where it was used by the missile and munitions school to train military personnel on rocket and booster loading procedures.[18] It was displayed for many years with a rocket partially loaded through the front clamshell doors. Unfortunately this rare and unique aircraft was not preserved and eventually broken up.

Turbine Test Bed JC-124C

Technically there were two turboprop C-124s. A second Globemaster, 52-1069 c/n 43978, was converted to a JC-124C by installing an experimental 16,000 shp P & W XT-57-P-1 turboprop engine with a 20-foot four blade propeller in the nose making it a five engine aircraft. It was actually a flying test bed for the XT-57 engine slated for the planned swept wing XC-132 Globemaster III. The new swept wing turboprop design was announced by Douglas in February 1957 as the successor to the C-124. Mounting the XT-57 engine in the nose of the C-124 was only for engine testing. It was too large to ever be considered for any upgraded or modified version of the C-124C.

The C-124B empennage was designed for higher stresses and is noticeably different. The horizontal stabilizer has considerable positive dihedral, and the vertical fin is nearly three feet taller. The area below the rudder extends straight down to a shorter tail cone. The dorsal fin extends at a greater angle to the taller vertical fin (Bob Williams collection of AAHS archives).

The Douglas JC-124C was tested and evaluated at Larson Air Force Base, Washington, which was a center for the development of aircraft and systems for the USAF. In 1954 the C-124C 52-1069 was flown to Logan Airport in Boston for transfer to the Pratt & Whitney Rentschler Plant in East Hartford. The Plant was named after engine developer Gordon

Rentschler. The XT-57 (PT5/T57) engine was an experimental project which came from the P & W J57 (JT3C) development program.

The J57 is an axial-flow turbojet developed in the 1950s as the first 10,000 pound force (lbf) thrust engine built in the United States. The engine was modified by adding a forward housing with bearings and an oiling system for the propeller. The four stage low turbine with a new low compressor and reduction gear drove a 20-foot diameter Hamilton Standard propeller developing 15,000 shaft horsepower. The propeller diameter is huge when compared to other aircraft. The jet thrust added another 5,000 horsepower. This combination made it the biggest turboprop engine ever built in the United States.

Probably one of the most bizarre flying test beds was the converting of production Globemaster 52-1069 to a JC-124C. A Pratt & Whitney XT-57-P-1 turboprop with a 20-foot propeller was installed in the nose. The 15,000 horsepower engine was intended for the huge XC-132 Globemaster III. The program was eventually cancelled, and the heavily modified C-124 test aircraft retired to the desert (Bob Williams collection of AAHS archives).

Major modifications were made to the nose of 52-1069 in order to install the large turboprop engine. The APS-42 radar dome was removed. The clamshell doors were sealed and a structure fabricated in the nose to support the engine mounting. The engine was mounted above the clamshell doors with a chin scoop under the XT-57 nacelle. To compensate for the extra weight of the engine being ahead of the nose, lead weights were installed in the rear of the cargo compartment as a counterbalance. The test pilots reported an increase in pitching making the C-124 fly like a dirigible.[19]

The exhaust was ducted downward through the nose and exited under the belly just forward of the nose gear well in "N" compartment bulkhead. An access panel was installed in the side of the nose directly behind the area where the nacelle of the turbine mates with the nose. The base airframe of 52-1069 was equipped with an APU unit in forward "P" compartment which vents out the lower side of the forward fuselage. The two forward lower fuselage windows were deleted. Several vents were installed on the side of the fuselage below the flight deck. In addition a large tube forming a ridge runs along the lower fuselage from the rear edge of the nose door seam area to just above the APU vents to the leading edge of the wing.

To test the large turboprop engine the JC-124C was flown to the test altitude on the four reciprocating engines with the turboprop feathered. After leveling off and maintaining an air speed of 242 mph, the inboard engines were shut down and props feathered. The T-57 was started and as soon as it was up to speed the outboard engines were shut down. The C-124 would then fly on the power of the single T-57 turboprop.[20]

When the test program was completed 52-1069 was flown to Davis-Monthan and retired. The XT-57 nacelle remained on the nose of the aircraft after the test engine and the four R-4360s were removed. Many parts were stripped before it was eventually scrapped. Testing a very large turboprop in the nose of a C-124 is probably not directly relative to the evolution of volumetric transports. However, the C-124 is a guppy aircraft which was used in the development of another outsize design which is relative in the evolution of guppy aircraft. It is unusual for an aircraft built from relatively primitive 1940s technology to be used as a test vehicle for the development of an advanced turbine. However, the size of the XT-57 turboprop required a very large aircraft for testing. Only so much data can be obtained from running an engine on a test stand. Actual in-flight performance data was needed to evaluate the ability and characteristics of the engine in flight.

7

Turboprop Transports

The XC-132 and C-133

The XC-132 is briefly reviewed as a sidebar in the development of volumetric aircraft. The C-124B was used as an engine test bed for both XC-132 and the C-133 which ultimately was produced. The proposed XC-132 is a volumetric aircraft but not considered a guppy because it was not developed by expanding an existing airframe. It did briefly play a part in the progression and development of hybrid and guppy airframes. The XC-132 was reviewed as a possible transport for rockets because it was considered to be the logical next step in large Douglas transports. Although it was planned to follow the C-74 and C-124 it did not share any airframe components. An upgraded XC-124 design was studied at the same time as the C-133 as a parallel competing project.

In January 1953 Phase I of the XC-132 project was under way at Long Beach. In February the Air Force instructed Douglas to begin development work on the smaller diameter proposed Model 1333 (C-133) Cargomaster. The XC-132 full scale mockup was started at the Douglas Long Beach factory a year later in February 1954. The XC-132 was planned in two versions, a transport with a 389,500-pound gross weight and an in-flight refueling tanker with a gross weight of 469,000-pounds. It was by any measure a very large aircraft at roughly two-thirds the size and lift of the Lockheed C-5A which did not come until 1968.

The mockup of the XC-132 brings into perspective its monster size. The cargo version had an estimated 100,000-pound payload at a gross weight of 389,500 pounds. It was on the scale of the XC-99 with a considerably higher gross weight. A larger tanker version was proposed with a take-off gross of 469,000 pounds. A month after the Air Force announced it to the press the project was cancelled in favor of the smaller C-133 Cargomaster (NARA collection of AAHS archives).

The XC-132 was a double deck configuration slightly smaller but on the scale of the Convair XC-99. It had a calculated gross of 80,000 pounds more than the Convair XC-99 and was fully pressurized. The transport version was projected with a span of 177.5 feet and length of 179 feet, 3 inches. The tanker version had a span of 193 feet, 3 inches with wing pods which housed hose reels for a probe and drogue system.

The C-133A project moved ahead with the first aircraft rolling out in January 1956. In March the Air Force requested a comparison study of the XC-132 and C-133 as a transporter for Atlas missiles from California to Florida. The study was relative to logistical problems NASA was beginning to experience. The search for a suitable missile transport ultimately resulted in the development of the first outsize volumetric transport which was created by Aero Spacelines B-377PG in 1962.

The XC-132 project continued with scale model wind tunnel test which were compared to the C-133. The test showed the XC-132 was considerably more aerodynamically efficient because of the wing configuration. Both were high wing but the XC-132 has a straight-line upper surface the entire length of the fuselage. Conversely the C-133 has an upsweep forward of the high wing which creates a resistance to airflow parallel to the fuselage adding aerodynamic penalties.

The C-133A could not replace the aging C-124 because Atlas and Titan ICBMs would not clear the two-piece hinged rear cargo door. The rear half retracted up into the fuselage, reducing clearance, and the forward half pivoted down to form a loading ramp. Despite shortcomings the second C-133A 54-0136 was loaned to NASA from 1965 to 1969 for Apollo test projects (Chuck Stewart collection of AAHS archives).

Additionally the flow along the C-133 landing gear housings mixed with the vortices from the top surface of the wing creating turbulence along the aft sides of the fuselage. The XC-132 airstream from the wings was smooth and the landing gear housings were slender in comparison creating an even flow with less drag. If the XC-132 had been built and the projected performance data was correct, it would have been a major advancement in military transports. The predicted advancement in efficiency was never proven. The Air Force officially announced to the press on 14 February 1957 the 460 mph XC-132 was under construction at Tulsa, Oklahoma. Despite the superior design and relative accurateness of data collected in wind tunnel test relating to airflow characteristics and reduced drag the XC-132 project was cancelled in March 1957 in favor of the smaller C-133.[1]

The KC-132 in-flight tanker version was estimated to have the ability to transfer 2,400 gallons of fuel per minute at 20,000 feet while cruising at 512 mph. This calculated cruising speed was 161 mph greater than a C-133. In addition the KC-132 tanker was projected to pass fuel at speeds nearly 50 percent higher than the Boeing KC-97 at greater altitudes.[2] The Boeing KC-97 tanker with Pratt & Whitney R-4360s struggled to fly fast enough to transfer fuel to jets. The faster receiving aircraft had to throttle back for the slower tanker. Ultimately auxiliary jet pods were added to the KC-97L to increase the speed.

Turboprop C-133 Cargomaster

The C-133 Cargomaster is relative to the evolution of hybrid and guppy aircraft for several reasons. It was built to take over the role of transporting missiles from the C-124 and a piggyback modification was studied to transport rocket boosters for NASA. When the C-133 piggyback idea was rejected by NASA, Douglas engineers became very receptive to Jack Conroy's proposed outsize modified B-377 transport. Until this time internal transport of rocket boosters was not considered possible.

Design work began on the Model 1333 (C-133) in February 1953 after the Air Force instructed Douglas to discontinue research on the circular fuselage Model 1324 known as the C-124X. Design concepts were changed to concentrate on the swept wing turboprop XC-132. The C-124 served well in transporting missiles up to a certain size; however, the military realized the need to upgrade to a turbine or jet transport fleet with greater capacity.

In September 1953 the Air Force made the decision to continue development of the C-133 as a primary logistical transport. The C-133 was more of a move to turbine power than a necessary volumetric transport. It could uplift more weight but the C-124 could still handle more volume of larger diameter. The C-133 first flew on 23 April 1956 powered by the Pratt & Whitney T-34 turboprop engines.

The T-34 engine was first tested by mounting it in the nose of a Boeing B-17G similar to testing of the XT-57 turbine in the nose of a C-124. The T-34 was put through an intense test program by mounting them on the YC-124B, two Lockheed YC-121F Constellations and two Boeing YC-97J Stratofreighter aircraft. It is quite ironic the two YC-97Js were eventually acquired by Jack Conroy for the T-34 turbines which were needed to power the B-377 Super Guppy. The C-133 was the only turboprop Air Force strategic airlifter to serve in active military service powered by the P & W T-34 engines.

The C-133 was an all new design with a circular fuselage. It could lift a heavier load and transport 96 percent of all equipment in U.S. Army inventory at the time. Its purpose was to be the primary transport of the Thor (IRBM) ballistic missile. Building the C-133 for missile transport was somewhat short sighted. It could not replace the C-124 which was still needed to transport Atlas and Titan ICBMs because they would not clear the door of the C-133A model.

Unlike its C-124 predecessor, the C-133 was produced in limited numbers. Only 35 "A" models were built from 1954 to 1957 and 15 "B" models between 1959 and 1961. The C-133 was produced in limited numbers to fill the need for a primary transport of ICMBs. Only 50 C-133 aircraft were produced which contributed to spares shortages in the later years of operation and reduced the operational life.

Design flaws in the C-133 contributed to severe vibration problems. They were withdrawn from service in 1971 after only 15 years for the "A" and 10 years for the "B" model. Interestingly the original C-74 Globemaster I suffered from parts shortages because of low production which also contributed to a short operational life. Except for the C-124, Douglas was never able to produce a large military transport in sufficient numbers to justify the development programs. In 1989 a similar problem occurred. The Aero Spacelines built NASA B-377SG Super Guppy N940NS which was powered by those same Pratt & Whitney T-34 engines and Curtiss Electric props used on the C-133 was grounded for lack of spares.

The C-133 "A" models were able to transport smaller missiles throughout the world; however, the two piece rear cargo doors prohibited the loading of the larger Titan ICBM. The aft section of the cargo loading door folded up inside the cargo compartment and the forward section folded down forming a loading ramp. In order to transport larger cargo, the rear fuselage was redesigned. The inward folding rear half of the cargo door was replaced with a pair of clamshell doors which opened outward. This new design provided more clearance at the aft upper extreme of the cargo compartment. Aircraft with the modification were re-designated as C-133B. The modification expanded the opening for easier loading and added three additional feet of cargo space. The last three "A" models were modified in production with the wider clamshell doors and re-designated as "B" models.

Douglas built a single C-133 "A" model with no construction number or Air Force identification which was retained at Long Beach for testing. After the redesign of the rear fuselage with clamshell doors a "B" model was also built without construction numbers. It was also retained at Long Beach by Douglas for testing and used for practice of loading missiles.

Convair officials observe the loading of an Atlas missile into a C-133B with redesigned clamshell cargo doors. The aircraft still experienced vibration problems which were never corrected. Because the missile was developed for the military it could withstand vibration without being damaged, unlike the larger Saturn stages (NASA).

Although the payload of the C-133 was 36,000 pounds more than the C-124, the fleet service life was shorter because of multiple problems including severe vibration and metal fatigue. The YC-124B Globemaster II test aircraft supplied considerable test data for the P & W T-34 engines used on the C-133. The low wing YC-124B test aircraft did not suffer from the same vibration problems with the same engines. The piston powered C-124C remained in service after the C-133 transporting outsize cargo until after the Lockheed C-5 came fully operational.

Ironically the C-124 transported many replacement props for stranded C-133s. The C-133 was built as a 10,000 hour 2,500 cycle airframe. The hours were extended to 15,000 then

17,000 and finally 19,000 hours; however, the airframes suffered from serious metal fatigue caused by vibration. The Cargomaster was evaluated to extend the service life to 25,000 hours but it was determined not to be cost effective. The airframes were virtually worn out from being shaken apart. They experienced a 20 percent loss rate over the life of service and were only kept active until the C-5 became operational in 1971.

The C-133 P & W T-34 engines were fitted with 18-foot Curtiss Electric CT735 propellers which were at the maximum limits of technology of the time. The reliability was improved over time but they remained a source of problems throughout the C-133 operational life. The chronic vibration problems were never overcome. All propellers had to be spinning at the exact same RPM or severe vibration set up creating a harmonic imbalance which could cause a catastrophic airframe failure. Even if they were all in sync, there was still considerable vibration and excessive noise inside the aircraft. When the prop tips exceeded Mach One the vibrations became so intense catastrophic failures were a distinct possibility anywhere on the airframe. Interestingly the Aero Spacelines B-377SG which was powered by the same Pratt & Whitney T-34 engines used on the C-133 had some vibration but did not suffer airframe fatigue.

Inside the C-133 prop plane while in flight the noise level easily exceeded 125 decibels (db). Noise above 90 decibels can cause hearing loss. Ear pain begins at a level of about 125 decibels. The vibration inside the C-133 cargo bay was so intense an airman would actually move across the floor while standing still. Because of the excessive noise the crew was limited to only five minutes of exposure in the cargo area. The noise was also excessive in the cockpit providing little relief for crew members. Military standards limit cockpit noise level to 113 decibels at maximum continuous power. Designers were able to keep it below 113 decibels at the pilot, engineer and navigator stations. The noise in the crew rest area was never resolved. It exceeded 113 decibels which was not comfortable for extended periods while at maximum power settings.[3]

In addition to noise and vibration problems, the T-34 engine nose case was subject to pinion bearing failures in the high-speed reduction gear. Dating back to 1960, the bearings would seize causing gear failures and ultimately nose case explosions. The compressor section would often ingest the debris resulting in complete engine failure. The problem became so severe the Air Force grounded the C-133B in December 1961 until the problems could be resolved.

They were returned to service in March 1962 after the engine nose modifications. In June 1962 the C-133A model

A perfect example of the C-133 "B" model limitations is the very tight fit of an Atlas missile. There are only inches of clearance on each side. Loading was extremely difficult and slow. Every precaution had to be taken not to damage the missile or the structure of the aircraft. Any mistakes could have been catastrophic. The problem intensified the need for a larger aircraft to transport increasingly larger rocket boosters for NASA (NASA).

was grounded pending engine upgrades. The upgrades reduced failures but they were not eliminated. Initially the propellers required a 500-hour time overhaul. After an exorbitant amount of costly testing and upgrades they were increased to 1000 hours. The engines went through five nose case upgrades through June of 1966 but failures continued.

The propeller electrical system suffered from multiple problems which caused at least one crash. In addition, the blade fairings were subject to cracks causing them to separate from the blade shanks. The blade problem was never resolved; however, installing fiberglass covers over the prop-blade cuffs prevented debris from entering the compressor section. Eliminating debris ingestion saved over $100,000 per incident. The engine failures persisted for multiple reasons throughout the C-133 operational life.

Despite many problems the C-133 Cargomaster was still able to lift heavy loads by setting many records. In 1960 it set a record by airlifting the Nike-Zeus air defense missile tracking system from Chicago to Tulsa which was loaded by Redstone Arsenal, Huntsville, Alabama, army personnel. In all probability some of those technicians gained experience from training on the turboprop YC-124B Globemaster II which was retired to Redstone for the purpose of training on loading procedures for rockets.

NASA negotiated an agreement with the Air Force to provide transport aircraft to support the space program in 1959. The Air Force utilized the C-133 to transport Thor and Minuteman (solid fuel) rockets for the military and NASA programs. The success of transporting these missiles prompted NASA to review proposed C-133 modification plans. An upgraded and modified C-133 was offered as a means of solving the logistical problems of transporting larger boosters for the space program.

In 1960 the C-133 was being used to transport the actual rocket motors which were produced by Rocketdyne for NASA. Space hardware which was too large to fit inside the C-133 or C-124 was being moved from manufacturers overland and by barge to Huntsville for testing and Cape Canaveral for launch. This was time-consuming and increased the possibility of damage. To remain on launch schedule NASA officials were searching for an immediate alternative to transporting the units by barge. The need for an alternative to barge transport of the S-IV rocket booster became paramount when the Wheeler locks on the Tennessee River collapsed on 02 June 1961.

Rocket boosters which were at Huntsville for testing were stranded above the dam. Although efforts were being made to move the booster around the dam by road, Wernher von Braun and NASA logistics manager Julian Hamilton discussed the possibility of rocket test fires for the space program being delayed. NASA had previously expressed concern about delivery schedules for S-IV and S-IVB from California. The units were too large to be transported overland or by rail leaving seagoing delivery as the only option. The possibility of severe weather or a political situation could shut down the Panama Canal affecting delivery and impacting the launch schedule.

External Cargo Transport

The Aerospace Industries Association of America conducted a study headed by experts P.F. Forderer and Vern W. Porter to review a concept named the "birdie back system." It is identified as being much like a car top carrier for aircraft to transport NASA hardware. The study also reviewed twin dirigibles with a center-wing tying them together. The dirigibles center-wing would have hard points to attach rocket boosters for transport.

The twin dirigible concept was revised and proposed to NASA again

Douglas proposed the "birdie back" system for transporting Saturn rocket stages on top of the C-133 with streamline cones on each end. Endplates were designed for the horizontal stabilizer for stability. The production C-133 suffered from severe vibration and interrupted air flow along the aft fuselage. NASA rejected the idea, which prompted Douglas to be more receptive to Jack Conroy's expanded fuselage B-377 proposal (NASA, MSFC).

in the 1970s by guppy builder Jack Conroy as a transporter for the space shuttle. He followed it up with a B-52 twin fuselage heavier than air vehicle named the Virtus. The design concept surfaced again in 2015 when Bill Allen and Burt Rutan began building a twin fuselage vehicle using Boeing 747 parts. The aircraft named the Stratolaunch will be used to transport space vehicles to altitude for launch into orbit.

Douglas proposed a second design split tail version C-133 modification with a completely new empennage. The design with a group of multiple size pods eliminated engineering mounting points for the actual hardware being transported. It was also rejected because of known C-133 problems plus the cost and time it would take to construct the modified aircraft (NASA, MSFC).

Douglas Aircraft reviewed the "birdie back" concept and made an unsolicited proposal to NASA for a piggyback configuration to transport Saturn S-IV rocket motors on top of the C-133. Marshall Space Flight Center (MSFC) at Huntsville, Alabama, let a contract to Douglas Aircraft in 1960 for a feasibility study of the proposal. In 1961 Douglas Aircraft met with a NASA project team at Langley Research Center to review proposed modifications to a C-133 for external transport outsize cargo. The proposal was considered problematic prompting other contractors to propose systems. Douglas had been working on multiple variations of the external concept for quite some time. Until Jack Conroy proposed the expand fuselage B-377 and convinced Douglas to let him make a presentation to NASA, it was not considered possible to transport anything as big as a rocket booster internally.

NASA could not wait for a new cargo aircraft to be developed which could transport booster stages internally. An interim method had to be found to keep the launch program on schedule. The results of the piggyback concept study were presented again to MSFC director Wernher von Braun at Tulsa, Oklahoma, on 26 September 1961. It consisted of multiple variations relative to the cargo load to be airlifted. The first design mounted the 68-foot long Saturn booster backwards on top of the aircraft with light weight nose cones attached to the ends to reduce drag. The proposal required a beefed up C-133 horizontal stabilizer with fixed vertical stabilizer fins mounted to the ends in a similar fashion to those on the later developed NASA B-747 Space Shuttle transporter (SCA). The end plates are supposed to compensate for interrupted air flow on the tail surfaces and reduce yaw. The C-133 already suffered from inherent vibration problems which were never resolved. In addition, a design flaw created turbulence along the aft sides of the fuselage. Vortices from the top of the wing mixed with the flow from the landing gear housings causing an unstable condition along the aft fuselage. Adding a booster or pod to the top of an aircraft which already has disrupted airflow in all likeliness would compound the problem. Also the aircraft was still suffering from unresolved engine and propeller in-flight failures.

A second and more interesting alternate proposal was offered with a completely new empennage featuring very large twin vertical stabilizers. By utilizing the twin tail design a series of at least seven different size pods relative to the cargo size could be mounted on top of the C-133 with the payload placed inside. The top half of the pods are removable with the cargo secured in a cradle built into the pod. The plan eliminated engineering multiple mounting points on the aircraft for actual hardware being transported. The aircraft had an estimated payload with either tail configuration of 124,985 pounds.

The added cost of attempting to resolve the existing problems in conjunction with the modification would be passed on in the build cost. NASA was already experiencing budget problem and could not justify funds for logistical aircraft development. The piggyback concept for the C-133 was considered a poor choice. The delicate nature of hardware designed specifically for space brought into question the viability of NASA utilizing the C-133. The C-133 would continue to transport rockets borrowed from military inventory and the dropping of test modules.

The C-133 piggyback proposal did not proceed past the preliminary artist renderings and demonstration models. The von Braun team evaluated the proposal but rejected it as too costly. The team indicated the best means of transport for space hardware would be internally because of the stresses exerted on external loads. It was determined the most practical use of the C-133 would be the modification of the rear cargo doors to expand the opening in order to transport Atlas, Titan and Minuteman rockets internally. The modified version was designated the C-133B. The military rockets could withstand the vibration during transport where as space hardware could not.[4]

The military was looking at transport problems from several perspectives. Not only was the Air Force supplying transport for NASA it was confronted with missile transport for national defense. The C-124 was aging and the C-133 suffered from multiple design flaws which were affecting performance. The C-133 was also limited by fuselage diameter. Air Force Headquarters sent the Material Air Transport Service (MATS) Qualitative Operational Requirement (QOR) in October 1961. It specified a successor to the C-133 was needed which could accommodate larger missiles. The operational life of the C-133 was already coming into question and the last one built would soon be leaving the assembly line.

Interestingly the C-133 piggyback concept never went away. Jack Conroy suggested it to NASA as a way of transporting the space shuttle on the back of a B-747. He began a design program and presentation to NASA prior to the agency purchase and conversion of an American Airlines B-747-100. The Soviet space program experimented with piggyback configurations years later by adding large twin vertical stabilizers to existing aircraft. In 1978 the Soviets modified a Myasishchev M4/3M bomber by stretching and changing the empennage to create the VM-T for external pod transport of space hardware. The 1980s Ukrainian built Antonov AN-225 designed to transport the Buran space vehicle had a split tail similar the second proposed C-133 piggyback design.

Alternative Transport Designs

NASA was open to any suggestion for the modification of an existing aircraft to practically transport a rocket booster externally or more desirable internally. The urgent need set up the conditions which led to aircraft broker Lee Mansdorf's idea of modifying B-377 airframes to transport large boosters internally. This eventually led to the Jack Conroy's revolutionary Aero Spacelines's B-377 conversions. The NASA logistics team only demonstrated cautious interest in Conroy's expanded fuselage proposal. Conversely, Wernher von Braun was excited about the concept and impressed with the design from the beginning. As a result, Aero Spacelines was incorporated in Norman, Oklahoma, on 15 November 1961 to begin production of an expanded fuselage design based on the B-377 Stratocruiser.

Douglas aircraft countered with a proposed jet powered modified C-133B designated the C-133X (model D-895) as a high capacity alternative to the rejected piggyback concept. The C-133X was proposed with a 17-foot outside diameter fuselage with a volume of 12,240 cubic feet and a 100,000-pound payload. Similar in appearance to the Lockheed C-141 the aircraft used an expanded C-133 fuselage combined with DC-8 wings, empennage and Pratt & Whitney TF33 turbofan engines. The hybrid would utilize the C-133B clam shell rear cargo door design with increased opening. It was believed this all jet version could be built at a lower cost by using off-the-shelf DC-8 components and would eliminate the vibration problems which plagued the C-133. However, there were many questions of stress problems when combining a Mach 0.41 fuselage with a Mach 0.75 wing.[5] It would have only been and interim transport at 17 feet in diameter.

Perhaps the most bizarre alternative proposed for moving outsize space hardware during the same period came from Texas Engineering and Manufacturing Corporation (TEMCO) of Dallas. The company which was established after World War II by Robert McCulloch and later merged into Ling-Temco-Vought suggested a glider concept. The Texas company proposed the Temco Air Trailer, which is a system using large gliders to transport outsize Saturn rocket boosters.

TEMCO proposed towing a glider containing the booster stage behind a Lockheed C-130A or a Douglas C-133A. The glider was not a full airframe. It was more of a cockpit module and backbone with wings and empennage which fits around the Saturn booster stage. The glider would be attached to the C-130 by a semi-rigid

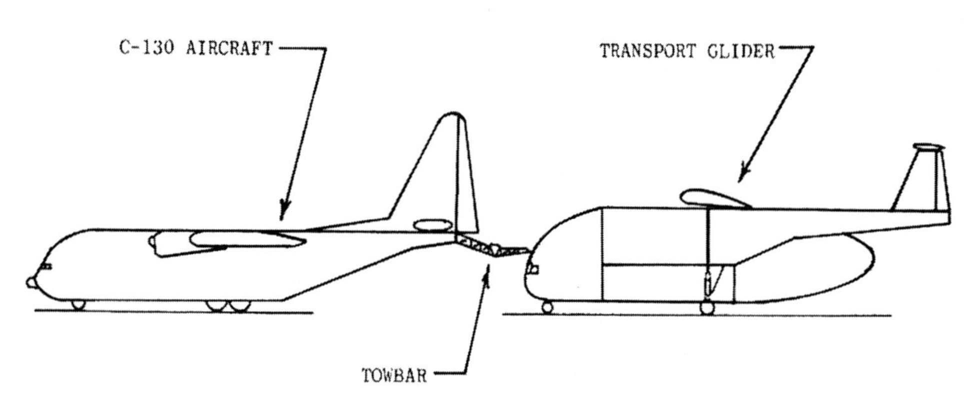

Probably the most unusual proposal to transport rocket boosters was the TEMCO Glider. It was a cockpit module with wings and empennage that would fit around the rocket booster to make it into a glider that would be towed behind a C-130 or C-133. NASA rejected the proposal as impractical because of potential of damage to the booster (NASA).

hydraulically dampened tow bar. It was considered feasible because a tractor aircraft towing a glider was a proven reality. The study concluded the C-130 could take off and land with the glider attached. The approximate travel time coast to coast was estimated at 21 hours. Because of the increased risk of an incident when landing to refuel, TEMCO proposed modifying the C-130 tow aircraft for in-flight refueling. This would allow the tow aircraft to make a cross country flight nonstop in a more efficient manner.[6] Not having to land multiple times to refuel the C-130 would also reduce stress and vibration on the rocket booster. The project was reviewed and soon rejected.

NASA C-133 Support Missions

The C-133 was designed as a logistical transport for military missiles which was viewed as a benefit to NASA transport needs. The military Thor IRBM was designed to sustain any vibration encountered in airborne or ground transport which resulted in them being airlifted on a regular basis by the C-124 Globemaster without damage. The opposite was true for outsize space components and rocket stages built for NASA. Boosters are designed as lightweight as possible to put hardware into orbit.[7]

Although the C-133 had serious vibration problems, there were no alternative aircraft in inventory which could perform the necessary capsule drop test required by NASA. The second built C-133A Cargomaster, 54-0136, was loaned to NASA from 09 June 1965 to May 1969 to support test projects related to the Apollo program. It was modified for Apollo command module test drops. These are the only air drops the C-133 ever made. Accompanying personnel were only allowed in the cargo compartment for short periods because of the noise level and vibration. It was never used for troop drops or transporting personnel because of the extreme noise and vibration which would have caused physical harm. Both sections of the rear cargo door were removed for this operation. Blast shields were installed on the sides of the fuselage to reduce the turbulence along the sides of the aircraft. An a-frame carrier was constructed and mounted in the rear extending out in the opening. It was designed to hold the Apollo command module for the test drops. To compensate for the weight of the command module two 400-gallon ballast tanks were installed forward.

The C-133 was based at Ellington Air Force Base, Texas, using the call number NASA 928 instead of Air Force identification. All of the test drops were made at El Centro Naval Air Station in California. The modified C-133 is reported to have made 46 successful drops to test the parachute system for the returning Apollo capsules.[8]

Once the test were completed Apollo missions 7 to 14 were supported by C-133B aircraft until they were decommissioned in 1971. The Air Force provided NASA multiple C-133B models for specific missions. Cargomaster 59-0532 transported the Apollo 7 command module from Norfolk to Long Beach on 25 October 1968 after it was recovered from the Atlantic.[9]

Apollo 8 was the first mission launched by a Saturn V booster which circled the moon and splashed down in the Pacific. In December 1968 the command module was picked up by C-133B aircraft 59-0531 at Hickam Air Force Base, Hawaii, and flown to Long Beach. The aircraft was retired in 1971. It was sold to Northern Air Cargo of Anchorage, Alaska, on 10 January 1975 and assigned civilian registration N2276V. It remained at Tucson for 25 years with the civilian registration. It was scrapped in January 2001 but the nose was still in the National Aircraft scrap yard at Tucson in October 2006.

Cargomaster 57-1613 used by NASA was originally built as a C-133A. The engines were changed and the fuselage was modified for clamshell doors re-designating it a C-133B on 14 March 1961. It was referred to as a C-133A/B because all other systems were "A" model. After Apollo 9 completed the objective of an Earth-orbital engineering test of the first crewed lunar module, the command module splashed down near Bermuda. It was picked up at Norfolk on 19 March 1969 by 57-1613 and delivered to Long Beach. In May 1969 Apollo 10 splashed down near American Samoa. The capsule was picked up by the next to last C-133B built (59-0535) at Hickam and flown to Long Beach.[10] That C-133 was also removed from service and delivered to Tucson on 22 March 1971. Both remained stored until they were scrapped in 1974.

The Apollo 11 module which took Neil Armstrong, Buzz Aldrin and Michael Collins to the moon came down near American Samoa in July 1969. Air Force dispatched C-133B 59-0526 to retrieve the capsule at Hickam Air Force Base, Hawaii. It was returned to Ellington Air Force Base, Texas, staging through Travis Air Force Base. It was also removed from service in July 1971 and scrapped in 1974. Apollo 12, the troubled Apollo 13 and Apollo 14 all returned to earth near American Samoa. The capsules were retrieved at Hickam by unidentified C-133s and transported to Ellington Air Force Base Texas.[11]

The NASA Memorandum of Understanding with the Air Force only specified aircraft would be supplied to support the NASA mission. The agreement never specified the use of C-133 aircraft. As problematic as the C-133 was, the Air Force had no other option because there were no other aircraft available with the range and lift. The C-133 was built in limited numbers and experienced many problems which contributed to a shortened service life, however, it served the NASA mission as required.

A total of 50 C-133s aircraft were constructed. Nine were lost in crashes; one was destroyed by fire on the ground and one destroyed by a tornado after retirement. Aircraft 57-1610 was the first C-133 to be retired on 11 June 1970. The 38 remaining were removed from Air Force inventory in 1971. Four of that group went to museums, seven were sold to civilian operators for work in Alaska and the balance was scrapped at Davis-Monthan Air Force Base, Tucson, in 1974.

8

DC-4 (C-54)/ATL-98 Carvair

The Outsize Car-Ferry Evolves

The number of outsize "guppy class" transports built other than the military Douglas C-124 is quite low compared to all other types of aircraft. Revolutionary outsize fuselage designs are purpose built aircraft which historically have come from small firms operating independently of each other. Two of the successful types, the Aero Spacelines B-377 conversions and the Aviation Traders Engineering ATL-98 Carvair are the end result of enterprising individuals seeking a way to repurpose aging airframes into cost effective solutions to a specific need. The DC-4 conversion to ATL-98 Carvair is a unique example of expanding an aircraft fuselage to produce a special purpose aircraft for passenger use.

The idea of outsize aircraft evolved from a post–World War II need for the military to move materials over long distances on short notice. The early years of the American space program relied on the military to transport rockets. In the 1960s NASA hardware was increasing in size far faster than aircraft manufacturers were designing transports. Just prior to Jack Conroy forming Aero Spacelines in America to build a transport to move outsize NASA hardware another guppy-type for commercial airline service took shape in England. There are surprising similarities in the construction methods used by Freddie Laker to build the Aviation Traders Model 98 and Jack Conroy's effort to build the B-377PG. In both cases parts were fabricated in small shops and a considerable amount of the work was done outside.

A small aircraft overhaul and maintenance company known as Aviation Traders Engineering Limited (ATEL) was tasked with building a commercial guppy type aircraft. The project was proposed to satisfy a need to increase market share in the British airline car-ferry service. The service at the time needed a more profitable second generation car-ferry aircraft to replace the Bristol Freighter. The company founder, Freddie Laker, conceived the idea to capitalize on a growing market to transport holiday travelers and their cars across the English Channel to mainland Europe.

The Model 98 Carvair conversion of the C-54/DC-4 was unique as the only specialized volumetric transport built for the commercial airline passenger market. All other purpose built volumetric guppy aircraft were conceived, designed and built for the military, NASA or aircraft manufacturers. The Model 98 Carvair was conceived by Freddie Laker for the Channel Air Bridge division of British United Airways (BUA), a sister company to Aviation Traders Engineering Limited.

British United had been moderately successful transporting passengers and their automobiles with the Bristol MK.31 freighter and later the modified long nose MK.32 aircraft. The Bristol MK.31 was only capable of transporting two automobiles making the profit margin very low. In order to be competitive in the market and show reasonable profit the carrier needed more reliable lift and capacity to expand further into mainland Europe market.

While the ATL-98 is unique as the only guppy type aircraft built specifically for an airline, adapting volumetric ex-military transports for passenger service was not a new idea. As previously described, Pan Am ordered 15 passenger versions of the giant double deck Convair XC-99 (B-36 guppy) in the 1940s then cancelled the order when cost analysis proved there was not a market in the world with enough sustained traffic to justify the operational cost. Convair proposed a second advanced C-99 to Pan Am with a drive on lower deck for vehicles. Laker's smaller ATL-98 transported cars and passengers on the main deck. Aero Spacelines proposed a B-377 with a drive on upper deck and passenger lower deck.

In 1959 officials at Channel Air Bridge reviewed multiple large aircraft which would be capable of transporting automobiles. The carrier needed a car-ferry aircraft with a standard airline cabin at a low operating cost which would produce a profit in a seasonal market. Although the carrier was planning to expand deeper into mainland Europe, it was not necessary to purchase fuel hungry giants to try and duplicate the elegant Clipper service of the past.

There is only a limited amount of market for special purpose outsize aircraft whether it is the relatively small DC-4 to ATL-98 Carvair conversions or B-377 to giant Super Guppy. Only 21 examples of the ATL-98 Carvair were produced. Studies were made and wind tunnel test conducted with the intent of producing a larger version utilizing DC-6 and DC-7B airframes but none were built. The DC-7 or Carvair 7 with an extended upper deck resembled the profile of the Boeing 747-400 with propellers. It was planned to be considerably longer than the production DC-4 Carvair conversion. The ATL-98 is the only guppy type produced with only the forward portion of the airframe expanded.

The ATL-98 is what is commonly known as a combi car-ferry aircraft which was designed for straight-in loading with provisions for up to five cars or bulk cargo and 22 passengers. The Carvair was built primarily for passenger transport. Five were delivered as pure freighters. Two of those were converted back to passenger car-ferry configuration after serving short duty on United Nations contracts during the war in the Congo in the 1960s. In later years all Carvairs built with passenger cabins which were acquired by second and third level carriers were converted to cargo configuration. The passenger cabins were removed and at least one had a roller floor system added.

All of the ATL-98 conversions outwardly appear the same. The enlarged forward fuselage nose units are built for straight in loading. The interior was designed to be adjusted in multiple combinations of passenger/cargo configurations with up to an 80-foot constant section in cargo configuration.

Two of the aircraft built for Aer Lingus (c/n 6 and 8) were returned to ATEL after the first season to have the area in the hump behind the cockpit modified. The control cable runs were rerouted up from behind the cockpit and along the ceiling. The modification increased the interior ceiling height giving a cathedral effect in the forward cargo compartment. Later conversions were built with the redesigned high ceiling forward cargo area after the stock of previously produced standard nose units was exhausted.

When introduced the Carvair generated strong interest among many airlines and foreign air force units because of the drive on-off capability. However, the commercial jet age was dawning in 1962, making airline executives skeptical of buying new propeller driven aircraft built from World War II era surplus airframes. Consequently, Aviation Traders sister company British United Airways was the only airline with an immediate need for the ATL-98. The carrier placed an initial order for only three aircraft with options for seven more for the Channel Air Bridge division.

There were no orders outside the British United family of companies. Aviation Traders production planners believed interest was strong and large orders were forthcoming. With only three orders it was deemed too expensive to purchase metal fabricating machines when the number of aircraft being produced was unknown. The large number of orders never materialized. As a result, all ATL-98s were hand built with individually fabricated parts.

The Aviation Traders ATL-98 Carvair was conceived to transport passengers with their cars across the English Channel to mainland Europe. Initially it was not identified as a guppy. One year after the Carvair entered service the Aero Spacelines B-377PG took flight in California and the guppy classification was created. Soon afterward aviation personnel began to identify the ATL-98 as a guppy (BAF, author's collection).

Only a few aircraft orders trickled in. Aircraft four and five were the first two ordered from a foreign company. Intercontinental the American sister company to Luxembourg based Interocean placed an order for two aircraft with special features for operation in remote areas of Africa. The charter airline had secured a United Nations contract for work in the Congo. Each aircraft from this time forward was ordered with custom features or special configurations.

The 21 Carvairs were converted from multiple series of C-54s. Some had been upgraded to commercial DC-4 standards after the war. Over the years of postwar service they had been modified to many different configurations. The special options ordered on the ATL-98 combined with multiple series base DC-4 airframes selected contributed to all of the ATL-98s being different in some way. Some of the Carvair special options were a folding ramp system, enlarged forward bulkhead for oversize cargo, wheel well crew access doors, nose door toilet, cockpit galley and moveable bulkhead for increased passenger seating configurations. None of these changes visibly affected the outside appearance except for window placement.

Aviation Traders built a single ATL-98 (c/n 11) as a long-range aircraft with eight fuel tanks and a central oil tank. Cockpit bunks and a galley were added in an effort to gain Ministry of Defense contracts to transport rockets or to receive an aircraft order from the British military. The modified long-range version was flown on a worldwide demonstration tour. After returning it was occasionally chartered by the Ministry of Defense to transport rockets to Woomera, Australia, for testing.

Running assembly line changes and upgrades are common place when building special purpose aircraft as they are refined to meet the mission. Not only was this the case with the ATL-98 Carvair but the enlarged fuselage Douglas C-124 Globemaster and Aero Spacelines B-377 guppy. Once the first C-124 Globemaster was converted from a C-74 all subsequent aircraft were continually upgraded with running changes. After the single XC-99 was developed from the B-36 base design it was in continuous upgrade which included a major redesigned of the landing gear and cargo loading openings. After the first B-377PG Pregnant Guppy, all subsequent models were upgraded with the extended center-wing, modified empennage and the lower fuselage profile change to produce a wider floor width.

The ATL-98 Carvair does not have an exceptionally large diameter fuselage compared to other volumetric type aircraft. The term "guppy" is assumed to indicate a very large fuselage aircraft. In most cases it does however, the term is relative to an aircraft which has the fuselage expanded from the original diameter. Soon after the term guppy became common to describe aircraft with expanded fuselages, control tower personnel routinely referred to the Carvair as a guppy. Compared to other volumetric aircraft, it is not really heavy lift with a cargo payload of only 18,000 pounds and an ATOG of just 73,800 pounds. It is small by today's standards; however, it is relative because the DC-4 base airframe was considered heavy lift in the late 1930s when the aircraft first came on the scene.

When Freddie Laker conceived the idea of a car-ferry conversion in 1959, the DC-4 became the prime candidate because it could carry a respectable wartime payload and operate into small unimproved airfields. Wartime C-54s were tough and well proven in the most adverse conditions. Those converted to commercial airline DC-4 standards after the war flew in regular service until replaced by the DC-6 and DC-7 which is a stretched version of the same airframe.

The DC-4 is not widely remembered or celebrated as a historical aircraft, yet, it is unique for multiple reasons. It is America's first mass produced four-engine land-based airliner. It can be argued the Douglas DC-4 is a contributing factor in two of the guppy types produced, the ATL-98 and Douglas C-124 Globemaster. The ATL-98 Carvair is obviously a DC-4 airframe with an expanded forward section and modified vertical fin. The C-124 is a little more difficult to visualize. It came from modifying the gangly looking monster known as the C-74 Globemaster. Douglas designers openly stated the C-74 was nothing more than a scaled-up DC-4. The C-74 Globemaster I debuted as one of the largest aircraft in the world in the 1940s. The fifth aircraft built was returned to Douglas to have the fuselage doubled in size to become the first C-124.

When Douglas proposed the 162,000 pound C-74 Globemaster in the 1940s, a commercial passenger version

Douglas produced approximately 1,170 C-54 aircraft for the military. After the war many were converted to commercial airliner standards. By the 1960s there was an abundance of DC-4s on the surplus aircraft market at bargain prices. Aviation Traders acquired 19 C-54s and two postwar DC-4-1009 airframes for ATL-98 conversion. Some were in such poor condition a tech team was deployed to make them airworthy enough for a ferry flight to England for conversion (AAHS archives).

was visualized as possibly replacing the DC-4 airliner. Pan Am made a second attempt to utilize outsize aircraft in the commercial market in the 1940s when Douglas offered the C-74 for commercial airliner service. The carrier placed an order in 1944 for 26 aircraft with lounges and private state rooms for service on the newly proposed New York to Rio de Janeiro and Buenos Aires service. This commercial version C-74 was designated the DC-7 by Douglas not to be confused with the later smaller production model Douglas DC-7. Pan Am selected the name Clipper Type 9 to describe the newly proposed airliner. Pan Am cancelled the order in 1945 after the unit cost rose to a prohibitive $1,412,500 per aircraft ending any chance of a commercial version.[1] Pan Am as well as other carriers opted for the smaller war surplus C-54/DC-4 airliners in the postwar years until the passenger market grew enough to need larger aircraft. By 1960 the larger airlines were phasing out the DC-4s for larger upgraded equipment.

Flyable C-54s were available on the open market from $100,000 to $110,000 in 1960. Some of those which had been poorly maintained were available for as low as $80,000. Aviation Traders purchased several believing they were perfect candidates because the aircraft would be completely disassembled and rebuilt as ATL-98s. Some were in such deplorable condition it required a team of engineers and considerable effort to get them airworthy to fly back to Southend, England. Three of the ex–Pan Am DC-4s found their way through several second level carriers before being purchased by Aviation Traders to become Carvairs 12, 15 and 21.

Interestingly, the ATL-98 Carvair shares some design features and characteristics with the much larger proposed advanced commercial version of the Convair C-99. Consolidated Aircraft (Convair) made multiple proposals to Pan Am in an attempt to sell a giant commercial airliner based on the XC-99. Designers proposed an advanced commercial version of the C-99. It was even larger than the military XC-99 and combined features of military cargo and troop transports with passenger airliners. The designers proposed a cargo/passenger combi version named the Clipper 9 to Pan Am. It was on a giant scale, but with design features similar to the much smaller ATL-98 Carvair. The cockpit was raised above the cargo compartment with frontloading doors for drive on vehicle transport. The nose wheel retracted into a pod outside and below the cargo floor. The commercial C-99 guppy never got past the drawing board because the production cost was too prohibitive.[2] Convair made a final effort with a third proposal of a commercial turboprop version. Similarly Aviation Traders proposed a larger turboprop ATL-98 Carvair built from the production DC-7 airframe. It was named the Carvair 7.

The idea of converting older piston aircraft to turboprops lingered for years because usable airframes with low hours were being retired for jets in large numbers. In 1965 Eastern Air Lines began phasing out its DC-7 fleet for newer more economical equipment. The DC-7s had been in service less than seven years yet Eastern was unable to sell them on the secondary market. The aircraft were valued between $100,000 and $150,000 each as opposed to the older and smaller DC-6 which was valued at up to $330,000.

The DC-7 suffered a similar fate as the Boeing B-377 Stratocruiser. They were very expensive to operate and maintain causing the first line carriers to move out of them as soon as jet equipment became available. There was no secondary market because air carriers could not afford the cost of operation. The DC-7 was powered by the gas guzzling plug-fouling Wright Duplex Cyclone R-3350 engines. Eastern Air Lines expressed a serious interest to Aviation Traders in converting the DC-7 fleet to the proposed turboprop powered Carvair 7 guppy cargo-liner. This was considered to be the best way to extend the operational life of the DC-7s which still had considerable hours left on the airframes.

Aviation Traders conducted studies and issued press releases stating a Dart turboprop DC-7B version with a 90-foot cargo compartment and 20-passenger upper deck was forthcoming. The fact a DC-7 conversion never materialized ultimately doomed the original DC-4 series ATL-98 Carvair to only 21 units being built. Many airlines considering the Carvair early on decided to postpone ordering the DC-4 conversion and wait for the DC-7 turboprop model. Aviation Traders eventually came to the conclusion the Pratt & Whitney R-2000 powered DC-4 Carvair was underpowered and a turbine guppy would be more efficient. However, too much time had passed. The airline industry had entered the jet age beginning a new era and was no longer interested in rebuilt prewar technology airframes.

The Eastern DC-7s are the same aircraft Dee Howard considered purchasing a few years later to build his proposed 40-foot diameter outsize transport for Saturn rocket stages. Howard's bi-wing giant consisted of a huge fuselage using DC-7 wings and fuselage components. It was powered by 10 Wright R-3350 engines. The design was wind tunnel tested but never got past the drawing board. NASA encouraged proposals for giant fuselage transports but did not have a large logistics budget. Jack Conroy already had the B-377PG in service and was building the larger B-377SG all with private funds.

DC-4/C-54 Development

The base C-54 airframe used for the Carvair originated from a 1930s vision of a long-range commercial transport. Twenty-six years before Freddie Laker visualized the ATL-98 Carvair, William Patterson proposed a four-engine transport. Prior to World War II air travel for the general public was

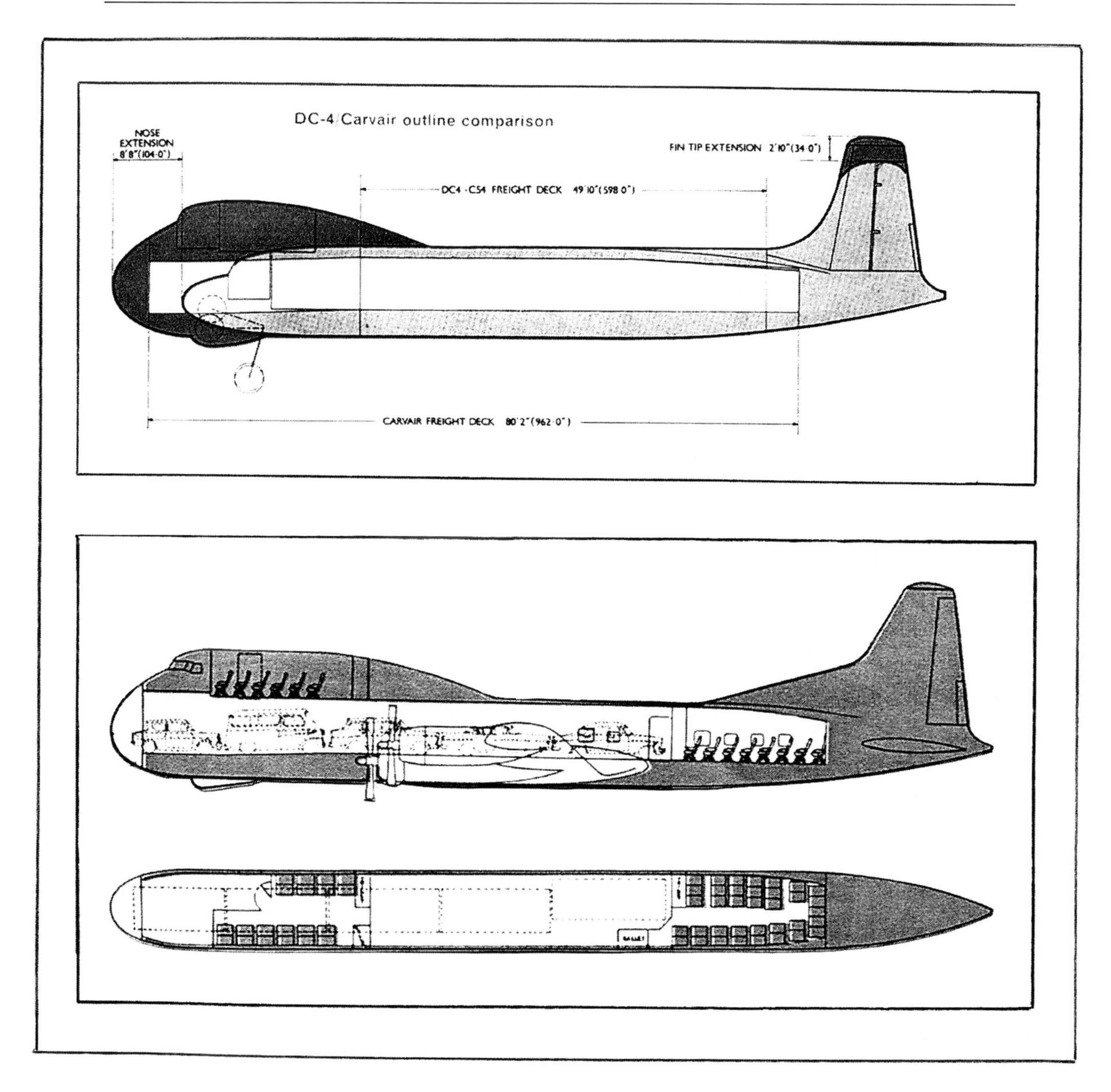

The Carvair is obviously a combi-configuration. Comparison profile shows the original DC-4 compared to the expanded Carvair conversion, which had a passenger cabin in the rear. The proposed turboprop DC-7 version had a passenger cabin in the rear and on the upper deck. The configuration bears a strong resemblance to the Boeing 747, which did not fly until seven years after the Carvair (Aviation Traders, author's collection).

considered novel and generally believed would never appeal to the masses. Patterson thought differently and envisioned a new era of air transport with large airlines flying the globe. His proposal of a 74,000 pound 175 mph airliner was extreme for the time but small by today's standards.

Patterson discussed his projections with Donald Douglas emphasizing the need for an airliner with a 2,200-mile range. Douglas estimated development cost at $3 million, which neither Douglas nor United Airlines could afford. Patterson proposed a group of airlines which were highly regulated could each put up $100,000 toward the collective development of a new aircraft.

Douglas enlisted G.T. Mead of United Aircraft and Jerome Hunsaker of MIT to do the basic engineering drawings. The result was the DC-4E with tri-finned empennage which resembled the later Lockheed Constellation. The 65,000 pound airliner first flew on 07 June 1948 powered by four 1450-horsepower Pratt & Whitney 14 cylinder R-2180

Twin Hornet engines. During the development Pan Am and TWA withdrew from the project in favor of the Boeing B-307 which is a bloated fuselage mated to B-17 wings and tail.

The tri-tail Douglas DC-4E was loaned to United Air lines for testing from May to August 1939. It suffered the same problems of all commercial and military large aircraft designs of the pre–World War II period. They were underpowered and too costly for the airlines or military to operate. The program was suspended and eventually the single DC-4E aircraft was sold to Japan.[3]

Patterson refused to give up still believing a four-engine airliner was the future. The three remaining carriers in the project, United, American and Eastern proposed a new scaled down version which would be more cost effective. In 1939 A.E. Raymond and E.F. Burton began designing a completely new airframe designated the DC-4. It was smaller with a 94-foot fuselage and proposed empty weight of 40,000 pounds with a 17,000-pound payload. The three airlines were please with the scaled down more efficient DC-4 design. On 14 February 1941, just two months after the bombing of Pearl Harbor, the first DC-4 flew from Santa Monica commanded by John F. Martin.

The Douglas DC-3 was the primary military transport when the war began because the military had not officially identified the need for long-range cargo aircraft to support combat operations. Some planners recognized the demand but it was not considered high priority to move men and materials across the Atlantic by any means other than by ships. Once Japan attacked Hawaii the war department realized the necessity of long-range global cargo aircraft to move men and materials over the vast area of the Pacific as quickly as possible.

Douglas had received airline orders for 61 of the new commercial DC-4s. The first 24 were in various stages of assembly or were just going into service with the airlines. The USAAF reevaluated the DC-4 as a military transport. Unlike the DC-3 and other aircraft of the time the DC-4 had tricycle landing gear. It provided a level constant cross-section fuselage from front to rear. The level cargo hold was far superior to the previous tail draggers which had tapering fuselage and angled floor making it difficult to load cargo up hill.

The success of the DC-4/C-54 constant cross-section fuselage in military service led to the development of other larger aircraft. The new DC-4 easily transitioned to military service with few changes. The level cargo floor was seen as a vast improvement to turn around time which was critical. The Army Air Force commandeered the undelivered DC-4 positions held by the airlines and pressed them into service. The 24 aircraft still at Douglas were revamped to Army specifications. All future DC-4 production was placed on hold while the military C-54 was put into production. The military version had an empty weight of 38,200 pounds and gross of 73,000. The design was so successful and rugged, a total of 1,163 C-54s were built for the military between 1942 and January 1946.

The British Car-Ferry

The Car-Ferry was a postwar era idea for a commercially profitable way to transport vacationers with their cars to mainland Europe. It was pioneered by British air carrier Silver City Airways as an attempt to expand services and increase market share. British Aviation Services (BAS) became the owner of Silver City Airways along with several other carriers in an airline consolidation move in 1947.

Commodore Griffith J. "Taffy" Powell was established and well respected in British aviation having flown with the prewar air carrier which became BOAC. He was the first British pilot to hold both land and marine aircraft license. He formed the Air Ferry Command to move military aircraft across the Atlantic during World War II. While serving in the war he set two transatlantic speed records. After the war he was offered the position of Managing Director of BAS and Director of Silver City Airways.

Following World War II the only means for British travelers to transport their personal cars across the channel to mainland Europe while on holiday or business was by ship. Car for hire business in mainland Europe was not available in many areas making travel difficult. There were no drive on-off ferries at the time resulting in cars being lifted onto ships with a derrick where they were frequently damaged. In addition it was time-consuming and required all the fuel be drained from the automobile.

Revenues at Silver City Airways were slipping in the ever changing postwar passenger airline market prompting management to look for new marketing ideas. It was suggested to Commodore Powell a niche market may exist in transporting cars and their owners by air the 47 miles between the two closest cross channel airfields. Although the cost would be higher than by boat it would eliminate the damage problem, fuel would not have to be drained and it would provide faster service creating a new source of revenue. The idea was rejected as pure folly. Transporting cars and their owners by air as a scheduled daily service was a speculative venture at best. British aviation officials were not amused when Powell continued to press his idea prompting them to enforce very restrictive regulations.

Commodore Powell met with strong opposition but was not one to yield if he believed there was potential in an idea. He maintained there was a market to transport cars from England to mainland Europe by air. If implemented Silver City Airways could control the service. As a publicity stunt Powell placed his personal Armstrong-Siddeley automobile in a Bristol 170 freighter and flew across the channel from Lympne, England, to Le Touquet, France.

The event drew the press and a large crowd to view a car being unloaded from an aircraft. It was enough to capture the imagination of a wealthy Le Touquet resident who also had a home at Southend, England. The gentleman approached Powell with a request to fly his Bentley back to England. Powell was delighted and the two worked out a fare. The car-ferry was born and the Bentley became the first revenue automobile at a cost of 22 pounds sterling ($61.62). The gentleman became a regular Silver City customer until his death in 1960.[4]

The service for holiday vacationers to fly with their cars to mainland Europe was initiated with the Bristol 170 on 06 July 1948. News of the novel service spread allowing individuals to drive to the airport, board their cars, ride in the rear cabin and drive away at their airport destination. The service caught on but was not a large revenue producer for Silver City Airways. Transporting only two cars per flight took up the capacity of the aircraft with no room for additional revenue. As other airlines introduced competing Bristol 170 service it became apparent a more cost effective larger capacity longer range aircraft was necessary to increase yield and remain competitive.

Air Charter and Freddie Laker

The car-ferry service was considered novel and expensive but it did not go unnoticed by other airline executives. The two-car Bristol 170 was noisy and uncomfortable without lavatory facilities. It produced revenue with some profit as long as Silver City could control the market. However, other carriers began competing with the inefficient Bristol. This diluted Silver City revenue in an already low revenue source service. Unlike sea transport the service had good selling points and the public wanted it. The problem was charging fares high enough to make a profit were prohibitive. The public liked the fact the automobile fuel tanks did not have to be drained which decreased loading and turnaround time. Once passengers reached their destination in mainland Europe the car was unloaded, cleared through customs and driven away on holiday.

In response to demand for more lift the Bristol Aircraft Company modified the design of the Bristol 170 or MK.31 aircraft by extending the nose and cargo hold ahead of the cockpit. The modified long nose version designated the Bristol MK.32 could transport three cars. However, it was still noisy and uncomfortable. Silver City Airways attempted to monopolize the car-ferry market by acquiring the rights to purchase all improved MK.32 three-car capacity car-ferry aircraft built. This would give the airline a competitive edge over the other carriers by increasing revenue by one third per aircraft. The other carriers were forced to operate service with the 2-car capacity Bristol 170 MK.31. The plan ultimately caused the demise of the airline because there was only so much traffic and the carrier was not financially strong enough to buy up all MK.32 production.[5]

Freddie Laker's Air Charter Airlines, an associate company of Aviation Traders Engineering Limited, introduced competing car-ferry service. His resume was not as formal as Commodore Powell; however, he was successful in building, flying, repairing and selling aircraft. His interest in aviation began as a child and carried into his youth when he worked for several aircraft companies. He was fascinated by aircraft, confident and loved a challenge. During World War II he became a flight engineer with the Air Transport Auxiliary ferrying aircraft. After the war he formed Aviation Traders Limited to salvage surplus military war planes. Aviation Traders became very profitable during the Berlin Airlift by servicing and supplying parts for aircraft used in the operation eventually making him a millionaire.

Laker's operation found him favor with British aviation authorities and positioned him for success when he proposed

Silver City Airways controlled production of the three car capacity Bristol MK.32 which prevented other carriers from acquiring them. Freddie Laker believed there was potential in the car-ferry market but a superior aircraft had to be found. After considering multiple options he was inspired to create the ATL-98 Carvair replacement (Silver City, author's collection).

a program to the British military to reduce cost and increase productivity of British troops being moved to assignments throughout the world. He determined troops deployed by ship to foreign assignments were in route up to one month in each direction resulting in them doing nothing for two months. He proposed his company, Air Charter, could move them on station by air which normally took less than three days and would increased productivity of each serviceman by a month per year.[6]

He was awarded a transport contract to fly troops in surplus Avro York bombers which were converted to passenger configuration. The plan was so successful he received a contract for all British trooping transports. In addition he was granted a seven year contract to transport nuclear material to Australia. His company Air Charter became a success by being the first private air operator to fly the longest air route in the world. The route of 13,319 miles from London staged through Malta, Nicosia, Bahrain, Karachi, Delhi, Calcutta, Rangoon, Singapore, Makassar, Darwin, Townsville and La Tontouta, ending in Fiji.[7] The British military was extremely pleased with his air operation gaining him favor over other air operators.

Laker was granted an unrestricted commercial passenger certificate in early 1954 to operate Air Charter as a small scheduled airline based at Croydon airport near London. He operated his first car-ferry proving flight in August. He began competing with Silver City with the two car capacity Bristol MK.31 Freighter marketing it as Air Bridge. The Bristol MK.31 was never suited for passenger or car-ferry transport. It was a military aircraft with fixed gear and a limited spar life. It was only capable of transporting two automobiles and 12 passengers in the small rear cabin. It was obvious to him an aircraft superior to the Silver City MK.32 was necessary to compete or expand in the market.

The prospect of profits because of the demand in the car-ferry market prompted the forming Jersey Airlines in 1956 which began car-ferry service to the Channel Islands. Silver City, Air Charter and now a third carrier Jersey Airlines were competing for a small niche market. A fourth carrier named BKS Transport had been formed by a group which included T.D. "Mike" Keegan. The carrier was created primarily to transport race horses by air with the Bristol MK.31 Freighter. In an effort to increase revenue Keegan and his group reviewed the possibility of also competing in the car-ferry market. Keegan would eventually acquire Freddie Laker's BAF Carvair operation and Jack Conroy's CL-44 guppy for Transmeridian Airways.

The Bristol MK.31 aircraft was slow, noisy and only flew at 2,000 feet. All the airlines were operating the MK.31 in competing car-ferry service from Lydd and Southend, England, to Ostend Belgium; Rotterdam Holland; Calais and Le Touquet, France. The vacation/holiday market was very seasonal leaving a fleet of aircraft idle during the winter months or at best transporting bulk freight occasionally. Laker eventually acquired the three-car long nose version Bristol MK.32 when Silver City could not absorb all production of the aircraft. However, the three-car capacity MK.32 was still not cost effective and had a short range. Laker believed there was a greater market in longer routes deeper into mainland Europe. If he could obtain an aircraft with greater range and a five-car capacity, he could increase market leaving Silver City Airways in the non-profitable short range market. He began reviewing existing new aircraft with more capacity and range but found them to be too costly. He concluded the only cost effective answer would be to modify an existing used aircraft to transport automobiles.

Laker was convinced a new aircraft capable of flying longer routes could be developed to increase market share and generate higher profits. He assigned Bob Batt and A.C. Leftley from Aviation Traders Engineering to find a suitable replacement for the aging Bristol. It was evident from the outset the development of a new aircraft would be too costly to produce in the small numbers needed. The alternative would be to modify an existing type if one could be found suitable.

A new turbine aircraft would cost between $1,400,500 and $1,960,700 resulting in amortization of at least $224 per hour on an aircraft operating 1000-hours per year over a minimum of seven years. The cost estimates were too prohibitive eliminating the possibility of purchasing new high bulk aircraft capable of handling automobiles. The group determined a new car-ferry aircraft could not exceed a cost of $560,000 amortized over seven years. The only acceptable option was to modify an existing airframe which could transport cars and passengers during the holiday season and bulk cargo in the off season.

Laker specified the new car-ferry aircraft must have at least a 70-foot cargo hold and be able to accommodate a 25-seat passenger cabin and toilet. Bob Batt and A.C. Leftley reviewed a number of candidates including the Blackburn Beverly and the proposed Armstrong Whitworth Argosy which first flew in 1959. The Argosy had an elevated cockpit and straight in loading but the cargo hold was only 47 feet long. The fuselage frontal area had a lower profile and the door opening was actually shorter than the Bristol MK.32 making it a marginal choice because larger cars would not clear the door.

If the Argosy had been more suitable some consideration may have been given to bringing the cost in line. Purchasing a new turbine powered Argosy for operation during only part of the year could not be justified. In addition it could not hold five cars. Laker also toyed with the idea of modifying a Handley Page Hermes. He had reviewed multiple airframe types and liked the concept of the frontloading

Bristol and the Argosy. It was reported he was considering frontloading designs and expanding the fuselage when the idea of grafting a new nose on the DC-4 came to him while taking a bath. It is easy to understand how he envisioned the concept. Channel Air Bridge had been operating the Bristol MK.32 with the elevated cockpit and the Argosy is a four-engine aircraft with a similar configuration to what was needed. He instructed A.C. Leftley to develop the idea of modifying a DC-4. Cost calculations were made along with detail engineering drawings and a wind tunnel model.

Laker's Carvair Design Controversy

The fuselage design of the Carvair very closely resembles the profile of a Boeing 747. Laker always maintained the position Boeing lifted his idea for the design of the B-747 fuselage configuration. It is a not an unreasonable claim when considering the B-747 did not fly until ten years after the ATL-98 Carvair was conceived. Something similar may have been on the Boeing drawing boards, however, nine years prior to being built is a long time. It is logical to assume, although not verified, that Boeing did study the drag calculations of the Carvair. The chances of the ATL-98 and B-747 having the same profile may be more than coincidence. The B-747 was originally designed as a freighter with a nose door for oversize bulk cargo and straight in loading. The cockpit is positioned eight feet above the main deck similar to the Carvair and blended into the fuselage with a long fairing creating a hump.

The sub-sonic B-747 became a passenger aircraft as a result of several factors. Increased air travel combined with smaller aircraft was contributing to severe airport congestion. It was believed the B-747 could relieve the congestion by carrying twice to three times more passengers on a single aircraft. The model 747 was announced in April 1966. Boeing had been researching a supersonic transport since 1952 but the study was not advanced to project status until 1958. In 1966 the B-2707 Super Sonic Transport (SST) transport division was formed at Boeing.[8] The B-747 which was being developed at the same time was widely accepted as a freighter design adapted to passenger configuration. When the supersonic B-2707 came to the forefront the aviation community believed it would render the B-747 obsolete and relegated to cargo work in the future.

Laker's case is further strengthened when considering at the time the ATL-98 Carvair was being tested; Boeing engineers were very interested in the drag and loss calculations of large diameter and high frontal fuselage aircraft. Up until this time most aerodynamicists did not believe it would be possible to fly an aircraft of a large diameter with a massive front. They were concerned the amount of drag of extremely large diameter aircraft would require excessive power from multiple large engines with high fuel consumption. The general consensus was it would also require very long wings. The weight of the fuel needed would displace payload. In theory there would be a point of diminishing return resulting in no gain in building an aircraft above a certain size.

Freddie Laker embarked on the development of the large front ATL-98 using A.C. Leftley's design and ATELs chief aerodynamicist Brian Kerry's calculations. It has not been conclusively proven but is quite plausible Laker's assertions are correct and Boeing did monitor the progress of the ATL-98 design. How much of the study was applied to the development of a scaled-up large commercial jet freighter is open to speculation. Boeing definitely monitored the development of the Aero Spacelines B-377PG being built in California.[9]

A year after the ATL-98 had flown the idea of Aero Spacelines developing an aircraft with complete disregard to the accepted size standards caused Boeing engineers to once again become very interested in a high frontal aircraft design. Conroy shared with Boeing the engineering theory and different drag formulas. Boeing closely monitored the building of Conroy's guppy and sent engineers to Van Nuys, California, on multiple occasions to observe the progress. They were not only interested in the build of the B-377 Pregnant Guppy but also the progression of flight testing.[10] Boeings interest in Conroy's expanded B-377 makes it easy to reason their engineering department would be interested in any expanded fuselage concept which was being developed or already flying.

Selecting the DC-4

Surplus wartime C-54 Skymasters were converted to commercial DC-4s and placed in airline service after the war ended. This was during a time when a new factory fresh Douglas DC-4-1009 cost $385,000 USD and an ex-military upgraded C-54 was around $100,000. As America began to apply wartime technology to commercial applications more efficient aircraft designs were developed. The DC-4s were slowly outclassed by the better equipment. By the time DC-4s were retired from third and fourth level carriers the price was approximately $80,000 each.

Laker's cost projections for DC-4 conversion made it an excellent car-ferry design. It had a 15 percent higher operating cost than the Bristol Freighter which is offset by a 100 percent higher capacity. Doubling revenue during peak traffic months and only increasing cost by 15 percent could drastically increase revenue yield during off periods. The final cost to build and attach a new forward fuselage, modify the existing vertical fin, move the rear bulkhead aft to accommodate a passenger cabin, install a uniform fuel plumbing system, fuel tank seal and mount four zero-timed P&W R-2000-7M4 engines and Hamilton Standard propellers came in at $560,200 USD, which included the $33,612 for the engines.

The British aviation industry was reformed and consolidated in 1958–60. As part of the consolidation Freddie Laker sold Air Charter and his aircraft salvage and maintenance company Aviation Traders to Airwork. As a result he became executive director of British United Airways. By 1960 the car-ferry market was struggling because of the low yields caused by market saturation. Channel Air Bridge car-ferry service was under the corporate umbrella of British United Airlines. After taking the helm at British united, Laker made it a priority to control the car-ferry market. He insisted on Channel Air Bridge being allowed to operate as a stand-alone car-ferry company within the Airwork corporate structure.

In 1961 Air Holdings was formed as part of the British aviation consolidation which included Airwork and many small airline companies. When the decision was made to convert surplus C-54s to car-ferry aircraft, it was natural the work would be accomplished by associate company Aviation Traders Engineering Limited. Laker always referred to his car-ferry operation as Air Bridge. The theme was carried forward with Channel Air Bridge and British United by giving each Carvair a special name: Golden Gate Bridge, Chelsea Bridge, Pont de l' Europe, Pont du Rhin, Maasbrug, Channel Bridge, Menai Bridge, and Pont d' Avignon.

ATL-98 Design and Engineering

The building of the ATL-98 Carvair was a multi-faceted redesign of an existing aircraft which is considerably more complex than grafting a new nose to an existing airframe. By modifying an existing airframe the majority of the engineering is eliminated which reduces cost considerably. Initially, Aviation Traders had to obtain permission from Douglas to perform modifications to the airframe. Douglas was receptive to the DC-4 conversion proposal because Aviation Traders also planned a DC-6 and DC-7 version. If successful the ATL-98 conversion could create a market for otherwise obsolete propeller driven aircraft. ATEL needed engineering drawings and Douglas stress and design data. Douglas had a favorable view of the project but wanted to be released from any liability after conversion.[11]

ATEL chief engineer Arthur Leftley was dispatched to Santa Monica, California, in March 1959 to meet with Douglas Aircraft engineers. He was to present the modifications and negotiate a reasonable price for the DC-4/C-54 drawings, stress data and manuals. ATEL officials set a budget of approximately £10,000 pounds sterling which amounted to about $28,000 U.S. to obtain the data. During the negotiations Douglas proposed a priced of "10,000." Leftley, who was thinking in terms of £10,000 pounds sterling, agreed on the amount thinking he was within budget. When he returned to Southend he was informed he had agreed on $10,000 U.S. (£3,570 pounds sterling). Leftley's error obtained the data and drawings for a third of what ATEL expected to pay resulting in a budget windfall for Aviation Traders.[12]

Most of the data immediately available from Douglas was on the postwar commercial DC-4-1009. Aviation Traders had purchased derelict military era C-54s with thousands of hours on the airframes. They had been upgraded to DC-4 standards after the war and were quite tired after many years minimal maintenance. Multiple changes, upgrades and patches had been preformed which added weight to the airframes. As agreed, Douglas retrieved the prewar C-54 data from the archives and forwarded it to ATEL. No one had considered the fact the Douglas test data was from a brand new 1942 model C-54. Flight test data was needed for comparison but it was determined it could only be used for reference. The flight test data was not the same on the nearly 20-year old aging airframes being used for conversion. The weary airframes responded quite different than new equipment. As a result, ATEL conducted more than 76 hours of testing on a C-54 to establish drag and lift characteristics on the forward fuselage.[13]

The new double deck nose production was set up in a shop at Southend Airport. The nose section was designed from scratch which required a full size mockup up. Smaller wind tunnel models were also constructed to determine the aerodynamic behavior of the aircraft with a greatly enlarged nose. In order to meet the required specifications of a 70-foot cargo hold and a rear passenger cabin the aft fuselage bulkhead had to be redesigned and moved 6 to 7 feet rearward. This affected the routing of the control surface actuating cables which had to be reengineered.

The routing of the cables in the tail had to coincide with the major changes of the control cables of the new raised cockpit. The layout of the original DC-4 cockpit was maintained and relocated using the same yokes, control pedestal and panel to reduce production cost. Because the new forward fuselage increased the length of the aircraft, the nose gear had to be relocated and modified to give unobstructed loading access. In addition to reengineering flight control systems a new passenger cabin, lavatory and galley was designed to meet the needs of the service.

Douglas built the C-54 in a number of different configurations to meet multiple wartime missions the military required. The modifications included double cargo doors, self-contained hoist and additional fuel tanks in center-wing and fuselage for increased range. The postwar C-54s had the hoist removed and rear half of the cargo doors permanently bolted shut when converted to postwar DC-4 commercial standards. Many of the DC-4s retained the multiple fuel tanks configuration. These aircraft had many different fuel line plumbing configurations with pumps and lines on both sides of the spar. After they were reconfigured to postwar DC-4 commercial service they were fitted with multiple series R-2000 engines, which added even more changes. The commercial airlines had overhauled and modified the systems over and over and in some cases the later "E" model wings were fitted to earlier aircraft. This resulted in a variety of problems which included in-flight incidents of fuel starvation and failures above certain altitudes. ATEL engineers recognized the problem from the beginning as the different series C-54 and DC-4 airframes began arriving at Stansted for conversion. Consequently the first task was to engineer a uniform fuel plumbing system for all ATL-98s.

When the ATL-98 design was debuted, more than 60 airlines and the air force of several countries expressed interest in the conversion. The interest was so strong Aviation Traders marketing estimated 100 aircraft would be built.[14] Most airlines wanted the planned turboprop version. As a result no orders were received. Additionally British United had not been granted long haul routes deep into mainland Europe. Because route authority had not been obtained, Channel Air Bridge placed an initial order for only three ATL-98s with an option for seven more. The aircraft had been specifically designed for Channel Air Bridge to increase its share of the car-ferry market. However, it was considered a multi use combi aircraft which would generate other orders. This would increase production which was needed to defray the cost. Without any deposits or orders outside the company it was not possible to estimate the number of aircraft which would be built.

The lack of orders inhibited the ability of ATEL planners to make cost estimates on tooling. This prevented any investment in expensive metal forming machines. Consequently all of the panels for the first nose units were individually hand formed on a wheeling horse by very skilled metal craftsman. The first nose for DC-4 G-ANYB was built in two sections split horizontally at floor level. Metal forming machines would be purchased once the expected orders materialized and a firm production number could be established.

Building the ATL-98 was not as straight forward as it appears. It is more than just cutting the off the C-54 airframe forward of the wing and grafting on a new double deck section with a nose door. Aviation Traders began the design work on modifying the first C-54 at Southend, England, in 1959. Within a year American aircraft broker Lee Mansdorf visualized expanding the fuselage of a Boeing B-377 in an effort to profit from NASA's logistics problem.

On 01 October 1960 ATEL engineers began the dismantling process of C-54B-1-DC c/n 10528 as the base airframe for the first expanded fuselage ATL-98 guppy. After the engines, outer wings, empennage and control surfaces were removed the nose was cut from the DC-4 forward of the wing at station X270. It was then notched back by cutting horizontally back to station X360 and cut around the upper half of the fuselage. The freshly stripped out airframe was placed in stands for stability and to prevent distortion in order to align and graft on the new nose.

The first ATL-98, G-ANYB, under construction at Southend, England. It is the only Carvair to have the new nose built in two separate sections then grafted together. The lower section was placed in a jig forward of the C-54 fuselage and upper section was suspended above. The area between the two sections was filled in before the nose was grafted to the existing fuselage (Guy Craven, courtesy Richard Vandervord).

After the ATL-98 structural design work and wind tunnel tests were completed, the full scale nose drawings were transferred to hard board. The pieces were cut and assembled to form a wood mock-up of the new nose. It was split longitudinally into an upper and lower section. The first expanded nose was not completed in a jig but on the aircraft.

The standards and procedures of the building process were still in the trial and error stage. The newly fabricated lower section of the nose was placed in a jig forward of the C-54 fuselage and lined up. It was grafted to the fuselage just forward of the wing. The upper cockpit half of the new nose was suspended above the lower section. The area between the upper and lower halves was then bridged with formers and ribs to complete the shape of the expanded forward fuselage. Stringers were attached to form the hump behind the cockpit to complete the grafting of the nose to the existing aft fuselage. The splice was then skinned over to complete the forward section. All subsequent conversions had nose sections built as a one piece unit before grafting to the C-54 airframes.

Nose Gear

The DC-4 nose landing gear was removed before the forward fuselage was cut away from the airframe. The new enlarged nose with straight in loading presented a clearance problem with the mounting and retraction of the nose gear. To reduce cost, ATEL retained and modified the original nose gear strut. The mounting position was moved forward in the new nose. Up-gear stowage presented a problem because the DC-4 strut has 94 degrees of travel to reach the full retracted position. This is 19 degrees above the horizontal line to up-lock placing it above floor level. In order to maintain an uninterrupted level cargo floor the ATL-98 nose gear was relocated forward and mounted four inches lower. The gear retraction arc is limited to 75-degrees stowing parallel to the fuselage horizontal datum in a set of blister type gear doors. This was achieved by placing a steel tube inside the strut piston to reduce travel.

Oil Tank

The base C-54 had a 22-gallon nacelle mounted oil tank behind each engine. Reciprocating engines have a tendency to consume or throw out a considerable amount of oil. The Pratt & Whitney R-2000 fitted to the DC-4 does not consume anywhere near the amount of oil of the R-4360 mounted on the XC-99, C-124 and B-377 guppies. However, the problem still exist which could limit range on extended C-54 flights. In 1942 when the military realized the need for a long-range transport, oil consumption on long overwater flights posed an unacceptable possibility of shutting down an engine for lack of oil. As part of the military C-54 design, the Skymasters were fitted with a 50-gallon central oil tank mounted in the lower fuselage. The tank has a four position selector which allowed the crew to transfer oil to any engine with excessive consumption. Oil could only be transferred in one direction to the engines. The individual engine oil tanks are always filled before departure. The engine oil tanks are never completely refilled when transferring from the central oil tank in-flight. Oil is transferred a few gallons at a time as needed to prevent having excess oil in an engine tank because it cannot be retransferred to another engine. If a high consuming engine fails and the tank is full that oil is unusable because it cannot be transferred to another engine to remain airborne. This was very important during World War II on extended flights across the Pacific.

The central oil tank was removed and supply lines capped by ATEL when the C-54 nose was removed during ATL-98 conversion. The tank was considered unnecessary because the Carvair was developed for short to minimum range cross channel service within Europe where there were multiple airfields to land. As a precaution the tank was shipped with the aircraft to be installed at the owner's option if the buyer anticipated long-range service. This did not become a problem until years later when third level operators operating in all cargo configurations realized range could be limited by oil consumption. At least two ATL-98s had central oil tanks added 25 years after they were converted to Carvair standards.

Vertical Fin and Rudder

Wind tunnel test determined the original vertical fin and rudder of the DC-4 was too short to deliver adequate yaw control because the new nose produced a change in airflow and turbulence along the rear fuselage. The problem occurs in designs where the fuselage is increased in size or external cargo is transported. The airflow problem required additional engineering of the tail section for proper yaw control. ATEL engineers initially designed endplates for mounting on the horizontal tail plane. Similar end plates were originally designed for the Aero Spaceline B-377PG but were found unnecessary with an extended vertical fin. The Airbus A300-600ST Beluga required end plates to compensate for airflow disruption created by the out size fuselage. The proposed piggyback Douglas C-133 configuration as well the NASA B-747 shuttle carrier aircraft (SCA) required end plates to compensate for turbulence set up by mounting the Space Shuttle on top of the fuselage.

Wind tunnel test determined excessive pressure would be exerted on the endplates if installed on the DC-4 horizontal stabilizer. This would require reengineering and reinforcing the horizontal tail plane to handle the load. Additional cost would be incurred and the end plates would add additional weight to the airframe reducing payload. A more cost effective solution was to reinforce the existing vertical fin and add 2 feet, 4 inches to the top. The new fin had a similar profile and is often thought to have been adapted from the production Douglas DC-7. Closer inspection will reveal it is a modified DC-4 fin with the same trim tab as the DC-4. The top of the new ATL-98 fin is 29 feet, 10 inches above the ground. This is halfway between the Douglas DC-7 and DC-7C which are 28 feet, 7 inches and 31 feet, 10 inches, respectively, which have straight fuselages without a forward hump.

Wing Bob Weights

Fuel is often used as ballast to reduce wing deflection and compensate for an extremely heavy payload which can overstress the wing root. The Carvair was expected to carry a heavy cargo load of five automobiles over a short distance which would not require an excessive fuel load. One of the more unique features incorporated in the ATL-98 is wing bob-weights. The new expanded forward fuselage and increased vertical fin added 2,300-pounds dry weight to the original fuselage. To increase the maximum zero fuel weight ATEL engineers came up with an ingenious idea. A set of removable 250-pound weights was installed in canisters in each wing tip. The weights were used to offset wing loading. Adding 500-pounds to the Zero Fuel Weight (ZFW) the farthest distance for the aircraft centerline allows the airframe to transport 1000-pounds more payload. The addition of wing bob-weights added 500-pounds more weight for a total of a 2,800-pounds increase over the original C-54 aircraft. The FAA reduced the ATOG from 73,800 to 72,900 pounds.

In simple terms the wings deflect as the aircraft lifts off picking up the fuselage. The weights in the wingtips reduce deflection by more evenly distributing the weight along the length of the wing maintaining the same stress loading with more cargo. The resultant stress on the wing root remains within limits. The only penalty would be fuel because the ATOG remains the same. Reduce fuel load is not an issue on the Carvair because it is designed for short to midrange service. All Carvairs fitted with the weights were designated as ATL-98A models.[15]

Windscreens

The cockpit window configuration is a slightly different contour from the original C-54. In order to reduce engineering, material and component cost ATEL reused many DC-4 original and off-the-shelf parts and assemblies. The cost of new cockpit windshield glass was considered prohibitive. The alternative was to adapt the original C-54/DC-4 glass to the new nose cockpit. After the DC-4 nose sections were removed from the C-54 being modified the windshield rivets were carefully drilled out. The glass was then cleaned and placed in new frames which were installed in the ATL-98 nose.

Aviation Traders did not have any spare cockpit glass in stock when Carvair production began. This presented a problem which slowed the completion of the new nose sections at Southend. Completion was delayed until the C-54 nose was cut away from the first donor airframe. The glass was salvaged then reconditioned and installed in new frames. As multiple aircraft were being brought in for conversion the supply of glass increased allowing the reconditioned panels to be completed and placed in parts inventory. As the new nose was built up the windshields were drawn from stock and installed.

Cockpit Configuration

The ATL-98 cockpit compares very closely to the original C-54 with some refinements. Neither the C-54 nor ATL-98 had a dedicated flight engineer's position. The ATL-98 was designed for a two-man cockpit crew with a third seat behind and between the pilots for a flight engineer

The cockpit of VH-INM, the last ATL-98 built. The configuration is actually a refined DC-4. Even the windscreens were salvaged from the old DC-4 and reinstalled in new frames for the Carvair. Although not required, a flight engineer sees the cockpit from this position. His seat is mounted on a track behind the console, which allows him to slide between this position and a work table behind the co-pilot (Aviation Traders, author's collection).

position when needed but no instrument panel. A flight engineer was not required, but often carried, on long flights to handle throttle settings and functions on the console. The pilots were often occupied with navigating and operating the radio while flying in unfamiliar areas which was quite common on Carvair charters. The engineer could only monitor the engine gauges on the forward panel but often made throttle settings and gave the pilots another pair of eyes.

When the DC-4s were built it was not considered necessary to have a dedicated flight engineer's position in a multi-engine transport. The larger postwar post–World War II (C-74 and B-377) transports were much more complex requiring a flight engineer as a necessity with an instrument panel to monitor engine functions along with throttle levers, mixture and prop controls. Commercial operators tried to prevent a flight engineer's position on all four-engine aircraft until the FAA stepped in on 15 April 1948 requiring a flight engineer on all four-engine aircraft weighing over 80,000 thousand pounds.

The ATL-98 has an engineer/navigator table mounted on the bulkhead behind the copilot. The engineer has a chair mounted on a track which slides sideways to a position behind the console. British Carvair operators did not require a flight engineer but often assigned the third crew member on long flights to assist the pilots and reduce the workload.

All 21 ATL-98s were virtually hand built. Depending on the options ordered by the original buyer there were many different main deck configurations as well as the upper area behind the flight deck. It can be safely stated that each Carvair is different. The options which could be ordered included two rear facing sleeperette seats, three crew seats, up to seven seats on cargo-liners, twin bunks and a galley. Configuration drawings also offered a cockpit lavatory; however, no records have been located of any aircraft being so equipped. All cargo-liners had a small portable toilet below the cockpit mounted in the nose door with a draw curtain for privacy.

It appears ATEL engineers attempted to stay as close to the DC-4 cockpit layout as possible in order to create a standard upper deck design. It should be noted many military DC-4s had crew quarters and bunks behind the cockpit for supplemental crew. These quarters proved necessary on extremely long flights operated across the Pacific. The crew quarters on the long-range modified ATL-98 can be compared with the DC-4s big brother C-74 and C-124 Globemasters. All were equipped with three bunks, a galley and lavatory in the area behind the flight deck.

Production Positions

During the British airline consolidation Freddie Laker demanded the Channel Air Bridge car-ferry operation be allowed to operate as a stand-alone division. Part of the 1960s British airline consolidation was to eliminate competition. Laker had placed an initial order for three Carvair conversions and took options for seven more to create work for sister company Aviation Traders. When the first three were delivered to Channel Air Bridge the additional conversions were not yet needed because route authority had not been granted for destinations further into Europe.

Carvair construction began with great enthusiasm. Almost 60 airlines expressed interest, prompting ATEL to estimate a production of 100 aircraft. Multiple aircraft were under construction at Stansted; however, no orders were received. Most of the interested carriers decided to wait for the proposed turboprop DC-7 model—which was never produced (Aviation Traders, author's collection).

It was obvious the failing Silver City Airways could not compete after Channel Air Bridge took delivery of the first of three ATL-98 Carvairs. Air Holdings purchased the major competitor British Aviation Services (BAS) in 1962 which was the parent company of Silver City Airways. The managing director of BAS was Commodore "Taffy" Powell who had created the original car-ferry service at Silver City Airways in 1948. At the time of the consolidation Silver City was still operating car-ferry service with the Bristol MK.32. On 01 January 1963 Silver City and Channel Air Bridge were merged to form British United Air Ferries (BUAF). The Channel Air Bridge name was dropped, bringing all companies under British United with a complete livery change.[16]

The next three Carvair conversions were started for BUAF on speculation. It was still believed many orders were forthcoming from the more than 60 interested carriers. The plan was to have the aircraft near completion leaving only certain configuration features to complete. When orders materialized BUAF could relinquish delivery positions making the aircraft immediately available to the buyer. Either way the aircraft were committed allowing BUAF to develop the market and not take delivery until aircraft were needed.[17]

Carvair Cargo-Liner

Initially the plan worked to have conversions in process when outside orders were received. The first aircraft ordered outside the Air Holdings/British United family of companies came from New York based charter operator Intercontinental in June 1962. The carrier requested the aircraft be completed as cargo aircraft with multiple special options for operation in remote areas. Carvair c/n 4 was slated for Channel Air Bridge and had already received British registration G-ARSH. Because of the urgency to acquire the aircraft for a United Nations contract Intercontinental stipulated it would only purchase the next two aircraft in production. Channel Air Bridge willingly gave up the delivery positions to allow Aviation Traders to make a sale. The aircraft were to be completed as pure freighters with special options and delivered to associate company Interocean based in Luxembourg. The ATL-98 was purpose built for car-ferry service. ATEL sales welcomed the possibility the order for cargo-liners could broaden the appeal and finally bring in orders for more aircraft.

Carvairs four and five were completed for Interocean with special options which included six seats behind the cockpit, a crew "engineer" seat behind the pedestal, two bunks plus a small galley with hot plate and beverage containers. A winch was installed at the rear of the cargo hold and a chemical toilet in the nose door. The forward bulkhead was enlarged to accept military trucks. A modified bulkhead was designed around the assumed measurements of military trucks used by U.N. forces. Loading tests were done by ATEL at Southend with Bedford military trucks to ensure they would clear. However once the aircraft reached the Congo it was discovered the U.N. vehicles had welded on tops which did not clear consequently none were transported.[18]

The most challenging special option was a self-contained portable four-ton capacity ramp system for loading vehicles at remote airfields with no support equipment. The design of the ramp system was quite a challenge. It had to be strong yet light enough it could be broken down and loaded by hand. Once the aircraft arrived in the Congo it was realized the ramps required considerable manpower to unload and were difficult to set up. Fortunately there were pallet loaders in Leopoldville and Albertville. The ramps were used only when necessary at remote airfields.

After the two Interocean aircraft returned from the 1963 United Nations contract work in the Congo, they were returned to Aviation Traders for overhaul and upgrade to car-ferry standards. Both were fitted with a 17-seat all first class passenger cabin for car-ferry charter service based in Luxembourg. The service was not successful prompting the sale of both aircraft to French carrier Cie Air Transport. Carvair four was lost when it crashed into a populated area at Karachi, Pakistan on 28 February 1967. Carvair five was acquired by another French carrier and leased back to British Air Ferries (BAF). It was eventually purchased back by BAF.[19] It spent considerable time leased out in the Middle East transporting construction materials in Oman and oil field service equipment.

Nose Loading Door

The ATL-98 Carvair nose loading door is designed with two hinges on the left opening through 170 degrees. The configuration is very similar to the one-piece nose door design which was conceived and patented by Charles H. Babb in the 1939. Babb who was previously mentioned designed a new frontloading cargo aircraft in the mid–1930s. It was considered revolutionary in an era when all cargo aircraft were side loading.[20] The aircraft was a high wing configuration with elevated cockpit which resembled the Bristol 170 MK.31 car-ferry aircraft. Babb also designed an overhead crane cargo loading system similar to the one fitted to the C-74 and C-124 Globemaster. Babb's designs were forgotten after his death when his company was acquired by Floyd Odlum's Atlas Corporation in 1952.

Eight years after the Babb Company was acquired by Atlas Corporation Transocean Airlines was acquired and the two companies were merged. Four of the DC-4s which Transocean had acquired in its early years were used to build ATL-98 Carvair two, three, seven and 13. It is also a coincidence the two DC-4s used to build Carvairs nine and 17 were purchased from the Babb Company. In 1957 the Babb Company was instrumental in Transocean acquiring 14 surplus B-377 Stratocruisers. These were the same Stratocruisers which Lee Mansdorf acquired in 1960 after the demise of Transocean. The Stratocruisers were eventually transferred to Aero Spacelines for airframe parts to build B-377 super guppies. The number of airframes which passed through the Babb Company and Transocean which became guppy ATL-98 and

B-377SG conversions is intriguing and yet it is all coincidental.

The first design ATL-98 nose door was all metal without a provision for radar or the twin landing lights mounted in the lower quarter. The landing lights were soon added to the nose door on Carvair c/n 1, G-ANYB and all subsequent models were so fitted. There were at least three different nose door configurations on the 21 aircraft built. The aircraft forward bulkhead dictated by the positioning of the alignment pins which called for design changes. Carvair one, two, three and six thru 18 had the same door although some had the dielectric patch for radar and some did not. Carvairs four and five which were ordered by New York based Intercontinental for Luxembourg based Interocean were different because the two aircraft had a notch forward bulkhead for military vehicles. The three Ansett Carvairs, 19, 20 and 21, were cargoliners. The nose door was different because the aircraft were fitted with a roller floor system to accommodate the universal 88 × 108 inch (P9) cargo pallets. To accept the pallets the bulkhead was squared at the bottom. Outwardly the nose doors generally appear the same. Eventually most were upgraded with a dielectric patch for radar. All nose doors were designed with a flat shelf area at the bottom. A small chemical toilet was mounted on the door shelf for the crew on cargo models.

ATL-98 First Flight

The Carvair name was not created until just prior to the ATL-98 going into service. The new Air Bridge car-ferry service was being vigorously promoted to the public in an effort to increase market share. The marketing department at Channel Air Bridge believed it needed a way to get the public excited about the introduction of extended European service. Freddie Laker seized the opportunity by proposing an aircraft naming contest to promote the new car-ferry. The prize for the promotion would be the transport of a car and two passengers to or from Calais, Ostend or Rotterdam. The contest was won by two gentlemen from Southend who proposed a most appropriate name of Car-via-Air which became shortened to Carvair. The name was applied in script to the lower front fuselage on the three Channel Air Bridge aircraft and later to the nose and tail when the livery was changed to British United.

Alastair Pugh, who was deputy editor of *Flight Magazine* in the 1950s before becoming product development manager at Channel Air Bridge, claimed some years later the Carvair logo was his handwriting. Apparently Pugh was working on a logo of the acronym "car-via-air" which had won the contest. He wrote out the name Carvair in his own hand.[21] The script is a very close match. Aviation Traders contracted all lettering to paint specialist Reginald Taylor. It can be safely assumed the paint shop took some liberty with Pugh's design of the Carvair logo and matched his hand as closely as possible.

Carvair G-ANYB was rolled out at Southend, England, on 17 June 1961 after 8.5 months in conversion. It flew on 21 June 1961 for two hours with Channel Air Bridge Chief pilot Don Cartlidge in command. He was assisted by Captain Bob Langley and flight test engineer Ken Smith. The chase plane was flown by senior Channel Air Bridge captain L.P. Griffith accompanied by Freddie Laker and ATEL chief engineer and designer A.C. Leftley. The crew reported the aircraft performed very well and handling was equal or superior to the DC-4. The landing and rollout was reported as greatly improved with the addition of DC-6 brakes.

The first double deck nose DC-4 conversion to ATL-98 took 8.5 months to complete. It was rolled out on 17 June 1961 as G-ANYB. There are no landing lights in the nose door. It only took Jack Conroy 45 days longer to build the much larger B-377 Super Guppy. The Carvair first flew on 21 June 1961 but did not go into service until February 1962 after 155 hours of testing (Brian Kerry, courtesy Richard Vandervord).

Engineers were very pleased the new nose only reduced the cruising speed by four knots. The rate of climb was actually better than the original DC-4. The new aircraft was tested for 155 hours before receiving the C of A on 30 January 1962. On 12 February 1962 it was christened Golden Gate Bridge by the wife of the Swiss Ambassador at Southend. After a festive ceremony it operated a demonstration flight to Switzerland and Ostend, Belgium, with a group of dignitaries on board.

During the next seven months,

Carvair one (G-ANYB) operated many promotional and scheduled flights to European destinations with great excitement. On 19 October 1962 G-ANYB was operated on a proving flight to Singapore for the Ministry of Aviation. The nature of the mission was not disclosed; however, it could have been live rockets, munitions or nuclear material. The mission was reported to be two-fold. It was an actual cargo carrying flight for the British government to prove aircraft viability to obtain Ministry of Defense contracts. Secondly it was an attempt to promote Carvair sales to carriers in Asia and the Pacific.[22]

The first three Carvairs entered service with Channel Air Bridge with few problems. The aircraft was large, heavier and more stable than the Bristol MK.32. The rear cabin and space for five cars was a vast improvement which increased the number of bookings. Carvairs four and five were completed for Interocean. With no more orders forthcoming it was assumed Channel Air Bridge would receive conversion number six. A corporate restructuring soon brought Channel Air Bridge under the British United family with a name change to British United Air Ferries. The change insured that only three Carvairs would ever wear the Channel Air Bridge livery.

Competition from other car-ferry carriers prompted Irish national carrier Aer Lingus to reluctantly order two ATL-98 conversions in January 1963 with an option for a third. The carrier placed the order under pressure from the Irish Government to provide car-ferry service and prevent encroachment from foreign carriers. It was projected the aircraft would operate at a loss to prevent a British carrier which attempted to obtain authority to fly Carvairs over Irish routes. The Irish order was received at ATEL with great fanfare because it came from a national scheduled "flag" carrier outside the British family of companies. Increased route authority had not yet been granted to British United Air Ferries prompting the carrier to exchange delivery position for Carvair six for a later delivery. Aviation Traders officials were extremely pleased to gain a sale to a foreign airline. Hopes were high at Aviation Traders that aircraft orders would begin to materialize from other carriers outside Great Britain.

The two Aer Lingus Carvairs began service in the summer of 1963. The carrier maintained they would only be marginally profitable as a car-ferry aircraft. In an effort to extract additional revenue they were flown as car-ferry service in the daytime and operated bulk overnight cargo service with mail and newspaper runs from London. Aer Lingus engineers began to review ways to modify the aircraft for more versatility and profitability.

Control Cables and Raised Ceiling

The re-routing of flight control cables for the new raised cockpit in order to match up with the original DC-4 cable runs presented ATEL with an engineering nightmare. The DC-4 control cables ran under the floor from the yokes and pedestal to the engines, wing and tail section. The cables from the new ATL-98 elevated cockpit had to be rerouted because they exit the cockpit floor into the cargo hold at ceiling level.

The Aviation Traders ATL-98 design team encountered similar cable routing problem as Douglas engineers at Santa Monica when converting the fifth C-74 to YC-124. The C-74 and DC-4 were similar with control runs under the cargo deck floor. Converting the Douglas DC-4 to ATL-98 was actually more complex because the aircraft was smaller with limited space. The C-74 flight deck was raised to create the straight in loading C-124. The cables were routed from under the cockpit floor over to the side walls and down at an angle along the outer edges of the cargo compartment and back to the wing.

The new cable routing for the Carvair required hundreds of new brackets and 600 additional pulleys. To alleviate the problem of getting them below the original cargo deck floor, the cables were re-routed 90 degrees through a pulley harp over to the side of the fuselage and 90 degrees down the sides of the cargo compartment walls. Another set of pulleys redirected them 90 degrees again under the floor to the center of the fuselage to re-connect to the original C-54 control cables and out to the wings.

The elevator and tab control cables on the original DC-4 ran from the cockpit under the cabin floor to the extreme rear behind the passenger compartment and up to the elevator pulleys. To build the required 22-seat passenger cabin the rear C-54 bulkhead had to be moved rearward four feet. This was necessary to produce a 70-foot cargo hold which would accommodate five cars. The repositioning of the rear bulkhead displaced the pulley mounting bars at the rear of the aircraft and restricted travel of the original cables to the elevator. To resolve the problem the cables to the tail exited the rear of the ATL-98 cockpit floor and were routed to the rear in fabricated trays along the cargo and passenger compartment ceiling. Pulleys were installed behind and above the repositioned rear passenger cabin bulkhead to direct the cables down to the elevator pulleys in the tail.

The ATL-98 flight control cable routing is remarkably similar to the Douglas C-124 Globemaster II. The elevator and trim cables on the military C-124 exit the rear of the cockpit and angle up to travel along the ceiling to the rear. The cable routing on the first twelve ATL-98s exited the cockpit floor and traveled straight back in cable trays and along the original C-54 ceiling. The setup worked quite well on the DC-4 to ATL-98 conversion. However, cable trays blocked access to the area in the hump behind the cockpit making it unusable space.

The early ATL-98 conversions did not have gill-liners in the cargo compartment leaving all the ribs, formers and framing exposed in plain view. The cockpit crew could open a hatch on the rear cockpit wall and look down into the cargo bay between the cable guide trays. This feature allowed the crew to monitor the security of the automobiles and check for any possibility of fire in the cargo bay.

Aer Lingus engineers proposed a design change to ATEL in 1963 shortly after the Irish Carvairs entered service. The carrier believed the empty space in the hump could be better utilized by rerouting the control cables up from under the flight deck floor. Aer Lingus had increased market share in the bloodstock race horse charter market after the demise of BKS Transport. The high strung horses were quite spirited and not comfortable with the low ceiling of the cargo hold which presented some problems in handling. It was suggested rerouting the cables behind the cockpit would create a cathedral type ceiling in the cargo compartment providing more headroom for the horses and possibly producing a calming effect. In addition Aer Lingus requested a Rolamat floor system to ease the loading of horse containers and bulk cargo pallets to decrease turnaround time.

ATEL engineers agreed and redesigned the rear bulkhead of the cockpit eliminating the inspection door. A frame with additional pulleys was installed on the back side of the flight deck rear bulkhead. The control cables were re-routed from under the flight deck floor up the back of the cockpit wall to the top of the fuselage. They followed the ceiling contour of the hump to tie in with the existing cable runs to the rear of the fuselage. The modification very closely matched the routing of elevator and trim cables on the Douglas C-124 from under the cockpit floor and upward along the roof to the tail.

The first two Aer Lingus Carvairs EI-AMP (c/n 6) and EI-AMR (c/n 8) operated the first holiday car-ferry season in a two-cabin 34-seat configuration with the standard cargo compartment ceiling. After the first season of operations ended the pair was scheduled to return to ATEL at Stansted in the winter of 1963–64. The parts had been fabricated to retrofit the two Irish Carvairs with the rerouted control cables system and the Rolamat floor system.

Carvair six arrived at ATEL, Stansted for ceiling modifications on 08 January 1964. Upon landing EI-AMP suffered a nose gear collapse causing severe damage to the lower forward fuselage and nose gear bay. After inspection ATEL engineers were not confident it could be repaired and the ceiling modification completed in time for the upcoming 1964 summer holiday season. Aer Lingus officials reviewed the situation and formed a backup plan. In the event it was not repaired in time the carrier would exercised a contingency option for a third conversion as a replacement for EI-AMP.

Coincidentally, in 1964 Carvair c/n 13 was already in production and scheduled for delivery to British United Air Ferries. The British United was anxious to delay delivery until market conditions improved. The delivery position was reassigned to Aer Lingus in anticipation of the substitution. Aviation Traders immediately changed the build specifications on Carvair c/n 13 to match the two previously built Aer Lingus ships. Carvair 13 became the first assembly line ship built with the raised ceiling configuration. The aircraft had already been assigned British registration G-ASKN for British Air Ferries. The registration was not cancelled pending the final outcome of EI-AMP repairs. G-ASKN was completed with the

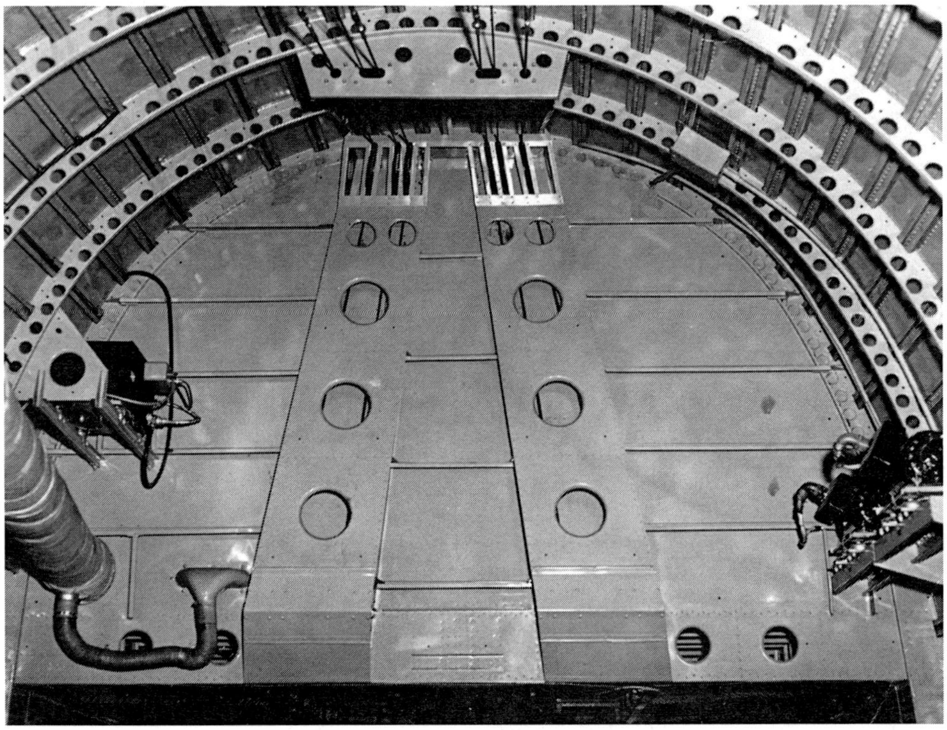

Aer Lingus was the first Carvair operator to utilize the space in the hump behind the cockpit. Looking forward from the cargo hold, this is the rear wall of the cockpit. The elevator and trim control cables exit under the floor and are redirected up the back of the cockpit wall then down the ceiling to the rear. This increased the ceiling height, which allowed for much more headroom for the transport of race horses (Aviation Traders, author's collection).

raised ceiling and rolled out on 19 December 1963 in anticipation of delivery to Aer Lingus.

Surprisingly, the nose gear damage on EI-AMP was repaired and the ceiling modification completed in record time allowing a return to Aer Lingus for the 1964 season. The change in cable routing to open up ceiling in the hump was a vast improvement. Once in operation it was reported the high ceiling greatly improved the transport of the horses. Handlers reported they appeared calm and were less likely to balk during loaded or in flight.

The delivery position of Carvair 13 reverted back to British Air Ferries as G-ASKN. By default the first production line ATL-98 to have the raised cargo bay ceiling became the only British operated Carvair built in this configuration. British United had agreed to the delivery delay because of market changes. Some of the Carvair destinations were receiving more supplemental passenger bookings causing BUAF to consider operating regular passenger aircraft on those routes. To meet the demand the seating configuration on G-ASKN was changed from a 22-seat single cabin to a twin cabin 55-seat configuration with only three car-ferry positions. The aircraft was placed in dedicated service on the Ostend, Belgium, route which carried more supplemental passengers than car-ferry traffic.

The area under the ATL-98 nose is considerably larger than the standard rear fuselage. The raised ceiling modification behind the cockpit of the new nose tapers down to the rear half of the cargo hold. The control cables are housed in the cover at the center top and are visible in the twin boxed covers at the extreme rear (author's collection).

As the season progressed Aer Lingus exercised its option for a third conversion as more of a backup aircraft because of dependability problems primarily related to engine failures. Carvair 14, EI-ANJ, first flew on 17 April 1964 and was delivered to Aer Lingus on 24 April after only four hours and 22 minutes of tests flights. It is the second production ship with raised ceiling and the first built with the Rolamat floor system for ease in loading palletized cargo.

After the raised ceiling design change in 1963, the nose production at Southend changed to incorporate the raised ceiling. All nose sections built from October 1960 to the fall of 1963 had the wasted space behind the cockpit because of the straight control cable runs. Determining the aircraft fitted with those early built nose units is quite confusing. The first nose unit for Carvair number one was fabricated in two pieces. All subsequent units were built as complete units in groups of four to speed production. Because it was unknown how many Carvairs would be built, the plan should have worked until the production order numbers were firm. However, multiple orders for the DC-4 Carvair never materialized because many interested airlines decided to wait for the more efficient proposed turboprop DC-7 version. Consequently the last block of nose sections was not completed.

Ultimately the lack of orders and limited production created six groups of noses which would have equaled 24 units plus the longitudinally split first nose for a total of twenty-five. In reality only 23 nose units were built including the two-piece on Carvair number one. Three blocks of four plus one (13 total) were built with the standard control runs before the raised ceiling design change. The units with standard control runs were used on Carvairs two through twelve leaving two surplus low ceiling nose units. After the design change for Aer Lingus two blocks of four plus one nose (9 aircraft) were built with high ceiling. Nose production was halted at that time because of the lack of aircraft orders. In order to use up the two remaining standard noses left over, they were grafted to ships 15 and 17 which were built on speculation without a specific need for a high ceiling. Carvair 15 took nearly 2.5 years to complete because it was only worked on when craftsman lacked other work. Carvair 17 was completed and stored without out engines and instruments for four years. It was removed from storage and completed for BAF in 1968 long after the assembly line at Stansted had been shut down. It first flew five months after the last Carvair, c/n 21, had been completed for Ansett and flown from Southend.

All Carvair nose assemblies were fabricated at Southend, England, and loaded on a special transporter for delivery by road to Stansted. They were grafted to the stripped down overhauled airframes with the DC-4 noses removed. All but three of the Carvairs were built at Stansted. Ships 1, 11 and 21 were converted at Southend (Aviation Traders, author's collection).

Carvair 16 is the first of two conversions built for Iberia (Aviaco) with raised ceilings. The carrier supplied two passenger DC-4s which were being phased out for newer equipment. After Carvair conversion they were delivered to associate company Aviaco. The last ATL-98s (18–20) constructed at Stansted were fitted with the raised ceilings noses. This left three unused high ceiling nose units stored at Stansted. When Ansett ordered a third conversion in 1968 long after the assembly line had been closed, the three high ceiling nose units were moved back to Southend. ATEL was still anticipating the remaining pair would be eventually used. After one was used for Ansett Carvair 21, the other two units were placed in storage. Parts from one of the stored units was used in 1972 to rebuild an ex–Aer Lingus Carvair owned by Eastern Provincial Airways after it was damaged in eastern Canada. The remaining unused nose was eventually scrapped. A total of nine aircraft were built with the raised ceiling option, seven on the production line and the two retro-modified for Aer Lingus.

Cabin Seating and Cargo Configuration

The Carvair was conceived as a car-ferry combi aircraft. It was initially designed to transport five-cars and 23-passengers. The capacity of the completed aircraft was reduced to 22-passengers. Unlike the Bristol predecessor the Carvair was fitted with a galley and lavatory for passenger comfort. The order by Intercontinental—Interocean for two custom built cargo transports prompted

ATEL to offer several additional configurations in addition to the standard car-ferry. A cargo version was designed for routine scheduled cargo service. Removal of the passenger cabin produced an 80-foot cargo compartment which was well suited to transport military rockets. This produced a second long-range cargo version with additional fuel tanks for government and military contracts.

Multiple seating configurations were offered to meet the needs of service in different markets. The three Aer Lingus aircraft were ordered with a four-car 34-seat configuration. They were fitted with the standard 22-seat rear cabin with an adjacent 12-seat separate compartment in the fifth car position. All car-ferry aircraft had the rear 17–22 seat cabin. Configurations with more than 22 seats had a separate cabin forward of the rear cabin with the additional seats. To adjust to changing bookings and market changes, ATEL created a moveable forward bulkhead in the forward passenger cabin. The bulkhead could be repositioned for a 34-seat four car or a 55-seat three car configuration with 33-seats in forward cabin. Padded snap in liners were created to prevent damage to the cabin walls when the bulkhead was moved rearward to transport more than three cars or bulk cargo. British United began using the 55-seat three-car configuration in 1964 for flights deep into mainland Europe. At least one aircraft was set up for 65-seats and bulk cargo with no car positions. Conversely, Ansett ordered three pure freighters without rear windows or passenger cabin. They are the only ATL-98s never fitted with passenger cabins.

British United experimented with a 55-seat three-car configuration for service deep into mainland Europe. The seating was separated into two cabins. The forward cabin had 33 seats. The door on the forward bulkhead allowed the flight attendant access to the cargo hold. It was her responsibility after takeoff to inspect the cars for leaking gasoline and verify they were securely lashed down. She would then report an all clear to the cockpit via intercom (Aviation Traders, author's collection).

Worldwide Marketing

In 1965, four years after the first ATL-98 was built only 18 aircraft had been completed. Two were built for Interocean, three for Aer Lingus and two for Iberia. The balance had been built for British United on a rather delayed schedule and yet the sales staff at Aviation Traders continued to believe more orders from outside the sister company were forthcoming. Earlier in production a demonstration tour was conducted through southern Asia and the Pacific. Indications are the tour was instrumental in Australia's Ansett Airlines decision to place an order for two Carvair freighters in early 1965. The order was more of a stopgap measure for Ansett to extract more time and utilization from the carriers DC-4 passenger aircraft. Converting them to Carvair cargo-liners would extend the life of the aircraft until economic conditions were right to upgrade to Lockheed L-188 Electra freighters. Up until this time all Carvair operators were in Europe with some charter work in the Middle East and occasional British Ministry of Defense flights to Australia.

The Royal Aeronautical Society conducted seminars in Asia and the South Pacific in an effort to generate sales to foreign carriers worldwide. Interest was high but no orders ever materialized. Proposals were also made to possibly lease Carvairs in Asia during the English winter months. In the off season the aircraft were relegated to ad-hoc operations transporting cargo. It was argued leasing them out would result in better utilization of the aircraft and better amortize the production cost. Possibly some of the carriers would realize the versatility of the ATL-98 and place orders. Aviation Traders was so confident Asian sales were forthcoming a production contingency plan was set up with Hong Kong Aircraft Engineering (HAECO) to produce the aircraft under contract. In addition plans were formed for HAECO to produce component kits. This would allow airlines with major overhaul facilities to convert DC-4s to Carvair standards in-house and reduce conversion cost and time. Ultimately Carvair orders never materialized in Asia or the Southern Hemisphere and HAECO never produced any ATL-98 components.[23]

Interestingly the Aviation Traders idea to produce component kits to modify aircraft to guppy configuration was successful years later by another aircraft builder. After Airbus found success with the first Aero Spacelines B-377 SGT-201s in France, Aero Spacelines was contacted with a request to produce component kits to build additional SGT-201 Super Guppies in France. A contract was signed in 1979 for the California based guppy builder to produced B-377 SGT-201 kit components at Santa Barbara for shipment to France. Ultimately only two Super Guppies were constructed by UTA for Airbus.

All ATL-98 enlarged forward nose sections were fabricated at Southend. All but the two were transported by road from Southend to the assembly line at Stansted. Three aircraft were built at Southend. The first Carvair, G-ANYB which was the operational prototype was built with a two piece nose split horizontally. The eleventh conversion, G-APNH, was the second Carvair built at Southend as a highly modified long-range aircraft with many special options for demonstration and sales promotion. It was built as a project to provide work for craftsmen at Southend who were idle. Carvair sales were very slow consequently workers were not needed on the assembly line at Stansted. They were bused to Southend to work on special projects which included Carvair modifications.

The third ATL-98 built at Southend was the final conversion. Carvair twenty-one was ordered by Ansett long after the assembly line at Stansted had been shut down in 1967. It was converted at Southend as more of an afterthought with left over parts which made it different from the other ships. There were three nose sections stored at Stansted. The last ATL-98 is the only Carvair to have its nose section moved from Southend to Stansted and back again without the rest of the airframe.

British Air Ferries/ Transmeridian

British Air Ferries marketing became concerned about the future of car-ferries in 1969. Car-ferry bookings had diminished from 29 percent of the airline's revenue in the mid–1960s to only 5 percent in 1970. At the end of the summer 1971 travel season T.D. (Mike) Keegan tendered an offer to Air Holdings to purchase British Air Ferries (BAF). The offer was accepted making BAF a subsidiary of Keegan's Transmeridian Air Cargo (TMAC) on 01 October. Keegan believed the car-ferry service still had some revenue producing life left. He had competed in both car-ferry and the bloodstock charter business years before as a co-founder of BKS Air Transport. BKS Transport and Channel Airways (not to be confused with Channel Air Bridge) were part of the reason Aer Lingus purchased the ATL-98 Carvair.

Keegan developed a new marketing plan for BAF with an image change and different livery. With fewer cars being transported and more bulk cargo the remaining aircraft were reconfigured to high density seating of 40, 55 or 65 seats with moveable bulkheads to meet changing markets and individual bookings. British aviation authorities assumed BAF would be absorbed into Transmeridian. However, Keegan operated BAF and TMAC as separate companies while jointly advertising and promoting the cargo operation as if it were one company.[24]

The purchase of BAF was only part of a major expansion program for Transmeridian. Keegan acquired the first of nine long-range Canadair CL-44 cargo-liners in 1969. During the same period Jack Conroy was seeking to start a new operation after leaving Aero Spacelines in 1967. In 1970, Conroy formed Santa Barbara based Conroy Aviation for a number of aircraft projects. His primary focus was to produce a new series of outsize CL-44 guppy aircraft. Conroy managed to

G-ASDC is the first of two BAF ATL-98s converted to a cargo-liner while still in first-line service. It was eventually acquired along with two others by U.S. cargo carrier Falcon Airways of Texas. It passed through multiple owners until finding work with Frank Moss servicing mines in Alaska. It was lost on 28 June 1977 when a fire in engine number two forced a crash landing in the Chandalar River near Venetie, Alaska (courtesy Paul J. Hooper).

acquire three Flying Tiger CL-44s from Robert Prescott. He expanded the fuselage of a single aircraft (N447T) to guppy standards creating the Conroy 103. One of the other CL-44s had crashed placing Conroy in financial difficulty. In the summer of 1970 Keegan worked out a three way lease agreement with Jack Conroy and Robert Prescott of Flying Tiger to obtain the CL-44 guppy aircraft for Transmeridian. Eventually Keegan was operating nine CL-44s, the Conroy 103 Skymonster and the remaining BAF Carvairs.

The Carvair car-ferry service continued from Southend until January 1975 when it was announced all passenger car-ferry service would be phased out. Keegan announced to the press the Carvair would no longer transport passengers but would be retained in cargo service. Within the month the rear cabin was removed from G-ASDC c/n 7 and it was converted to an all cargo configuration. In July 1977 G-ASHZ c/n 9 was also converted to an all cargo configuration.[25] After years of British-European car-ferry service the remaining Carvairs were sold off to second level carriers eventually serving throughout the world as cargo-liners.

ATL-98 Carvair Production Summary

Car-Ferry Guppy Individual History

Each of the 21 ATL-98 Carvairs has its own story. While the majority were in British airlines service, some of them were operated on missions to many remote areas of the world. By the time the Carvair was passed on to third and fourth level carriers the fleet had been on every continent except for Antarctica. The Carvair was a work horse living up to the reputation of its C-54 wartime predecessor by transporting virtually anything which would fit through the door. The Carvairs which survived were virtually worn out when they were finally withdrawn from service. Ten were broken up and seven crashed. Two were lost in war, one to explosives and the other was abandoned to enemy forces. Two still survive in 2017, one in Texas and the other in South Africa. The following brief accounting of each airframe demonstrates how these 21 aircraft virtually covered to globe as DC-4s and again after Carvair conversion.

Carvair c/n 1, G-ANYB

The first ATL-98 was built from a C-54B which was delivered to the Air Force on 22 January 1945. It was returned to Douglas in 1946 for conversion to DC-4 commercial airliner standards. Braniff Airways purchased it for South American service. It was fitted with a JATO system for high altitude service for routes to La Paz, Bolivia and other similar high altitude destinations. As a DC-4 it passed through several carries before being purchased by Aviation Traders for ATL-98 conversion. It first flew as Carvair on 21 June 1961 as G-ANYB going into service with Channel Air Bridge. The company became British United Air Ferries in January 1963. The first Carvair was retired in March 1967 and placed in flyable storage at Lydd, Kent, England. It never returned to service and was allowed to deteriorate until broken up in August 1970.[26]

Carvair c/n 2, G-ARSD

The second ATL-98 was built from a C-54A which was delivered to the Air Force on 27 May 1944. In September 1949 it was declared surplus and sold to California Eastern Airlines where it saw Pacific service and DEW Line contract flights. It was leased to Orvis Nelson's Transocean to fly military contract service in the Pacific. Aviation Traders acquired it in June 1961 for ATL-98 conversion. It first flew as a Carvair on 25 March 1962 and went into service with Channel Air Bridge and subsequently passed to British United Air Ferries and British United Airways. Four days after the company became British Air Ferries it was withdrawn from service and stored. It was broken up on 26 August 1970 at Lydd, Kent, England.[27]

Carvair c/n 3, G-ARSF

The third ATL-98 was built from a C-54B which was delivered to the Air Force on 10 July 1944. It was declared surplus in 1945 and sold to Braniff Airways and upgraded to DC-4 standards. It was leased to Northwest Airlines then subleased to Orvis Nelson's Transocean Airlines for military contract flying in the Pacific. The DC-4 was eventually purchased by British carrier East Anglian Flying Services for Carvair conversion. The carrier cancelled the order and it was acquired by Aviation Traders in June 1961 for ATL-98 conversion. It first flew as a Carvair on 28 June 1962.

It is the shortest lived Carvair. After only six months of service it crashed on the morning of 28 December 1962 at Rotterdam, Netherlands. The field had experienced heavy snow which covered the field and surrounding area leaving no definition. The aircraft passed over the field in a white-out before attempting to land. Because of heavy clouds and low visibility the aircraft approached below the normal ILS glide path landing short clipping a dike. One of the wings contacted the ground cart-wheeling the aircraft. One wing broke off with the aircraft rolling upside down. The cockpit collapsed down to the top of the forward panel. The Captain perished and the co-pilot was seriously injured. The passengers escaped with only scrapes and bruises. The aircraft was broken up and salvaged back to Southend, England.[28]

Carvair c/n 4, G-41-2 (N9758F)

The fourth ATL-98 was built from a C-54A which was delivered to the Air Force on 30 June 1944. Braniff Airways purchased it in November 1945 for conversion to DC-4 standards. It was also fitted with a JATO system for service to high altitude fields in South America. Channel Air Bridge acquired it for Carvair conversion on 06 June 1961. In order for Aviation Traders to secure a sale to a foreign air carrier the delivery position was switched. It was special built with many special options without the passenger cabin for Intercontinental—Interocean Airways. The aircraft was ordered specifically for a United Nations contract in the Congo. After the mission was completed it was returned to Aviation Traders for overhaul and conversion to car-ferry standards with a passenger cabin. After an unsuccessful car-ferry and charter service out of Luxembourg, it was sold off to French carrier Cie Air Transport. While operating a charter flight it crashed at Karachi, Pakistan, 08 March 1967. The aircraft took off overweight in high temperature and humidity. Just after becoming airborne it suffered a catastrophic engine failure crashing on to a road in a populated area. It was one of the worst Carvair losses. Four crew members perished along with seven on the ground with 15 civilians suffering serious injuries.[29]

Carvair c/n 5, N9757F

Carvair Five was built from a C-54A which was transferred to the Navy as a R5D-1 on 03 August 1944. On 10 April 1946 it was declared surplus and purchased by postwar startup Veterans Air Express. It was transferred to Matson Aviation Services for conversion to DC-4 standards. The carrier defaulted on payment and it was acquired by Matson Airlines & Steamship Company and placed in Hawaii service. Matson was forced to drop the service after the CAB granted route authority to Northwest and United Air lines. In 1956 Matson reviewed purchasing Orvis Nelson's Transocean to reinstitute Pacific service with the DC-4. It transited multiple carriers before being acquired by Aviation Traders on 27 December 1961. It became the second cargo-liner built for Interocean. It possibly transited more carriers as a DC-4 and ATL-98 than any other Carvair. It was purchased from BAF in April 1979 by Ruth May in Dallas, Texas, and registered in America as N83FA for lease to Falcon Airways. It transited several more owners before being acquired by Bob McSwiggan's Custom Air Service for ad hoc charter work based in Griffin, Georgia. The primary charters were transporting automobile parts and aircraft spare engine AOG flights.

On the night of 03 April 1997 N83FA crashed on a positioning flight from Griffin when the crew attempted to take-off with the control surfaces locked. Tragically the two man crew perished.[30]

Carvair c/n 6, G-ARZV (EI-AMP)

Carvair Six was built from a C-54A which was delivered to the Air Force on 11 April 1944. It was sold to American Airlines on 10 April 1946 and converted to DC-4 commercial configuration. The DC-4 passed through multiple carriers before being purchased by Channel Air Bridge in May 1962. It was transferred to Aviation Traders and first flew as a Carvair in bare metal from Stansted to Southend on 21 December 1962. Channel Air Bridge had given up the delivery position in order for Aviation Traders to make a sale to Aer Lingus. It became the first ATL-98 to have the raised ceiling and Rolamat floor system for palletized cargo loading which were retrofitted by Aviation Traders after the first season of service. When phased out by Aer Lingus it was acquired by Eastern Provincial for work in Canada supporting a massive hydroelectric project. It crashed on 28 September 1968 while attempting to land on a gravel field at Twins Falls, Labrador, Canada. The crew and all passengers survived without serious injuries. The aircraft was scrapped on the scene.[31]

Carvair c/n 7 G-ASDC

The seventh ATL-98 was built from a C-54A which was delivered to the Air Force on 06 November 1943. On 03 May 1946 it was transferred to the War Assets Administration for disposal. It was purchased by California Eastern Airlines and upgraded to DC-4 commercial standards. The DC-4 transited several carriers before being acquired by Interocean for operation in the Congo. It was traded in to Aviation Traders as partial payment for Carvairs four and five. After conversion it first flew as a Carvair on 19 March 1963 going into service with British Untied Air Ferries. It appeared in the classic James Bond movie *Goldfinger* and was eventually converted to a cargo-liner in 1975. Falcon Airways of Dallas, Texas acquire it with Carvair nine in April 1979.[32]

It transited multiple carriers flying in Hawaii and eventually flying under Hondu Carib in Alaska. Captain Frank Moss was in command on 28 June 1997 when departing the gold mine operation at Venetie, Alaska. The number two engine suffered a catastrophic failure and caught fire on climb out. The engine mounting became engulfed in flames and the engine separated from the wing falling into a wooded area and starting a forest fire. Moss stated it was not possible to make it back to the gold mine airstrip. He was able to crash-land on a sandbar in the Chandalar River. All three crew members escaped with minor injuries. The aircraft was destroyed.[33]

Carvair c/n 8, EI-AMR

The eighth ATL-98 was built from a C-54B which was delivered to the Air Force on 15 November 1944. The C-54 was built in an airline configuration as a military staff transport. It was one of the few C-54s fitted with single cabin entry door and no cargo hoist. The Air Force maintained it in active military service until 1950 when it was sold to Twentieth Century Airlines. It passed through multiple charter operators before being acquired by Aviation Traders on 23 October 1962. Interestingly, it had to be retrofitted with the double cargo door during conversion. It first flew as a Carvair from Stansted on 18 April 1962. In the winter of 1963–64 it was returned to Aviation Traders along with Carvair six to be retrofitted with the raised cargo compartment ceiling and Rolamat floor system. Eastern Provincial purchased it along with the other two Aer Lingus Carvairs in 1968 for work in Eastern Canada. It suffered major damage on 03 May 1972 at Gander, Newfoundland, when the nose gear collapsed. It was repaired with parts from one of the surplus nose sections stored by Aviation Traders at Southend. In 1973 it was sold to a Norwegian operator and ferried back to ATEL at Southend for overhaul and upgraded to Wright R-2600 engines. The charges were never paid and it remained in storage until broken up on 13 September 1978.[34]

Carvair c/n 9, G-ASHZ

The ninth ATL-98 was built from a C-54B which was delivered to the Air Force on 11 January 1945. After only nine months, it was transferred to the Reconstruction Finance Corporation on 15 November 1945 for disposal. It was returned to Douglas for conversion to DC-4 commercial standards and sold to Western Airlines. After 10 years with Western Airlines it went to carriers in Mexico and Panama before being acquired by Aviation traders on 24 October 1962 for ATL-98 conversion. It went into British car-ferry service in 1963 where it remained until June 1976. It was placed up for sale. After no buyers came forward it was leased out in Africa transporting building materials for a large French construction company and returned to Southend after the lease. In 1979 it was purchased with Carvair seven by Falcon Airways of Dallas, Texas. After Falcon ceased operations, it passed through several other owners including Jim Blumenthal, who painted it in the colors it still wears today. It was purchased by Custom Air Service of Griffin, Georgia, for ad hoc charter work but stored for nearly ten years. It was brought back in service in 1997. In 2016 it was based at the old Perrin Air Force Base at Dennison near Sherman, Texas. It is in the better condition of the last two surviving ATL-98s in 2017.[35]

Carvair c/n 10 G-ASKG

Carvair 10 was built from a C-54A which was delivered to the Air Force on 24 August 1944. It was transferred to the Reconstruction Finance Corporation for disposal and eventually converted to DC-4 commercial standards but was not immediately sold. It was purchased by Norwegian charter carrier Braathens SAFE on 13 March 1947. Seaboard & Western acquired it in October 1950 for conversion back to cargo configuration to operate MATS charters in the Pacific. Interocean acquired it in 1961and traded it to Aviation Traders as partial payment for Carvair four and five. It first flew as a Carvair on 29 July 1963. British United was set to take delivery, however, it was leased out for a short time to Italian carrier Alisud. After return to BUAF it continued in car-ferry service until 1969 when it was sold to French carrier Transports Aeriens Reunis. Eventually a construction company based in France purchased it to transport building supplies in Africa. A small African operator purchased it for oil field service work. It was withdrawn from service in October 1989 in need of overhaul. The engines were periodically run up to possibly put it back in service. Éclair purchased it in 1993 and brought back to flying status to transport tobacco products. On 05 September 1995 it was destroyed with explosives under unknown circumstances because of military conflict.[36]

Carvair c/n 11, G-APNH

The eleventh ATL-98 was built from a C-54B which was delivered to the Air Force on 22 June 1944. After six months it was returned to Douglas for conversion to DC-4 commercial standards. Delta Airlines purchased it on 27 April 1946 operating it for eight years. North American Airlines acquired it in 1954 for charter work. The DC-4 passed through multiple American and European carries until acquired by Freddie Laker in 1960 when it was transferred to Aviation Traders on 01 February 1963 for ATL-98 conversion. There were no immediate orders prompting the decision to build it as a long-range cargo transport. It first flew 04 January 1965 after two years in construction. The British government chartered it for evaluation on worldwide military supply missions and to transport rockets to Australia. The titles were changed to Air Ferry in 1965 for service during the Rhodesian U.D.I. in Africa. After the mission it was returned to British Air Ferries. It operated a number of recovery missions transporting damaged aircraft. On 18 March 1971 in an attempt to land in a strong crosswind at Le Touquet, France, G-APNH crash landed sustaining serious damage. Engineers accessed the damage and considered it not cost effective to repair. It was broken up for parts and transported back to England.[37]

Carvair c/n 12, G-AOFW

The twelfth ATL-98 was built from a C-54A which was delivered to the Air Force on 15 July 1944. It was transferred to the Reconstruction Finance Corporation on 17 June 1946 and returned to Douglas for upgrade to DC-4 commercial standards. Pan American Airways acquire it on 15 March 1947. After six months of service it was damaged in a landing in Alaska. It was repaired and sold to Twentieth Century Airlines. The DC-4 was acquired in 1950 by Alitalia for

European service. In 1954 it was acquired by California Eastern and returned to America for operation on MATS contracts in the Pacific. A year later in December 1955 it was acquired by Freddie Laker and returned to Europe. Air Charter Airline operated it for many years before conversion to ATL-98 in April 1963. It first flew as a Carvair on 11 February 1964. It was slated for service with British United but it was leased for seven months to Spanish carrier Aviaco. After the lease it returned to service with British United flying for nearly 12 years. It made the last British car-ferry flight in January 1977 and was withdrawn from service and placed in open storage. Multiple groups wanted it preserved as a historical reminder of car-ferry days and several BAF captains made unsuccessful purchase offers. It was placed on display at Southend and allowed to deteriorate ultimately being broken up in December 1983.[38]

Carvair c/n 13, G-ASKN

The thirteenth ATL-98 was built from a C-54-DO which was delivered to the Air Force on 08 February 1943. It is the ninth DC-4 built and the oldest airframe converted to Carvair standards. After three years with the Air Force, it was transferred to the Civil Aeronautics Administration (CAA). Eastern Airlines acquired it in1951operating it in scheduled service until May 1955. It passed through multiple carriers until acquired by ATEL in a default sale in July 1963. It is the first assembly line Carvair to be built with the raised cargo bay ceiling flying as a Carvair on 08 February 1964. After 12 years with British Air Ferries, it was parked at Southend on 03 June 1976. An international construction company purchased it to transport building materials from France to Central Africa based out of Libreville, Gabon. In 1977 it was impounded at Brazzaville, Congo, and in 1979 it was confiscated by the Empire of Central African government for non-payment of servicing contracts. African operator Aero Services purchased it in 1982. It was scrapped at Brazzaville in 1986 for lack of spare parts.[39]

Carvair c/n 14, EI-ANJ

The fourteenth ATL-98 was built from a C-54B which was delivered to the Air Force 27 November 1944. It was sold back to Douglas in 1946 for conversion to DC-4 commercial standards. Western Airlines flew it for nine years before selling it to SOBELAIR (Sabena) on 12 January 1956. It passed through several carriers until purchased by ATEL on 09 July 1963. Carvair conversion work began for British United but it was completed for Aer Lingus and first flew on 17 April 1964. After Irish service, it was acquired by Eastern Provincial in February 1968. After service in northern Canada it was purchased by a small Norwegian contract carrier and leased back to British Air Ferries in October 1971. The Norwegians then operated Red Cross contract relief missions to war torn Cambodia. While operating in Red Cross markings LN-NAA was grounded at Bangkok for monies owed and allowed to deteriorate. Unverified sources from an engine salvage company reported it was broken up in 1987–88.[40]

Carvair c/n 15, G-ATRV

The fifteenth ATL-98 was built from a C-54E with a quick change interior for cargo or passengers. The C-54 was delivered to the Air Force 05 April 1945 then transferred to the Reconstruction Finance Corporation and returned to Douglas for conversion to DC-4 commercial standards. Pan American Airways purchased it on 14 November 1945. After 16 years with Pan Am, it was acquired by a Beirut-based cargo carrier in 1961. After two years of cargo service in the Middle East with minimum maintenance, it was in poor condition. Aviation Traders acquired it in November 1963. Conversion was started but with no orders it was stored partially complete. After two and a half years it was finished first flying as a Carvair on 23 March 1966. British United flew it for a year before selling it to French carrier Cie Air Transport to replace Carvair Four which was lost at Karachi. Cie leased it out for oil field work in Nigeria. In August 1970 it was grounded in France where it remained until November 1972. British Air Ferries purchased it for spares and it was broken up and returned to England.[41]

Carvair c/n 16, EC-AXI

The sixteenth ATL-98 was built from a C-54B which was delivered to the Air Force on 23 December 1944. It was surplused in 1946 and purchased by American Airlines and converted to DC-4 commercial standards. In 1949 Spanish carrier Iberia acquired it from American. On 15 February 1964 it was flown to Stansted for conversion for Iberia subcompany Aviaco. It flew as a Carvair on 04 June 1964. After four years of service in 1968 Aviaco leased it to Caribbean carrier Dominicana as part of an economic aid program for struggling Latin countries. On 23 June 1969 while operating a return flight from Miami to Santo Domingo the Dominican Carvair suffered a catastrophic engine failure on takeoff from Miami. The overweight aircraft was flown by an inexperienced captain who pulled back the wrong engine while attempting to return to the field. It crashed into a densely populated area of Miami. The aircraft carrying a max fuel load took out an entire city block. The crew of three and a cockpit passenger perished along with six civilians on the ground. There were multiple injuries on the ground with twelve very serious.[42]

Carvair c/n 17, G-AXAI

The seventeenth ATL-98 was built from a C-54B which was delivered to the Air Force on 19 July 1944. It was transferred to the Reconstruction Finance Corporation in October 1945 for return to Douglas for upgrade to DC-4 commercial standards. United Airlines purchased it on 12 February 1946 for commercial service. United operated it in passenger configuration until it was replaced by newer equipment. The DC-4 was converted to a cargo-liner for the balance of United Air lines service. It was acquired by Orvis Nelson's Transocean in December 1956 but was only in service two years before being leased to Lufthansa. United Air lines purchased it back in 1959 and re-sold it to the Babb Company. It was sold to New York based Intercontinental (Interocean). Aviation Traders acquired it in February 1964. It was converted to ATL-98 and placed in storage at Stansted less engines and instruments. After four years the decision was made in December 1968 to complete the conversion. It first flew as a Carvair on 02 April 1969 and went into service with British Air Ferries. In December 1975 it was purchased by French carrier SF Air for operations out of Nice, France. It was subsequently acquired by American operator Aero Union as a support aircraft for a fleet of fire suppression tanker aircraft. The Carvair transited multiple U.S. small operators before being grounded at Naples, Florida. Custom Air Service which owned Carvair c/n 5 and c/n 9 purchased it in June 1992 expecting to put it in service. It was found to have extreme corrosion and was broken up for spares and transported back to Griffin, Georgia, where the cockpit remains.[43]

Carvair c/n 18, EC-AZA

The eighteenth ATL-98 was built from a C-54B which was delivered to the Air Force on 12 July 1944. It is the sister ship to the C-54 which became Carvair three. The two had consecutive C-54 construction numbers (c/n 18339–18340) and U.S Air Force serial numbers (s/n 43-17139 and 43-17140). It is quite interesting when considering the two aircraft went to different carriers after military service then years later both were brought back together and converted to ATL-98s. The USAAF declared the C-54 surplus in 1944. It was upgraded to DC-4 commercial standards and sold to American Airlines in 1945. Spanish carrier Iberia purchased it in 1949. It was transferred to Aviation Traders on 05 November 1964 for ATL-98 conversion for Iberia sub-company Aviaco. It first flew as a Carvair on 12 March 1965. After only four years of service it was leased to Dominicana along with Carvair 16 as part of a Latin economic aid program. After the crash of Dominicana Carvair c/n 16, the company experienced serious morale problems with crews not wanting to fly the Carvair. It was withdrawn from service on 12 January 1974 and placed in flyable storage. In November 1978 it was disassembled and moved to downtown Santo Domingo, Dominican Republic, to be converted into a restaurant. It never opened and became a piano bar and lounge.[44]

Carvair c/n 19, VH-INJ

The nineteenth ATL-98 was built from a postwar DC-4-1009 commercial airliner which never saw military service. It was delivered to Scandinavian carrier SILA on 17 May 1946 and subsequently transferred to SAS. It remained in Scandinavian service until January 1954 when it was purchased by Japan Airlines. Ansett Australian purchased it in August 1963. In May 1965 it was ferried to Aviation Traders at Stansted, England, for Carvair cargo-liner conversion. Because it was built as a postwar passenger airliner it had a single cabin entry door which had to be changed to the double side cargo door. It flew as an ATL-98 cargo-liner on 14 September 1965. It was only tested for nine days before making the 50 hour flight home to Australia. After seven years with Ansett, it was withdrawn from service in June 1972 and placed up f or sale. On 23 June 1973 it was purchased by South East Asia Air Transport (SEAAT) and subsequently leased to Cambodian carrier Air Cambodge. It operated out of Phnom Penh under war zone conditions until 13 April 1975. A week earlier it had arrived in Phnom Penh which was under siege with one engine out and a critical load of supplies. Because of mortar fire and snipers crews were unable to change the engine. When the field was overrun by rebels it was left behind. In the 1980s it was reported parked off the side of the field with locals living in it. There were unconfirmed reports of it still being there as late as 1997. The final fate is unknown.[45]

Carvair c/n 20, VH-INK

The twentieth ATL-98 is the second of only two Carvairs built from a postwar DC-4-1009 commercial airliner with a single cabin entry door. It is the latest model DC-4 airframe used for ATL-98 conversion. The DC-4 was delivered to a Norwegian carrier on 24 June 1946 which was later merged into SAS. On 27 October 1956 it was acquired by Japan Airlines and leased for a short time to Korean Airlines. The DC-4 was declared surplus in February 1964 and acquired by Ansett and converted in Hong Kong to a cargo-liner. The DC-4 was operated for about a year before ferrying in June 1965 to Aviation Traders at Stansted for ATL-98 conversion. It was the last assembly line Carvair built first flying after conversion on 27 October 1965. After testing it departed Southend in November 1965 for the 50 hour return to Australia. After nearly eight years Ansett withdrew it from service in 1973 and placed up for sale. It transited multiple owners and operators in Australia, New Zealand and Indonesia before arriving in Hawaii in 1982. After a failed Hawaiian cargo service it was purchased in 1996 for relief work in the Congo. The Carvair had been parked in open storage at Honolulu for quite some time. It was made airworthy and ferried to Griffin, Georgia, for additional work by Custom Air Service which was the only Carvair operator and service company. Because of mechanical problems in Florida it returned to Griffin, Georgia, and the sale was switched to Carvair 21. Canadian contract operator Hawk Air based at Terrace, British Columbia, Canada, purchased Carvair 20 for work servicing gold mines in remote Alaska. On 29 May 2007 it was destroyed on landing at Nixon Fork, Alaska, during a test flight by new owner Brooks Fuel. The aircraft was forced to land downwind on a field that only has one approach. The aircraft landed short when a wind shear slammed it into an embankment at the threshold. The two man crew was not seriously injured. The aircraft was destroyed.[46]

Carvair c/n 21, VH-INM

The twenty-first ATL-98 was built from a C-54E which was delivered to the Air Force on 07 April 1945. On 30 August it was transferred to the Reconstruction Finance Corporation for disposal. Douglas Aircraft purchased it back and upgraded it to DC-4 commercial standards. It was purchased by Pan American Airways in November 1945 and was operated until February 1958 when it was sold to Japan Airlines. Ansett Australia acquired it in March 1965 operating cargo service for three years. The decision was made to have it converted to Carvair standards in 1968. In March it was ferried to Aviation Traders at Southend because the Stansted production line had been shut down several years earlier. The last Carvair (c/n 21) was built using one of the three extra nose units left over in storage. It flew as the last ATL-98 on 12 July 1968 and retained the DC-4 registration VH-INM. After testing it departed Southend on 19 July for a 50 hour return to Australia. At Damascus, Syria, a cockpit window cracked and blew out. After some delay a new window was located and installed. Several days later the fire warning system malfunctioned at Bangkok. The last ATL-98 arrived home in Australia on 26 July 1968 after 168 hours en route.

It was withdrawn from service in February 1973 after five years and placed in storage. Ansett sold it to Australian Aircraft Sales on 01 January 1974. The aircraft passed through at least six owners or operators before being operated in Hawaii. It along with Carvair c/n 20 was acquired by an aircraft broker for unspecified work in Africa. After Carvair 20 suffered mechanical problems in Florida en route to the African buyer it was substituted for the sale. The details of the new owner are vague and the Congo relief program collapsed leaving the last Carvair grounded in South Africa. In March 2002 it was acquired by DC-4 operator Phoebus Apollo based in South Africa. As of 2017 it was grounded in non flyable condition with an uncertain future.[47]

The End of DC-4/ATL-98 Carvair Era

The DC-4/C-54 are a unique aircraft with multiple operational lives. America's first four-engine transport began as a dream in the 1930s for a long-range commercial airliner. Shortly after production began it was transformed into a war machine reverting back to a commercial airliner years later when hostilities ended. The DC-4 was operated under the most adverse conditions and proven to be an outstanding design. It fostered the more powerful and stretched DC-6 and DC-7 versions and inspired the giant C-74 Globemaster I. When the DC-4 commercial life ended a handful of airframes were transformed into a new type ATL-98 Carvair guppy to repeat the cycle.

There are noted parallels between the Aviation Traders founder Freddie Laker and Aero Spacelines founder Jack Conroy. Both Laker and Conroy believed a commercial market existed for outsize volumetric transports. Laker had been awarded British government contracts which were operated with standard type aircraft. At his direction, a long-range ATL-98 (G-APNH) was built on speculation to acquire Ministry of Defense contracts transporting military rockets. Almost immediately a more powerful and efficient stretched turboprop DC-7 or Carvair 7 conversion was proposed. By comparison Conroy built the B-377 Pregnant Guppy to obtain government (NASA) contracts to transport rocket stages. Before his first B-377 guppy was completed he was designing a larger turboprop version Super Guppy and planning a commercial version which became the Mini Guppy. Both men were developing completely different guppy designs in the same time period independent of each other and yet they were seeking similar markets on different continents.

Freddie Laker was a man of ideas and action who became a British aviation legend. He created the ATL-98 Carvair, a purpose built aircraft for a niche market where it served well. The design even got the attention of Boeing. Laker also attempted to replace the Douglas DC-3 by building the ATL-90 Accountant, a small twin turboprop which was not well accepted. There was an alternate swing nose car-ferry version of the ATL-90. While Laker was attempting to replace the DC-3, Conroy converted two DC-3s to turboprops in an effort to extend the life of the aircraft.

Between the costly failure of the ATL-90 and the bureaucracy of the corporate structure of Air Holdings Laker became disenchanted. He resigned his position with Air Holdings in 1965 and went on to form Laker Airways and Skytrain, a trans–Atlantic low fare discount service. The company filed for bankruptcy in 1982 for multiple reasons. Laker moved to the Bahamas and reorganized operating a smaller version of Laker Airways out of the Bahamas until 2005.

Laker died on 06 February 2006 at the age of 83 in Hollywood, Florida.

The ATL-98 Carvair is a stunning tribute to Freddie Laker. Although it was produced initially for short distance car-ferry service it turned out to be a cargo workhorse. These 21 enlarged fuselage aircraft were eventually owned or operated by 75 different carriers and registered in 16 countries. The ATL-98 was built as a car-ferry aircraft for holiday travelers to mainland Europe. It had the ability to operate into short unimproved fields with an 18,000-pound payload. Only 21 were built and yet the Carvair has seen combat in the Congo, Laos and Cambodia transporting military equipment and extracting wounded personnel. Besides automobiles, it has transported royalty, gold shipments, celebrities, rockets, nuclear material, military vehicles, recovered aircraft, circus animals, rock groups, sports teams, a whale and oil drilling equipment. It has seen every weather condition the world offers from the Sahara desert to the Arctic. It has operated to the farthest and most remote places on earth as far south as Comodoro Rivadavia, Argentina, transporting a radio tower and as far north as Iceland and Umnak Island, Alaska, transporting reindeer. The only other expanded fuselage guppy which can match this achievement is the military C-124 Globemaster II which was built in large numbers for the United States Air Force for global military operations which include both the Arctic and Antarctic

A comprehensive review of the entire ATL-98 fleet with complete engineering details combined with in depth operational history of each individual aircraft is available by the author in the book *The ATL-98 Carvair.*

9

Boeing B-29 and B-50 Superfortress

A Revolutionary Concept

The Aero Spacelines guppy series of modified B-377 and C-97 airframes began with an idea for a special purpose volumetric aircraft to transport NASA outsize rocket boosters. John M. "Jack" Conroy had been shopping for surplus Boeing Stratocruisers to start a VIP airline to Hawaii when he was introduced to aircraft broker Lee Mansdorf. The possibility of him creating his own "Conroy Class" of giant aircraft was not envisioned and purely a chance circumstance. It was a moment in aviation history when Conroy gave up a dream of owning his own passenger airline to change conventional aircraft design forever. After being challenged with the idea by aircraft broker Lee Mansdorf, Conroy and a circle of friends were gathered at Johnny's Skytrails Restaurant adjacent to the airport at Van Nuys, California. The group of legendary pilots began tossing out preposterous suggestions on Conroy's idea of how to profit from the lack of NASA logistics and the availability of surplus Boeing 377 aircraft.

The commercial Boeing B-377 was plagued with problems throughout its short life in passenger service. The airlines found it to be expensive to operate and temperamental. It was the ultimate propeller driven aircraft for passengers providing the prewar romance of luxury travel to the postwar era. However, it was built to standards of a past era and could not compete with the jet transports which were coming on line. The very short life in commercial service placed them on the secondary market with considerable time left on the airframes. The B-377 was destined for scrap because it was not economically practical for commercial freighter conversions and it was too costly for passenger charter operators. By all measures it was not considered to be a successful airliner. Very few B-377 flew again after 1960. They were broken up and sections of the airframes used to construct giant specialized transports known today as guppies. The unique ultra-large transports flew for 30 years with the last example still flying for NASA in 2018.

The Pregnant (B-377PG) and Super Guppy (B-377SG) are the benchmarks of specialized outsize aircraft. Jack Conroy's first B-377 conversion was not originally named or had a model number. It was considered a modified B-377 and identified as a booster transport. The concept was considered pure folly when the project was first conceived. The first Aero Spacelines conversion was so successful it changed the perception of previous attempts at volumetric transports. Only eight Aero Spacelines designed aircraft in five different configurations were ever built. They set the standard and became the inspiration for all other giant transports and wide body airliners which have followed.

The first Aero Spacelines B-377 Pregnant Guppy conversion built from a Pan American Boeing B-377. The airframe was stretched by adding a 16-foot section from another B-377 behind the wing before expanding the upper fuselage. The separation joint is visible behind the wing, which allows the rear half to be rolled away for straight in-loading of NASA boosters (NASA, USAF).

Prior to the Aero Spacelines B-377 Pregnant Guppy, other expanded fuselage aircraft had been constructed using existing airframes or components from other aircraft. The XC-99 and C-124 had larger fuselage diameters than the base airframes from which they evolved but the interiors were small in comparison. Prior to the refining of Lee Mansdorf's outsize design, the mention of building an aircraft with a cargo hold 20 feet in diameter was not thought possible and often considered laughable. Jack Conroy introduced the era of guppy class transports in 1962. As revolutionary as the B-377PG was, it is small compared to the outsized aircraft built to support aerospace manufacturing today.

The first Aero Spacelines booster transport was a purpose built design for transporting the second stage S-IV of the Saturn Space Vehicle and related oversize cargo from the U.S. West Coast to Huntsville, Alabama, and Cape Canaveral, Florida. The B-377PG and all the non-military volumetric types which followed including the Airbus A300-600ST Beluga, Conroy 103 CL-44 and Boeing 747LCF are all purpose built for aerospace hardware and aircraft sub-assembly transport. The only enlarged fuselage exception is the previously profiled smaller ATL-98 Carvair which was built as a single purpose commercial airliner for Channel Air Bridge (British United Air Ferries). The Carvair was chartered on occasion to transport aerospace and military hardware. A single special long-range version was built in an attempt to gain British Ministry of Defense contracts to transport rockets.

The cost of creating a small number of giant aircraft from scratch is prohibitive. While aircraft manufacturers in 1962 were debating whether it was possible to transport boosters internally, Conroy was grafting together portions of existing surplus airframes which had reached the end of commercial service life. Engineering cost could be greatly reduced by mating a new outsize fuselage to the wings, gear, cockpit and empennage of an existing aircraft. The B-377PG guppy was built exclusively for oversize NASA space hardware. However, a presumed need in the civilian commercial market was a major factor in planning and designing a third generation smaller diameter B-377MG Mini Guppy. Conroy proposed a car-ferry B-377 guppy commercial cargo/passenger combination (combi) version as either a replacement or a competing design to the Freddie Laker ATL-98 Carvair car-ferry.

Reliability problems and high operating cost had contributed to the B-377 Stratocruiser demise. It is difficult to envision anyone considering surplus Stratocruisers to transport delicate space hardware. The evolution from Boeing B-29 to Super Guppy Turbine-201 is a unique story which began in 1939 with the first designs of a heavy bomber for the Army. The defeat of Germany in World War II combined with the B-29 action over Japan set the stage for an incredible chain of events which produced the need for an outsize transport aircraft.

It would have been inconceivable to predict a German rocket scientist who worked for the Nazis would one day direct the American space program which put a man on the moon. Even more unbelievable is the story of an American B-17 pilot who became a prisoner of war and eventually built an aircraft to transport rocket assemblies designed by the German scientist for the U.S. space program. Fate put rocket scientist Wernher von Braun and American bomber pilot Jack Conroy together. As a result of unplanned and unforeseen events, surplus Boeing airframes and the ingenuity of Conroy a unique outsize aircraft was created to keep NASA Apollo launches on schedule.

In an effort to simplify aircraft identity and for continuity the Aero Spacelines guppy aircraft are identified throughout the text as the following: Pregnant Guppy as PG or B-377PG, Super Guppy as SG or B-377SG, Mini Guppy as MG or B-377MG, Mini Guppy Turbine as MGT or MGT-101, Super Guppy Turbine as SGT or SGT-201.

Boeing B-29 Superfortress

The operational history and service record of the Boeing B-29 model 345 is well represented and recorded in numerous books and sources. For this reason it will not be fully covered in this accounting. The primary focus is to relate the design and engineering evolution of the B-29 (B-50) and how the unpopular B-377 airliner derived from it became the base airframe for an expanded fuselage transport built by Aero Spacelines which greatly assisted in putting man on the moon.

Boeing was working on a new large bomber design in 1934 which resulted in the building of the Boeing XB-15. It first flew in 1937 setting load and altitude records; however, it was big with large wings cumbersome and slow. In 1935 Douglas Aircraft designed a competing giant, the XB-19, but once again it was slow and underpowered. The USAAC was not impressed with either of these aircraft. The war had not yet begun and the Army was having trouble obtaining funds for the smaller B-17. Secretary of War Harry Hines Woodring cancelled the XB-15 project. Boeing was not deterred and continued developing a new design without government funds which was the Model 334A (B-29).

This new design was little more than a pressurized B-17 with tricycle gear but it is a direct predecessor to the B-29. The cockpit still had a stepped up windscreen design with a glass nose similar in appearance to the B-17. Engineers improved the design of the Model 341 by developing the Model 345. Boeing submitted plans of the proposed B-29 prototype to the Army in 1939. The design was more clean and streamlined than any aircraft ever built before. The stepped up windshield was eliminated with the cockpit glass fitting the contour of the nose. The advance design high wing, pressurized, tricycle gear heavy bomber was reviewed by USAAC in May 1940. The Army had earlier issued requirements for a 400 mph bomber with a 5,300 mile range capable of carrying 2,000 pounds of bombs to the target. On 11 May 1940 the Army appropriated $85,652 for wind-tunnel test on the Model 345. The results were impressive enough for the Army to appropriate $3,615,095 on 24 August 1940 for the construction of two XB-29 prototypes. The contract was amended on 14 December for a flyable third aircraft and the evolution began.[1]

The new B-29 design was introduced as the world's heaviest bomber with a wing load which far exceeded anything previously produced. Army technicians and aeronautical engineers were extremely concerned with this increased wing loading and feared catastrophic consequences during takeoff and landing. The Army pressured Boeing to increase the wing area for lighter loading and to reduced

Top: **Mockup of original Boeing B-29 cockpit design resembled the B-17 with standard windscreen configuration. It was quite different from the final production B-29 nose. If this design had been adopted, the B-377/C-97, which is a double deck B-29, may have had a drastically different appearance. The B-377 Pregnant Guppy possibly would have had a traditional aircraft nose. Sitting behind the nose on the right is a constant section of the C-97 fuselage mockup (Bill Norton collection of AAHS archives).**

Bottom: **The B-29 was introduced as the world's heaviest bomber, with a revolutionary wing design. It did not have a centerwing and was made up of four sections which bolted together at the centerline of the fuselage. The new configuration with Fowler Flaps had the highest wing load of any aircraft ever produced. The military and most aeronautical engineers believed it would be disastrous. Boeing stated it was necessary to reduce drag and still produce the lift desired. The center-wing was added on the B-29B and carried through to all subsequent models including the B-377 (Bob Williams collection of AAHS archives).**

stress. Boeing protested strongly pointing out an increased wing would increase drag and reduce performance, which was critical. Previous aircraft either did not have flaps or they were hinged rear portions of the wing. Boeing's solution to the dilemma was an extension of the trailing edge of the wing which could be extended or retracted as needed. The concept was known as Fowler flaps.

This unique design changes the camber and increases wing area generating the required lift at lower speeds. In addition it reduced wing loading and stall speed. The effect is similar to increasing the lift coefficient by increasing wing area. In reality, the wing is not changed and the lift coefficient is dependent on the camber. The Fowler flap is hinged and retracts into the wing. When deployed, it moves rearward from the trailing edge increasing the wing area and simultaneously moves down increasing the camber. The XB-29 with the all new wing design first flew on 21 September 1942.

The revolutionary wing design with a span of 141.23 feet was used on the B-29, B-50, C-97 and B-377. Eventually it was incorporated into the Aero Spacelines B-377 PG and subsequent models. This is remarkable when considering the next revolution in wing design did not come until the swept wing for the jet age. The wings of the Boeing "Strato" series only differed slightly from model to model. The B-29 and YC-97 had a wing area of 1,739 square feet, B-50 at 1,720 and the KC-97 at 1,769 square feet. The slight differences can be accounted for in wing tip, flap and aileron configuration. The early B-29 wings were the same on the outside but were structurally different from all subsequent models.

The early B-29 design did not have a center-wing. It was made up of four sections, a two-piece wing including the nacelles which bolted together at the fuselage center line and acted as a single unit. The outer wings were attached to the ends. The design was changed on the B-29A and B models to five sections consisting of a short center-wing which protruded on the outside edge of the fuselage. A separate wing section carrying the pair of nacelles was bolted to it and the outer wings attached to the ends. This later center-wing design greatly increased the strength and reliability allowing safer and higher wing loading. If Boeing had not changed the design to a center-wing it is most likely the guppy series of aircraft would not have been built. The center-wing design was a factor in increasing the span of the Aero Spacelines B-377 Super Guppy. A completely new wider unit was cost effective because it provided a way to increase the span with minimum engineering and continue using the original wings. The span for all subsequent B-377/ C-97 guppy conversions was increased by 15 feet from 141 feet, 3 inches to 156 feet, 3 inches.

The B-29 was fitted with the new 18-cylinder twin-row Wright R-3350 Duplex Cyclone engines. The engines immediately had problems with lubrication and cooling which resulted in engine fires. The new bomber program was almost cancelled when the second XB-29 (41-1003) was tragically lost on 08 February 1943. Boeing's chief test pilot, Eddie Allen, and ten crew members perished when it crashed into the Frye Meat Packing plant at Seattle, Washington. Nineteen workers in the meat plant also perished. The aircraft experienced an engine fire which Allen reported as extinguished. A second fire erupted causing two crewmen to bail out. Both perished when their chutes did not open. The other eight crew and Allen were on board when it pan-caked on to the roof of the packing plant and exploded.[2]

The Army was desperate for the B-29s to fight the war in Europe. Production delays resulted in it not being completed in time and only seeing service in the Pacific. The revolutionary design combined with engine problems, airframe quality control and the deadly crash of the test aircraft placed the B-29 bomber program in jeopardy.

Ninety-seven B-29s had been built at Boeing Wichita by January 1944 but only 16 were flyable and most of them were grounded for modifications. The engines were overheating with the potential of fire on every test flight. There were problems with the cockpit windows which had distortions and were not properly sealed. The aircraft could not be pressurized because of poor fitting parts and seals. There were wiring problems and electrical fires. Problems were expected because it was the most sophisticated aircraft ever built. The design and engineering was sound; however, quality control was extremely poor or nonexistent. Parts from vendors were not uniform and the B-29 required a standard of manufacturing and quality which was not previously known. Bomber aircraft constructed up until this time were non-pressurized manual machines with little electronics or remote control armament. The workers had never built anything which required this degree of skill.

General Henry H. "Hap" Arnold visited the Wichita plant on 11 January 1944. He wanted thousands of aircraft and expected 175 of the new bombers to form the new bomber command. After inspecting the plant he challenged the workers by signing the 175th fuselage section which was waiting assembly. Arnold declared he wanted this airplane complete and ready to fly before the first of March. Two months later he was told the aircraft was not complete and in addition none of the B-29s were combat ready. Arnold was livid and accepted no excuses.

The entire B-29 program was a disaster. In addition to the tragic crash of the XB-29 Boeing had lost control of the production program. Arnold dispatched General Bennett Meyers with full authority to do whatever it took to get the aircraft completed. It was an all-out effort because the military desperately needed a heavy super bomber. It was either fix the aircraft or cancel the program. If the program had been cancelled the outcome of World War II could have been quite different. Even if America and the allied forces had been successful there would not have been a Stratocruiser airframe to form the base of the Aero Spacelines guppies. This could have created a domino effect changing the outcome of America's efforts to land on the moon.

Seventy-five of the new airplanes required the cockpit glass removed and replaced. All 586,000 electrical connections in the completed aircraft were removed, resoldered and replaced. The R-3350 engines were removed and upgraded with airflow direction baffles for proper cooling. The cowl flaps were reduced in size to better control air flow. To correct oil flow problems the engine sumps were modified, rocker arms drilled for oil passages and the exhaust valves were replaced with improved hardened steel units. The engine nose casings were modified. The engines were considerably improved but despite the efforts engine fires were still a problem throughout the life of the B-29. The aircraft skin was removed from portions of the wings to modify and strengthen weak areas. The rudders were removed and replaced with modified structurally stronger units. The program amounted to completely overhauling brand new aircraft.

The USAAF originally ordered 250 B-29 aircraft. Seven months later the order was amended to 500 which were to be built at Wichita. To fill demand construction was contracted to Bell Aircraft in Marietta, Georgia and Martin Aircraft in Omaha, Nebraska. The three XB-29s built at Seattle and fourteen YB-29s built at Wichita had been fitted with three blade props. All production models were fitted with Hamilton-Standard four bladed props. A total of 2,537 B-29s were built, 1,644 at Boeing Wichita, 536 at Martin Omaha and 357 (total 668) at Bell Marietta. A total of 1,122 B-29A models were built at Boeing Renton, Washington and 311 B-29B models at Bell Marietta. In all a total of 3,970 Boeing Model 345 (B-29) were built plus three XB-29 and 14 of the YB-29s.[3]

Sixty-five B-29s were assigned to Operation Silverplate, which was the code name for USAAF aircraft participating in the Manhattan Project. The 509th Composite Group

Top: Final B-29 cockpit configuration was a departure from conventional designs to make it more streamlined and reduce drag. The pilots sat far apart with a split forward panel and an open view through the nose. The pedestal was located on the left next to the pilot with all the primary engine controls mounted on the engineer's panel behind the pilot. The B-29 was a flight engineer's aircraft (AAHS archives).

Bottom: Ninety-seven of the 250 B-29s ordered had been built by January 1944 at Boeing Wichita but only 16 were flyable. There were serious problems with the engines overheating and catching on fire. Quality control was non-existent. No one had ever built an aircraft that required this amount of precision. After corrections were made the order was amended to 500 aircraft. Many were built by Martin at Omaha (Martin collection of AAHS archives).

became the first nuclear weapons unit and received 53 of the 65 modified aircraft. Oddly 46 of the B-29s in the program were not built by Boeing but by Martin at Omaha. On 06 August 1945 B-29-40-MO c/n 44-86292 Enola Gay dropped the first atomic bomb on Hiroshima, Japan, becoming the most famous warplane in history. When hostilities ended many aircraft plants were shut down. The Boeing Wichita plant was reopened in 1948 to convert twenty-nine B-29s to KB-29M in-flight tankers by installing 2,300 gallon fuel tanks in each bomb bay.

Boeing B-50

The Army placed an order with Boeing in July 1945 for 200 of the more powerful model 345-2 aircraft which were designated B-29D. They were powered by 28-cylinder 3,500 horsepower Pratt & Whitney R-4360-35 engines. After the surrendered of Japan the order was reduced to 60 aircraft. In December 1945 the B-29D model 345-2 was re-designated B-50 to win government approval. Congress was reluctant to appropriated funds for an existing war machine after hostilities had ended. Boeing changed the designation to B-50 in order to circumvent funding of an existing aircraft. The B-50 was ordered as a new stop-gap long-range bomber until proposed jet powered models could be developed.

The major change to the B-50 was the construction material of the wing. It was identical to the B-29 wing design but fabricated using a stronger yet lighter grade of aluminum. The improved metal made it 600 pounds lighter and 16 percent stronger. This superior wing would eventually be incorporated into the C-97 Stratocruiser transport and ultimately the B-377 guppies. The installation of larger R-4360 engines increased the power by 59 percent necessitating a taller vertical fin and larger flaps. Because of the slow development of jet bombers, the military reconsidered the role of the B-50 by ordering a total of 371aircraft which were in production until 1953.The larger flaps, taller folding vertical fins and R-4360 engines with chin scoops of the B-50 are the same as those adapted to the C-97.

The B-50B was also tested with the V-belt track type landing gear for dirt field landings. A single aircraft was fitted with the Goodyear V-belt track type landing gear which was also tested on the XB-36, C-82 and C-47. The experimental track-type landing gear system was developed for heavy aircraft to reduce the footprint. The system tested reduced the footprint to just one third of the psi placed on the runway by a multiple wheel undercarriage. The purpose was to allow heavy aircraft to land on unprepared landing strips. The gear was extremely heavy even with the use of light weight materials. A series of taxi test were done which produced an extremely loud screeching. After considerable testing the system was determined to be impractical and never considered for production on the B-50.

Two B-50D models were modified as tanker prototypes. After considerable testing it was determined early model B-50s could be modified to KB-50J and KB-50K series for in-flight refueling. The modification of 112 B-50As and RB-50Bs to KB50J tankers was contracted to Hayes International at Birmingham, Alabama. Speed was a problem when the piston powered KB-50s refueled the faster jets. The jets had to throttle back to speeds which were just above falling out of the sky and the KB-50 was at full power. To increase speed for refueling jet aircraft, 24 additional TB-50H aircraft were converted to KB-50K models by adding a pair of GE J-47 jets at the outboard auxiliary wing fuel tank positions. The addition of the jets increased the speed to 444 mph at 17,000 feet at a gross of 179,500 pounds on the J and K models. In a rather unique twist, as the KB-50s were being phased out the pumping equipment and jet pods were transferred to the KC-97 Stratocruiser tankers.[4]

A total of 371 B-50s were produced from 1947 to 1952. They were modified in a number of configurations. Hayes International converted 136 B-50s to hose tankers for use by the Tactical Air Command for refueling fighters. The hose-and-probe system could refuel three aircraft simultaneously. The KB-50s were old and scheduled to be phased out. However, these tankers saw action in Vietnam during 1964–65 assisting fighters which were low on fuel over enemy territory. The flight crews were placed in perilous danger often refueling at low altitudes while taking ground fire. The last operational aircraft a WB-50D was retired in 1967 when replaced by a Boeing WB-47.[5]

10

C-97 and B-377 Stratocruiser Series

C-97 Stratofreighter and KC-97 Stratotanker

As the war began to turn in 1944 military planners believed Germany could not hold on much longer and the conflict in Europe would be ending. When hostilities ceased American production of bomber aircraft would be scaled back. After the victory in Europe all orders for bombers were reduced except for the longer range Boeing's B-29. It was being built in increasing numbers with the anticipation the war with Japan would drag on for some time. After the atomic bombs were dropped on Hiroshima and Nagasaki the war ended abruptly, catching aircraft manufacturers by surprise. Military aircraft production was reduced shortly after VJ day placing many companies in financial peril. Most manufacturers had no contingency plans for commercial aircraft on the drawing boards. Boeing did have the model 367 (C-97) transport designed in 1942 for the military. Up until this time the Army had no experience in dedicated cargo aircraft. Military transports had always been civilian aircraft adopted by the military and modified. This was the case with the DC-3 and the DC-4 which became the C-47 and C-54 respectively. The Army ordered three Boeing XC-97s (model 367-1-1) in January 1942 as dedicated military transports. Boeing presented it to the press as the transport version of the wartime B-29. It was not offered in a commercial configuration to the airlines until after the military placed large orders.

The C-97 ultimately became a successful and versatile military transport but the civilian B-377 Stratocruiser was only a marginal commercial airliner. When B-29 production was cancelled with the provision only those airframes at the Renton plant which were near completion would be finished, the Model 377 transport design was presented to the major airlines. The plant was closed in 1946 when the last B-29s were completed. Only the first of the three XC-97s ordered by the USAAF in 1942 was a model 367-1-1. The second and third aircraft were designated Model 367-1-2. These first model 367 transports were truly double deck B-29s (Boeing description "double-lobed") because they utilized the same gear, wings, empennage and troublesome yet improved R-3350 power plants. The lower half of the fuselage was the same diameter of the B-29 and the upper was 11 feet, 9 inches in diameter. The floor was 8 feet wide with a 75-foot long cargo bay producing a total fuselage volume of 6,600 cubic feet.

The first XC-97 flew on 09 November 1944 with Boeing test pilot Elliot A. Merrill at the controls assisted by John B. Fornasero and engineer Kenneth J. Duplow. Also on board was radioman Clifford Dorman and second engineer-observer William Elfendahl. Fornasero would later command the first flight of the commercial B-377 with co-pilot Robert Lamson. Although built for the USAAF and considered a success, none of the XC-97s saw military service. On 09 January 1945 at 11:38 a.m. the first military XC-97 flew a 20,000

The Boeing XC-97 was a new aircraft with the appearance of fat B-29. It had the same wings and systems as the B-29. The engines were the troublesome Curtiss-Wright R-3350s. The B-29 tail assembly with short vertical fin was simply repositioned to a higher mounting on the larger fuselage. The XC-97s did not have a flight engineer's station. The first three built never saw active military service (Bob Williams collection of AAHS archives).

The Air Force ordered a single YC-97B as VIP transport. It was configured for 80 seats and a lounge essentially becoming a C-97 forerunner of the commercial B-377. The airframe was substantially improved with the taller vertical fin and more powerful Pratt & Whitney R-4360-35A engines with the chin scoop and a flight engineer's panel. Interestingly it did not have rear clamshell doors. Instead it had the B-377 type section 45, which was so important to the production of the Aero Spacelines B-377 guppies (E. Stoltz collection of AAHS archives).

***Left:* The C-97 clamshell doors and self-contained ramp system allowed the C-97 to transport most military vehicles as demonstrated by unloading a bulldozer. The military praised the C-97 as the ideal military transport and yet only 77 were built. The Douglas C-124 soon became the primary military transport (USAF).**

pound payload demonstration flight from Seattle to Washington, D.C., in six hours and three minutes. The new transport covered 2,323 miles at an average speed of 383 mph to set a new record for transport aircraft.

Lessons learned early in the war of not having sufficient transports prompted the 1945 postwar orders of, six YC-97, three YC-97A and one YC-97B for a total of ten 367-5-5 models. The YC-97s were identical to the XC-97 except they had the small "Andy Gump" chin scoop nacelles which had been developed for the last B-29s with the Wright R-3350 engines. The three YC-97A models were fitted with larger Pratt &Whitney R-4360 engines. They were the first to have a flight engineer's station in the cockpit. Initially the C-97s were planned with only a pilot and co-pilot. The work load for a large four-engine transport proved to be more than a two man crew could handle.

The single YC-97B was built as a VIP transport in airline configuration more like the B-377 commercial version. It had 80 seats, a lounge and airliner type windows. It was powered by the new P & W R-4360-35A engines with the large chin scoop. There were no C-97 rear clamshell doors. It had the same section 45 as the commercial version and fitted with a larger tail fin which became standard on all subsequent models.[1]

The AN/APS-42 radar dome did not become standard until the C-97A model. Between 1947 and 1958 Boeing produced 888 C-97s. As the need increased 219 were converted to KC-97E and F tankers. Another 592 were built as KC-97G tanker models by adding a pair of 700-gallon external wing tanks for a total of 8,513 gallons of tanker fuel.

Near the end of the Berlin Airlift a C-97 was called into action as an operational exercise. It made ten round trips into Berlin transporting over 200 tons of supplies. It was considered a live test because military planners needed operational data under adverse conditions to evaluate future operations. The Air Force confirmed the 14,000 horsepower Stratocruiser operated without incident using only two-thirds of the 5,300-foot Tempelhof runway. This operational test validated the Air Force decision to order 50 more improved C-97A model aircraft.

In 1947 the Army Air Force (AAF) stated the C-97A with Pratt & Whitney R-4360 engines was the only self-sufficient transport in air force inventory with a 26.5 ton cargo capacity. The AAF stated, "It requires no special ground handling equipment and has a self-contained ramp system for easy loading of military vehicles. It can haul more ton-miles of cargo than any other transport in service. In addition it can transport 107 litter patients or 135 fully equipped troops." Despite the Air Force viewing the C-97 as the ideal transport, only 77 were built as C-97s and 811 were built were built as KC-97 in-flight refueling tankers. By 1948 the C-124 assumed the duty of America's primary military transport.

B-377 Stratocruiser Commercial Development

The commercial version B-377 is based on the YC-97B model 377-4-7. All commercial Stratocruisers were designated B-377-10 with the 10 indicating the exterior configuration. The last number designates the carrier and interior configuration, for example a Pan Am aircraft was 377-10-26. Pan Am agreed to purchase twenty B-377-10-26 Strato Clippers on 28 November 1945 with very strict conditions. The contract stipulated Boeing could not deliver any B-377s other than to the United States Government until Pan Am received its first aircraft. Boeing designed many B-377 configurations which began with model number 377-1 and went to at least 377-28.[2]

Pan Am also made a request to Boeing for a six-engine model designated B-377-28-26. The study produced a configuration with a Wright R-1820 engine with three bladed 11-foot, 7 inch propeller mounted outboard of number one and four engines. The configuration also called for two external under the wing fuel tanks mounted between the P & W R-4360 engines. The KC-97G had external tanks mounted outboard of the number three and four engines. Pan Am reasoning for these extra engines is not clear but it does pose several questions. Was Pan Am officials concerned the smaller more reliable engines were needed because the R-4360s were prone to failures? It is a recognized fact after the B-377s entered regular service with the four Pratt &

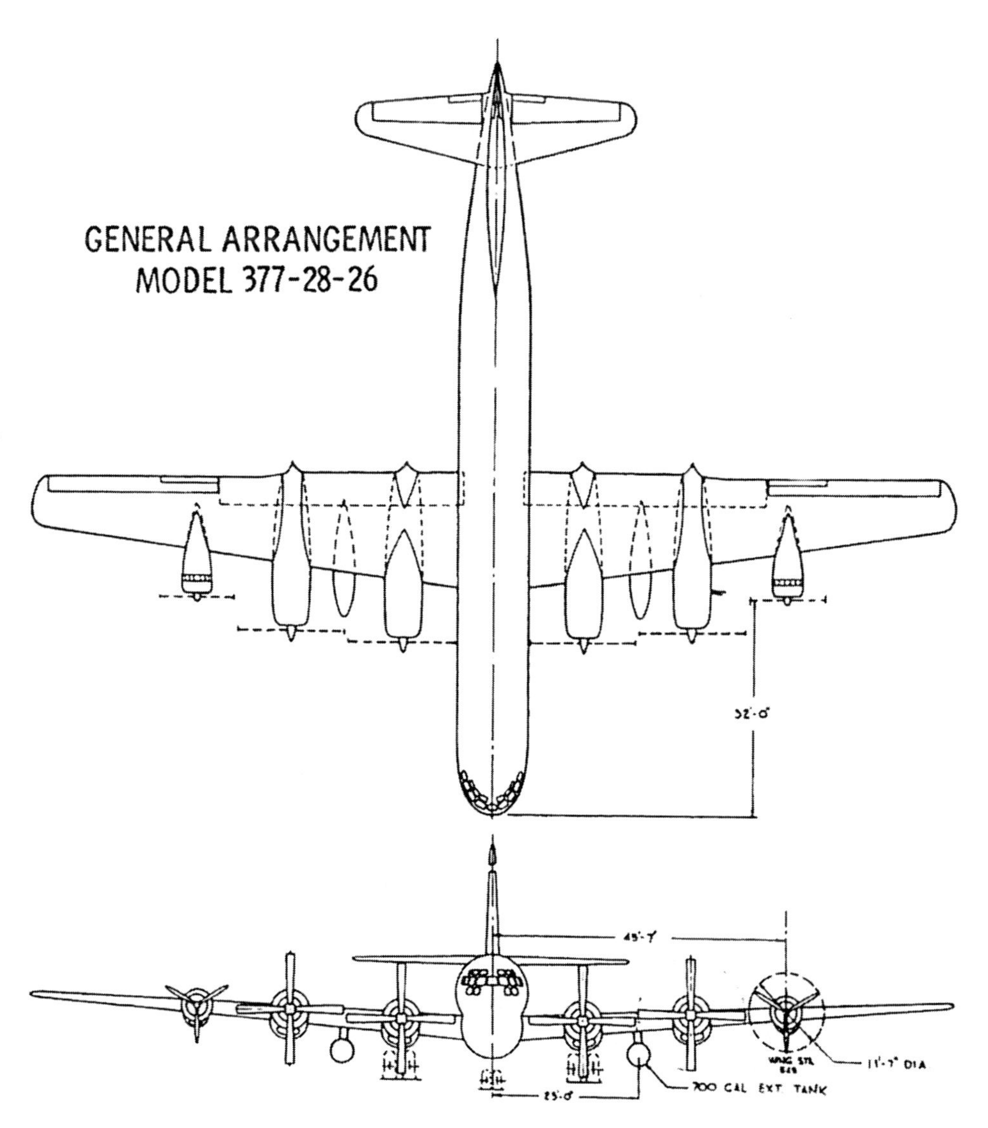

Pan Am requested a design study for a six-engine B-377 with Wright R-1820 engines mounted outboard of number one and four Pratt & Whitney R-4360 power plants. The plan called for a pair of 600-gallon external fuel tanks mounted between the primary engines (Boeing, courtesy Lloyd S. Jones).

Whitney R-4360 power plants, three engine landings were common place. Possibly Pan Am proposed extra engines as a safety precaution for extremely long over water flights where if an engine failed the aircraft had sufficient power to continue on. Also the B-377 struggled to become airborne on hot days requiring a lot of runway. Profile drawings were made but no model or wind tunnel test were conducted.

Pan Am had incurred considerable expense on interior design and promoting the six-engine C-99 commercial version of the Consolidated-Vultee XC-99. The commercial C-99 order was cancelled after it was determined the passenger capacity exceeded the market and it was too expensive to operate. Possibly a six-engine C-97 was an attempt to gain prestige and capitalize on previous Pan American promotions of a six-engine airliner. A six-engine B-377 was not practical but could have given Pan Am the desired worldwide edge. It would help recover from highly advertising the six-engine C-99 then canceling the contract. The presumed concern about public image is highly speculative because in later years Pan Am had no problem selling seats on the first commercial flight to the moon which still has not taken place.

The B-377 Stratocruiser was loved by passengers because of the comfortable accommodations associated with the by-gone era of luxury air travel which compared with steam ships. The major airlines were initially impressed having become accustomed to lesser ex-military equipment built from prewar technology. The B-377, although big, offered features to reduce ground time and servicing. It was the first commercial airliner to offer under-wing pressure refueling. The system enabled the carriers to pump on up to 7,620 gallons in 15 minutes. This was a major improvement producing a considerable advantage over the Douglas DC-4 and DC-6 which fueled over-the-wing.

The commercial Boeing 377 had many advance features; however, it was built too late in the era of large propeller driven aircraft and was soon outclassed by jets. The Stratocruiser was troublesome and very expensive to operate prompting airlines to unload them as soon as newer more efficient turbine equipment became available. There was no secondary market for the 377s leaving most sitting idle waiting for the scrap man. Aircraft broker Lee Mansdorf acquired almost the entire existing fleet for a tenth of its value. A few were sold off, but the majority was supplied over time to Aero Spacelines for outsized guppy transport production. It was necessary to combine fuselage sections of multiple aircraft to lengthen the base B-377 before expanding the upper half to create the guppy. Aero Spacelines was not the first to stretch the Stratocruiser. Boeing proposed the B-377-17-39 turboprop version with Allison XT40-A-6 engines. The planned 153,000 pound design had 80 inches added before the wing and 40 inches aft for a high density 133 passenger configuration. The Boeing customer list indicates the advanced high density version was being considered by Cubana Airlines.

Even more intriguing is an earlier proposal by Boeing to add jet pods outboard of standard engines on the commercial test aircraft NX1039V to increase the Max ATOG to 165,000 pounds with an estimated cruising speed of 360 mph. Much later the USAF installed J47 jet pods on KC-97G tankers re-designating them as KC-97L models. One can only speculate if Conroy was aware of the proposed stretched B-377 and C-97 and alternate configurations.

Jack Conroy had many friends in the aviation industry including test pilots and engineers. He flew military Stratocruisers in the Air National Guard and participated in cocktail hour brain storming sessions with legends of aviation making anything possible. He voiced concerns of a need for more powerful engines for the Aero Spacelines B-377 Pregnant Guppy before it was completed and flew for the first time. He and Wernher von Braun discussed a larger more powerful turbine version while the B-377PG was being developed.

Aero Spacelines suggested in press releases it was considering installing jet pods on the B-377 Pregnant Guppy to increase power. The reason it was never attempted can be explained by the fact the B-377 commercial wings (PG and MG) were not engineered with hard points like the C-97 which were used to mount external fuel tanks and later jet pods. The second Aero Spacelines Super Guppy built with C-97 wings had hard points for external fuel tanks. Boeing had proposed jet pods on commercial B-377s NX1039V which was never taken seriously by Pan Am. The airlines were in a constant struggle to reduce operational cost and not burn more fuel. Aircraft NX1039V was eventually delivered to Pan Am in 1949 and named Clipper Good Hope. It disappeared on 29 April 1952 over the Brazilian rainforest and became one of the most bizarre aircraft losses in commercial aviation.

Both the model B-367 (C-97) and B-377 were designed with only a two man pilot and co-pilot cockpit crew. Early cockpit configurations of the three XC- and six YC-97s did not have an engineer position or panel. The military C-97 was built with a separate position for the flight engineer. The B-377 two-man cockpit was even approved by the Civil Aeronautics Administration (CAA). Pan Am vice president and chief engineer Andre Priester protested and had Boeing raise the center control stand and install spur throttles and mixture controls for the engineer similar to the C-97. United intended to fly the B-377 without an engineer but had an office chair with a seat belt installed in the cockpit for the position.

Northwest tried to eliminate the engineer by requesting instrument panels on sliding tracks which the pilots could swing into position. United began flying the Douglas DC-6 aircraft with only two pilots in the cockpit. Four months after the B-377 first flew the Douglas DC-6 was grounded after two in-flight fires, one resulting in the loss of 52 lives. The CAA spent months trying to decide if the engineer was necessary. Almost immediately pilots began to protest, pointing out the workload of large advance transports and potential problems of only a two man crew in the event of an in-flight emergency. Those perceived problems proved to be frequent on the commercial B-377.

Crew complement hearings were held on 6–8 October 1947. As a result the CAA administrator reversed his earlier decision requiring Boeing to add a flight engineer's station. The CAB amended Civil Air Regulations on 15 April 1948 requiring a flight engineer on all four-engine aircraft over 80,000 pounds. Beginning with the YC-97A, all military models and all commercial B-377 had the flight engineer position. The B-377 flight engineer proved invaluable as the commercial carriers suffered many runaway props and engine failures which required the attention of all three cockpit crew.

On 08 July 1947 the prototype commercial B-377-10-19 Stratocruiser NX90700 flew for the first time becoming the Boeing test ship. It was not delivered to Pan Am until three years later when it was modified to a B-377-10-26 and registered an N1022V before delivery on 24 October 1950. It received the Pan Am name Clipper Nightingale flying until 1961 when it was traded in for Boeing 707s with 27,538 hours on the airframe. It was eventually acquired by RANSA and scrapped in Miami in 1961.

Pan Am early cockpit after the carrier had Boeing install a third seat with spur throttles for the flight engineer. Only the first three military XC-97s were built without a flight engineer's position. The airlines attempted to fly the B-377 with only a pilot and co-pilot. United put in an office chair for the engineer and Northwest moved the panels out of his reach (Boeing, author's collection).

Right: **After several incidents the CAB required a flight engineer on all aircraft over 80,000 pounds. Pan American went to a dedicated flight engineer's station as seen on Pan Am B-377 number 14. The edge of a full engineer's panel is visible on the right. The engineer had his own set of controls for the throttles and propellers visible in the center (Boeing, courtesy Glenn Iverson).**

By 1947 the Boeing Seattle plant was busy building the upgraded B-50 bomber version of the B-29. Wartime production had ended and Boeing was concentrating on the B-377 Stratocruiser airliner as the state-of-the-art commercial aircraft. Labor tension slowed production causing aircraft scheduled for delivery in 1947 to slip until 1948. In April 1948 Boeing union employees went on a 140 day strike pushing Stratocruiser delivery until 1949. By 1948 the military YC-97A was in production and the Boeing Renton plant was reopened to take the overflow. The Air Force placed orders for the tanker modification of the B-29 to KB-29 but by 1950 it was obvious the Curtiss-Wright R-3350 powered KB-29s were not powerful enough for in-flight refueling duty. The KB-29 was being replaced by the KB-50 with the more powerful R-4360 engines. Even with the larger engines the KB-50 needed jet pods to increase the speed for refueling jets. The next step was a new higher gross tanker version of the larger capacity C-97 with the more powerful later series Pratt & Whitney R-4360 engines.

B-377 Commercial Operations

Pan Am was considered to be on the cutting edge of commercial world travel prior to World War II. After the war Pan Am founder Juan Trippe intended to re-establish the carrier's premier status. He placed an order for 20 Boeing B-377 Stratocruisers on 28 November 1945. Noted designer Walter Dorwin Teague was enlisted by Pan Am to create a plush interior for the sophisticated traveler. The Pan Am contract with Boeing for $25,000,000 worth of aircraft got the attention of other long haul carriers. Northwest placed an order for ten (10) at a cost of $15,000,000. Other carriers followed with Scandinavian carrier SILA ordering (4) for $6,000,000, American Overseas (8), United Air lines seven (7) for $11,000,000 and BOAC (6) aircraft for a total of 35 additional B-377s on order. The other carriers also engaged industrial designer Teague to create the Stratocruiser interiors. Transcontinental and Western (TWA) expressed interest in the B-377 prompting the allocation of an additional ten construction numbers 15984–15993 but they were never built. SILA was absorbed into SAS opting out of delivery prompting BOAC to assume the carriers order in April 1949. In an effort to sell more aircraft and help the airlines increase revenue, Boeing proposed a 95-passenger high density Stratocoach version in July 1949. Ultimately no airlines were interested in the high density coach version and the concept was dropped.

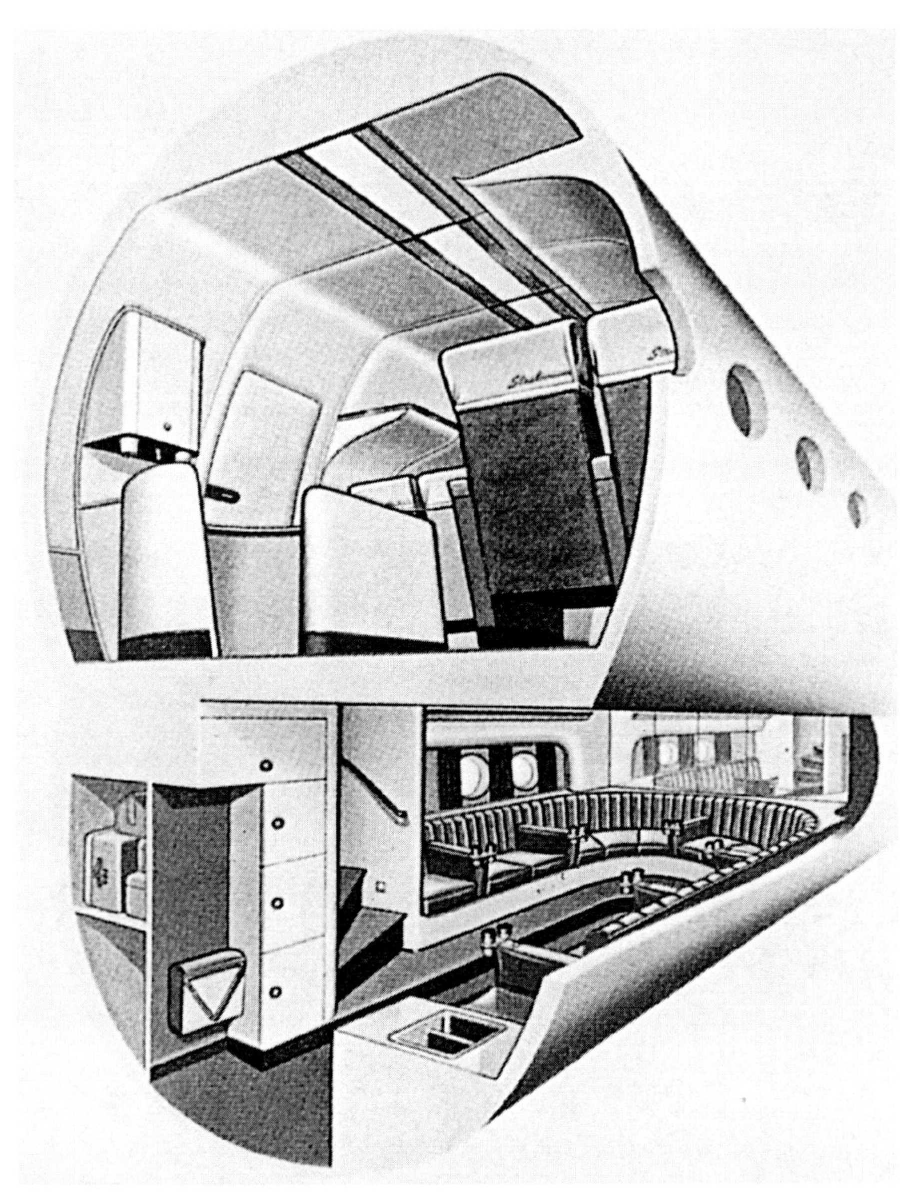

Designer Walter Dorwin Teague was engaged by Pan Am to design plush interiors for the sophisticated traveler. His designs included lounges, staterooms and berths. Looking forward, this Pan Am B-377 lower rear lounge interior is located behind the wing spar (Pan American Airways, author's collection).

Pan Am had previously announced in December 1948 it would take over American Overseas assets and absorb the carriers eight Stratocruisers. The acquisition ultimately made Pan Am the operator of the most B-377 airliners. The fifth Stratocruiser N1025V was actually the first delivered to Pan Am because the first four were in the

Boeing flight test program. The B-377 airliner was the epitome of propeller driven luxury travel combining the standards of luxury ocean liners with the speed and convenience of air travel. No other piston engine airliner came close to the service it provided. It was big, fast, heavy and stable with comfortable seating and roomy lounges, berths and state rooms. It truly was the last airliner of the romantic era of luxury air travel.

Boeing experienced considerable difficulty and expense in getting the B-377s certified for commercial service. The CAA placed several requirements on the aircraft which later proved to be problematic. In order to obtain the certificate of airworthiness (C of A) the CAA required the installation of leading edge spoilers on the wings between the fuselage and the inboard nacelle. The low-speed control designed into the aircraft when coupled with a mild stall produced a condition which allowed abnormally low airspeeds in power-on flaps up stalls. This produced extreme unwanted conditions during the attempt to recover from a stall. The wing tended to drop when the aircraft stalled. The idea of installing the spoilers was to make the nose drop first. The spoilers were designed to slow the onset of a stall at high angles of attack. The device worked when the aircraft was in a steep dive but at a cost. The Air Force, in typical government fashion, attempted to solve the problem on the C-97 version by installing a placard in the cockpit with a stall warning range of 20 mph. The spoilers in some configurations did make handling a stall much easier. However, on takeoff and landing the spoilers seriously compromised the handling characteristics by destroying an area of lift near the fuselage requiring operations at higher speeds.

As a condition of certifying the commercial B-377, the CAA required leading edge spoilers between the fuselage and inboard nacelles, which are visible as dark lines. The spoilers were intended to slow the onset of a stall during high angles of attack. On takeoff and landing the spoilers compromised the handling of the aircraft, reducing lift near the fuselage. As a result higher speeds were required when the aircraft already had problems getting off with heavy loads in high temperatures (Leonard Opdyck collection of AAHS archives).

The CAA finally agreed in 1952 to a flight test program to review the pilot complaints using a Pan Am B-377. The objective was to verify what the pilots and airline officials had contended for years, the spoilers were not necessary creating unsafe conditions. The pilots argued no flight crew would ever place a large commercial airliner into a steep dive situation where the spoilers were an asset; therefore, they are not necessary. The spoilers were eventually removed but it was almost at the end of the B-377 commercial operational service. After the spoilers were removed nose wheel landings were nearly eliminated and the aircraft handling characteristics were greatly improved.[3]

The spoilers were only part of the problem. Almost immediately after the B-377 entered service pilots began to report it did not meet performance standards and was underpowered. It was slow to climb and cruise speeds were deficient. Actually the biggest problem was runway length. Most runways of the 1940s and 50s were around 5,000 feet. The Stratocruiser required 7,000 feet to get off at maximum ATOG on a cool day (59-degrees) at sea level. This was compounded on hot days at sea level, which was the condition at many of the Pan Am Pacific destinations.

Commercial aviation was still not fully aware of the adverse affect temperature has on power settings and lift. The heavy trans–Pacific west bound flights out of Los Angeles only took off to the west. This presented a problem with the ocean on one end and a high traffic boulevard on the other where the runway began. In those days a rolling fence was fabricated to stop automobile traffic on Sepulveda Boulevard in order to extend the runway length for the B-377 takeoff. The roll-out and first quarter of takeoff was on the east side of the road to provide enough distance to obtain takeoff speed by the time the ocean was coming up on the end of the runway. Eventually the road was re-routed to permanently extend the runway.[4]

A total production of 888 (77 C-97 and 811 KC-97) were built for the military and only 56 commercial B-377s for six airlines. Most sources report there were 55 commercial Stratocruiser aircraft built. However, the prototype NX90700 was later sold to Pan Am and converted to passenger configuration in the carriers own shops to become commercial B-377 number fifty-six. Pan Am originally purchased 20 Stratocruisers fitted with a rear galley and round windows. When the prototype was reconditioned and converted to an airliner for Pan Am in 1950, it brought the total to 21. The Pan Am takeover of American Overseas Airlines in 1950 added eight more to aircraft to the carrier's fleet. This made Pan Am the largest operator with 29 B-377s which is over half of the production. The Stratocruiser had bladder type fuel tanks which were positioned in compartments in the wings. Pan Am fitted ten aircraft with five additional fuel tanks totaling 450 gallons for a total capacity of 8,240 gallons. They were designated as B-377-10-26S Super Stratocruisers for the trans–Atlantic route. Eventually the entire fleet was fitted with turbo-chargers resulting in an additional 50 horsepower per engine.

The last six Stratocruisers built were ordered by BOAC with mid-ship galleys and round windows. All six were eventually acquired by Aero Spacelines for guppy production. BOAC acquired four more in the same configuration with round windows which were originally ordered by Scandinavian Airlines (SILA) but never delivered. Three of those were also used in Aero Spacelines guppy production.

BAOC began phase out of the B-377s after receiving De Havilland Comet 1A in 1950. When the Comets were grounded the carrier took a step backwards from the jet age by purchasing the six piston driven United Air lines Stratocruisers which were being phased out. BOAC contracted to Lockheed Air Services in New York to change the Curtiss Electric propellers to Hamilton Standard before the United B-377s were delivered to BOAC. This made BOAC the second largest operator of the type. The window configurations of the different carriers proved helpful in identifying which derelict fuselage sections were used on the construction of the Aero Spacelines guppies.

Northwest purchased 10 B-377s with aft galleys and all square windows. After ten years of operations the eight remaining were traded into Lockheed for Lockheed L-188 Electras powered by Allison 501 turbines. This group of B-377s was also purchased by Lee Mansdorf with six eventually used in the construction of Aero Spacelines guppies.

United Air lines purchased seven B-377s with square windows in the upper main cabin and round windows on the lower deck. The first four delivered to United survived to find their way to other carriers before storage at Mojave and purchase by Lee Mansdorf and transfer to Aero Spacelines.

American Overseas ordered eight with round windows in the main cabin with rectangular windows below and in the aft galleys. They were eventually acquired by Pan Am with only one (N402Q) finding its way to Transocean. After the carrier went bankrupt it was purchased by Mansdorf and eventually used in the guppy program.

Stratocruiser Reliability

The idea of converting B-377s to produce volumetric aircraft for transporting outsize NASA space hardware is most intriguing when considering reliability issues during commercial service with the airlines. The Stratocruiser set the standard in airline luxury yet had a very short commercial life plagued with high operational cost and catastrophic losses. Decades later the B-377 airframes were redeemed by becoming the base for the Pregnant and Super Guppy which transported critical hardware to help put a man on the moon. The Aero Spacelines guppies both under contract and later owned by NASA have a service record of never damaging a piece of NASA hardware. However, there were a few close calls.

Pan American was the B-377 launch customer operating the largest fleet. The third built, N1025V, was the first delivered to Pan Am on 31 January 1949. With great fanfare the press stated it was the fastest and most luxurious airliner ever built. In worldwide service the public perceived it as the ultimate in luxury air travel, even though it was problematic and prone to catastrophic engine and propeller failures. The selection of B-377s for conversion to outsize cargo-liners by Aero Spacelines was completely chance circumstance because they were unwanted surplus aircraft which were cheap, powerful and big. Lee Mansdorf believed they would provide a good platform for airframe expansion.

A number of B-377s damaged in commercial operations were rebuilt and returned to service. All carriers experienced problems; however, Pan Am appears to have more than its share of failures and losses. The first indication there was a problem came on 24 January 1950 when a Pan Am B-377 lost the number four engine over the Pacific en route to Tokyo. A day later Northwest Airlines had the number one engine separate from the wing near Chicago. Within a few weeks the CAA ordered inspections of the Hamilton Standard model 2J17F3 propeller blades. Pan Am wrote off seven Stratocruisers, or one a year, between 1952 to 1959. This is not a record which instills confidence with the traveling public. There were other bizarre incidents of doors blowing open in-flight with passengers and/or crew being sucked out in addition to fires and engine failures. Between January 1950 and June 1952 Pan Am suffered 60 engine failures.[5] These failures are in addition to the disastrous problems and losses caused by propeller failures.

The most severe problems encountered were incidents of runaway props and blade failures. The problem was twofold. Propellers disintegrated throwing off metal pieces or an entire single blade would separate causing an out of balance condition. This occurred in both normal operation or during an engine over-speed. The loss of a blade causes an unbalanced condition which virtually shakes the aircraft apart if the engine is not shut down and brought under control immediately. Engine failure problems were compounded by propeller blades being slung through the fuselage. Frequently crews were unable to feather the remaining runaway prop resulting in engines being ripped from the wing or extreme vibration causing structural failure of the wing.

The six original operators had the option of either Curtiss Electric C644S-BS02 prop hubs with 16-foot, 8-inch diameter 1052-20C4-30 blades with round tips or Hamilton Standard Hydromatic 24260 hubs with 16-foot 6-inch 2J17F3-8W square tip blades. United, Sila (SAS) and American Overseas ordered the Curtiss Electric props which weighed 864 pounds. To reduce weight Pan Am, BOAC and Northwest selected the Hamilton Standard Hydromatic propellers which are 100 pounds lighter. Pan Am also did not like the Curtiss Electric units because they were slower to go into reverse. If the pilot was not careful and paused as the prop went through flat pitch it would over speed. To prevent the problem a throttle stop was installed requiring the pilot to lift up the gate before moving the lever into reverse.

Despite some idiosyncrasies, the Curtiss Electric propellers preformed well without serious problems throughout the commercial operational life of the Stratocruiser. The Hamilton Standard props had an engineering defect resulting in numerous catastrophic failures causing the loss of aircraft and lives. To save weight the steel Hamilton Standard blades were hollow with a black neoprene core. After a period of operation the core tended to crack and break down. The centrifugal force caused the core to be forced toward the tip creating an out-of-balance condition. Within a few weeks after the first incident in 1950 the CAA ordered inspections of the Hamilton 2J17 blades. Engineers originally surmised the problem was with the surface of the blades and ordered daily inspections for cracks. In an effort to correct the disintegrating problem, Hamilton Standard began plating the blades in 1954. The CAA certified the nickel plated blades but the failures continued because it did not address the problem.

Engineers came to the conclusion the problem may initiate in the core. Inspections were required daily using a tone-tap test to determine if the rubber was separating leaving voids inside the prop blades. The filler was changed to a nylon substance but the problem persisted. The hub and four blades weighed 761 pounds. The force created by the out-of-balance condition of losing a blade was extreme which caused catastrophic failures. The blades would fly off or disintegrate which often damaged adjacent engines causing loss of power and decreasing any chance of recovery. The CAB issued a letter on 28 June 1955 directing the removal of the Hamilton Standard 2J17 blades and replacing them with solid blades as they became available. The lack of urgency to fix the problem is apparent because Airworthiness directive AD 58-19-02 was not issued until late 1958. It required all Hamilton Standard 2J17 hollow blades be removed from service no later than 01 August 1959. This was twelve years after the first Boeing B-377 flew.

During the same period the props were disintegrating, carriers were also experiencing "prop overspeed," or uncontrolled runaway. The over-speed condition also caused props to disintegrate but it was not relative to the original blade problem which confused the investigations. Two separate problems were causing the same failures. An over-speeding propeller would disintegrate as the RPMs became extremely high but still no one realized this was because of the over-speed and not internal blade failure. The only defense to control an over-speed condition is to feather the prop. The problem was compounded in a number of instances when the crew was unable to feather the runaway prop. The end result was the same, either slinging a blade or the engine separating from the wing causing additional damage or loss of the aircraft.

Eventually, it was determined the inability to feather the propeller after a failure was a result of the bearing seizing in the nose transfer case for lack of oil. The nose case oil supply line was brazed rather than bolted into place. It was surmised the vibration during an out-of-balance condition because of the propeller failure especially with a runaway would cause the oil supply tube to break. With the oil supply cut off the bearing would fail and the crew would be unable to feather the prop. In early 1957 the FAA issued an Emergency Airworthiness Directive, stating, "As a result of propeller shaft oil transfer bearing failures, several cases of loss of propeller control occurred which make it impossible to feather the affected propellers." The directive ordered all engines to be inspected and the oil supply line replaced no later than 31 May 1957.

The Stratocruiser was the epitome of piston commercial aircraft but relative to the numbers built it had more problems than the military C-97. Pan Am runaway propellers resulted in the loss of 25 percent of the carriers B-377 fleet. The reasons are not clear why these problems were allowed to continue for so long. Either the investigation techniques of the time were not advanced enough or in a heavily regulated environment the wheels of bureaucracy were slow to act. In addition, it is not possible to assess the influence on authorities by airlines and manufacturers for financial reasons. As a result some of these losses are unique in aviation history. The reliability problems in commercial service could have had an impact on the dependability of Aero Spacelines guppy aircraft built from B-377s. Reviewing the history emphasizes the irony of a derivative of such a problematic aircraft being a major asset in putting a man on the moon.

United Air Lines was the first airline to lose a commercial B-377. United Stratocruiser N31230 c/n 15970 was on a crew training flight on 12 September 1951 near Redwood City, California. The pilot was in the process of his semi-annual instrument check. The captain had conducted several ILS approaches and was in the process of an engine out procedure before returning to San Francisco. The number four engine was feathered when the aircraft stalled and went inverted at approximately 300 feet. It crashed near Redwood City into the mud flats in about five feet of water. The three United crew members perished in the crash. The aircraft was fitted with the inboard spoilers which were known to be problematic, however, it was not stated if they contributed to the crash. The United loss was not directly related to propeller problems experienced by Pan American Stratocruisers. The Pan Am aircraft had Hamilton Standard propellers which were plagued with blade failures and out of balance conditions. The United aircraft was fitted with the relative trouble free Curtiss Electric Propellers. Although United suffered the first loss, Pan Am was the first carrier to lose a Stratocruiser which resulted in commercial passenger fatalities.

Pan American B-377 Disappears

One of the more bizarre B-377 losses of a Pan American airliner became an international incident. On 29 April 1952 N1039V c/n 15939 Clipper Good Hope suffered a catastrophic in-flight failure at 14,500 feet when one of the blades separated from the number two propeller. The out of balance vibration set up a series of failures causing the engine to separate from the aircraft. The airframe suffered catastrophic failure, breaking up in-flight and falling into the Amazon jungle.

Clipper Flight 202, El Presidente Especial, originated in Buenos Aires on a multi-city route to New York. The flight was to make stops at Montevideo, Uruguay; Rio de Janeiro, Brazil; and Port of Spain, Trinidad. The flight arrived at Rio de Janerio from Montevideo at 01:05 GMT. After a crew change at Rio, Captain Albert Grossarth was in command. He was assisted by First Officer L. A. Penn Jr., Navigator John T. Powell, Flight Engineer Paul L. Stilphen and Radio Operator Leroy R. Holtzclaw. The Stewards were Anthony L. Urda, Anthony Nasco, Julio Hansen and Stewardess Patricia Monaghan.[6]

Captain Grossarth requested and received clearance to fly direct Rio—Port of Spain rather than the regular more circuitous route. The estimated flying time was 10 hours and 30 minutes. The flight was cleared from Rio to checkpoint Barreiras at 12,500 feet, to checkpoint Santarem at 14,500 feet and to Port of Spain at 18,500 feet. The actual fuel on board was 7,400 gallons, which put the aircraft 3,550 pounds below Max ATOG. The aircraft taxied out at 23:17 local time (02:17 GMT). It proceeded to Runway 32 and stopped short to run up all engines with no noticeable malfunctions. At 23:30 local the flight advised it was returning to the ramp for maintenance. However, after ten minutes in position it turned on the runway headed northwest and took off.[7]

The Stratocruiser crew reported back the off the ground time at 23:43 local time on 28 April 1952. The B-377 had a crew of nine with 41 passengers for the 1,200 mile leg over the Amazon basin to Port of Spain. The passenger list included Brazilian Attorney General Jorge Goddoy and his wife, Washington Secretary to the Brazilian Embassy Luiz Felipe Damorim Antony, Pan Am Vice President Daniel Radersma, General Foods Corporate Officer Lockett Coleman and a number of other influential American and Brazilian citizens.[8]

The flight was at 12,500 feet when the crew checked in at 01:16 local over Barreiras, Brazil. They reported, "abeam of Barreiras cleared to 14,500 feet." This was the last transmission from Flight 202. It never arrived at Port of Spain, prompting a search of over 300,000 square miles of Amazon jungle. The search resulted in one of the most bizarre recovery missions ever attempted flavored with international intrigue, renegade soldiers-of-fortune, hostages and casualties among the members of the rescue team.

On the first day the area where the Stratocruiser could have gone down was searched by 21 American and Brazilian aircraft. By the end of the second day over 40 aircraft were in the area. On 01 May 1952, the third day of the search, a Pan Am C-46 freighter commanded by Miami based Captain James Kowing spotted an 80-foot burned hole in the jungle canopy in an extremely remote area at latitude 9 degrees 45'36" south, latitude 50 degrees 47'18" west. This area of wreckage was calculated at 282 nautical miles N-NW of Barreiras or 887 nautical miles from Rio. The coordinates placed the wreckage approximately 40 miles from any known civilization. The wreckage of the fuselage and full right wing and part of the left wing with number two nacelle were in the charred crater. The rest of the left wing with number one engine was located in another burned area and the empennage in a third area. A nine man United States Air Force rescue team was dispatched from Ramey Airfield in Puerto Rico with instructions to parachute into the crash site. Upon arrival at the scene aerial photographs were taken as the scene was accessed. After orbiting above the scene and observing no sign of survivors, the decision was made to return to Ramey Air Force Base to reassess and organize a recovery operation.[9]

Boeing B-377 NX1039V was used as a test and demonstration aircraft prior to delivery to Pan American Airways. It was lost on 29 April 1952 over the Amazon basin with no survivors. The resulting search and recovery mission became an international incident. It was determined a propeller blade had separated from an engine. The out of balance condition virtually shook the aircraft apart in midair (Boeing, courtesy Glenn Iverson).

The complete absence of roads and thick jungle canopy made it impossible for anyone to access the area. The crash area was so remote there was no record of it ever having been explored. It was thought to be inhabited by hostile tribesman. The crash was more than 35 miles from the nearest point of any known civilization which was a very primitive Indian village on the Rio Araguaia River. The logistics of putting together such a recovery appeared overwhelming. A plan was put forward to fly a land plane to the nearest airstrip at Araguacema, 523 miles south of Belém. An amphibian would also be flown in. The team would transfer to the PBY amphibian which would fly the 85 miles to Lago Grande and land in the river. This could be done because it was the rainy season and the river would be high enough to land and taxi to high ground. A staging camp would be set up from which a passable foot trail could be cut to the crash site. It was soon determined that approximately 26 miles of trail would have to be cut through extreme dense jungle.[10]

A joint Brazilian-American recovery mission was organized to try to reach the site. The Brazilian government assigned Brigadeiro Rodrigues Coelho, commander of Forcas Aereas Brasaileiras (FAB) to head the investigation. He in turn assigned Major Jose Carlos de Miranda Correa to head the mission. The mission was jointly led by Major Correa and Humphrey H. Toomey. Correa was a respected military officer who had been awarded the U.S. Distinguished Flying Cross in World War II and Toomey, an Annapolis graduate, was the manager of the Pan Am Latin American division. They engaged Brazilian cattle rancher Aloisión Solino who had used the gravel field which was some distance away at Araguacema to fly out dressed beef. Solino was known to the Carajai and Taipirape local native tribes at Lago Grande and spoke their language. However, the wreckage was in an area believed to be occupied by the hostile Ciapos Indian tribes. The team was advised to be heavily armed and not attempt to engage the Ciapos tribe. They were also advised the area was inhabited by wild boars, black leopards, jaguars, and boa constrictor and viper species of snakes.[11] Finally the team was briefed on infectious insets in the forest and told to stay hydrated because of the extreme humidity.

A United States Air Force C-82 Packet and C-47 was dispatched from MacDill Air Force Base at Tampa, Florida, to Belém in northeast Brazil. The plan was to fly a 27 man team and equipment from Belém to the field at Araguacema. The area was so remote the dirt airstrip at Araguacema had to be improved before the C-82 and C-47 could land. From there the team and equipment would be transported by a Panair do Brasil PBY-5A sea plane to the village of Lago Grande which was still a 35-mile walk to the crash site.

The plan was to set up a base camp at Lago Grande and cut out a 200-foot clearing for a helicopter. The Air Force disassembled a Sikorsky S-51 helicopter and loaded it into the Fairchild C-82. It was transported in pieces on 12 May to the gravel airstrip at Araguacema. The helicopter was reassembled there using a small portable crane which was transported in by the Douglas C-47. After assembly the plan was to fly it to Lago Grande to support the search team. During reassembly the helicopter tail rotor shaft was discovered to have been damaged in shipment requiring a primitive field repair. The two pilots advised the team flights would be limited because of the damaged rotor.[12]

Meanwhile the victim's families became impatient with the slow progress of the recovery mission. The press being totally unaware of the logistical problems contributed to public outrage by speculating one of the passengers was carrying a fortune in diamonds. Another story reported a shipment of gold bullion on board. As usual, politicians exploited the situation. Ademar de Barros, former governor of the São Paulo state who was a candidate running for the presidency of Brazil, instigated his own plan without regard to the investigation. Barros stated to the press he would put together his own recovery mission. He organized and dispatched a group of rag-tag mercenary types headed by a character identified as Lino de Matos. The group parachuted into the jungle approximately four miles from the crash site with only minimum supplies and no plan for extraction. They cleared an area large enough for a two place helicopter to bring in Matos the next day.

The official recovery team was not aware the mercenary group was ahead of them in the jungle. While the operation was being coordinated at Araguacema, a message was received from Lago Grande. It stated the supply flight observed parachutes in the tops of trees and persons on the ground in a clearing approximately four miles from the crash site. Each man in the official Pan Am expedition had been issued a side arm and ammunition on the advice of the Brazilian government in the event they encountered hostile tribesmen who were known not to be friendly to the outside world. The situation was now compounded with the possibility of hostile soldiers of fortune in the area. Experts in the investigating team included CAB Airframe Specialist Martyn Clarke, Brazilian Military Medical officer Captain Alcino Nova da Costa, Pratt & Whitney expert John Cogswell and Pan Am investigators Charles Manly, power plant specialist Ralph Dobbins and aircraft maintenance specialist Charles Boaz.[13]

The 27 man expedition team with native guides cutting an access path found the foliage so thick they had to split into alternate relief teams to hack through the jungle. Friendly local tribesmen indicated there were watering holes en route but after 16 miles through the jungle they found them dried up. The helicopter attempted to drop supplies and containers of water, but most were lost in the jungle and not recovered. The ground team set off smoke bomb signals for the PBY to drop daily water and supplies with disappointing results. The supply bags that were recovered had notes attached with compass headings to the wreckage. It was not until the team was deep into the recovery expedition they were informed by radio of the mercenary team ahead of them. The investigators were concerned the crash site would be contaminated by these ne'er-do-wells who thought there were diamonds or gold in the wreckage. It was also feared they might scavenge or loot the personal belonging of the deceased.[14]

The ground expedition team cutting through the jungle was still 10 miles from the crash site. Lieutenant Ferreira who was representing Major Correa at Lago Grande made the decision to use the helicopter to establish Brazilian government security at the site. Lieutenant Ferreira, a Para State Police officer and FAB medical officer were the first transported to the clearing thought to be occupied by Matos and the mercenaries. When they arrived it was determined the parachutists had left the area where they had landed. It was assumed they headed for the crash site. After Ferreira had secured the clearing, Major Correa and a medical officer were transported by helicopter to the parachutists' clearing. The PBY and later a second FAB PBY aircraft dispatched for Major Correa's use provided air cover for the helicopter and the dropping of supplies. By 15 May all the investigating team including power plant and airframe experts plus Pan Am observers had been transported by helicopter to the clearing used by the parachutists.[15]

At the same time, one of the groups cutting the trail was ordered back to Lago Grande. The other half of the group was ordered to discontinue cutting a trail and make their way expeditiously to the crash site. A plan was considered to transport the returning trail cutting team to the clearing to begin clearing a trail from the other end. There was concern if the investigating team transported to the clearing could not be extracted they would have to walk out and would have no way of locating the 16 miles of trail cut by the ground team.[16]

The investigating team at the clearing was now confronted with multiple problems. The decision was made to use the clearing as a base camp to obtain air drop supplies. A second camp could be set up at the crash site. On the afternoon of the 15th a select group set out for what was expected to be a four mile walk to the crash site. The wreckage was to the northwest over the rugged Tomanacu Mountain Range. A very circuitous route led the party over extremely rugged terrain requiring ropes to descend the cliffs. It was now apparent the four air miles to the crash site was approximately seven miles overland. The environment was so hostile only eight of the original investigating team would be able to proceed to the crash site. The decision was made for the group to overnight on the trail. The group would be split with those able proceeding forward and the rest returning to the clearing. The medical officer suffered a heart attack, several collapsed from fatigue, many suffered serious injuries in falls and one suffered a severe wound when he was impaled on a sharp branch.[17]

The returning group made camp on the trail for the night and radioed for an emergency water drop by the helicopter the next day which would be marked by smoke bombs. On the morning of 16 May the injured began the return back to the base camp. The remaining eight members of the team headed by Major Correa would proceed to the crash site. The returning party encountered six members of the parachutists group with Linos de Matos on the trail. They were out of water and supplies.[18]

The group asked to join the official party returning to the parachute base camp. It was learned they were terrified of the night. The group was so afraid of the dark and wild animals they fired their weapons and tossed grenades at any sound in the brush. The official team negotiated with the fortune hunters in an effort to all work together. During the last few miles the combined group relied on the local guides to cut vines for drinking water.[19]

After descending an 800-foot cliff the investigating party found the wreckage on the afternoon of 16 May. The main area of wreckage was at an altitude of approximately 1,300 feet on the side of the mountain. It was apparent that someone had preceded them, as evidenced by disturbance in the wreckage. The main wreckage area consisted of a 100-foot hole burned in the jungle canopy. Indications were the aircraft made a nearly vertical descent upside down in a horizontal attitude, it pancaked in. This was evidenced by trees which were damaged from overhead and pierced the aircraft structure. The resulting fire had been so intense very little could be determined. A number of large trees were smoldering at the base and could be heard cracking as they collapsed in the area. They could not be extinguished and were expected to smolder until the next rainy season.[20]

On 17 May the group who had been instructed to abandon cutting the trail from Lago Grande and proceed forward singly to the scene arrived at the parachutists' base camp. There were 25 persons in the group which included newspapermen. The group was also out of water. During the night some of the canteens of water belonging to the official search group disappeared.[21] It was not determined if it had been taken by stragglers from Matos and his group of fortune seekers or the desperate newspapermen.

The search team at the crash site was never able to locate the number two engine. It was deduced one of the prop blades failed and separated possibly causing damage to the adjacent engine or penetrating the fuselage. However, no cut marks were found on the fuselage from prop debris. The bodies were recovered for burial in the jungle. Humphrey Toomey and Major Correa conducted a joint Protestant-Catholic burial service. The group was then informed by radio the helicopter was experiencing severe vibration problems and could only be used for extreme emergency.

After the dead were buried the crash scene was measured and photographed. The lack of water and casualties in the rescue group prompted the decision to shorten the two week investigation to only two days. At the base camp the fortune hunting mercenaries became hostile and desperate when no diamonds or gold were found. The group still believed there had been treasure on board the Stratocruiser and demanded it for themselves as reward for participating in the recovery.

Further complicating matters, presidential candidate Ademar de Barros, who organized he mercenaries, had made no plans to supply his group or extract the band of rag-tag soldiers-of-fortune. When emergency water was dropped by the helicopter they seized it and refused to share. Then the helicopter was sent in with only one pilot to try and extract the injured of the recovery team one at a time. When Matos and his band of thugs realized they were being left behind. They turned their guns on the investigating team and held them hostage. They demanded the helicopter bring a chainsaw and fuel to cut an airstrip in the jungle for an aircraft to land.

The tense standoff wore on until finally the American ambassador demanded intervention by the Brazilian forces. The Brazilian military swung into action preparing to dispatch military paratroopers to rescue the investigation team and newspaper party. Back in the jungle Major Correa radioed to Lago Grande that they had not been harmed and he had finally negotiated a peaceful solution with the mercenaries. A gun battle in the jungle was avoided prompting the Brazilian military to stand down. A small landing strip was cut in the jungle for a light aircraft to extract the injured. Eventually the able crew would all be rescued one-by-one.[22]

On 23 May information was received at Lago Grande that a privately owned Brazilian registered light aircraft was involved in an accident while landing on the airstrip which had been cut at the parachutists' clearing. The following day a FAB aircraft dropped parts to repair the aircraft. On May 28, a full month after Pan Am Flight 202 was lost in the jungle, crash team investigators Major Correa and CAA Advisor Scott Magness were the last men evacuated.[23]

While the investigation continued a sufficient airstrip was built at Lago Grande and a 35 mile road cut to the crash site. Four and a half months after the crash on 18 August 1952 the bodies of the victims were exhumed from the jungle graves and transported to Rio de Janeiro for burial.[24]

By 23 August, the base camp at Lago Grande and the advanced camp near the wreckage were completed. A Pan American C-46 was used to transport Jeeps with trailers along with power plant and structure investigators to Lago Grande. From there they were transported 35 miles through the jungle to the advance camp. The old trees on the trail were still actively on fire. They had been smoldering since the crash and occasionally could be heard crashing to the ground as their bases collapsed.[25]

Conclusion

The investigators determined a propeller blade separated from the number two engine. The out of balance condition set up a vibration so severe it could not be brought under control. The number two engine separated and sometime during the event the airframe broke apart. The wreckage consisted of the main area with the fuselage from the nose back to the dorsal fin and the right wing with number three and four engines with propellers still attached. A portion of the left wing with the number two nacelle was also attached. The number two engine and propeller were missing.

The outboard portion of the left wing with the number one engine and propeller was found 2,300 feet away. The tail section was found 2,500 feet away and consisted of the vertical fin, right horizontal stabilizer and aft section of the fuselage. Examination of the tail determined it failed in flight. There was a film of oil on the left side of the tail and traces of zinc chromate paint. There was also an indentation caused by the number two engine cowling. There are three emergency shut-off valves on each engine for hydraulic fluid, fuel and engine oil. The hydraulic fluid valve was found to be in the closed position. The other two valves were never found.[26]

It was concluded the propeller on the number two engine suffered a catastrophic failure with a blade separating, placing the engine in an extreme out of balance condition. The extreme vibration caused the engine to separate from the aircraft. The engineer managed to shut off the fluid valves but not before oil was sprayed on the left side of the fuselage and tail. At the same time the outer left wing broke off outboard of number two engine. The engine cowling and debris struck the left horizontal stabilizer. The G forces were so great on the out-of-control aircraft that the tail section broke off. At this point there was no control. The two engines on the right were still turning causing the aircraft to roll over. It fell vertically upside down in a horizontal position into the jungle below.[27]

A definite timeline could not be established from the time of the failure to time of impact. However, it was probably less than a minute from the time the blade separated and the complete engine left the wing. From that point the aircraft was doomed with no chance of recovery. It was only seconds before the tail separated and the aircraft began an inverted spin downward.

Propeller Problems Continue

Despite the tragic loss in the Amazon the propeller problem was not resolved. A similar incident occurred on 06 December 1953 when Pan Am Stratocruiser Queen of the Pacific N90947 c/n 15963 was at 10,000 feet, 350 miles

Pan Am B-377 N90947 "Queen of the Pacific" suffered a catastrophic propeller failure over the Pacific between Honolulu and Wake Island en route to Tokyo. An out of balance condition caused the engine to separate from the wing. The crew struggled with the aircraft flying at only 2,200 feet. They were intercepted by an Air Force SA-16, which escorted them to Johnston Island. Notice the luggage on the ground (Pan American, courtesy Glenn Iverson).

northwest of Johnston Island halfway between Honolulu and Wake Island en route to Tokyo with 35 passengers and a crew of seven. The number four propeller suffered an out of balance condition ripping the engine from the wing. Debris from the propeller fouled the rudder causing control problems. The crew was able to recover after considerable loss of altitude and maintain control of the aircraft. An Air Force SA-16 seaplane on patrol intercepted the Stratocruiser 150 miles northwest of Johnston Island flying very slowly at 2,200 feet. The B-377 was escorted to Johnston Island where it made an emergency landing on three engines. This time the passengers were fortunate and there were no injuries. The aircraft remained at Johnston some time while being repaired for a ferry flight out. It was rebuilt and returned to service operating until 18 December 1960 when it was retired.

On 26 March 1955 N1032V c/n 15932 Pan Am Clipper United States was lost 42 minutes out of Seattle when the propeller on number three engine went into overspeed and ran away. The aircraft ditched in the ocean with four fatalities, one injury and 18 rescued.

On 16 October 1956 N90943 c/n 15959 which was acquired by Pan Am from American Overseas in 1950, was forced to ditch in the Pacific. The B-377 Pan Am Clipper Sovereign of the Skies departed Honolulu as Flight 6 for the overnight 8 hour 54 minute flight to San Francisco commanded by Captain Richard N. Ogg. He was assisted by First Officer George Haaker, Flight Engineer Frank Garcia and Navigator Richard L. (Dick) Brown. The aircraft had arrived in Honolulu from Asia and was completing the last leg of an around-the-world flight. The Stratocruiser departed at 20:26 with 24 passengers and a crew of four on the flight deck. Chief Purser Patricia Reynolds was assisted by Mary Ellen Daniel and Katherine Araki in the Cabin.

At 21,000 feet approximately 1,150 miles SW of San Francisco the aircraft lurched and the number one engine began a shrill whine as the propeller exceeded 2,000 RPMs as it ran away. The RPMs were rapidly climbing as First Officer Haaker pressed the propeller feathering button but nothing happened. The needle hit the peg on the gauge and stayed there. He lowered the flaps to 30 degrees to slow the aircraft. Engineer Frank Garcia shut the fuel off to number one but the RPM continued to increase. Captain Ogg had been up and scrambled to his seat. He radioed the *Pontchartrain*, a Coast Guard Ocean Station Vessel permanently stationed midway between Hawaii and

Captain Richard Ogg manages to land Pan Am Stratocruiser N90943 next to a Coast Guard Cutter in the Pacific on 16 October 1956 at approximately 1,150 miles southwest of San Francisco. The B-377 suffered a runaway propeller on the number one engine resulting in a catastrophic failure. The 24 passengers and crew of seven only suffered minor injuries (William Simpson, U.S. Coast Guard).

the U.S. mainland, and called for a bearing on the cutter discovering it was only 40 miles away. Captain Ogg told Garcia to cut the oil supply to force the engine to seize. In a few minutes the RPMs decreased as the engine seized but the prop continued to windmill. They were now at 5,000 feet. Captain Ogg called for more power on the other three engines to try and level off. The engineer reported the number four engine was not responding and was only at half power. The *Pontchartrain* was now visible in front of them. At this altitude the two good engines were consuming fuel at a rate which could only take them 750 miles. San Francisco was over 1,000 ahead and they were too far from Honolulu to return.

It was the middle of the night and there were no options left. Captain Ogg made the decision to attempt a landing in the Pacific alongside the Coast Guard Cutter. He orbited the ship on two good engines and one at half power until daylight. All the passengers were moved forward of the wing in the event a prop should penetrate the fuselage on landing. The sea was calm as he made the last turn at 08:16 and lined up beside the ship with the gear up and full flaps. As the aircraft hit the water it twisted to the left breaking off the tail. The forward section with all passengers and crew floated for twenty minutes, allowing all to escape with only five slight injuries. This was possibly the most publicized of all Pan Am crashes and was covered by the press for days. It actually helped repair the tarnished imagine of Stratocruiser losses by showing the skill of Captain Ogg and the crew.

On 09 November 1957 Pan Am N90944 c/n 15960, which was also acquired from American Overseas in 1950 and re-named Clipper Romance of the Skies, departed San Francisco as Flight 7 to Honolulu. It was on the first leg of an around-the-world flight commanded by Captain Gordon Brown when it disappeared over the Pacific. The last report was approximately 1,000 miles from Honolulu. It was determined to have been 90 miles off course when searchers found burned wreckage floating in the water. All 38 passengers and crew of six perished. The search team recovered 19 bodies. The CAB was never able to determine a cause.

It was common on Pan Am Clipper flights for many of the passengers to be accomplished and well-known. On board Flight 7 was H. Lee Clack, the Dow Chemical Tokyo manager and his family, prominent surgeon William Hagan and his wife, World War II French ace and vice president of Renault Motor Cars Robert La Masion, Thomas McGrail, attaché to the American embassy in Rangoon, Burma, and fashion designer Soledad Mercado. Also on board was Air Force Major Harold Sunderland of the 1,134th Special Activities Squadron. It was believed he was involved in a clandestine international black bag operation for which his destination was never revealed. There were several other questionable and nefarious characters identified as both passengers and crew members.

The flight was at 10,000 feet when a routine position report was made to the *Pontchartrain* Coast Guard cutter which was on station at the point of no return. This is the same cutter which Pan Am Flight 6 ditched beside in the water just a year before. Only twenty-two minutes after passing the *Pontchartrain* Flight 7 impacted the water. No distress call was ever given. Some of the recovered bodies exhibited impact trauma and most had drowned. There was fire damage above the waterline on the wreckage indicating there may have been a fire prior to impact. Most of the bodies had on life jackets indicating they had some time before the crash. Both Pan Am and the FBI suspected foul play.

The two individuals most considered to be suspicious were Major Sunderland and the Pan Am purser Eugene Crosthwaite. Absolutely nothing was disclosed about Major Sunderland and his mission. The 46-year-old Crosthwaite was known as a disgruntled employee and had been depressed since his wife passed away of cancer there months earlier. He had a strained relationship with his 16-year-old daughter from a previous marriage. He had amended his will the day before and left a copy in the glove box of his car. In addition Crosthwaite had shown his father-in-law a hand full of blasting caps a few weeks before the crash.

The FBI abruptly bowed out of the investigation with little explanation and the CAB was ready to put the blame on Crosthwaite when the activities of another listed passenger, William Harrison Payne, stood out as suspicious. His body was never recovered and he had stated he was on the way to Honolulu to recover a debt. However, the debt was less than the $300 ($1968 today) fare to Honolulu. He had taken out three life insurance policies on himself and one was only a few weeks before the crash. Payne was known to be of questionable character. He had previously been in several scams and confrontations with authorities. He owned a hunting lodge in Northern California which was heavily mortgaged and was in serious financial trouble. Payne had been a Navy frogman and demolition expert which made the insurance investigators very suspicious.

The direction of the investigation changed again when it was revealed this same aircraft, N90944, operating as Flight 7 on 18 June 1957, suffered a runaway propeller on number three engine. Captain Clancy Mead was in command of the flight from San Francisco to Honolulu. The crew was unable to feather the runaway prop. Captain Mead declared an emergency as they struggled to maintain control of the Stratocruiser. The crew barely got the crippled aircraft back to San Francisco, fighting to keep it level and losing altitude all the way.

This incident was two weeks after the FAA issued an Airworthiness Directive to inspect and replace the oil transfer tubes on the prop hub no later than 31 May 1957. The directive had not been complied with then and it is strongly suspected it had not been followed when Flight 7 went down with no warning. After years of speculation, it is now suspected the aircraft suffered the same fate as Pan Am Flight 202 which went down over the Amazon. A catastrophic propeller failure created an out of balance condition which could not be brought under control. The failure literally shook the aircraft apart in flight, possibly breaking off the wing. Even today there is an effort to recover the airframe using a deep water submersible to finally close the case.[28]

There were other Pan Am B-377 airframe losses which were not related to propeller problems:

On 02 June 1958 NX1023V c/n 15923 Clipper Golden Gate made a heavy landing at Manila in a severe rainstorm. There were two casualties among 46 passengers with all 8 crew surviving. The aircraft was written off.

On 10 April 1959 N1033V c/n 15933 Clipper Midnight Sun crashed into an embankment short of the runway at Juneau, Alaska, and burned. Five passengers and five crewmembers survived.

On 09 July 1959 N90941 c/n 15957, which was acquired from American Overseas in 1950 and re-named Clipper Australia, made a belly landing at Tokyo with 53 passengers and crew of six. All survived and aircraft was written off.

The other commercial B-377 operators suffered in-flight failures but only one was because of a propeller blade loss and the crew managed to land safely. In comparison, Pan Am suffered a disproportionate amount of Stratocruiser losses than other airlines. There have been a number of reasons suggested including the fact Pan Am was the launch carrier which historically experience problems when a new type is introduced into regular commercial service. However, it appears the propeller problems and errors in proper corrective action contributed to multiple losses.

On 25 January 1950 Northwest N74608 c/n 15954 number one engine failed and separated from the wing. The aircraft made successful emergency landing at Chicago–Midway Airport. It was eventually lost on 02 April 1956 when it crashed in Puget Sound on takeoff from Seattle resulting in five fatalities. The crew lost control due to severe buffeting.

On 12 September 1951 United Air Lines N31230 c/n 15970 crashed into San Francisco Bay on a training flight. The aircraft stalled at low altitude with one engine feathered while practicing engine out procedures. All three crew perished including pilot Fred Angstadt, Pacific flight manager for United.

On 25 December 1957 BOAC suffered and engine failure and fire after loss of number four propeller. The flight made successful landing at Sydney, Nova Scotia, with no casualties.

B-377/C-97 Configurations and Service Reliability

The preceding accounts give some idea of the irony of Pan Am B-377 airframes becoming the candidate for the Aero Spacelines Pregnant and Super Guppy series of aircraft. The number of in-flight failures and losses by Pan Am and other carriers relative to the numbers of aircraft produced made reliability a concern. The high operating cost combined with beginning service too late for piston aircraft contributed to a brief airline life. The airframes with considerable service hours left in them were relegated to scrap because no carrier wanted them. Lee Mansdorf actually purchased them for less than the scrap value because Boeing, Lockheed and others wanted to get them off the books.

The 56 commercial B-377 Stratocruisers were configured for 55 to 100 passengers or 28 sleeper berths and equipped with a lower level bar and lounge. Boeing produced a total of 888 B-367 (C-97) military versions for the Air Force. Only 60 of those were C-97 cargo-troop transport aircraft which were significantly different from the airline version. The rest of the production included 816 KC-97s tankers and twelve other variations. The military C-97 version was only operated by the U.S. and Israeli Air Forces. In 1952 in an effort to resurrect the Stratocruiser and keep production of the C-97 alive, Boeing offered 25 P & W R-4360-75 powered freighter aircraft to commercial airlines as the C-97F cargo-liners at $1.2 million each. Northwest, Pan Am, BOAC, Slick and Flying Tigers expressed an interest relative to CAA certification. Boeing agreed to the certification work with an estimated cost of $500,000 to get them approved. Multiple problems with the certification program prevented Boeing from marketing the aircraft. The past expenses and delays of getting the original civil B-377 Stratocruiser certified played a part in Boeing eventually giving up the program of building a civilian C-97F commercial versions.[29]

Although the B-377 commercial version was plagued with many problems in airline service, the later model military C-97 was very reliable and served the Air Force well. During the Air Force phase out of the Douglas HC-54 Skymaster Air-Sea Rescue aircraft for transition to the Lockheed HC-130s, the C-97 was called on to fill an interim role because of C-130 production delays. A group of 29 KC-97G tanker aircraft with considerable hours remaining on the airframes were converted by Fairchild-Stratos corporation in 1963–64 for rescue work. The refueling boom was removed and rear clamshell doors were retrofitted. All of the lower deck and four of the upper deck fuel tanks were retained and converted from holding JP-4 jet fuel to 115/145 gasoline. Instead of in-flight refueling they became an internal source for extended range. Up to 12 seats were installed in the cargo bay behind the cockpit along with a galley for extra crew and Para-rescue personnel.

Each engine of all series military C-97 and commercial B-377s was fitted with a 35-gallon (32.5 usable) tank in each engine nacelle. The 2.5 gallons unusable oil was contained in a standpipe to supply the propeller feathering governor for Hamilton Standard props.[30] The engine oil tank on the commercial B-377 had the same capacity of 35 gallons.[31] However, the 2.5 gallons was not needed in reserve to activate the pitch change mechanism for the United, Sila-SAS and American Overseas B-377s fitted with Curtiss Electric propellers.

The range of large reciprocating engine aircraft is often reduced by oil supply. Boeing sought to eliminate the problem on the C-97/B-377s by installing a 56-gallon central oil tank with 2-gallon expansion in the forward belly which could replenish the engine oil tanks in-flight. The central oil system was critical on the KC-97 and HC-97 which were required to stay aloft for extended operations. An additional 50-gallon tank was installed on the HC-97 to feed the 56-gallon central oil tank. The Pratt & Whitney R-4360s consume considerably more oil when operating at low altitude on search and rescue missions. For this reason and because the HC-97 could remain aloft for up to 30 hours extra 55 gallon drums of oil were often carried to replenish the central oil tank.

The Curtiss propellers on the United, Sila and AOS planes also had another advantage of being able to change pitch without running the engine because they did not use engine oil. This feature was appreciated by maintenance technicians on the ground. The units were much cleaner and easier to maintain because they never required de-sludging of the prop dome. They did contain grease but it was never mixed with the dirty engine oil like the Hamilton Standards which created sludge with the consistency of modeling clay.

The nacelle engine oil tanks were not serviced over 28 gallons in-flight when flying above 15,000 feet. A space of 7 gallons was recommended for expansion and to handle oil foaming or fuel dilution when needed in cold weather. Leaving this space prevented excess oil being forced out the breather. The Pratt & Whitney R-4360s are known to normally consume 1.5 to 4 gallon of oil per hour which could affect aircraft range. High time engines could consume 5 to 6 gallons per hour. Even low-time engines would consume more oil at lower altitude operation.

On extremely long flights the engine nacelle tanks were often run down to 7 gallons before replenishing from the central tank. Even then only about 3 gallons would be replenished at a time. The reason being if you lost an engine your oil consumption could go up on the remaining three and any oil in the tank of the shut down engine could not be pumped back out. In extreme cases when oil levels became critical, a high consuming engine would be shut down at cruise for low oil then restarted for decent and approach.

Crew comfort on extended missions was always an issue. On low altitude HC-97 search and rescue missions the cockpit became unbearably hot for the crew. A supplemental air conditioning system was installed in the cockpit comprised of a series of individual units produced by Vornado which closely resembled the after-market units installed in automobiles of the 1950s.[32]

The HC-97s rescue aircraft were also used worldwide to support the NASA manned space flight programs. They were reliable enough for NASA to depend on them to be positioned on station for launch or re-entry malfunction. The HC-97s were removed from service in 1972. The author had the pleasure of being temporarily assigned to HC-97s of the 305th while serving in the U.S. Air Force in the 1960s.

The Last B-377 Operator

In the early 1960s the B-377 commercial fleet was outclassed by the new generation of jet airliners. The surviving fleet of B-377 aircraft was left idle after being traded in on newer turbine or jet equipment. Although there were a lot of hours left on the airframes, Transocean was the only carrier which attempted to revitalize the Stratocruiser for commercial transport.

Transocean evolved from secondary carrier Orvis Nelson Air Transport (ONAT). It was formed on 09 March 1946 under the stewardship Orvis M. Nelson and seven of his friends. Nelson had been a senior pilot for United Air Lines from 1941 to 1943 flying under contract with the Air Transport Command. On 01 June 1946 ONAT was incorporated with a name change to Transocean (TAL). Nelson acquired 14 Stratocruisers from BOAC in three blocks. Six of them were never delivered and the registrations not applied. The group included the last four originally delivered to United Air Lines in 1949 and subsequently sold to BOAC in the mid–1950s. They were acquired by Transocean in 1959 with help from the Babb Company. These aircraft were transferred to Transocean but some did not go into service and never flew again.

Transocean had established itself as having an excellent safety record. This is quite a contrast to Nelson's later adventures with surplus C-74s acquired for Air Systems and flown under Aeronaves de Panama registration (see chapter on C-74s). Transocean bankrupt and shut down in 1960 after corporate raiders stripped the company of its assets. Orvis M. Nelson had a rather unique passive role in the evolution of guppy aircraft. The Stratocruisers were brokered by Airline Equipment Company and sold for a mere $105,000. Thirteen of the Transocean aircraft were acquired by Lee Mansdorf.

At least three were moved to Mojave an ultimately used for the Aero Spacelines guppy program while the rest remained at Oakland, California. It has been verified that at least six were broken up for scrap at Oakland with selected parts going to Aero Spacelines.

Lee Mansdorf managed to acquire a total of 34 of surviving B-377s. The Venezuelan based Rutas Aereas Nacionales SA (RANSA) acquired eleven B-377s for cargo operations, one from Alex Rozawek, three from Alfred Equipment and the balance from Lee Mansdorf. It is not clear how many were actually operated because many were used for parts and shared with Linea Internacional Aerea. The remains of the RANSA B-377 fleet were scrapped at Miami after the company ceased operations in June 1966. Ecuadorian based Linea Internacional Aerea purchased N1040V c/n 15940 and ferried it to Quito Ecuador where it was impounded. The carrier acquired parts of N1041V c/n 15941 and N90945 c/n 15961 from RANSA but none of these aircraft were ever placed in service.

It has been reported Lee Mansdorf purchased a group of surplus C-97 around the same time as the B-377s. At a later date a group of C-97s were purchased for the Aero Spacelines guppy program. However, they were not part of the B-377s originally purchased which were parked at Oakland, Long Beach and Mojave and subsequently transferred to Aero Spacelines.

It was chance circumstance which put the B-377 in the important role of transporting NASA hardware in the race for space. Both the B-29 and B-377 shared the same wing and structural design and both had questionable operating records. The B-29 program was almost cancelled because of engine fires and the crash of the second XB-29 with the Wright R-3350 engines. The C-97 was improved with the Pratt & Whitney R-4360 engine but the B-377 commercial version suffered catastrophic propeller failures resulting in major losses. The Boeing B-377 is not remembered as a success while in commercial airline service. The Pan Am Stratocruiser had a tarnished safety reputation as a commercial airliner, yet the carrier's B-377 aircraft became the primary airframes used to build the Aero Spacelines guppies. In a unique twist of circumstances the Stratocruiser eventually redeemed itself by transporting rocket boosters for NASA without any damage.

11

Aero Spacelines

NASA Logistics

In any large project logistics are critical to meet both production and delivery dates. When Congress passed the Space Act in 1958, logistics for transporting rockets for space was primarily a function of the military. In the early Mercury and Gemini programs NASA relied on Thor, Atlas and Titan missiles from military inventory which were transported by Air Force C-124 Globemaster aircraft. As the program progressed the rocket boosters required for a mission to the moon would have to be of a size which had not been built before. The program called for a multi-stage launch vehicle which dwarfed military rockets. The logistics problem became serious because contractors built the components all over America. They relied on NASA to provide a transportation system for delivery to the launch site. One of the most glaring examples is the Saturn launch vehicle which was developed without any concern for logistics.

In the early phase of the program planners felt there was no need for a logistics program similar to the military. Missiles for national defense were needed in great numbers all over the world. NASA did not need a large number of rockets dispersed to multiple launch site locations. It was theorized NASA would have a limited number of launches at scheduled times at one location as opposed to the military need for unscheduled launches from multiple sites. It was reasoned only the national defense required an elaborate logistical system.[1]

In the space program, especially the lunar flights, the launch windows are critical using much larger hardware which requires considerable advance planning. The early false assumptions of NASA not requiring a logistical program created a mammoth problem. The size and complexity of the Saturn vehicle, the numbers of suppliers around the country and multiple test and launch sites proved NASA planners had made a major mistake assuming logistics were not necessary. The problem was complicated by lack of funds which caused program managers to divert needed funds from other departments. NASA had to find a way to fund a logistical program, yet stay within budget.[2]

President Kennedy announced on 25 May 1961 America would be on the moon by the end of the decade. His optimism and desire to beat the Russians in space handed America's space program an enormous challenge. With only nine years to build and test a rocket with payloads measured in tons and not pounds, NASA was suddenly confronted with an engineering nightmare of enormous proportions. No one really knew what it would take to manufacture and gather all the parts to assemble a vehicle capable of getting to the moon. Marshall Space Flight Center (MSFC) at Huntsville, Alabama, let a contract in 1960 to Douglas Aircraft to study the feasibility of an oversize transporter aircraft to move the needed components. Douglas had considerable experience with the military transporting Thor IRBM rockets with the C-124 Globemaster. The success of the C-124 demonstrated the most practical system for transporting rockets was by air.

Military rockets are designed for air transport and can take considerable shock and vibration. However, space boosters, because of size, were designed for sea-going transport because there was no alternative by air. After three months of study Douglas proposed a piggyback method using the C-133 Cargomaster (see Chapter 7). Up until this time Douglas engineers had not considered internal air transport as a viable option for anything which did not fit inside the Douglas C-124 or C-133.

Transport time was critical because the boosters were never designed to be laid down on the side. Every hour it is laying down increase the chances of stress damage. They had to be light weight and were designed to stand on end. When booster stages were filled with fuel they were quite strong. However, they could not be shipped standing up or full of fuel. To reduce weight for launch into space they were designed to minimum structural standards which barely allowed it to support itself. A good comparison is an aluminum soda can. It is quite strong when it is full standing upright but as any tavern owner can tell you on Saturday night it can be crushed quite easily by hand when empty.

Astronaut Alan Shepard told Aero Spacelines Super Guppy pilot Jack Pedesky the rocket boosters were fine on lift off. However, the worst part of the flight was just before separation when the boosters were empty. The entire assembly would shake and vibrate violently. All the structural integrity had been expended by burning fuel, leaving only an empty shell of the rocket booster.

At the beginning of the space program, the only accepted means of transport for the enormous boosters was by sea. Even then NASA engineers feared the delicate boosters would sustain damage during the extended and sometimes hazardous sea voyage. The boosters had to be transported 6,578 miles from California through the Panama Canal, up the Mississippi and down the Tennessee River to Huntsville, Alabama, for testing because they were too large to transport by air. The delivery time from California to Marshall Space Flight Center (MSFC) at Huntsville became weeks and possibly months. Even after the boosters moved from California through the Panama Canal across the Gulf of Mexico, it was still a 14-day trip up the Mississippi, Ohio and Tennessee rivers to Huntsville. The voyage of 1,240 miles from New Orleans has multiple segments with transfers to multiple tugboat operators en route. The first segment of 867 miles up the Mississippi to Cairo, Illinois, can take ten days. The barge is then transferred to another tugboat for the 50 mile trip up the Ohio River east to Paducah, Kentucky, to the outlet of the Tennessee River. The barge is transferred again for another 325 miles up the Tennessee River through the Kentucky, Pickwick, Wilson and Wheeler locks before arriving at the MSFC docks on the south end of Redstone Arsenal in Huntsville, Alabama. After testing the components were then moved by barge 2,175 miles by the same river route to the Gulf of Mexico and around to Cape Canaveral, Florida. The logistics of the entire American space program was at the whims of the Mississippi, which is subject to flooding in the spring and freezing in the winter. In addition there were reports of moonshiners in the remote areas and young boys with their first rifle who often took shots at barge traffic.[3]

These potential problems combined with the extended delivery time left no doubt to von Braun and NASA officials the launch schedule was in jeopardy. It was imperative to find alternate transportation methods. Aircraft broker Lee Mansdorf was intrigued when he first read about the logistical problems in an aviation publication. He discussed the problems with his many contacts in the aviation community as he formulated a plan to modify an existing aircraft. NASA solicitations for outsize transports proposals motivated him to entertain a possible way to profit from the situation but NASA did not have a logistics budget. Mansdorf was not interested in building an aircraft himself but could supply airframe components, support and resources if the right team could be put together. Once John M. (Jack) Conroy was introduced to Mansdorf, the idea of building an outsized transport was set in motion. Conroy was convinced NASA was in a situation which would force them to find the funds if he created the solution. He and Mansdorf combined their creative ambitions to produce America's first non-military volumetric transport. Additionally, Douglas Aircraft engineers were actively working on external piggyback air transport delivery systems using the C-133 to move booster components produced in California. Douglas engineering became very interested when Conroy presented his outsized internal payload transport aircraft as an alternative to the external piggyback proposals presented to NASA.

While air transport proposals were being studied NASA was still relying on sea-going barges to transport the booster components to MSFC and Cape Canaveral. Transport problems compounded to a critical level on 02 June 1961 when

After the locks collapsed at Wheeler Dam on the Tennessee River, an alternative transport method had to be found. The Corps of Engineers was dispatched immediately to build roads and docking facilities above and below the dam. The boosters were offloaded on special transporters and towed by road eight miles around the dam then reloaded on a barge to complete the delivery (NASA).

the Wheeler Locks west of Huntsville on the Tennessee River collapsed, leaving the Palaemon transport barge above the dam. With all river traffic halted, the barge containing a dummy S-IV and the first SA-1 flight stage was stranded above the dam in Alabama with no alternate transportation route. Von Braun had promised the president America would be on the moon by the end of the decade. The launch schedule was now in serious jeopardy. In 1962 Marshall Space Flight Center managers conducted studies on the hardware logistics of the Saturn V program but still did not have a firm alternative plan.

Although the rocket stages are built to the highest manufacturing standards, they are still subject to vibration damage. It was almost impossible to control movement or vibration during the long voyage from California to the Gulf of Mexico and Huntsville via the Tennessee River waterway. If any damage occurred there was not time to repair or build more components and remain on launch schedule. The military proposed building an oversized amphibious transporter at an estimated cost of $5,000,000. It was estimated to take four years to build. The space program did not have four years if America was going to make it to the moon. NASA was left with no options for the boosters. Conroy's outsize aircraft was still in the idea stage and the boosters were too large for existing aircraft.

As an immediate emergency fix, military engineers were dispatched to build a road as quickly as possible around the collapsed lock but the boosters had to be offloaded on to special built transporters, moved miles around the dam then reloaded on another barge. This was time-consuming and increased the possibility of damage to the rocket boosters.

Alternate means were used to move some of the smaller components. The adapter unit which connects the instrument unit to the service module was transported from Tulsa, Oklahoma, to Huntsville, Alabama, slung under a CH-47A Army helicopter. It was successful for a one-time fix but not acceptable for regular transport. A transport system was needed to move the upper stages of the Saturn S-IV and S-IVB as well as other outsize space hardware. Two of the most unlikely entrepreneurs, a swimming pool salesman and an aircraft salvage broker, were ready to make aviation history.

Jack Conroy, Clay Lacy and Lee Mansdorf

Jack Conroy sold swimming pools when he was not flying in the California Air National Guard or on call for several non-schedule airlines. He had toyed with the idea of starting a VIP commercial airline of his own with B-377s in late 1959. His plans suddenly changed after a meeting with Lee Mansdorf to purchase several Stratocruisers. Clay Lacy accompanied Conroy to the 1960 meeting with Mansdorf to propose a deal to start an airline with the surplus B-377s. Neither of them had any idea Conroy would be building a giant expanded fuselage aircraft using the Stratocruisers as a base airframe.

After that meeting Conroy began planning the construction of the largest airplane in the world. He was so taken with the idea after learning of NASA's logistical problems; he began the project using his own money on pure speculation. Jack Conroy was already a decorated World War II American airman having been shot down while flying a B-17 over Germany. Conroy believed he could accomplish anything he set his mind to do. The challenge to be a part of getting to the moon before the Soviet Union was tailor made for his ego. This was a chance to make a profit while following his passion of starting his own airline cargo charter operation. Don Stratman who worked for Conroy for years said, "I don't know about profit. He just loved to fly and build airplanes."[4]

To say "Smiling" Jack Conroy led an interesting life is a gross understatement. John M. (Jack) Conroy was born in New York on 14 December 1920 but spent his youth in Oklahoma where his father opened the first steel mill west of the Mississippi. He was very bright, graduating from high school in just three years. His dad wanted him to become an engineer but he had other ambitions.

Conroy left Oklahoma for Hollywood with hopes of becoming an actor. He trained at the Pasadena Playhouse and was reported in industry tabloids as being good at acting. He actually found light success and appeared in nine movies as a member of the Tough Little Guys, the rival studio group of the Dead End Kids. At the same time he was learning to fly on the weekends, making his first solo flight in 1940. He became somewhat disenchanted with his slow start in the movie business and left Hollywood for work in Hawaii. He took a job in Honolulu excavating for underground fuel tanks at Pearl Harbor. Conroy was on Oahu on 07 December 1941 during the Japanese attack. Within days he enlisted in the Army.

The Army assigned him to the Air Corps because he already held a private license. Conroy proved to be a natural pilot and in short time was commanding a B-17 with the 8th AF, 379th Bomb Group in England. On 30 November 1944 the allied forces dispatched 1,250 bombers to targets in Germany. Jack Conroy's aircraft Landa was one of the eighty-six which did not return. He was on his 19th mission in a B-17 with co-pilot Milton Fox. Their target was the synthetic oil plant at Leipzig. The B-17 was hit by flak, forcing the crew to bail out over Zeitz, just 25 miles from the target.

Conroy and the crew survived being shot down only to be captured. They became POWs for the last six months of the war at Stalag 1. The POW camp was only 19 miles from where Wernher von Braun was developing rockets for Germany. A strange twist of fate would one day bring the two men together on America's space program, where they would become friends. Von Braun would become head of NASA and Conroy would develop the aircraft which would transport the Saturn launch vehicle hardware needed to put America on the moon.

Conroy was a decorated airman receiving the DFC, Air Metal and Purple Heart for his military service. After the war he remained on Air Force active duty as a special mission pilot and instructor until 1948. He joined the Air National Guard where he remained for the next 12 years while flying for more than a dozen non-schedule airlines logging considerable hours in Central America. In the 1950s Conroy also operated a small fixed base operation in Blyth, California. While serving in the Air National Guard he was instrumental in creating Operation Boomerang, which was a plan to fly round trip coast to coast in the same day. On 21 May 1955 Conroy set the first round trip record from Los Angeles to New York to Los Angeles between sunrise and sunset in an Air National Guard F-86 Saber Jet. Less than a year after the F-86 flight, Jack Conroy met Clay Lacy, who was also serving in the Air National Guard.

In 2015 Lacy related the story of his first meeting with Jack Conroy. Lacy had already established a reputation as an outstanding pilot. Conroy was known in the local aviation community and had gained recognition for his coast to coast record flight. Clay was in the Air Guard restroom when he heard Jack yelling, "This toilet paper is like sandpaper! Can't the government buy anything decent?" Conroy was washing his hands while standing next to Lacy and said, "I'm Jack Conroy." Clay said, "I'm Clay Lacy" and they shook hands. Jack said, "I need to talk to you." They went to Johnny's Skytrails bar and began a friendship which lasted the rest of Jack's life. Lacy said "Jack taught me a lot and got me in airplanes which I would not have gotten to fly as a National Guard lieutenant. I flew to four continents with him in a T-33, B-25, C-47, F-86, C-97, DC-4 and Tri-Turbo DC-3. We had great fun together; he was my best friend."[5] As he told the story there was no question of the great admiration the two men had for each other.

Lacy also tells a story of being in a Hong Kong tailor shop which specialized in flight crew uniforms. The tailor asked Lacy if he knew Conroy. Lacy said yes. The tailor said, "He very strange man. He order suits for five different airlines. He real pilot of just dress like one?" The tailor was very curious and did not understand how Conroy could be flying for five different non-scheduled airlines at the same time. He had thoughts of Conroy being some kind of "great imposter."[6]

In time Jack introduced Clay to aviation legend Bill Lear and Lockheed test pilots Herman "Fish" Salmon and Tony LeVier. Conroy also introduced him to Allen Paulson, who owned California Airmotive. The company was mostly a parts business but Paulson began buying and selling surplus aircraft. Soon Lacy joined the business with Paulson. By 1958 there was a glut of used commercial aircraft available. They purchased all the surplus TWA Martin 404s, which were then converted to corporate planes. Paulson eventually became the founder of Gulfstream Aerospace. In 1974 Paulson's American Jet acquired the Pregnant and Mini Guppies after Aero Spacelines was absorbed into Unexcelled Chemical.

In 1966 Conroy and Lacy duplicated the F-86 coast to coast Operation Boomerang record in a Lear jet with three of Jack's children. Jack Conroy loved to hang out with his flying friends at the local airport Skytrails watering hole in Van Nuys.[7] He was a natural pilot and aviation tinkerer who often expressed his ideas on improving aircraft with this legendary group of aviators. Lacy was in business with Paulson buying and selling aircraft. He knew Lee Mansdorf and offered to introduce them. Lacy told the story of going with Conroy to see Lee Mansdorf after he decided he wanted to buy or lease three Stratocruisers to start his own airline.[8]

Lee Mansdorf was the oldest of four brothers (Lee, Harry, Fred and Norman) who were in the family business of buying surplus army aircraft for parts. Brother Fred was a talented engine mechanic who oversaw the rebuilding and resale of aircraft engines. Lee was a test pilot and Harry was a World War II B-24 pilot who was shot down twice becoming a becoming a POW. The four Mansdorf brothers were creative businessmen building a worldwide aircraft trading business.

The brothers pioneered skywriting and towing advertising banners behind aircraft. They traded a fleet of aircraft to the Venezuelan government for 75,000 acres of land. Lee was chased out of the country after giving 10,000 acres to the local people. The brothers were known for purchasing horses in Brazil and trading them throughout the world. Lee was a regular visitor to Cuba, flying there numerous times even after Castro came to power. The flights were discontinued once the Cuban trade embargo was implemented. The Mansdorf family accumulated considerable wealth by using their fortunes to purchase large blocks of land including nearly 1,700 acres in Malibu, California, which included three linear miles of beachfront.

Lee Mansdorf was well established in trading, rebuilding and modifying airframes prompting him to seek opportunities to purchase fleets of aircraft. In 1947 Grumman ended production of the G-44A Widgeon sea plane. The next year production was licensed to a French manufacturer and tooling was shipped to the company in France. After the company produced 41 of the new model amphibians designated the Scan 30, production was shut down because of sluggish sales and the failure to obtain a contract with the French navy. Mansdorf purchased 31 unfinished airframes in 1952 and had them shipped to California. They were upgraded and completed with better engines by associate company Pacific Aircraft Engineering as Gannet Super Widgeons and advertised by Mansdorf.

Mansdorf supplied aircraft to many of the postwar independent air operators. Some were rebuilt and placed in service while others were used for parts. A unique example is a surplus World War II ATC DC-4 which had been refurbished for Argentine President Peron. It had been disassembled and stored in Argentina when purchased by Mansdorf. Transocean's Orvis Nelson knew Mansdorf and learned the airframe was in excellent condition. Nelson purchased it from Mansdorf in 1953 in three pieces. He shipped the DC-4 to TAL's Hangar 28 at Oakland, California, where it was rebuilt and returned to service as the Argentine Queen. This was the first time an airline rebuilt an aircraft in company shops from surplus parts.

Mansdorf acquired a Navy R5D-3 (C-54 c/n 50877) on 22 August 1957 which he re-sold to Stanley Weiss of Twentieth Century Airlines. Weiss is known for introducing coach airfares to the airline industry. Weiss was president of four different airlines and was considered one of the most honest and astute individuals in the airline industry. Because he was so well thought of in the airline industry, Eddie Rickenbacker recommended Weiss as his replacement when he retired from Eastern Airlines.[9]

Jack Conroy and Clay Lacy were both airline pilots in the California Air National Guard (ANG) at the same time Lee Mansdorf was brokering aircraft. Lacy was flying for United Airlines and Conroy for non-schedule charter airlines. Conroy was always looking for opportunities. He was moderately successful with the swimming pool business after setting a coast to coast speed record in an Air National Guard F-86. He was receiving a lot of publicity in the local media from the record setting flight and appeared on the television show *What's My Line?* He even built a swimming pool for Lucile Ball. Lacy said he was having fun with the publicity opening airports and making speeches. However, he was not making much money from the appearances and his vision was to operate his own airline.[10]

Mansdorf had acquired the existing group of surplus B-377 airframes on speculation he could resell them for considerable profit because there was considerable time left on the airframes. As early as March 1956 rumors were circulated the Babb Company had negotiated with Boeing to purchase 14 of the 16 Stratocruisers scheduled for trade-in. Pan Am began pulling the B-377 from service in 1958 as trades on the Boeing 707s. The Pan Am Stratocruisers were completely phased out by 1960. The Babb Company sold six of them to associate company Transocean Airlines and the carrier purchased additional B-377s from BOAC. Prior to ceasing operations the Transocean fleet of 14 B-377s had been valued at $14 million. They were now idle after the demise of the carrier and there were no buyers or offers. Mansdorf who had previous dealings with Nelson acquired all 14 of the aircraft through Airline Equipment Company for a bid of only $105,000.[11]

Mansdorf also acquired a portion of the Northwest B-377 fleet which was taken on trade by Lockheed as part of the purchase of new L-188 Electra aircraft. The Northwest B-377 aircraft were parked at Burbank with an uncertain fate. Jack Conroy had left the National Guard to pursue other flying interest but returned when the unit received C-97s. It was the perfect opportunity to become familiar with the aircraft and build time on them at the government's expense.

After seeing the commercial B-377s stored at Burbank and Oakland, Conroy concocted a plan to acquire the surplus aircraft for a VIP airline between California and Hawaii. He did not have any money but he was a deal maker. It has been said when Jack Conroy got an idea it never occurred to him he would not succeed. Clay Lacy introduced him to aircraft broker Lee Mansdorf who owned the Stratocruisers. Conroy believed he could convince Mansdorf to sell or lease him the aircraft with no collateral. Lacy went with Conroy to pitch his proposal to Mansdorf in an attempt to obtain several aircraft. He expected Mansdorf to finance the deal because no one else was interested in purchasing the airliners.

Mansdorf had been seeking buyers for the surplus B-377s when he came across a magazine article describing NASA's logistical problems of moving the huge rocket boosters. Multiple proposals had been put forth on how to move outsize hardware for the space program around California and to MSFC at Huntsville and Cape Canaveral. Only two months before President Kennedy had announced to the nation the United States would be on the moon by the end of the decade. Mansdorf realized the financial potential in providing NASA with a transport system. He believe the surplus Stratocruisers were the perfect airframe for a transporter. He had a monopoly on the supply of aircraft with relatively little invested in them. This would allow him to produce a volumetric transport cheaper than anyone else. He had several ideas sketched out and was looking for the right person to convert the surplus aircraft. While Conroy was expressing his ideas and the desire to acquire Stratocruisers, Mansdorf decided to toss out his idea. He showed Conroy the article describing NASA's logistical problems. He explained that running an airline and dealing with the public was labor intensive and low yield. Transporting cargo for NASA under a government contract could be very lucrative. Besides, cargo does not talk back, require meals or lose luggage.[12]

Mansdorf then pulled out a sketch of an enlarged fuselage Stratocruiser and slid it across the desk to Conroy. He said, "What do you think about this?" He envisioned taking the Stratocruiser and building a giant fuselage which would open from the top. It was an extreme concept using the principle of a top loading cargo door. He proposed lowering the rocket boosters with a crane into a giant fuselage. Conroy was immediately interested in the project. He and Lacy left Mansdorf's office without a deal to purchase the B-377s. Lacy said, "You could see Jack was already thinking about how to modify the aircraft into an outsize transport."[13] The idea of a VIP airline would soon be a forgotten dream.

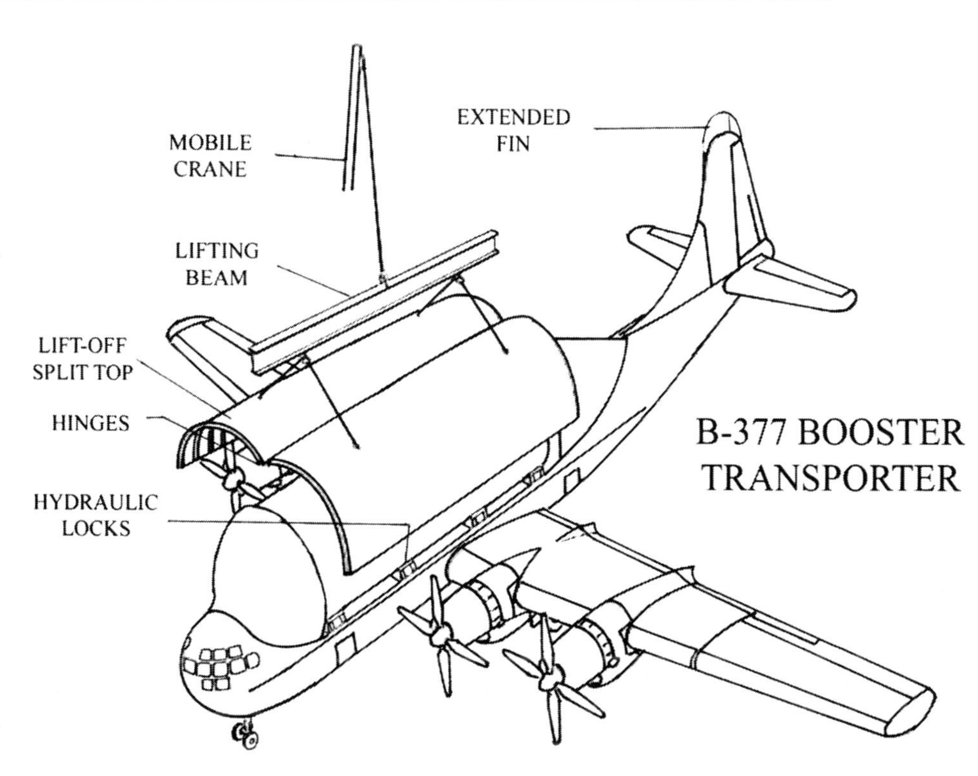

Lee Mansdorf toyed with the idea of building an enlarged fuselage B-377 with a bi-fold lift-off top to transport NASA boosters. He proposed the idea to Jack Conroy, who attempted to purchase some of the surplus Stratocruisers for a VIP airline (ASI, courtesy Tom Smothermon).

During a cocktail hour brainstorming session with his friends at Johnny's Skytrails bar, Conroy began to explain why he was shelving the VIP airline idea. He showed the open top B-377 drawing around the table. The conversation changed from starting an airline to the proposed concept of transporting large rocket boosters and pieces of space hardware for NASA. The agency was seeking an alternative to sea-going transport of rocket boosters and Douglas was evaluating a piggyback transport. All agreed the military did not have aircraft capable of outsize cargo. Douglas Aircraft was working on an external transport design because their engineers did not believe it was possible to build and aircraft big enough to transport a booster internally. There were other proposals of strapping a tail and wings on a booster and towing it as a glider behind another aircraft. NASA was struggling for an alternative to move the rocket boosters from the factories in California to MSFC and Cape Canaveral. The problem was becoming critical because NASA planners had not made any commitment to logistics in the early years of the space program.

Douglas Aircraft proposed the C-133 piggyback idea to NASA on 26 September 1961. NASA was not receptive because the aircraft already had known stress and vibration problems. The field was open to any entrepreneurial proposals. Mansdorf 's idea of modifying Stratocruisers for outsize cargo seemed to be a viable alternative to the proposed C-133 external pods and faster than NASA's sea-going transport of rocket boosters.

NASA relied on the Air Force for transport of smaller rockets. The C-124 Globemaster had performed well transporting the missiles but was also built from World War II technology and showing its age. As large as it was it could not transport the outsize boosters. The C-133 was designed and developed primarily for missile transport but it had a smaller fuselage. It could handle lengthy cargo but could not accommodate large diameter outsize bulk. In October 1961 the Material Air Transport Service (MATS) received a Qualitative Operational Requirement (QOR) from Headquarters Air Force. It specified a successor to the C-133 was needed which could accommodate larger missiles.

Conroy's informal cocktail hour group agreed the government would take years to figure out how to transport the rockets. Mansdorf convinced Conroy there would be far more opportunity, stability and profit in a contract with NASA than forming an airline and dealing with the CAB. However, he had no desire to convert the aircraft himself. The project had potential but he was a deal maker and aircraft broker not interested in managing a day-to-day aircraft manufacturing operation. It was Mansdorf's idea but it was obvious Jack was really convinced. Mansdorf was still thinking in terms of an open top outsize Stratocruiser when he said to Jack, "You think you can handle it?"[14]

As the group tossed out ideas around a big table at Skytrails, Conroy was busy sketching out variations of Mansdorf's design on cocktail napkins. Conroy was not an engineer but was full of ideas and acutely aware of how aircraft are designed. As they discussed the concept of an outsize aircraft with doors which opened on top, Clay Lacy was watching Jack frantically drawing on napkins. Conroy had initially sketched out an idea to build an outsize fuselage with a swing nose similar to the Super Guppy. The group pointed out this would involve developing a hydraulic system, hinges and require reinforced structure adding weight. After a moment, Conroy produced a new drawing. Lacy jumped in saying, "Jack has a better idea!" Conroy said this design will cut weight, reduce engineering and decrease production time. He had drawn an expanded fuselage Stratocruiser with a detachable tale which rolled away for easy straight in loading. Lacy pointed out Conroy's idea of taking the tail off an aircraft then slide the rocket booster in and bolt the tail back on was by far the easiest to accomplish and added the least amount of weight to the airframe.[15]

By the end of an evening of cocktail fueled creativity, the not yet named guppy class aircraft took life from Mansdorf's idea. Jack Conroy's original passenger airline proposal lay under a pile of cocktail napkins scrawled with ideas of a way to save NASA and make aviation history. Conroy was so motivated he could hardly sleep that night. His creative imagination was awakened as he began expressing ideas of how to create a series of oversized special purpose aircraft. Within weeks he envisioned aircraft with up to forty-foot diameter fuselages based on almost every large multi-engine aircraft which existed.

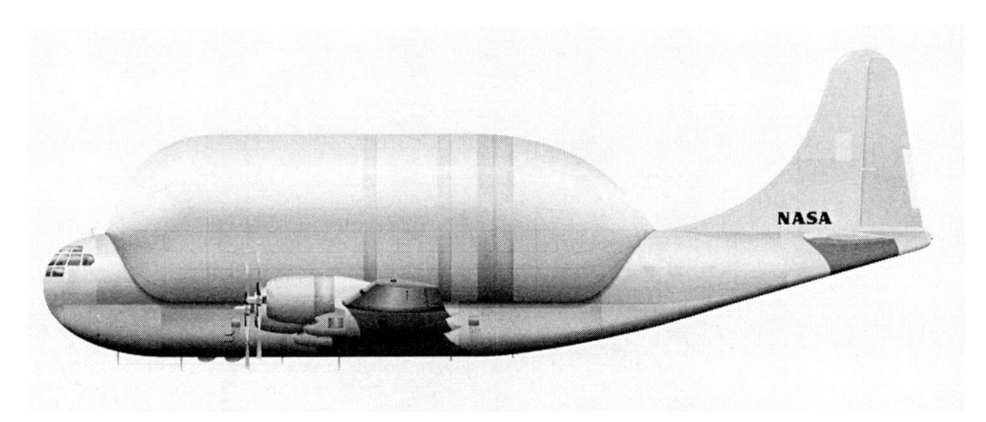

The first expanded B-377 design with a rollaway tail was a huge cylinder with teardrop ends mounted on a Stratocruiser airframe. The forward section was not blended into the B-377 nose and did not have a dorsal fin. In reality it probably would have been very unstable and generated turbulence at the rear rendering the control surfaces marginal (courtesy Michael Zoeller).

Conroy envisioned a fleet of outsize aircraft to move rocket stages for NASA. Aviation artist Doug Ettridge was engaged to produce the drawings of the proposed outsized booster transporter. The detail drawings of the unnamed aerial transporter were depicted with the rollaway tail concept, which impressed Wernher von Braun at the very first meeting (Doug Ettridge, courtesy Angelee M. Conroy).

Only days after Conroy was introduced to the idea he began to set up a new company. Lee Mansdorf agreed to supply several Boeing B-377s as his financial contribution to the project but would not invest any cash. If Conroy wanted to do it he would have to come up with the money to build them. In the following days, Conroy reviewed the problems NASA was experiencing in moving the giant rocket boosters and associated space hardware by barges. He discussed the internal transport design with Douglas engineers who were skeptical but a few other aerodynamicists agreed it might be possible. Conroy decided the logistical need was great enough for a volumetric transport and NASA would buy it if he could build it. He managed to put together a small group of engineers who believed building such an aircraft was possible.

Furman Associates, headed by Dottie Furman, was selected as the public relations firm to handle all press releases for the newly formed venture. British Aviation artist Douglas Ettridge was engaged to produce a series of drawings of the proposed outsize transporter. Ettridge had only been drawing and painting aircraft for about five years beginning in 1957. His ability to express Conroy's ideas in fascinating detail was exceptional. Ettridge became friends with Los Angeles television newsman Clete Roberts, who often took Ettridge flying. Roberts was fascinated by Conroy's adventures. It is not clear how the trio met and who was first to introduced them. It is possible Dottie Furman was instrumental in putting the group together. Whatever the situation Furman handled all the press releases and Roberts was always included in every Conroy announcement. Ettridge produced the presentation drawings for probably everything Conroy dreamed up. The drawings used at the meetings with NASA director Wernher von Braun were done by Ettridge.

Conroy mortgaged his house and only came up with $20,000 but it was enough to get the conversion project started. He contacted everyone he could think of who may be willing to invest in the project. In the process of setting up the new company he discussed the project with Robert "Bobby" Siegfried of Oklahoma who handled all his aviation insurance. Siegfried suggested Conroy should meet his friend Lloyd Dorsett of Dorsett Electronics who already had contracts with NASA. He emphasized that Dorsett was a very shrewd businessman with outstanding credentials and a good understanding of the space program.

Lloyd Dorsett founded Dorsett Electronics in 1946. The Norman, Oklahoma, based company was the perfect choice to partner in the project. Dorsett was a pilot with degrees in physics, electronic engineering and a Ph.D. in industrial engineering. His company produced space and missile telemetry equipment which was being used by the military and NASA.[16] Dorsett was aware of the NASA logistical problems through his dealings with the government. This put him in the position to provide needed contacts and lend credibility to the aircraft building project. Jack Conroy and Lloyd Dorsett agreed to form a new company together.

Aero Spacelines was incorporated (ID 1900172981) in Norman, Oklahoma, on 15 November 1961. Dorsett provided considerable startup capital and assumed the position of chairman of the board. Conroy took the position of president and was named director along with Gene Hopkins and Lee Mansdorf. Kenneth Healy became the VP of operations. The technical advisors were listed as Herman "Fish" Salmon and Alvin M. "Tex" Johnston. It is interesting that Lee Mansdorf was also a vice president of U.S. Missile Corporation at the same time.

Conroy selected "Fish" Salmon because they were close friends and he was an outstanding test pilot. Salmon had been part of the brain storming sessions at the Skytrails bar when the idea was first kicked around. Salmon and Conroy would make many test flights together after the Pregnant Guppy was built. Salmon was an extremely calm personality. His easy going nature was the object of a standing inside joke among the group because he was able to sleep anywhere. Multiple photos were located of him sleeping in airplanes. He had been chief engineering test pilot for Lockheed-California. His reputation as a Lockheed test pilot is legendary with several close calls.

Ten years earlier in April 1955 Salmon was at the controls of the second prototype XF-104A 53-7787 during a test firing of a 20 mm gun. Vibration caused the bottom ejection seat hatch to separate from the aircraft causing an explosive decompression. The aircraft began to breakup as Salmon managed to eject. He was injured when he landed in a remote desert area.

Aero Spacelines Incorporates in California

The Aero Spacelines Oklahoma corporate filing was followed up on 10 January 1962 when Aero Spacelines filled corporation papers (ID CO426078) as an outside company moving to California. Conroy struggled to raise money and was continually underfunded. While many in the banking community believed the guppy to be true folly, Carl Schatz, who was the president of the Independence Bank of Encino, was intrigued with the guppy project. He liked Jack and was very creative in securing financing for the project on many occasions.

The California Corporations Commission was just the opposite monitoring virtually everything Conroy did. He wanted to build airplanes; however, the idea was so revolutionary the commission saw the entire project as an investment scheme. As soon as the first guppy began to take shape his efforts got the attention of the engineering departments at the major aircraft manufacturers. Douglas Aircraft had participated in some of the initial design aspects and assisted in presentations to Marshall Space Flight Center officials. Boeing became very interested in drag calculations dispatching engineers to Van Nuys to review Conroy's progress. Conroy developed Mansdorf's idea of an outsized transport, struggled to raise capital, built the aircraft and gambled on getting a NASA contract and all with private money. It is questionable if any of the early investors ever saw a return on their investment.

Clay Lacy said you could tell from across the field if Jack had money by the sound of rivet guns. If he had money you could hear multiple power tools and rivet guns thrashing away. If he was broke you could hear one guy every few minutes installing a rivet.[17]

There is some indication financial speculators became aware of the potential of Conroy's aircraft project very early but would not consider investing capital in such a speculative venture. Aero Spacelines always operated on the financial edge making it attractive for speculators in the shadows to just wait until the development of the aircraft was successful then make a move on the company. At the time the government was spending billions to put a man on the moon and the general opinion was it would last forever. Space was the future and America was committed to beat the Soviets by landing a man on the moon by the end of the 1960s. A long term NASA contract to transport space hardware could be worth $100 million over time if America could conquer space.

We now know Conroy was building his airplanes in a sea of financial alligators as he struggled to secure financing. Once he obtained a contract with NASA several individuals with questionable motives saw Aero Spacelines as a potential takeover target. Unexcelled Chemical was being transformed into a small conglomerate. By 1965 the Wall Street appetite for emerging growth companies and the lack of capital made Aero Spacelines the perfect candidate. The opportunity for stock manipulation and insider trading placed Aero Spacelines at the mercy of the manipulators of Unexcelled Chemical stock.

12

B-377PG Pregnant Guppy

The First Conroy Class Volumetric Aircraft

Jack Conroy had an impressive resume and many friends in the aviation community. His coast to coast speed record in 1955 brought him name recognition. The meeting of Clay Lacy in the 1950s began a close friendship and produced many contacts in aviation. Lacy was already on the way to becoming an aviation legend, eventually flying for 40 years with United Airlines. During his career he acquired 30 type ratings and 29 world speed records. In 2017 Clay Lacy was reported to have logged over 50,000 hours in 300 types of aircraft, which is more than any other pilot in history.[1]

Conroy was not an aerospace engineer or even an aircraft mechanic. Flying was easy for him because he had a natural ability to feel an aircraft when at the controls. He was excited about NASA's request for transport feasibility designs because of the major logistical problem in getting components to Huntsville. Von Braun and his team had rejected the idea of moving rocket boosters on the back of an existing aircraft leaving the field open for additional proposals including internal transport.

Conroy needed a way to get NASA to review his proposal of an outsize aircraft capable of transporting rocket boosters internally. He contacted R.W. Prentice, who managed the S-IV logistics program at Douglas, and presented his idea. Most aircraft engineers and aerodynamicists believed an external canister mounted on top of an existing aircraft was the only alternative for NASA hardware transport. Conroy continued to press his design of an extreme diameter "volumetric" airplane despite the respected opposing opinion of aircraft engineers and aerodynamicists. They believed no airframe could be distorted enough to transport internally a S-IV rocket stage and still be structurally sound enough to fly or have enough power to be efficient.

Designing an aircraft which will efficiently fly depends on the drag/lift coefficient. Drag is the summation of all forces which resist the aircraft forward motion. Most aircraft designers were very skeptical of increasing the frontal area because any change in the external configuration will change the drag coefficient which affects power required and the operating cost. The wetted area is the total area which has high speed air flowing along it. A greater wetted area increases the drag; however a greater wetted area can be overcome with longer wings. The amount of frontal area was believed to be relative to length of the fuselage. With this in mind, engineers were skeptical Conroy's aircraft with extreme frontal area could fly efficiently. The accepted opinion of building a fuselage in the diameter required theoretically would make it so long and with such a great wingspan it would be structurally unsound. In addition the size and number of power plants required to get it off the ground would be prohibitive. In retrospect one can see the B-377PG was on the verge of being underpowered. In contrast, the Super Guppy with extended wingspan and turboprop engines was adequately powered. Furthermore Conroy's outsize fuselage design inadvertently took on some characteristics of a lifting body which was not a concept fully understood at the time. It is interesting that frontal area was theoretically limited and yet mounting a large pod on top with a large frontal area was thought feasible.

Conroy enlisted Irv Culver who was a senior aerodynamicist at Lockheed. He is said to have coined the name "Skunkworks" for Kelly Johnson's secret aircraft project facility. Culver said an enlarged fuselage aircraft was possible and would be stable. In the beginning Conroy did not have the funds to rent wind tunnel time for testing. Culver put a Stratocruiser model on a stand and attached silk strings to it. He then took cardboard and modeling clay to develop a shape for the fuselage. He placed the primitive model in front of a fan and adjusted the shape until he felt comfortable with the results. From this experiment he did the initial drag and lift calculations for Conroy. In 1961 Lee Mansdorf hired structural engineer Abraham Kaplan of Strato Engineering to draw the proposed outsize "transporter" dictated by Culver's design and calculations.

Abraham Kaplan founded Strato Engineering in 1952 providing complete aircraft design engineering. He had an excellent rapport with the FAA and was confident he could get the transporter (guppy) certified. The head of drafting for Strato, Eugene "Gene" Stanley, did many of the design drawings under Kaplan's direction. Strato created some of the early design drawings for Lee Mansdorf's idea of a proposed booster transport. Some of those early renderings had been used to capture Jack Conroy's interest in building an outsize transport in the first place.

Douglas Aircraft had been working on external transport systems for NASA without much progress. Conroy's gift of persuasion convinced Douglas executives it was possible to build a giant transport and he was the man to do it. Douglas was looking at Conroy as a sub-contractor or consultant. However, Conroy was so convincing he persuaded Douglas officials to let him make a presentation to NASA officials in Washington which included the team from Marshall Space Flight Center headed by Wernher von Braun. Chief of MSFC Logistics John Goodrum later stated, "Conroy touched a chord with the agency's visionary director Wernher von Braun who warmed to the idea from the start."[2]

Conroy gave Abe Kaplan only three days to complete the drawings based on Culver's calculations. The drawings and data along with a model built by Lloyd Jones and artist renderings were presented in Washington to the NASA MSFC team, House Budget Committee representatives and Pentagon officials. The majority of the NASA managers and engineers were skeptical at best. The group of bureaucrats listened to Conroy with icy skepticism. Aeronautical engineers from NASA expressed doubt it could fly because of the outsize shape. NASA did not have the funds for such a project but von Braun loved innovative concepts and was impressed with the idea. The House Appropriations Committee told von Braun, you have one mission and it is to put a man on the moon and this does not involve developing airplanes.

The group was less than enthusiastic with Conroy's proposal. One of the officials brushed the entire presentation aside by saying that no one had found a way to move big components around economically and that if Conroy wanted to develop such an aircraft and it flies the government may be interested but would not be funding any such project. During a break in the meeting Conroy stepped out of the room. He overheard one of the skeptics say rather loudly: "The damn thing looks like a pregnant guppy. It will never fly!"[3]

Jack Conroy was convinced it would fly. He had discussed it at length with Irv Culver and Abe Kaplan who staked their professional credibility on it. Conroy was not one to challenge when he believed in something. He had convinced engineers at Douglas to allow him to present the project to NASA and the budget committee. After all Douglas had failed with the proposal to transport the boosters on the back of a C-133 aircraft. Conroy lived on optimism and always set unreachable goals. He was so convinced it would fly he believed the trip to Washington was a formality to obtain funding. The committee saw it as folly and told him so. Jack Conroy took his inspiration from many sources. He had a quote from inventor Charles F. Kettering which hung in his office: "If I could only get across to people the extreme difficulty of starting anything new."

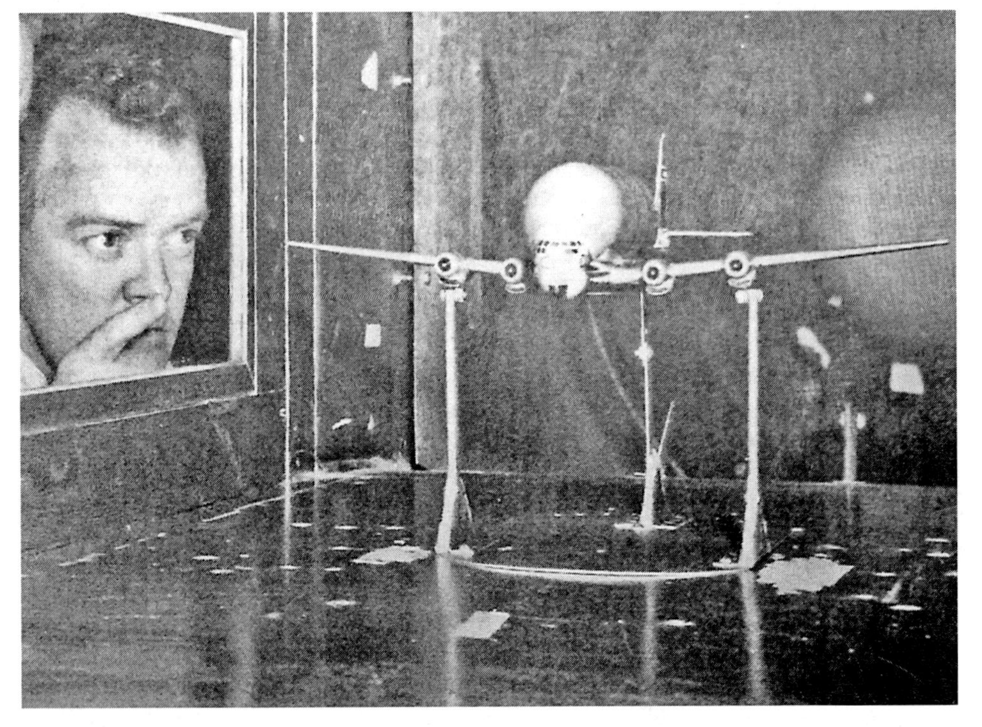

Initially Jack Conroy did not have funds for wind tunnel testing. Irv Culver put a model on a stand in front of a fan. He used cardboard and modeling clay to produce a shape he felt comfortable would fly. Eventually wind tunnel testing was done at the University of California under the close review of Conroy. He was hands-on involved in every aspect of the development and building of the aircraft (Dottie Furman, courtesy Angelee Conroy).

Jack Conroy did not take the "pregnant guppy" comment lightly. He would get his revenge by not only building and selling the aircraft to NASA but he would christen it the Pregnant Guppy. Von Braun had not sided with the skeptics. He was not only a scientist, engineer and pilot but a visionary. He identified with Conroy's enthusiasm of thinking bigger with new concepts. After all von Braun was in the process of doing something that had never been attempted before. He was building the biggest rockets ever attempted to explore space and no one was telling him "no." He was immediately impressed by Conroy's idea and believed it was the solution to the booster transport problem. Von

Braun's approval was all the incentive Conroy needed. With no contract and only the enthusiasm of Wernher von Braun, Conroy was convinced he could sell the airplane to NASA once it was built.

The first design was not streamlined and had the enlarged fuselage rise up abruptly behind the B-377 cockpit. In all subsequent designs the enlarged fuselage was more streamlined and blended above the cockpit. The aft tapered to the tail behind the enlarged constant section. A model with the redesigned frontal area was wind tunnel tested for two weeks at the University of Southern California. To begin production Robert W. (Bob) Lillibridge was hired as vice president of manufacturing and engineering. A team was assembled for the project which was contracted to On Mark Engineering at Van Nuys, California. Conroy began construction using his own funds plus an initial investment by Lloyd Dorsett and whatever he could convince friends to invest in the project. In his mind he was not building one aircraft but the first of a fleet of imagined giant transports. He had Doug Ettridge producing drawings of almost every large aircraft which existed with enlarged fuselages before the Pregnant Guppy was even finished.

The B-377 Stratocruiser N1024V in the first livery. It was repainted in the last Pan Am livery about three months before being phased out in 1958. Aero Spacelines moved it to Van Nuys to have the fuselage lengthened before the construction of the expanded upper fuselage (Henry W. Arnold collection of AAHS archives).

The future need for larger transports was becoming apparent to the military, NASA and even commercial airlines by 1962. Aircraft companies were studying heavy jet transports for the military prompting the Boeing CX-4 (Cargo-X) program in August 1962. The program produced multiple designs for larger aircraft but almost all had six-engines which was unacceptable. The existing theory still required the wing to be lengthened proportional to diameter of the fuselage. Boeing, Douglas and Lockheed were competing with design studies of large military transports but they adhered to established design theory of wing length proportional to fuselage. If a practical cost effective design could be found it could bring a major military contract. Boeing began looking at the aerodynamic characteristics of Conroy's modified outsize transport design using the standard B-377 wing. It was essentially the B-29 wing, which was revolutionary when first introduced. Now Conroy was building an extreme fuselage aircraft and utilizing their technology to cut cost. Aeronautical engineers were becoming very interested because he had expanded the fuselage but did not increase the wingspan.

Boeing engineers encouraged Conroy to share his drag calculations of large diameter transports with them because military design studies were being conducted at the same time. Boeing volunteered wind tunnel test for Conroy's proposed expanded B-377 Stratocruiser to determine the aerodynamic feasibility of the oversize fuselage modification for other applications. A lot of the data compiled from Conroy's guppy design was eventually used by Boeing in the design of the B-747 which was independent of the Air Force call for a new outsize transport.

With satisfactory wind tunnel results on the expanded B-377 fuselage design, the next step was to build it. Mansdorf had acquired most of the surplus fleet of Stratocruisers. Fourteen of them were ex–Transocean that he acquired through Airline Equipment Corporation along with 20 spare engines. The Northwest aircraft N74601, N74602, N74603, N74604, N74605, N74606, N74609 and N74610 were acquired from Lockheed and moved to Mojave along with the only four airworthy Transocean aircraft N402Q, N403Q, N404Q (N9600H) and N413Q. Pan Am B-377s N90942 and N1042V were also moved to Mojave and N1024V to Van Nuys. The balance of 10 Transocean Stratocruisers, N401Q, N405Q to N412Q and N414Q were parked at Oakland and broken up by ASI fabricator Don Stratman for parts to support the guppy program. Six of B-377s acquired by Transocean never had the U.S. registration or livery applied and remained in BOAC colors. Conroy released a statement in April 1962 stating Aero Spacelines owned or had options on 12 Boeing B-377s, eight ex–Northwest, two ex–Pan Am and two ex–Transocean aircraft. In addition he had acquired the Northwest stock of B-377 spares which included engines, (the troublesome) Hamilton Standard props and engineering drawings.

The conversion project began in January 1962. Conroy did not have any production facilities or experience in building aircraft. Consequently the construction was contracted to aircraft modification company On Mark Engineering at Van Nuys. Bob Lillibridge was hired by Conroy as the project engineer for the conversion.

On Mark was founded in 1954 by Robert O. Denny and was recognized by the FAA as an unlimited airframe repair facility. For over 15 years the facility overhauled and converted Douglas A-26 aircraft. Some were converted to executive transports but the primary business was the modification of A-26 aircraft to B-26K counterinsurgency work for the CIA and Air Force. On Mark had completed approximately forty B-26K models. The company also performed overhaul and upgrades for 105 Navy and Air Force C-54 transports. On Mark had considerable experience in airframe work on smaller aircraft. The aircraft modification company was very capable. However, the building of the Aero Spacelines outsize transports was a challenge for even the most experienced facility.

Construction began outside at Van Nuys in January 1962 because there were no large hangars available for the project. The conversion was contracted to On Mark Engineering. Scaffolding was erected with only narrow boards for the workmen to stand on. Work progressed slowly on a pay-as-you-go basis as evidenced by this view in March 1962 (author's

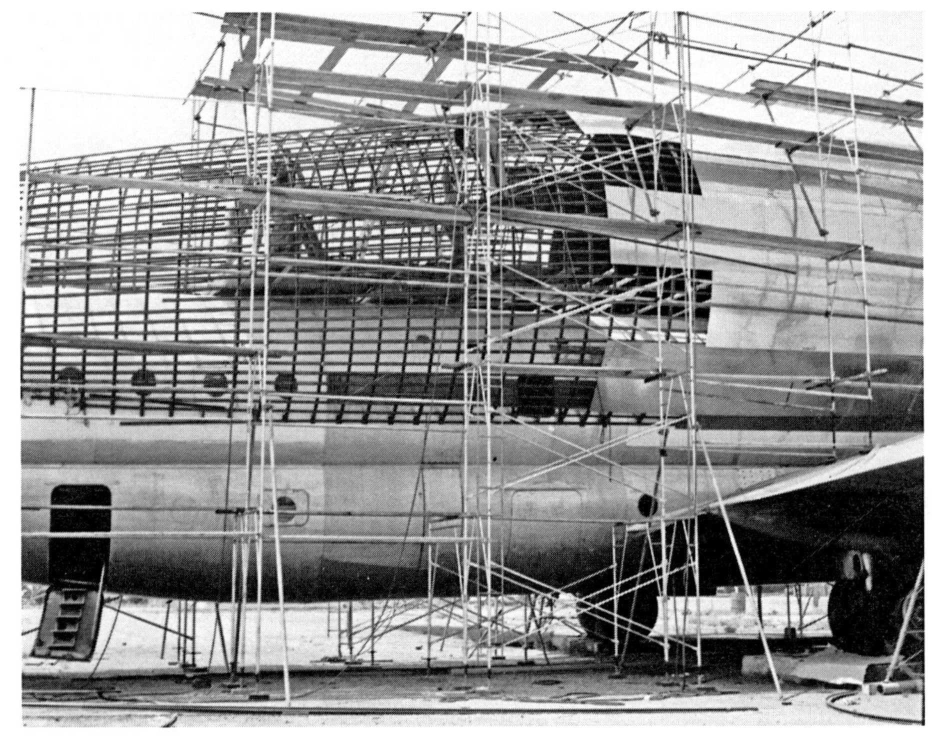

The fuselage of N1024V was stretched by splicing in behind the wings of a 16-foot forward section 42 from BOAC B-377 G-AKGJ. It is clearly visible in darker color between the two windows as the ribs and stringers are grafted in to the existing B-377 airframe approximately 24 inches above the original floor level (Lloyd S. Jones collection of AAHS archives).

Aero Spacelines applied to the FAA for a certificate of airworthiness for the outsized Stratocruiser on 24 January 1962 under part 49 U.S.C.A. Many shortcuts were being taken which are only acceptable on military aircraft. In order to obtain certification with all the shortcuts and modifications, NASA proposed it should be considered a "public aircraft." The plan was changed to have it certified as a public aircraft operating under exclusive contract to NASA. For this reason Aero Spacelines did not immediately apply for a Commercial Operators Certificate, assuming it was not necessary.

The entire guppy aircraft conversion was to be done outside. Conroy developed a three phase plan. They would lengthen the B-377 fuselage and conduct test flights. Phase two consisted of building the enlarged upper fuselage. If it flew and was stable the inner B-377 fuselage would be cut away and the tail separation joint would be added.

The airframe of Pan Am Stratocruiser N1024V (c/n 15924) was separated just behind the wing. The straight Section 42 forward of the wing from station 230 to 430 measuring 16 feet, 8 inches was removed from ex–BOAC B-377 G-AKGJ (Transocean N407Q) (c/n 15976) which only had 18,985 hours when scrapped. The section was spliced in behind the wing of N1024V stretching the fuselage from 110 feet to 127 feet. Many adjustments and changes had to be made for the relocated center of gravity. After completing the structural modification and calculating a new CG, it was flown to determine the handling characteristics and the difference in power. Although Conroy insisted on photographing the project daily to record the progress and document the modifications, no photos of the lengthened B-377 with shortened inboard propellers were located during the entire research for this project.

Workmen fabricate the large upper frontal area while perilously perched 18 feet above the ground on a single plank suspended between two maintenance stands. The appearance of the scaffolding and the entire process did indeed resemble an amateur backyard project (Lloyd S. Jones collection of AAHS archives).

The stretched B-377 was flown only twice, once with original props and once with the shortened inboard props which would be required to clear the expanded fuselage. Clay Lacy stated they only needed to cut about eight inches off the inboard propellers. However, the technicians got a little carried away and cut off nearly 15 inches. Despite the shortened props there was no discernible difference in speed.[4]

Additional flight test were needed with the stretched B-377 but Conroy was running out of money. The project moved forward with the limited data collected. With little flight data construction began on an unproven large non-structural 20-foot outside diameter fuselage. It was built up above the existing fuselage because it was not determined if the outsize unit could support itself.

These were possibly the most primitive conditions under which any large aircraft had ever been built. It was less of an aircraft manufacturing operation and more of an amateur backyard Frankenstein airplane project. There was no hangar or cover available consequently all work was done outside with scaffolding. Tarps were strung over the scaffolding to provide shade and protection for workers. The reciprocating Pratt & Whitney R-4360 engines remained on the aircraft. In April 1962 Conroy told the press he planned to install small 3,000-pound thrust Westinghouse J-34 jet engines between the inboard and out board engines. At this point it was still believed the size of the new fuselage would reduce the speed to a point where additional power would be needed. The stretched B-377 sat on the gear while jacks were used to support and stabilized the airframe at critical points with one at the rear of the fuselage to prevent tipping.

A fuselage inside diameter of over 19 feet would create a new aircraft capable of transporting the upper stages of the Apollo boosters however, the larger lower booster sections would still have to be transported by barge. At this point Conroy was already planning larger conversions because he was aware the Pregnant Guppy would only be a stop gap aircraft. As the space program progressed larger aircraft would have to be built to keep up with NASA transportation need.

The new upper half was built up by attaching large diameter ribs above the existing fuselage. Rib attachment was spliced in above and just behind the rear cockpit bulkhead. With the old fuselage still intact the lower ends of the rings were secured through the skin to the existing rib of the old fuselage. The first five forward rings were each larger than the preceding until the sixth and seventh were over 19 feet in diameter at the beginning of the constant section.

The constant section ribs were attached above floor level to create a 30-foot long constant section in the 80-foot cargo bay. At the end of the constant section 38 more ribs each progressively smaller were added back to the waist. Ten additional smaller ribs were added on each side of the existing vertical fin to just forward of the horizontal stabilizer. All were tied together with longerons back to the fuselage waist and were allowed to sag over the sides of the fuselage. The stringers and longerons were then trimmed with the ends attached through the skin in the same manner as the ribs. Temporary supports were added across the top of the old fuselage and attached on each side to the new ribs. Sheets of plywood were laid across the supports to allow workers to walk on top of the old fuselage and work on attaching the longerons and stringers to the ribs.

The 30-foot constant section of the fuselage was skinned first with alternating panels for strength. Longerons and alternating stringers were attached from the first constant ribs forward and down across the five ribs over the flight deck and grafted in just above the cockpit windows. The eyebrow cockpit windows were eliminated. The contour of the fairing also covers half of the rear cockpit window on each side. An extended dorsal fin was built up and added forward of the vertical fin to enhance stability and yaw control.

The original 8-foot wide B-377 cargo floor was left intact. The new upper fuselage expansion is grafted in along a line approximately 24 inches above the floor level. It was spliced in at this height to raise the huge new fuselage high enough to allow clearance of the inboard props. This is another example of design short cuts to complete the aircraft as quickly as possible. When Conroy's second design B-377SG was built the center-wing was extended to allow the grafting in of the expanded fuselage at floor level. After the framework was built up on the B-377PG, Aero Spacelines ran out of funds again. Because the entire project was privately financed, when finances ran short work ceased. Production was shut down for months until Conroy could find more investors.[5]

The method of building and supporting the massive top was quite primitive. The roof of the original Stratocruiser was left in place and even two-by-four lumber was used as bracing for the massive outsized fuselage. One of the reasons the old fuselage was left in place for flight testing was fear the structure may not be able to support itself in-flight. The engineering calculation showed it would support the weight and could withstand the stress of flight, however, nothing this size had ever been built before.[6]

Parts were often fabricated by hand from sketches without engineering drawings. In many instances the engineering drawing was done after the part was made and installed. Strato-Engineering draftsman Jim Luff would be dispatched to Van Nuys airport to examine the part, take photos and make notes then draw the print to match. Conroy was known for saying, "It's amazing what you can do without paperwork."[7] Despite the appearance of Conroy's engineering being quite cavalier, he was very thorough insuring the aircraft met the required design and safety standards. An FAA inspector stated, "Jack paid attention to FAA requirements and cooperated fully with the certification process. His methods seemed haphazard but his engineering was done by FAA approved designers who were well-known to the western region engineering staff."[8]

To obtain airflow data for comparison to the guppy, Conroy used ex–Pan Am B-377 N1038V which was still airworthy at the time.[9] Stratocruiser N1038V would eventually contribute enough parts to provide the registration to second built outsize super guppy built using YC-97J components and turbine engines. A camera was fitted to the left wingtip and side of Stratocruiser N1038V. The fuselage was tufted to obtain critical airflow data for comparison to guppy N1024V.

Construction progressed slowly as 38 frame rings, each progressively smaller, were added behind the fuselage constant section. The original registration, N1024V, can be seen under the last rib. The dorsal fin, which was installed prior to the first flight, had not yet been fabricated (Lloyd S. Jones collection of AAHS archives).

The so called B-377 test flights were conducted with Conroy in the left seat and Herman "Fish" Salmon in the right. Lloyd Jones sat behind them and recorded the numbers as Conroy called them out. There was not enough money to buy fuel for any long test flights or even short flights. To reduce cost Conroy would run the engines up to takeoff power, race the B-377 down the runway then just as it started to lift off he would chop the power, reverse props and stand on the brakes. This was done a number of times until the brakes got so hot they were smoking. They would have to let the brakes cool down before attempting it again.[10] As soon as the minimum amount of data was compiled, Conroy declared the test complete.

Von Braun was enthusiastic about the aircraft after Conroy's first presentation in Washington when a bureaucrat declared it looked like a "pregnant guppy." He became very hands-on with the project. As the building of the guppy progressed von Braun made numerous visits to Van Nuys with his engineering team and NASA's chief of Saturn logistics Julian Hamilton to consult with Conroy and review the progress. By the time the fuselage began to take shape von Braun knew it was going to happen and was completely sold on the idea. He had faith in Conroy but others at NASA expressed doubt, especially when reviewing the concept of unbolting and removing of the rear fuselage for loading. Some of the doubters were aware of comments which had been made by the project review group during the first presentation at the Pentagon. Those same skeptics expressed doubt by continuing to refer to it in a negative manner as a pregnant guppy. Like it or not Conroy's newly created class of outsize volumetric aircraft would forever be known as "guppies."

Conroy was not operating in a vacuum and news of his progress was being reviewed at Boeing and Douglas. The engineering department at Boeing had taken the data from the guppy model wind tunnel test, the numbers from the B-377 high speed test data and their own data from the B-377 wing characteristics to produced drag calculations. All related material was considered in an attempt to calculate loss of speed because of the high frontal fuselage. Up until this time, no one, including the most forward thinking designers, believed it would be possible to fly an aircraft of this diameter without increasing the wing. The Air Force was consulting with manufacturers for a new large transport aircraft. Boeing had two designs on the drawing board, the military transport which was initially designated as the CX-4 program and a large commercial passenger aircraft which would become the B-747. Boeing was concerned about the amount of power needed and speed loss for an aircraft with a massive frontal fuselage. The guppy was a flying test bed. Boeing engineers reviewed the different drag formulas calculating what they believed would be a speed loss between 22 and 40 knots. They were very surprised after it flew and the loss was only 15 knots.[11] Boeing is a giant in aircraft manufacturing with departments of engineers and aerodynamicist and here was a guy with a handful of employees and a couple of consultants building an aircraft outside with hand tools setting the standard for outsize transports.

First Guppy Flight

FAA inspector James Bugbee, who years later would pilot the near disaster Conroy 103 CL-44 guppy test flight, was present for the guppy practice runs the night before and the first official flight on 19 September 1962. It has been reported the first flight was tarnished with an aborted takeoff. However, Bugbee stated the supposedly aborted takeoff was actually a handling test monitored by him as requested by the FAA. The FAA in Washington was skeptical from the beginning and reluctant to give the western region the okay for Conroy to takeoff from Van Nuys with a million citizens living in the San Fernando Valley. Bugbee was told by his superiors at the FAA the test flight would be approved however, if there was an incident he would be held accountable and dismissed.

Bugbee's primary concern was about aircraft handling if an engine failed during takeoff. As a result of pressure from his superiors, he insisted on an additional test of the Vmc drift the morning of the first flight in the day light. The FAA feared and expected handling problems because the fuselage was so greatly expanded and the rear section sloping sharply to the tail. Conroy had been concerned and agreed to conduct the test by accelerating the Pregnant Guppy down the runway at Van Nuys, getting airborne between 10 and 50 feet then pull the power back on an outboard engine to ensure there was enough rudder to correct the yaw. The power to the other three engines was then cut with aircraft settling back down and rolling out to the end of the runway.[12] Clay Lacy verified the test were done then smiled and said, "We had already done it more than once." He was referring to the night before.[13]

Up until the last minute the FAA was not comfortable granting permission for any flight testing of the aircraft. Conroy was able to convince them high-speed taxi text would not be a safety risk to the community. He was told by the FAA not to advertise anything. Behind the scenes Lloyd Dorsett had been frantically working his FAA contacts in Washington in an attempt to gain permission to fly it. On the morning of the first flight the press had mysteriously been invited to breakfast at the Skytrails airport restaurant. Conroy made a speech and briefed the press on what was going to happen. The group was not aware the FAA still had not granted permission for a test flight. Suddenly, two liveried attendants threw open the curtains to reveal the B-377PG in all its glory sitting on the ramp.

The Los Angeles TV newsman Clete Roberts was reporting live along with numerous radio stations. The coverage resulted in hundreds of people gathering at the airport to witness the takeoff. Dorsett was successful in getting the FAA to grant permission for the flight. A typed letter from the local FAA office was hand delivered at 10:30 a.m. As FAA representative James Bugbee walked out to the aircraft with Conroy a very excited man ran up to Conroy with a jeroboam of champagne trying to thrust it into his hands while encouraging him to celebrate the event. Bugbee said, "I did not know if he wanted to drink it or christen the aircraft. I quickly grabbed the guy and shoved him toward a ground handler telling him to get the champagne out of sight before the press got a photo." Years later Bugbee stated, "That was close, if the press had gotten a photo and there was a serious incident with the aircraft it would have been blamed on alcohol consumption with the FAA present. It would have forever ended my career."[14]

James Bugbee and flight test engineer Jim Ashley later went on to fly FAA test flights for certification on the Aero Spacelines PG, SG and MG.[15] Test engineer Ashley was always a bit concerned on engine out climb test flights. The guppies were fitted with very tired Pratt & Whitney R-4360s. Some of the engines were from the batch of old Stratocruisers obtained from Mansdorf and the rest from the Northwest B-377 spares inventory which was acquired by Aero Spacelines. The engines did not restart well in flight. Engine-out test often resulted in engine-out landing test.

On the morning of 19 September 1962 Conroy took the controls with Clay Lacy in the right seat, Bob d'Agostini as flight engineer and Bill Cuffe as flight mechanic. Conroy, Lacy and d'Agostini had all flown C-97 Stratocruisers together while serving in the 146th Transport Wing of the California Air National Guard. Just before the flight Conroy told Lloyd Jones, who built the guppy model used in the original presentation to von Braun, "I couldn't stand to wait, so I flew it last night!" Years later Jones confirmed the story when he asked Clay Lacy in 2012 if it actually happened. Lacy said the night before they were doing taxi test and took it up to about 50 feet then settled it back down on the runway. They did it multiple times taking it up to 80 feet.[16] Lacy stated in 2015 he recalled it was more like five or six times because they wanted to determine handling with an engine pulled back.

The delays on the day of the first flight were caused by waiting for the FAA to grant permission for the flight. Conroy was convinced the awkward monstrosity would fly with ease. After all he had already gotten it in the air multiple times the evening before. When the FAA required a yaw test the morning before any flight could be made, Conroy in his showman fashion teased the press and spectators. He brought the engines up to takeoff power while rolling down the runway then suddenly cutting the power as if something had malfunctioned. Conroy had toyed with the airport authority by saying they were not sure it would be flown. Either Conroy forgot to tell the tower or they did not expect him to obtain approval for the first flight. Whatever the reason the tower was not advised and had not requested crash trucks to standby. Conroy taxied back to starting position as if he were going to do another high speed run. He brought the engines up to max power while standing on the brakes. The Van Nuys traffic control suddenly realized Conroy was taking off and alerted police and crash trucks to take up positions. The takeoff was as smooth as probably any Stratocruiser had ever made as it lifted off nose down and headed for Mojave with a string of chase planes occupied by the press. After landing at Mojave, Jack Conroy told the press the plane far surpassed their fondest hopes. The controllability and stability was better than the original B-377.[17]

The Pregnant Guppy roars down the runway and lifts off for the first time on 19 September 1962 with Jack Conroy and Clay Lacy at the controls. The takeoff was a smooth as any Stratocruiser had ever made as they gained altitude and turned toward Mojave. Interestingly there is no wedge-shaped fairing at the wing trailing edge to direct airflow to the tail. The red stripe bare metal color scheme was only used on the first aircraft (Dottie Furman, courtesy Angelee M. Conroy).

There is some doubt if it was better than the B-377; however, it flew straight and level and was controllable. Jack Conroy was a showman. He had just completed and flown the first aircraft with a giant fuselage without incident. His entire operation was based on NASA being impressed enough to appropriate funds for a contract to transport launch vehicle rocket boosters and the lunar excursion module.

Although Conroy stated it flew smoothly on the first flight, the B-377PG was subject to buffeting as the flaps were retracted. When taking off with the flaps set at 25 degrees the aircraft behaved in a normal manner. As the flaps were retracted the buffeting increased with severe intensity. Irv Culver suggested it needed a directional fairing on the side of the fuselage at the wing root. Clay Lacy experimented with different flap settings and determined the buffeting was very light at 10 degrees and smoothed out as they were retracted. Consequently, a standard was set requiring the flaps not be deployed past 10 degrees pending the installation of a wedge shaped fillet on the sides of the fuselage.

The B-377PG was subject to buffeting when the flaps were retracted. NASA needed the aircraft so desperately that there was not time to correct the problem. Clay Lacy determined it could be flown safely by only using 10 degrees of flaps. It flew for nearly five years until a wedge-shaped fairing was designed to correct the problem. It was not installed until about 1967 during an overhaul (ASI, courtesy Tom Smothermon).

The aircraft was so critical to the space program the PG was not taken out of service long enough for the modification to be made because it would have required additional testing. In addition, the project was out of cash and struggling to move forward. It was originally believed the depression the length of the fuselage created by the B-377 figure eight cross section would aptly control airflow when the aircraft was at altitude in cruise. It flew for years without the fuselage airflow directing fillet although in certain configurations there was a problem with the airflow washing up the sides of the fuselage reducing control.

There have been numerous false reports over the years of the dorsal fin being added after returning from the demonstration flight to MSFC for von Braun's review. This is obviously not true because the dorsal fin is quite visible in first flight photographs. The references most likely are in respect to the wing trailing edge air deflectors installed on the fuselage sides to prevent buffeting during flap retraction. A wedge shaped device was eventually designed by Irv Culver and added to the side of the fuselage above the wing trailing edge. The purpose was to improve the handling characteristics with flaps deployed by more evenly directing airflow. The exact date of installation is not absolutely determined. It was later than 1965 after the Super Guppy entered service and most likely in November 1967 when the Pregnant Guppy was withdrawn from service for a major overhaul. It flew for at least three years without the fairing and possibly as much as five years. The fillets were present when it was scrapped in 1979.

When considering aircraft manufacturers allow five years from design to first flight, building the Pregnant Guppy was an amazing accomplishment. Conroy did it in two years and almost six months of the time work was at a standstill for lack of capital. The FAA had only approved the one test flight to Mojave. Conroy was not supposed to be flying it around anywhere without careful scrutiny from the FAA. It still needed considerable testing and Jack was calling everyone at the FAA he could get to listen to get permission for more flights.[18]

Somehow Conroy was able to get FAA observers to agree to fly in a 400 horsepower Comanche alongside the Pregnant Guppy. The guppy crew conducted simulations of loss of critical engine (#1) at V-1, continuation of takeoff with inoperative engine wind-milling, wing and cowl flaps in takeoff configuration, gear down and wings within five degrees of level. Six takeoffs were made with no difficulties. Four landings, three of which were full stall in gusty cross winds of approximately 90 degrees with velocities of approximately 15 knots were made. Two of the takeoffs and all four landings were observed by FAA engineering flight test personnel. In the next five days of testing a total of eight hours and 40 minutes were logged. Lacy said, amazingly the FAA was satisfied by 6 of the last evening.[19]

Conroy was aware the Air Force could no longer provide transportation to NASA because the rockets were too large. He was desperate to show off his creation which was being promoted as the world's largest airplane. In an effort to influence NASA and secure a contract it would be in his interest to show it off to the Air Force. Somehow he got permission from the FAA to fly a cross country test flight as long as he avoided any major population centers. Sandbags were loaded to give it a payload of 27,000 pounds. The next day (20 September 1962) the partially completed outsized flying contraption left Mojave for the Air Force Annual National Convention and National Aerospace Panorama Exhibition in Las Vegas. The B-377 fuselage was still intact inside the outsize outer fuselage which was being supported by a structural framework and wooden supports. Conroy hyped the aircraft and made a speech at the convention. The PG was displayed in Las Vegas then returned to Mojave for additional testing. By 25 September a total of 10 hours, 10 minutes flying had been completed.

Demonstration Flight to Huntsville

The NASA budget allocated very little for logistics and nothing for development of a new aircraft transporter. Conroy based everything on von Braun's enthusiasm. He was convinced NASA needed a volumetric transport and would commit to a contract once it was built. Von Braun had assured Conroy he would find the funding if the aircraft became operational. With no government funds available, Conroy had mortgaged everything he owned and squeezed all his friends for financial support in order to build the airplane. He was out of money after spending over $1 million. He desperately needed a commitment or a letter of intent from NASA stating the agency would negotiate a lease. This would get the creditors off his back and help raise more cash to complete testing, remove the inner B-377 fuselage and install the separation joint.

Conroy was convinced if he could just get the guppy to MSFC at Huntsville and demonstrate its potential to von Braun, he could secure a commitment from NASA to continue with the project. Von Braun dreamed big and was enthusiastic from the first moment he was shown the model and design drawings. After two visits to Van Nuys to observe the progress of the build, NASA officials still had not provided funds for the project. All along von Braun had assured Conroy if he could get the airplane completed and certified he would get him a contract. The project was so short on cash there was not enough money to purchase fuel for a one way trip the Marshall Space Flight Center. On 28 September 1962 with Jack Conroy in command, Jack Richards as copilot and Bob d'Agostini as flight engineer the Pregnant Guppy N1024V departed Mojave at 9:45 a.m. with a planned arrival at Oklahoma City of 3:15 p.m. The aircraft remained in Oklahoma for a day while Conroy negotiated an agreement to borrow fuel guaranteed by Lloyd Dorsett.

On Sunday 30 September they departed Oklahoma City staging through Perrin Air Force Base at Sherman, Texas. The stop at Sherman was to prevent landing in Dallas as part of the restriction not allowing them to operate in populated areas. Today the Perrin Air Base is Gainesville Field where ironically the last flyable ATL-98 Carvair guppy is stored. The Pregnant Guppy proving flight was originally planned to stop at Houston for a demonstration to personnel at the Manned Spacecraft Center and a test flight with NASA representative W. Graves on board.[20]

The demonstration flight itinerary was changed to fly directly to Redstone Airfield (KHUA) at Marshall Space Flight Center in Huntsville, Alabama. Conroy arrived low on fuel and out of money to a less than enthusiastic crowd. NASA engineers from Marshall still questioned it being airworthy enough to be a serious contender to transport space hardware. NASA Chief of Logistics John Goodrum was present along with many ranking NASA officials. There were the usual guppy jokes with several skeptics referring to it as a misshapen blimp with wings.

In contrast, Wernher von Braun was eagerly awaiting the arrival along with team members who had been with him since the V-2 program in Germany. The group included ex–Peenemunde scientist Karl Heimburg, ex–German engineer Konrad Dannenberg and ex–Luftwaffe test pilot Hermann Kroeger. Von Braun was enthusiastic about the project. He desperately needed a transportation alternative to shipping the boosters by barge. The test program was falling behind schedule because of the time required to transport components. To prove the airworthiness and stability of the aircraft Conroy offered a demonstration flight to any of the NASA team. There was not a lot of enthusiasm in the crowd except for a few of von Braun's group and MSFC Saturn logistics manager Julian Hamilton. Conroy requested a full load of fuel from the army facilities at Redstone for "ballast." His request was twofold. He was low on fuel and financially broke. If he could not sell von Braun and NASA the airplane he would be stranded in Alabama because he did not have funds to purchase fuel to get home.[21]

Julian Hamilton along with German team member Hermann Kroeger, who was deputy director of the Guidance Control Division at MSFC, eagerly volunteered to ride on the demonstration flight. Kroeger was a trusted von Braun team member. It was a strange reunion because Kroeger had worked with von Braun on German rocket programs at Peenemunde near where Conroy had been a World War II prisoner of war.[22]

Redstone Field is 7,300 feet with no parallel taxiway

and only a turnaround area at each end of the runway. After some effort to get into position, the huge aircraft took off as von Braun and the others watched. As they gained altitude over the rolling hills and cotton fields of Alabama and leveled out, Conroy prepared to dazzle Kroeger. He demonstrated the ability of the aircraft to maintain altitude and course with number one engine out. When they landed Kroeger, who was a former Luftwaffe test pilot, was so excited he could not contain his enthusiasm as he rambled on in German and English.

Von Braun became so excited with Kroeger's report he also decided to take a test flight. This was not well accepted by many NASA and military officials who were present. This was not a proven aircraft and the top was being supported by two-by-fours. Von Braun and the entire future of the U.S. space program would be at the mercy of Jack Conroy and his semi-complete volumetric guppy flying contraption. If anything happened to von Braun, America's space program would be in severe jeopardy or perhaps collapse.[23]

Von Braun, who was a rated pilot himself, boarded the aircraft against strong objections of NASA officials. The men took off. Conroy did the same maneuvers and demonstrations which had so impressed Hermann Kroeger. Conroy was later quoted as saying, "After we got airborne, I turned the controls over to him. Then I signaled the flight engineer [Bob d'Agostini] to slowly reduce power and then feather [shut down] the left outboard engine [number one]. Doctor von Braun didn't even notice it. Then the flight engineer reduced power and feathered the left inboard engine." Von Braun was aware something was different; however, the plane was still flying well on two engines with a heavy load of fuel. When von Braun expressed his surprise of it flying on two engines, it is reported Conroy responded, "We do it all the time to save fuel." Years later Conroy stated, "no one has ever told me since then the Pregnant Guppy wouldn't fly."[24]

One must take into consideration there is a lot of Conroy hype in the story. If the plane was full of fuel and had more than 20,000 pounds of ballast as reported, it flew on two engines at more than 141,000 pounds, which was the calculated new ATOG. This is far in excess of the reduced 133,000 max operating weight which would be determined after the first all up weight test flight in May 1963. That flight, nearly eight months after visiting Huntsville, almost ended in disaster because it would not fly at 141,000 pounds. It is more likely the army provided enough fuel to bring it up to a reasonable ATOG. Flight engineer d'Agostini confirmed the story; however, he did not state the actual takeoff weight. He said, "in addition the oil level was so low I could not unfeather the engines. I had to replenish 15 quarts from the central oil tank while in a panic. I don't think the good doctor realized what trouble we were in. Jack later said, 'I almost gave myself a hernia. I was standing on the rudder with everything I had to keep the aircraft flying straight while Bob frantically got them restarted.'"[25]

Upon the return to Redstone Field von Braun was as equally excited as Kroeger.[26] Von Braun stated he was amazed at the ease with which the aircraft could be controlled maintaining 4,000 feet at 170 knots with two engines shut down in normal climb power on engines three and four with the trim tabs in neutral position.[27]

NASA officials continued to express concerns of the completed aircraft not having any hinges or swing tail at the separation joint. The detaching of the entire rear of the aircraft for loading had not been done before. Conroy explained rolling the tail away on a carrier which is transported within the aircraft eliminates any clearance problems when loading a rocket booster. Then the rear half of the aircraft is reattached. He countered NASA concerns by saying the mechanic who torques the bolts is required to fly with the aircraft, which actually seem to satisfy the skeptics.

The demonstration flight to MSFC was a success. Conroy impressed von Braun and secured his approval. The two

Flight tests were conducted at NASA's Dryden Flight Research Center at Edwards Air Force Base before the fuselage separation joint was installed. Close inspection reveals tufts along the rear fuselage and dorsal fin to determine air flow characteristics. The remains of the PAA stripe from when it was a commercial airliner are still visible on the vertical fin (NASA, courtesy Bob Burns Collection).

men began informal contract talks later in the day and closed the deal. Conroy had gambled everything on this trip and accomplished his mission. However, he did not have sufficient funds to purchase fuel to get home. He needed to top off the tanks with a full load to fly back to California. The army supplied fuel for a second time at the request of von Braun. Conroy departed Huntsville with a handshake on a good faith agreement with von Braun. Within a year of the first flight the modified N1024V B-377PG would make a cargo flight for NASA.

The guppy was flown back to Dryden Flight Research Center at Edwards Air Force Base without incident. The FAA did stall test before turning it over to the NASA pilots for further testing. During one of the flights, FAA test engineer Jim Ashley walked to the rear of the aircraft. He stated it was very scary experience with the rear of the aircraft twisting and creaking like a basket in a storm. "That was enough for me. I returned to the cockpit and tried not to think about it for the rest of the flight."[28]

Further testing was placed under the direction of Paul Bickel. In October of 1962 NASA pilots Joe Vensel and Stan Butchart flew the B-377PG for evaluation. The original B-377 fuselage was still in one piece inside the new expanded top and the separation joint had not been fabricated. Over 50 hours of test flights were logged including 35 takeoff and landings. The test included flights with two engines inoperative in many combinations, i.e., one and two out, three and four out, one and four out, etc. Approaches to stall to the point of mild buffeting were accomplished in various configurations. The FAA accepted the results and it was determined to be aerodynamically sound. A report of their observations was forwarded to NASA officials in Washington and Marshall Space Flight Center at Huntsville.

B-377PG Certification

Abe Kaplan of Strato-Engineering wrote to Jack Conroy on 24 October 1962 to update him on the certification process with the FAA. He stated a preliminary type board was held with the FAA. The drawings of the typical splice ring, loft information, skin contour, dorsal fin structure and attachment to the existing fuselage structure were accepted as sufficient. In addition the preliminary stress analysis was made and analyzed. The data was considered satisfactory and would only require editing. Kaplan further stated the fuselage separation rings were analyzed for air loads and found quite satisfactory. However, the rings were analyzed for cargo loads which turned out to be critical. The rings needed to be strengthened considerably where they attached to the existing structure for continuity. The loads and structural requirements for the wing support bulkheads were determined, although, they would not be installed until the present old fuselage structure is removed.[29]

In the event the detachable tale section was not acceptable by the FAA, a contingency design had been engineered by Strato Engineering. Kaplan stated, "although drawings were not submitted on control systems for a swing tail, considerable thought has been given to the subject. The FAA has looked at the system and considered it satisfactory. We are confident a Supplemental Type Certificate (STC) will be issued on the Pregnant Guppy once the data is edited and submitted to the FAA and the flight test program is completed."[30]

The outsize giant was flown to Templeton Field in Denver on 24 October 1962, which was the same day Kaplan submitted his report to Conroy on the FAA review of the separation joint. Conroy attended the meeting of Aerospace Traffic Managers to receive their approval and assure them the guppy was on schedule for FAA certification. Interestingly, the preliminary work Kaplan did for a swing tail was used in the development of the Mini Guppy.

Jack Conroy wrote a letter to von Braun on 29 October 1962 stating ASI had conducted initial test flights on the modified B-377 and it exceeded expectations. He also stated it was equipped with radar and thermal deicing and was capable of worldwide service operating from 4,000-foot fields with a guaranteed cruise or 235 mph. It was described as having a cabin for four passengers fitted with reclining seats, galley and toilet facilities with access to the flight deck. Once again this was Conroy optimism. The cockpit was virtually unchanged from B-377 days. The letter optimistically stated the aircraft will be capable of cabin pressurization. (However all pressurization equipment had been removed for weight reduction during construction.)

Von Braun had expressed interest in an even larger aircraft from the beginning. He knew a larger aircraft would be needed as the space program progressed. This was just the first step for testing and development to determine the viability of such extreme fuselage aircraft. The general opinion was the space program would grow. After the moon the next step would be Mars and a fleet of larger transports would be needed. Conroy's letter stated upon acceptance of the contract for service of the first aircraft construction would start immediately on a second conversion. The next aircraft would be ready in 180 days with the same terms except the guaranteed mileage will be half of the first. The second aircraft "may" be a turboprop configuration. It would be capable of transporting the larger S-IVB. The mileage of the second aircraft would be approximately double the rate of the first aircraft.[31] It is of interest to note Conroy stated it "may be turboprop." The fact is Conroy was only speculating it would be turboprop. He knew it would be underpowered. It was initially planned with the same Pratt & Whitney R-4360 piston engines because he had not been able to obtain suitable turbines. Consideration was given to using C-97 wings and augmenting the reciprocating engines with jet pods.

Deputy Director of NASA at MSFC Harry H. Gorman wrote back to Conroy stating the airplane he had built was impressive. "We at Marshall have an interest in its value to transport large vehicle stages and propulsion units. We are interested in your air service to transport the S-IV stage. Your aircraft must have the ability to transport cargo 18 feet, 6 inches in diameter and 60 feet long at weights of 40,000 pounds and operate into airfields of 4,400 feet. After reviewing your letter of 29 October and viewing the aircraft while here in Huntsville, we believe it will meet the requirements. However, it must be certified by the FAA under Part 8 of the CAR and approved by NASA. Our headquarters in Washington will review your proposal and make comparisons of other methods of transportation before a commitment can be made. If approved we will contact you for further discussions."[32]

Conroy knew from the letter he had a contract if the aircraft was certified. NASA needed it and there was no other viable alternative. The proposed Douglas C-133 piggyback proposal was no longer competition. It had been determined not practical because of structural and vibration problems. Conroy had the only completed flying volumetric transport in the world.

A three year contract was proposed stating NASA would provide all loading equipment, make arrangements for Aero Spacelines to purchase fuel at NASA and military facilities. In addition NASA would be responsible for insurance on all cargo. The following cargo rates were proposed:

> "The basic rate for the aircraft for the first 500,000 miles will be $6.35 per statute mile. Mileage will be computed direct from airport to airport.... Mileage rate will be the same for live or ferry flights. During the first 150,000 miles of operation there will be a surcharge of $3.00 per mile ($9.35 per mile). The $3.00 per mile will be credited to the Government in the last 150,000 miles of the 500,000 miles aforementioned."[33]

1st Year 150,000 miles @ $6.35 + $3.00	$9.35/statute mile
2nd Year 175,000 miles @ $6.35	$6.35
3rd Year 25,000 miles @ $6.35	$6.35
3rd Year 150,000 miles @ $6.35–$3.00	$3.35

> "The mileage may be accelerated so the entire 500,000 miles could be completed by the first or second year. All mileage over 500,000 miles would be at the rate of $5.50 per statute mile. Monthly or quarterly guarantee and schedule of payments will be negotiated."[34]

Conroy was very optimistic believing a NASA contract would allow him to build an entire fleet of guppies. Not only could he monopolize the transport of NASA hardware because he had the only outsize aircraft, he could exploit the commercial market. The only problem was he was completely out of money as usual. Lloyd Dorsett's original investment was gone and the company stock in practical terms was worthless. He tried to enlist investors but their reaction was, "You have flown the plane for NASA and you can't get a contract?" Conroy had not even been able to raise enough money to remove the old Stratocruiser fuselage from inside the new guppy and complete the separation joint of the tail section.

Banking on von Braun's enthusiasm Conroy believed a NASA contract was forthcoming and formed a holding company. During one of von Braun's visits to Van Nuys to review the progress, Conroy introduced him to his banker Carl Schatz in an effort to influence the bank to extend him more credit. Von Braun repeatedly told them there was no money in the NASA budget for the airplane but if they got it certified there would be a contract. Conroy reasoned NASA's credibility was on the line and the bank should provide the needed funds to get the aircraft completed and certified.

Von Braun was committed to the president's plan for America to be on the moon by the end of the decade. There were huge logistical problems moving hardware to the test site and the Cape which could jeopardize the program. Conroy had the aircraft which could reduce transport time from weeks to hours on multiple components. He pointed out NASA had no choice but to offer him a contract.

In an effort to produce cash Jack Conroy created a second company, Hayvenhurst Van Nuys Corporation, aided by venture capitalist William M. Ballon. It was incorporated on 15 February 1963 (ID C0446085) as a Domestic Stock Business to solicit investors and to sell stock. Aero Spacelines would be held by Hayvenhurst. All NASA contracts would be with Hayvenhurst Van Nuys Corporation leaving Aero Spacelines as an aircraft manufacturer with a flight operations division. The Hayvenhurst name was selected because Aero Spacelines offices were at the corner of Hayvenhurst and Arminta Avenue in Van Nuys, California.

Conroy was able to leverage the belief he would soon have a contract with NASA in order to raise capital. NASA announced the contract two weeks later. The U.S. Government was spending billions on the space programs. Once there was a NASA contract it was much easier to convince investors and sell stock. Unfortunately, investors were not always as optimistic as Jack. On one occasion they barred him from entering his own office. He managed to get a court order to allow him access but for a while there was an armed guard on the door. Eventually the stockholders gave in, allowing him more time to make good on their investment.[35]

The original Aero Spacelines, an Oklahoma corporation, was purchased by Aero Spacelines, a California corporation, in February 1963 for $500,000. The corporation was reorganized retaining directors Mansdorf, Conroy and Dorsett. Also listed as directors was Conroy's friend, World War II veteran and venture capitalist William Ballon, along with Mitchell Howe, F. George Humiston and Robert M.

Left: After the fuselage was cut, the rear section was moved back to install the mating rings. Jack Conroy demonstrated the size of the Pregnant Guppy by standing on top of the old B-377 fuselage inside the expanded 19-foot guppy. The size of the new fuselage is impressive (Dottie Furman, courtesy Angelee Conroy).

Below: Boeing promoted the size of the B-377 Stratocruiser fuselage by having flight attendants from the six original carriers—United, Northwest, American Overseas, Pan American, BOAC and Scandinavian—pose in the upper half. The contrast is enormous when considering Jack Conroy was photographed standing on top of this fuselage, which was inside the Pregnant Guppy (Boeing, author's collection).

McGuire. William Ballon was a new investor in the project. Dorsett agreed to purchase ex–BOAC B-377 G-ANTX c/n 15965 from Lee Mansdorf in June 1963. J.H. Olverholser eventually replaced Dorsett as chairman of the board.

Certification of the Pregnant Guppy

The next phase was to separate the rear fuselage for installation of the bulkhead ring structure and locking devices. The circumference of the fuselage was cut behind the wing and the rear portion separated and rolled back approximately six feet. The original B-377 fuselage was still intact inside the 127-foot long aircraft. Conroy held a news conference and posed for photographers while standing on top of the old Stratocruiser fuselage inside the new outsize fuselage. To establish a perspective as to how large the Pregnant Guppy fuselage was, there was a complete B-377commercial airliner fuselage below his feet and the new top was nearly 12 feet above.

No data existed on stress levels which could possibly occur at the separation joint because an aircraft of this diameter had never been built before. Removal of the tail section to load cargo was a totally untried concept. Abe Kaplan of Stratos Engineering had previously analyzed the design of the rings and recommended they be strengthened considerably where they attached to the existing structure for continuity of stress on the airframe. In an effort to obtain any data on circumferential joints the engineering staff at Aero Spacelines consulted with the engineers at On Mark where it was built. Although much smaller, On Mark had dealt with designing a circumferential "ring spar" for the A-26 Marksman airframes which were modified to increase cabin space. The design for the guppy fuselage ring structures were loosely based on the On Mark A-26 calculations.

A pair of structural rings measuring slightly more than 19 feet in diameter was designed to fit on each side of the fuselage separation joint. The rings were fabricated by laminating curved sections of 4130 steel with each lamination approximately ½ inch thick. Several laminations sections were staggered and bolted together forming the two rings which were the full diameter of the figure eight circumference of the upper and lower fuselage. These rings were huge and extremely heavy, literally thousands of pounds of steel.[36]

The heavy steel mating rings emphasize the question of the Pregnant Guppy being tail heavy and how to deal with it. The airframe had been stretched by adding 16 feet of constant B-377 fuselage behind the wing. This also added considerable weight behind the main gear when considering the front was not extended. When the stretched guppy is compared empty, the center-of-gravity (CG) is considerably farther back than a standard B-337 or C-97. It is logical to assume once a Saturn booster or other cargo is onboard the weight would be concentrated from the center-wing forward. However, it is most likely tail heavy when empty and rumored to be difficult to handle.

When the B-377PG was first flown to Redstone Field at Huntsville for NASA review it was reported to have sand bags loaded forward for ballast. Conroy arrived with minimum fuel. He convinced NASA officials it was necessary to take on a load of fuel for ballast before a demonstration flight for von Braun.[37] Conroy was able to fly there without incident although he most likely arrived in an out of balance condition. The smaller B-377 fuselage was still intact inside the larger fuselage concentrating some weight forward but the rear was extended behind the wing adding more weight aft drastically changing the CG. The balance was expected to improve once the inner fuselage was removed.

After the inner and outer fuselage was cut around the circumference behind the wing and separated to insert the separation rings. On Mark Engineering assigned Roland Hipley to remove the old B-377 fuselage. Don Stratman who worked for Conroy from the beginning stated, "Hipley was one of the best craftsmen I ever worked with who skillfully cut the old fuselage out by using only hand held power tools."[38]

Removing the B-377 inner fuselage reduced the weight behind the wing. When the steel separation bulkhead rings were installed weight gained by removing the inner B-377 fuselage was added back to the total airframe weight; however it was at end of the constant section and closer to the main gear. Either way it appears to still have been tail heavy when empty. The weight of the steel rings became less of an issue in all subsequent guppies except the Mini. The others were fitted with a swing nose placing the weight of the fuselage opening joint forward. It has been suggested one of the reasons for the swing nose was not only for better access but better weight distribution.

After the top of the smaller original B-377 fuselage was cut away, the outer oversize fuselage connections were secured at each joint rib along the length of the lower B-377 airframe giving it structural integrity. With the old fuselage cut away a new 19-foot, 9 inch high, 80-foot long cargo bay with a volume of 29,187 cubic feet was fully open. With the fuselage rear section in place, the two rings fit face to face and secured with 32 bolts which are 1.25 inches in diameter and 10 inches long. Each time the aircraft is loaded or unloaded all the bolts have to be removed and the control cables disconnected to separate the tail. When the tail is rolled back into position the bolts are reinstalled and torqued before flight. It is a time-consuming exercise but still reduced the delivery time of a rocket booster from California from weeks to hours.

The original Boeing design B-377 lower half of the figure eight fuselage with eight foot floor is quite evident with the tail rolled away. The fuselage cross section shows the contrast of the much larger upper half dwarfing the smaller bottom. The new expanded upper half is spliced in about 24 inches above the floor level indentation of the original Stratocruiser figure eight cross section which runs the full length of the fuselage. The 8-foot main deck floor remained intact leaving a very narrow area for the cargo to be secured. It is easy to understand how large cargo could make the vertical CG far above the floor creating a top heavy condition.

To help reduce weight and achieve a 91,000 pound empty weight all the interior and pressurization system had been removed. The primary weight saving feature of the rear section modification came from not hinging the tail. A swing tail would have required adding a hydraulic actuator system and hinges of approximately 600 pounds each. The design used was simple; unbolt the rear section of the aircraft and roll it out of the way for loading. Interestingly, as aforementioned a hinged system was designed in the event the FAA would not accept the complete detachment of the tail.

Even with all the weight reduction measures, the completed Pregnant Guppy weighed approximately 7,500 pounds more than the original Stratocruiser with a maximum ATOG set at that time of 141,000 pounds. Unlike later models which were fitted with a pressurized cockpit, the B-377PG was altitude restricted and never refitted with a pressurization system. It normally cruised at 235 mph with a max cruise of 250 mph compared to 301 mph of the original B-377.

Aero Spacelines had not applied for a commercial operator's certificate because it was assumed not to be needed. The Pregnant Guppy was considered a public aircraft. This was done to cover all the construction shortcuts which would have never been approved on a commercial transport. It was built for an exclusive NASA contract which placed it in a category with military aircraft. After considerable pressure from the FAA, Hayvenhurst DBA as Aero Spacelines was forced to apply for Commercial Operators Certificate WE68 (c) on 20 February 1963 under Air Regulation 45.2. A little over a month later on 12 March Aero Spacelines applied to the FAA for an exemption from Parts 1 and 8 (special purpose) of Civil Air Regulations regarding the Airworthiness Certificate. The FAA is allowed to grant exception if it deems it is in the public interest.

It is obvious NASA was putting the pressure on the FAA to facilitate the process because of an urgent need for the outsize transporter. The exception was granted and a supplemental Type Certificate relative to the Civil Air Regulations was issued on 02 May 1963 clearing the way for Conroy to begin transporting components for NASA. The maximum payload was set at 33,000 pounds.

On 16 May 1963 N1024V was flown for the first time as a completely modified aircraft with the inner B-377 fuselage removed and the rear separation joint installed. NASA officials had been closely monitoring the certification process. Von Braun and a group from MSFC visited Van Nuys to inspect the completed aircraft with the tail separation joint installed. Von Braun inspected every part of the airframe. Conroy and von Braun posed for photos along with the NASA team ASI officers and financial back-

Top: **The cargo area and cockpit was inspected by the NASA team after the separation joint was installed. Notice there is no reinforcement in the upper structure above the cockpit. The B-377PG is the only guppy with the expanded upper fuselage grafted on two feet above the cargo floor, which required NASA to fabricate a loading fixture for wide cargo (Dottie Furman, courtesy Angelee M. Conroy).**

Bottom: **Wernher von Braun and a group from MSFC visited Van Nuys to inspect the completed Pregnant Guppy. The tail section was rolled back to demonstrate the loading process. He inspected all areas of the completed aircraft in detail. Von Braun reminded Conroy of the urgency to get the aircraft certified and in service as soon as possible (courtesy Angelee M. Conroy).**

ers. The group posed for the press under the letters "World's Largest Airplane." Von Braun was very pleased, urging Conroy to complete the certification as quickly as possible in order to begin transporting rocket boosters.

Flight certification test were conducted at Mojave. The first time the B-377PG was flown at an all up maximum weight almost turned into a disaster. The aircraft was loaded to 141,000 pounds which had been calculated to be a reasonable max ATOG after conversion. The original B-377 max ATOG was 145,800 pounds. The improved later model military KC-97G models have a max takeoff of 175,000 pounds.

Conroy was at the controls as they began the takeoff roll. The aircraft made a long ground run but was not rising. It was skipping along prompting him to pull the gear up to reduce drag and increase the speed. It remained airborne but it would not climb. The aircraft was struggling but the air speed was constant at 148 mph (128 knots) with rolling terrain still in front of it. The screaming P&W R-4360s were maxed out and it still would not climb.[39]

The aircraft was barely clearing the rises in terrain with the mining town of Boron dead ahead. A turn could cause side slipping putting a wing into the ground which most definitely would cause it to cartwheel and crash. Number three engine was overheating, flight engineer d'Agostini feared a catastrophic failure and requested to pull it back. The aircraft was straight and level but not climbing. Conroy feared any loss of speed would insure disaster. D'Agostini was watching the temperature gauge reporting the overheating condition. Conroy shouted, "Don't touch it!" The engineer responded, "It will burn up." Conroy shouted back, "Let it burn, we need all it's got!" The R-4360s were at max power and burning fuel at a rate in excess of 16 gallons per minute which is slightly more than 100 pounds. It was becoming slightly lighter each minute. The aircraft was at the critical point when a few hundred pounds was just enough reduction in weight to make it start to climb. After the test flight the guppy returned to Mojave. During the debriefing and discussion of the overweight situation everyone agreed the weights needed to be recalculated. The decision was made to reduce the max ATOG by 8,000 pounds to 133,000 pounds.[40]

These near disaster situations occurred with some regularity in the guppy flight test program. This was the one of the most unique aircraft since aviation began. It was so revolutionary one would expect a sophisticated and well planned test program. In reality it appeared more like "Laurel and Hardy build an airplane." Conroy actually had the best engineers and technicians available; however, they were out of money and nothing like this had ever been attempted before. They were truly making it up as they went along. NASA was very impatient because the aircraft was needed to transport boosters to stay on launch schedule. There was never enough money to complete any test program. Conroy and his engineers did the bare minimum and in many cases projected the end result. The workers would often joke, "We build it to fit, draw it to match and paint it to cover."

The NASA space flight scheduled called for the first two-stage launch of the SA-5 Saturn vehicle in May of 1963. This would be the first test launch with both stages live. The program had experienced many delays because of the 18 to 21 day trip by boat from California to the cape. To stay on schedule transfer by barge was no longer an option. The guppy could do it in 12 to 18 hours. Brainerd Homes, director of Manned Space Flight Program, made sure the FAA was advised of NASA's critical need for the aircraft to transport the S-IV-5 stage of the vehicle. It is of note the first two gup-

Jack Conroy (center) and Wernher von Braun with NASA and ASI group pose for the press beside the B-377PG at Van Nuys. Included is MSFC logistics manager Julian Hamilton, far right with camera, who flew on the guppy at Huntsville when it was demonstrated. Next to von Braun is ASI chairman J. H. Olverholser (courtesy Angelee M. Conroy).

pies (B-377PG and SG) were not formally type certified by the FAA. Both the Pregnant and later the Super Guppy were considered extreme modifications of B-377/ C-97 airframes with a built up upper fuselage half. FAA advisors and NASA flight test staff worked with ASI and offered advice during flight testing to keep the aircraft within the parameters of the FAA requirements.

The B-377PG was privately owned; however, NASA sponsorship and exclusive operation provided a way to circumvent certification problems. FAA officials followed the development of the aircraft and NASA test pilots had flown it giving them the ability to attest to it meeting airworthiness standards. This was not the case with the later built six commercial guppies.

Once the aircraft flew and appeared to be as represented, NASA-MSFC offered Aero Spacelines a contract for a period of two months contingent on a successful completion of the FAA certification test which include the transport of the inert S-IV booster. NASA signed the interim contract on 28 May 1963 in the amount of $194,850 ending on 31 July with conditions it could be extended until 31 August. The contract covered only the test period to determine if it was airworthy and if it could transport space hardware. Not only did the test flights transport a dummy S-IV third stage across the country, active modules and components relative to the moon program were transported. However, the FAA certificate was not effective until 13 November 1963. The need for the aircraft was so critical it was actually flying components before it was certified.

NASA wanted the aircraft online in July and August 1963. It had to be capable of transporting Saturn moon launch boosters from Douglas at Huntington Beach, Apollo modules from North American at Downey and F-1 engines from Rocketdyne in Canoga Park to Huntsville and Cape Canaveral. Possibly the most important certification test was a flight with the inert stage of the S-IV from Douglas Aircraft. On 13 June 1963 the PG was loaded at Los Angeles International with the dummy S-IV stage of the Saturn I space vehicle for a planned test flight to Edwards Air Force Base the next day. The cargo weighed 20,379 pounds which is the equivalent weight of a live rocket stage. Crews from Douglas aircraft installed an instrument package to measure vibration, temperature and pressure for NASA to determine if the delicate cargo was compatible with the guppy aircraft. The flight served as a part of the FAA certification and a simulated booster transport for NASA.

On the morning of 14 June 1963 guppy N1024V departed LAX piloted by Jack Conroy with co-pilot Don Walker, flight engineer Bob d'Agostini and ASI director of maintenance Bill Cuffe. On board was FAA pilot Robert E. Bear and FAA engineer William A. Bryde to conduct certification maneuvers with a full cargo load. Douglas test engineer Richard Trudell was also on board to monitor the instruments fitted to the S-IV booster.

All test proved successful prompting the FAA to issue a certificate of airworthiness on 10 July 1963. An Aircraft Type Certificates (ATC) is usually issued after the manufacturer's test program but before being delivered to the customer. The original B-377 ATC was issued to Boeing in the 1940s after it met the licensing requirements for unlimited commercial operation. Aero Spacelines as the company which modified the B-377 to B-377PG received ATC A-912 under a Supplemental Type Certificate under Part 8 to the original ATC. Conroy continued to press for licensing under 4B transport category because he wanted to produce aircraft for the commercial outsize cargo market.

The reduction in Max ATOG because of the Mojave test flight incident resulted in a decreased range with extreme loads. The empty weight of the B-377PG was increased from the base stretched B-377 pre-conversion weight of 83,500 to the guppy weight of 91,000 pounds. A standard B-377 Stratocruiser empty weight is 78,920 pounds. The B-377PG range was set at 1,500 miles; however, it was drastically reduced by cargo weight. In order to meet the transport needs of NASA hardware the payload was set at 33,000 pounds, giving an aircraft dry weight of 124,000 pounds. This leaves only 9,000 pounds (1,500 gallons) for fuel at 6.0 pounds per gallon with a maximum 33,000 pound payload. The fuel capacity of the original Stratocruiser is 7,790 gallons (46,740 pounds). With a four-engine fuel burn of 540 gph, considering hold and payload fuel and a cruise of 235 mph the range of the 377PG with maximum 33,000 pound payload is limited to less than two hours or 500 miles depending on weather. It is rare the aircraft would be operating at maximum payload thus the range is increased as cargo load is reduced. The undeniable fact is the aircraft was underpowered. Even when operating at the reduced max ATOG it struggled to get off the ground and had to make frequent fuel stops.

The B-377PG was built primarily to transport the S-IVB rocket stage which weighs 23,000 pounds. At an all up weight of 133,000 pounds with 19,000 pounds (3,166 gallons) of fuel the maximum range can be up to five hours. Range is also limited by engine oil consumption. The P & W R-4360 engine is a real oil burner which consumes or throws out 1.5 to 5 gallons per hour per engine. Each engine has a 35-gallon tank which can be replenished from a 56-gallon central tank located in the lower forward nose. A high oil consumption engine could affect range.

The B-377PG Certificate of Airworthiness was very restrictive stating: "It will be Certified only for the special purpose of carrying spacecraft modules, persons and cargo for compensation or hire for the National Aeronautics and Space Administration." The restrictions contained in the operating limitations formed a part of the certificate, stating:

1. The *only* flights authorized are for the special purpose of carrying spacecraft modules, persons and cargo for compensation or hire for the National Aeronautics and Space Administration, subject to the following conditions (Ref. FAA Exemption # 258, dated May 2, 1963—Regulatory Docket No. 1679).
   a. The cargo shall consist solely of S-IV Saturn and Apollo spacecraft modules and related cargo;
   b. The persons carried shall be restricted to the technicians designated by the National Aeronautics and Space Administration and carried to insure security and monitor loads to which cargo components are subjected in transit;
   c. Cargo and persons may be carried only between locations as prescribed by the National Aeronautics and Space Administration.[41]

The Pregnant Guppy unloads at the Cape Canaveral Skid Strip (XMR) for the first time on 11 July 1963. The "dummy" Saturn S-IV booster was transported from Los Angeles to the Cape as a cross-country test of aircraft performance and reliability. The FAA certificate was not in effect until 13 November (NASA, USAF).

During the B-377PG limited evaluation contract with NASA from May thru July in 1963 a trial run test was conducted. It transported the first and second stage Titan II rockets of the Gemini program from Martin Aviation in Baltimore to Cape Canaveral. The first major mission took place before the effective date the FAA awarded an airworthiness certificate was affective. On 11 July NASA sent the non-flight Saturn S-IV Battleship dummy across the country to Cape Canaveral as a test of guppy performance and reliability. As part of the program NASA assisted by Douglas Aircraft (DAC) set up very specific procedures for loading to minimize any chance of damage. Douglas assigned Loy Wallain to design a loading procedures manual and manage the operation. The preparation of the aircraft to accept the cargo load was assigned to Aero Spacelines director of maintenance Bill Cuffe who coordinated the procedure with loadmaster Fred Austin. He was responsible for the following procedure:

1. Secure the aircraft in loading area.
2. Install stabilizing jacks on aircraft; install jacks and handling dolly on removable tail section.
3. Remove bolts; disconnect control cables, separate tail section from aircraft.
4. Move tail section away from the loading area with locally supplied tug and secure.
5. After loading is completed ASI crew will reposition tail section, install and torque bolts, attach control cables.
6. Dollies and stabilizing jacks will be removed in reverse order and stored on aircraft.[42]

Problems with assembling the tail dolly and removing the rear fuselage section and reattaching began with the first test load. Up until this time the detached tail had only been moved straight back in practice and demonstration. The mockup of the S-IV booster was scheduled to be loaded at Los Angeles International. The dollies were attached to the rear section and it was moved back and to the side. Once the booster test load was secured, Cuffe and Austin attempted to reattach the tail. It took multiple attempts to get the tail lined up because there was no provision for controlling the caster wheels on the dolly. After the first flight the decision was made to fabricate a tiller for each castor in order to better align the tail for reattachment. In addition it was also found the dolly parts were difficult to identify for assembly once they were unloaded from the belly of the aircraft. The decision was made to color code the parts to make it easier to assemble the tail dolly. Those first few cargo flights looked like the Keystone Cops loads an aircraft. In their defense it should be noted this was the largest item ever loaded on an aircraft. Nothing like this had ever been attempted before. While researching this work several crew members openly stated they had to make it up as the operation progressed because parts were fabricated then modified on a trial and error basis.

To circumvent the certification process for a commercial aircraft the Pregnant Guppy was declared a "public aircraft" on 10 July 1963. This is a convenient way to approve the aircraft because it would fall in the same category as military aircraft and not subject to FAA regulations. An additional reason the FAA did not type certify the Pregnant Guppy is because the propellers have no in-flight low pitch stops (21.3 degrees). If the prop drops below 20 degrees at cruise the reverse thrust (drag) is so great it could tear the engine from the wing.[43]

On 06 September 1963 NASA and Hayvenhurst Van Nuys Corporation DBA as Aero Spacelines entered into a second NASA contract for "air transportation services" for the period 01 September 1963 to 30 June 1964. The contract stated Aero Spacelines would furnish B-377PG N1024V to NASA for exclusive use to transport S-IV stages, large booster components, tools and fixtures. The contract also stipulated all movement would be at the direction of NASA and the aircraft would be available on 48-hour notice. NASA agreed to a minimum usage of 7,000 miles per month not to exceed 23,000 miles with a guaranteed minimum of $120,000 per month.[44] The contract totals $995,884 for the minimum amount of use planned by NASA with an option for an additional two-year period.

While operating under the contract on 10 September 1963 N1024V was parked at Los Angeles International. At 11:15 a.m. a commercial Pan Am 707 (N705PA) was taxing when the pilot misjudged the distance between aircraft. The right wing of the 707 struck the tail of N1024V, causing substantial damage to the guppy horizontal stabilizer.[45] Parts were brought in from the supply of Stratocruisers parked at Mojave. The aircraft was successfully repaired in record time to continue the NASA evaluation schedule. Later in the month the guppy made the first payload flight transporting a Saturn S-IV stage from Sacramento to Cape Kennedy. At a cost of $16 per mile the flight reduced transit time via barge by three weeks. Surprisingly this was only one year after the first test flight.

During an early flight transporting an S-IV booster stage from California to Huntsville, N1024V encountered severe weather forcing a diversion off course to avoid turbulence. High fuel consumption forced an unplanned fuel stop at a SAC base which the crew believed was acceptable under the NASA agreement. The night landing caught SAC security by surprise. The guppy was met by armed security and impounded. The crew was forced to remain on the aircraft until morning when Marshall (MSFC) officials could sort out the details and explain the situation to SAC security. It was then refueled by the military and the flight continued. Apparently, details agreed to in the contract had not been passed on to SAC. By the end of summer 1963 the B-377PG was making regular cargo flights for NASA transporting Saturn S-IV rocket stages and associated hardware from California to MSFC in Huntsville and the Cape Canaveral in Florida.

The need for new larger transports was being evaluated by the military in late 1962. The Air Force let a contract for a study identified as "Large Strategic Transport for Outsize Cargo" or the CX-4 program. The military was reviewing larger diameter fuselage configurations. Conversely the major aircraft designers were still making some of the old assumptions of fuselage diameter dictating wingspan. All of the new aircraft designs submitted called for six engines. They were rejected because they were not superior to the four-engine Lockheed C-141 which was scheduled to fly for the first time in a few months. Up until 1962 the Air Force provided all air transport of boosters and hardware for NASA. Conroy had sidestepped the development of large transports by major manufacturers for the military by providing NASA a volumetric alternative. The military was interested in new "wide body" configurations and now commercial airlines were pressing aircraft manufacturers for a larger replacement for the Boeing 707.

Maneuvering and reattaching the tail on the first few flights was quite difficult. Bill Cuffe and Fred Austin modified the tail dolly by color coding the parts for easily assembly. They added a tiller on the forward castors to more easily maneuver the tail into place after loading (Dottie Furman, courtesy Angelee Conroy).

The Air Force came back to aircraft manufacturers in October 1963 with a request for a conceptual design proposal designated CX-X. It required an aircraft of approximately 550,000 pounds with an 180,000 pound pay-

The 148-foot Saunders-Roe SR.45 had a 219-foot wingspan and was powered by 10 Proteus 600 engines. Only one of the three built, G-ALUN, ever flew, making 47 test flights. It was placed in storage along with the two other partially completed airframes in the summer of 1954. Conroy discussed with NASA officials a plan to acquire and convert a SR.45 to a land based guppy transport (BAS via Dottie Furman, courtesy Angelee M. Conroy).

load an at least 17.5 feet high by 13.5 feet wide and 100-foot long cargo bay. Suddenly aircraft manufacturers were very interested in reviewing the Aero Spacelines guppy performance numbers and drag calculations for an outsize fuselage airframe.

The rocket engine developed for the Apollo flight to the moon was scheduled for testing in Huntsville. The Pregnant Guppy was dispatched to pick up the first Rocketdyne F-1 engine built at Canoga Park, California, for transport to MSFC on 30 October 1963. The FAA was not pleased with Aero Spacelines operating the B-377PG for NASA because the transport certification was not effective until 13 November. The FAA forced ASI to suspend flights until the certification was effective. Subsequent F-1 engines were transported by the Air Force in Douglas C-133 Cargomaster aircraft. The FAA filed suit against Aero Spacelines contending the aircraft was operated illegally resulting in a fine of $1000 for each flight. Aero Spacelines contested the Commercial Operators Certificate was not necessary because it was being operated as a "public aircraft" for an exclusive NASA contract. After a lengthy legal process of almost two years, the court ruled on 30 April 1965 the guppy was indeed a public aircraft used exclusively in service to the government and was not subject to the fines.[46] The court made the point of there being no other type aircraft in existence which could transport the outsize cargo, so why penalize the one provider the government needed.

Conroy Plans Larger Guppies

The British Saunders-Roe SR.45 flying boat was the inspiration of Arthur Gouge. He had risen to chief designer and general manager of Short Brothers in the 1930s. His knowledge of flying boat design and hydrodynamics brought him to Saunders-Roe where he became vice-chairman in 1943. He teamed up with Saunders-Roe designer Henry Knowler to submit design proposals to the Ministry of Supply for a long-range commercial flying boat. The design was very unique for a flying boat of this size because it had retractable wingtip floats as opposed to sponsons. In July 1945 the British Ministry of Supply reviewed and endorsed a need for long-range flying boats. As a result the in 1946 Ministry Director George Strauss authorized the construction of three Saunders-Roe SR.45 Princes aircraft. BOAC expressed interest in the flying boat for its London to New York service. By 1951 BOAC had moved on to an exclusive land plane service. As a result, it was announced in March 1952 only a single ten-engine Proteus 600 powered aircraft would be completed. The others would wait for more powerful 700 series engines to be developed.[47]

The 148-foot long flying boat with a 219-foot wingspan first flew on 22 August 1952 commanded by Saunders chief test pilot Geoffrey Tyson with a 14-man crew. This was the largest British flying boat ever built but it was nine years after the initial design was conceived. The postwar commercial market had changed drastically. The romantic flying boat era had ended leaving very little demand for such service. The aircraft was flown 47 times totaling 98 hours during testing and up to 30,000 feet which was quite impressive at the time for an aircraft weighing 190,000 pounds empty and a max ATOG of 330,000 pounds. By 1954 it was obvious more powerful engines were needed but they were still not available. The only flyable SR.45 G-ALUN was flown for the final time on 27 May 1954. The three airframes were cocooned and placed in storage. The flyable G-ALUN was stored at West Cowes and the other two at Calshot, England. They remained dormant with only one offer from Aquila Airways which was owned by British Aviation Services. Aquila offered £1,000,000 pounds each for the flying boats which was rejected. Multiple other offers were made by startup airlines into the 1960s but none materialized.

Beginning in 1951 the U.S. Air Force began studies on a nuclear powered aircraft. During the mid–1950s, Convair converted a standard B-36 to an NB-36 to study the possibility of nuclear powered aircraft. The B-36 was never nuclear powered but developed concepts for protection of the crew from radiation. Not to be outdone the U.S. Navy conducted preliminary studies and contracted to the Glenn L. Martin Company in 1956 to build models of a nuclear powered conversion of the Saunders-Roe SR.45. In 1958 the U.S. Navy in concert with Convair which had already studied nuclear powered aircraft, held talks with Saunders Roe (SARO) officials to consider purchasing the three giant SR.45 flying boats for conversion to nuclear power. The theory was the weight of a reactor in a land base plane would overstress the airframe and runway. However, a seaplane the size of the SR.45 could sustain the stress of landing because of the distribution of weight in the water. The proposed design mounted a Pratt & Whitney liquid-metal-cooled reactor with a heat exchanger below the center-wing. The superheated air would be directed through a pair of turbines mounted on top of the wing. The aircraft would take off on four conventional jet engines. When cruising altitude was reached the turbines would be started. The plans did not progress past the drawing board and the Navy suspended the project.[48]

The three flying boat airframes remained in storage in England. Jack Conroy became aware of the availability of three giant aircraft in 1963 while reviewing possible large airframes for future guppy conversion.[49] Von Braun expressed interest in a larger guppy shortly after the first ASI presentation. The success of the B-377PG prompted NASA to express a renewed interest in a larger version to transport all of the Saturn rocket stages for the Apollo missions. The B-377 Very Pregnant Guppy was already being developed. Conroy had considered multiple large airframes for possible conversion. Two of SR.45s had not been completed and the airframes were cocooned and stored awaiting possible completion.

A land based conversion of the SR.45 flying boat had been proposed in 1957. The design called for removing the lower boat hull and retractable wing floats. A multi-wheel landing gear was incorporated into each side of the fuselage similar to designs seen on the Lockheed C-5 Galaxy. The Proteus 600 engines would have been replaced with Rolls-Royce Tynes. It also incorporated a rear belly loading ramp similar to the Douglas C-133. Jack Conroy reviewed the British Saunders-Roe Princes SR.45 flying boats for conversion to booster transports. Preliminary design work was done for a SR.45 guppy designated PG-3. It was basically a forward section of cockpit, backbone, wings and empennage mated to an all new giant fuselage. There was speculation the project may have been an effort to block other competing potential guppy designs.[50]

It has not been determined how much data Conroy was able to obtain on the SR.45 land version. However, his expanded fuselage design very closely resembles the Saunders-Roe version. As a result of research conducted by the Aero Spacelines engineering department, he took an option on the two incomplete flying boats on 30 June 1964. Artist Doug Ettridge created operational renderings of the potential conversion. Preliminary drawings were done by Stratos Engineering to convert the SR.45 flying boats to land based turbofan transports by removing the lower boat hull. Only one had been completed and flown which was beached at West Cowes, England. The other two unfinished airframes were stored without engines. It was believed this would be an excellent opportunity to obtain two zero time new airframes at bargain prices.

Conroy was not able to raise the capital to purchase the two partial flying boats. The success of the B-377PG gave him enough leverage with NASA

Conroy and von Braun discussed larger guppies before the B-377PG was completed. The plan was to eliminate all barge transport for the large boosters. Conroy proposed acquiring the two incomplete SR.45 airframes for conversion to 40-foot diameter land planes. The ASI PG-3 design eliminated the 10 Proteus engines in favor of six Rolls-Royce turboprops (Doug Ettridge, courtesy Angelee M. Conroy).

to propose a joint venture to produce a larger volumetric transporter. In 1964 Congress considered a NASA request for funding of $9 million to purchase all three 12-year old Saunders-Roe SR.45 flying boats for conversion to transport the larger sections of the Saturn V rockets. The actual construction would be contracted out to Aero Spacelines and NASA would purchase the finished aircraft.

Senator J. William Fulbright of Arkansas was outraged. He was no fan of the space program, frequently pointing out the enormous cost. He said, "The fact that NASA plans to convert the flying boats exemplifies quite well the folly of the crash program to place a man on the moon by 1970." He further stated, "It is difficult for me to believe a proposal of this nature would ever be considered by a national government official." He wanted to slow down the space program to a more realistic target date. Fulbright stated, "There is so much waste in this space program, I am ashamed to try to include all of them in the [*Congressional*] *Record*."[51]

A redesigned jet version SR.45 guppy was proposed using paired Pratt & Whitney J-57 turbojets from surplus B-52s. A wind tunnel model was produced and test data collected. The wing floats were replaced with tip tanks to extend the range for the 150-foot long monster. The 219-foot wingspan was deemed sufficient for the nearly 40-foot diameter fuselage with approximately 34 feet useable diameter inside (NASA, courtesy Larry J. Glenn).

Although larger, the front area of the cockpit configuration of the ten-engine, six-propeller 330,000 pound ATOG monster physically resembled the Boeing B-377 Stratocruiser. Conroy believed he could modify the SR.45s to transport the 82-foot long Saturn V second stage which was 33 feet in diameter. Eoin Mekie, a British lawyer, became the chairman of British Aviation Services Group (BAS) in 1950. Mekie purchased the three flying boats for a group of investors in 1964. It is interesting the BAS Group was the parent company of Silver City Airways where Commodore "Taffy" Powell originated car-ferry service across the English Channel in 1948. During a British airline consolidation in 1962 British Air Services was merged into Air Holding the parent company of Channel Air Bridge and Aviation Traders Engineering Limited (ATEL) builder of the Freddie Laker designed ATL-98 Carvair guppy.

The intent of Mekie and his investors was to resell the SR.45s and make a tidy profit by providing the aircraft to either NASA or Aero Spacelines. Several other groups had previously attempted to purchase them for airline service. The three aircraft were originally built at a cost of $28 million (£10 million pounds sterling). Only G-ALUN had flown a total of 98 hours. Ultimately the three airframes were cocooned and stored waiting for engine technology to catch up. It was believed they would eventually be sold to an investor and completed. However, the flying boat era had long passed and no reasonable offers were received. No inspections or periodic maintenance were conducted while the three aircraft were in storage adjacent to the water. When the protective covering was removed they were found to be severely corroded because of the damp climate where they had been stored.

During the time negotiations were being conducted to purchase them, Conroy also developed a jet powered design using four of the paired B-52 jet pods for an eight engine configuration. The floats were replaced with wingtip fuel tanks. In anticipation of acquiring them he went as far as having a model built for wind tunnel testing. Congress failed to appropriate funds for the purchase. After the airframes were found to have severe corrosion Conroy did not exercise his option to purchase. On 12 April 1967 G-ALUN was towed to Southampton breakers yard on the river Itchen. After years of storage the Saunders-Roe flying boats were scrapped in 1967 for $49,000 (£18,000 pounds sterling). The forward hull and cockpit from the wing forward of G-ALUN was acquired by a salvage company on the Itchen River and used as an office and workshop until the mid–1970s.[52]

Large Transport Aircraft

The development of an outsize transport for Air Force and Army planners became critical in 1964. Multiple aircraft design programs and projects were taking place simultaneously. The Air Force was seeking design proposals while NASA was letting study contracts for transports with up to 40-foot diameter cargo holds which would accommodate all stages of the Saturn V rocket. In March 1964 Air Force Headquarters set requirements for the CX–Heavy Logistics System (CX-HLS). As with the 1963 CX-X design it had to be a four-engine aircraft with an 180,000 pound payload and at least 17.5 feet high and 13.5 feet wide cargo hold. Boeing, Douglas, General Dynamics, Lockheed and Martin-Marietta submitted outsize design proposals on 18 March. The Air Force let design contracts to Boeing, Douglas and Lockheed for detailed proposals. The request specified a design with front end loading and a raised cockpit similar to the configuration of the smaller ATL-98 Carvair. All three manufacturers submitted high wing designs. The Boeing and Douglas designs had a standard tail but Lockheed submitted a T-tail configuration. The Boeing design resembled the Soviet Antonov AN-124.

Studies concluded the B-377PG Pregnant Guppy with 19-foot diameter fuselage would be able to meet the NASA flying schedule for the time being but not without some problems. Considering the mission and amount of flying being done by a one-of-a-kind experimental aircraft it performed quite well. It had been built outside under marginal conditions and considered a primitive modification of an obsolete airliner which had reliability issues. After eighteen months of exclusive NASA service the U.S. military was also confronted with logistical problems requiring a volumetric transport.

The Douglas C-133 Cargomaster fleet was grounded in January 1965 preventing the Air Force from flying DOD missile transport and NASA cargo missions. The U.S. Air Force requested permission from NASA to engage Aero Spacelines to transport Atlas, Titan and Minuteman ICBMs for military purposes. The C-133s had been transporting the Atlas and Titan rockets to the cape for NASA use as launch boosters in the Gemini and Mercury programs. The B-377PG came to the rescue when it was dispatched to Baltimore on 23 January to transport the 71-foot first stage of the GT-3 Gemini Titan to Cape Kennedy. Again, on 16 May 1965, it returned to Baltimore to transport the first stage of the Gemini IV to the cape. Two days later it returned to transport the second stage for the second manned space flight in the Gemini program.

This would be the tenth American manned space flight which included two preliminary X-15 flights. The Gemini launch on 03 June is significant as the first space walk by an American when astronaut Edward White stayed outside the capsule for 20 minutes. The significance of the mission prompted NASA officials to be concerned with the need for additional transport services. There was no other aircraft available. NASA could not fall back on the Air Force if the Aero Spacelines guppy was out of service because the military was relying on NASA to release time on the guppy after the grounding of the Air Force C-133 aircraft. The B-377PG kept the component delivery on schedule and was now considered critical to the space program and the military.

The increased flying of DOD and NASA flights came at a cost of unscheduled maintenance delays. On 23 October 1965 N1024V suffered an engine failure and was grounded for

The Air Force C-133 was the primary transport for Atlas and Titan rockets until it was grounded. The military requested assistance from NASA to allow the Aero Spacelines to assume the transport duties until the situation was resolved. The B-377PG uplifted a 71-foot Gemini-Titan stage at Baltimore for delivery to the Cape in 1965. Two days later it returned to pick up the next stage (NASA, courtesy Angelee M. Conroy).

three days at Ellington Air Force Base in Houston waiting for a spare to arrive. As a result of this incident, NASA issued new demands requiring the positioning of spare engines across the country at SAC bases where ASI was also provided fuel. This was some insurance against delays; however, the space program was still relying on a single aircraft.

The Pregnant Guppy was transporting the LM test model TM-6 and test article LTA-10 from Grumman at Bethpage, New York, to Cape Kennedy on 27 September 1966 when a serious incident occurred. The aircraft made a refueling stop at Dover Air Force Base when a fire broke out inside the aircraft. It was quickly extinguished preventing any damage to the NASA cargo but it was a stark reminder of how fragile the logistics program was.

The detachable rear fuselage section also proved to be an occasional problem when rolled away from the aircraft during loading operations. Weather conditions were critical when the guppy was on the ground. Any wind more than a few knots created concern because it has a tendency to move the detached rear fuselage like a large sail. An incident occurred at Ellington Air Force Base, Texas, which could have jeopardized the NASA delivery schedule.

Pregnant Guppy Damage

During an unloading operation at Ellington Air Force Base, the tail of N1024V was rolled aside to unload critical NASA hardware. While the aircraft was being unloaded the detached tail was swept away by a sudden gust of wind causing considerable damage. Weather conditions were not considered to be threatening when the operation began. An intense freak gust suddenly came up catching the rear section as it sat in the dolly. Because the wind direction was straight on the open end of the fuselage, it formed a sail. The rear began rolling picking up speed until it spun around and crashed into a building.

The horizontal stabilizer was severely damaged along with the bulkhead mating ring which matches the other half of the aircraft. An inspection determined the horizontal stabilizer could not be repaired and would have to be replaced. Aero Spacelines home office at Van Nuys was contacted to set up the repairs and delivery of parts to Houston. ASI engineer and senior fabricator Don Stratman was dispatched to Mojave to remove a horizontal stabilizer from another Stratocruiser and make it ready for transport to Houston. A Fairchild C-82 Packett was chartered from Steward-Davis at Long Beach to transport the stabilizer. Stratman got the assembly removed and loaded on a flatbed truck for the transfer to Long Beach. When the stabilizer was loaded into the C-82 it was much longer than the cargo compartment leaving it sticking out the back. The delivery schedule at NASA was critical. The guppy had to be back in service as soon as possible. Once again, creativity born of necessity came to the rescue.

Stratman and the C-82 crew partially closed the clamshell doors leaving a gap of about four feet with the stabilizer sticking out the back. They took military style cargo tie-down straps and laced them through the door structural support and cinched them up tight. It obviously changed the CG of the aircraft but the pilot said he would taxi it around and if it felt stable and not too tail heavy he would try it. The C-82 took off and did not appear to handle any different although the partially open rear clamshell doors made the shape of rear fuselage like a flying wedge creating drag.[53]

While Stratman and his crew were retrieving the stabilizer at Mojave, Aero Spacelines vice president of engineering Bob Lillibridge along with mechanics and sheet metal specialist had been dispatched to Houston to remove the damaged parts. The steel ring which mates the rear to the forward half of the aircraft was sprung and obviously would not fit back in position to lock in place. In a bold move to make emergency repairs Lillibridge took a large sledge hammer and managed to pound the mating ring back into position.[54]

Meanwhile, the C-82 with the horizontal stabilizer and Stratman on board were slowly cruising at less than 200 mph en route to Ellington Air Force Base. The 1,400 mile trip progressed into the evening. The pilot got lost and overflew Houston into Louisiana. Fuel consumption was higher than normal because of the extra drag of the rear clamshell doors being partially open. The C-82 landed at Lafayette, Louisiana, to be greeted by curious onlookers. The aircraft was quickly refueled then flew back to Houston. By this time the Pregnant Guppy had been moved into a NASA hangar out of the wind for repairs. The removable tail section with repaired mating rings was reattached to the forward half of the aircraft to check the fit and provide stability. The horizontal stabilizer was unloaded from the C-82 and mounted on N1024V in only three days and returned to service.[55]

This was a stark reminder to NASA logistics at MSFC the space program was completely dependent on a single outsize B-377PG aircraft for component delivery. Not only was NASA dependent on it for missile transport but the Air Force was experiencing transport problems as well. On 30 September 1965 the Air Force selected the proposed Lockheed C-5 Galaxy as the winner of the CX-HLS Heavy Transport design competition. Boeing almost immediately shifted focus from large military transports to forming a design group for the development of the B-747 outsize passenger aircraft.

Instead of attempting to modify the configuration of the proposed CX-HLS military transport, the Boeing design team scaled-up the B-707 to create the new low wing expanded fuselage B-747. The idea of the CX-HLS raised cockpit was retained. Two passenger aircraft designs were evaluated. One had a full upper deck resembling the Airbus A-380 and the other had the cockpit on the lower deck with a high frontal fuselage resembling a scaled down Airbus A-300-600ST Beluga. Eventually the raised cockpit design of the production B-747 was selected with a shortened upper deck. Pan American's forward thinking Juan Trippe surprised all the airlines when he placed an order for 25 Boeing 747s in 1966.

The space program progressed through the 1960s relying heavily on the B-377PG and the B-377 Super Guppy after 1965. The Pregnant Guppy was becoming quite weary and in serious need of upgrades although it was still in high demand. Also, it did not have the wedge fairings at the trailing wing root to prevent buffeting during flap retraction. The moon launch was drawing close and it was evident the PG could not be pulled out of service leaving the B-377 Super Guppy as the single outsize transport to handle the busy NASA schedule. The Mini Guppy first flew in May 1967 and was expected to provide limited backup for the B-377SG super guppy. In mid–1967 Conroy announced to the press the Pregnant Guppy would be removed from service in October for a much needed overhaul.

When built at Van Nuys in 1962 the original wiring and fuel plumbing was left in place and spliced into or repaired as needed. At the time NASA was desperate for an outsize transport and Conroy was financially exhausted prompting cost cutting and shortcuts anywhere possible to get the aircraft completed. The B-377PG made it possible for NASA to fulfill von Braun's commitment to land on the moon by the end of the decade. However, in 1967, two years before landing on the moon the B-377PG was tired and obsolete. The plan was to upgrade all systems and correct the short cuts taken when originally built in order for it to receive an FAA type certification. Conroy was planning on building larger guppies for the space program and utilizing the overhauled B-377PG for commercial cargo.

Getting it out from under the restrictions of a "public aircraft" would allow ASI to operate commercial charters after the exclusive NASA contract ended. Plans were made to build and install a new 23-foot center-wing identical to the Super Guppy which would extend 7.5 feet on each side of the fuselage. This would increase the span to 156 feet, 8 inches like the SG. It would improve the weight distribution on the ground by moving the main gear farther apart. The extended center-wing would also place the mountings far enough away from the fuselage to allow the inboard engines to be fitted with the same 16-foot, 8 inch props as the outboards.

Work began on the ramp near the old hangar at Santa Barbara. The outer wings, rudder, props and cowlings were removed. The paint was stripped including all lettering and stripe with spear on the tail. On the last day of October it was towed from the airport ramp across Hollister Avenue to the same area where the Mini was built. After being secured at the off airport facility the engines were removed and overhauled. Rewiring was completed but the plans to install a new center-wing never came to pass.[56] Although not positively verified it is assumed the wing trailing edge fuselage fillet designed by Irv Culver was installed at this time.

Engineers were faced with the reality of the first two guppy conversions having a limited service life. The shortcuts taken to get the first guppy certified in 1963 for immediate NASA work were significant. It was becoming very apparent the expense of the upgrades amounted to a complete rebuild which could equal the original cost of construction. Work progressed to a point but plans for a complete update and modifications lingered. The costs of updating it to the same standards as the planned next generation wide-floor model was not based on sound economic reasoning. It would be far more practical to build a new super guppy. ASI did propose a more advanced turbine powered Mini SGT-101 (see Chapter 15) for commercial cargo. It was followed by a proposed Super Guppy SGT-201 design for NASA and the military (Chapter 16). An overhaul of the B-377PG's existing systems without additional modifications was completed in December returning it to service in January 1968.

Jack Conroy never publicly acknowledge the Pregnant Guppy was underpowered. Even with a reduced maximum ATOG it struggled on hot days to take off with heavy loads. In an effort to add needed power when at all up maximum weight a plan was devised in 1968 to supplement takeoff power with external short term rocket bottles (RATO) mounted to the rear of the aircraft. While it appeared to be a simple solution, the test flight nearly ended in disaster. RATO or JATO was developed by the military during World War II and first tested on 16 August 1941 prior to the Japanese attack on Pearl Harbor. It was designed for short field takeoff at high gross weights. Braniff Airways had successfully used the JATO bottle system on commercial DC-4s for takeoff from high altitude fields with high humidity in South America in the late 1940s.

To test the assisted takeoff RATO system, four Aerojet General 15KS-1000-A1 rockets were attached to the guppy fuselage. The system consists of paired mountings of 200-pound steel rocket bottles with an electric ignition. Once lit, it cannot be shut off burning for only twelve seconds. Each rocket bottle supplies a constant 1,000 pounds of thrust or approximately 350 horsepower each. When mounted in multiples, RATO can provide considerable assistance in getting a heavy payload airborne from short fields, high altitude airfields or in extreme heat. It is open to speculation if NASA would have ever agreed to place a piece of critical space hardware onboard if the guppy was using RATO assist.

Aero Spacelines pilot Bruce Stratton was in command

of the B-377PG when flight tests were conducted using the RATO system. He confirmed the boost measurably improved takeoff but there were problems. The most feared situation which can occur during flight is an onboard fire. Stratton recounted the event in 2015. "As the aircraft lifted off the bottles were activated. We felt the aircraft rise up from the added power. The rockets were so hot they ignited the paint on the inside compartment of the aircraft. It was a scary moment for the crew as smoke began to accumulate in the cargo bay. The heat was so intense it could have ignited the skin and burned the bottom out of the aircraft and brought us down. Fortunately, one of the crew was able to extinguish the fire with a hand held unit." After evaluating the test results and reviewing the possibility of fire, the use of RATO was dropped as an alternative means of increased takeoff power and never used on operational flights. Captain Stratton eventually accumulated over 300 hours flying the PG, SG and MG.[57]

The B-377PG was built with a planned obsolescence. As the space program progressed, the rocket components increased in size consequently the parts outgrew the aircraft. The B-377PG N1024V operated under contract to MSFC past the completion of the Apollo program. It transported NASA and DOD cargo for 11 years logging over 6,000 hours. The vast list of space hardware transported includes Apollo modules, F-1 rocket engines, Gemini launch vehicles, Pegasus meteoroid detection satellite hardware and Saturn I instrument unit.[58] In addition, Atlas, Titan and Minuteman rockets were transported for both NASA and the military.

Unexcelled Chemical acquired control of Aero Spacelines in 1965. Cutbacks in the space program reduced aircraft use resulting in considerable loss of revenue at ASI. The crash of the MGT-101 in 1970, the lack of serious buyers for the SGT-201 and loss of stock value because of devious stock manipulators brought the company to its knees. By 1973 Unexcelled transferred control of ASI to subsidiary Twin Fair discount department and grocery store chain of New York. Aero Spacelines under the new company ceased flight operations on 31 December.[59] Conroy had not been with the company he created since August of 1967. Aero Spacelines had changed focus from heavy airframe modification and construction of guppy transports for exclusive government contracts to production and marketing of the aircraft for the commercial market.

Last Pregnant Guppy Flight

In order to divest itself of guppy flight operations Aero Spacelines placed the Pregnant and Mini Guppy up for sale. They were acquired by Conroy's old friend Allen Paulson in 1974 via American Jet Industries. The B-377PG was not in the best condition. It had been hastily built and flown to the limits servicing NASA contracts. The future of the aircraft was in limbo as to whether it could be overhauled and upgraded to acquire a type certification. Paulson asked his friend and business partner Clay Lacy to ferry the Pregnant Guppy from Santa Barbara to Van Nuys along with engineer John Kinzer. Jack Conroy and Clay Lacy had enjoyed a long friendship. Conroy had selected Lacy to co-pilot the aircraft on the first flight at Van Nuys. Lacy felt he and Conroy should fly it together on the last flight because they were at the controls on the first flight. Lacy said, "I called Jack and asked if he wanted to fly the last flight with me. Jack said he had something else to do and seemed uninterested. I knew if Jack did not do it he would regret it later." Lacy continued to pressure

Jack Conroy wearing his lucky flight suit moments before he and Clay Lacy flew the Pregnant Guppy for the first time on 19 September 1962. Twelve years later he arrived at the flight line wearing the same flight suit to pilot the final flight of the B-377PG from Santa Barbara to Van Nuys with Lacy. He asked Lacy if he recognized it. Lacy said, "I knew this flight meant a lot to him" (courtesy Angelee M. Conroy).

Conroy but he continued to say he was not interested. Lacy called him again and saying you need to do this. "If you don't do this you will regret it." Finally, after a lot of persuasion Conroy reluctantly agreed.[60]

Lacy said, "Jack showed up on the ramp wearing an old tattered and faded flight suit." He asked Clay, "Where is your flight suit?" He wanted to know if Lacy recognized the one he was wearing. Jack said, "This is the flight suit I wore on the first flight, where is yours? Go get it!" Clay said, "I had no idea where my old flight suit was but I could tell this meant something to Jack and he wanted to fly it one last time. I was really glad I kept after him until he agreed. Jack was the best friend I ever had!"[61]

Twenty-six years after it rolled out at Boeing as a B-377 Stratocruiser in April 1948 the Pregnant Guppy registration N1024V was cancelled. It was registered to American Jet as N126AJ on 23 October 1974. The Aero Spacelines titles were painted out but the American Jet lettering was not applied. The Pregnant Guppy had logged over 6,000 hours since it was converted from a Pan Am Stratocruiser. It is doubtful this landmark aircraft ever flew again because there is no record of any contract flights with American Jet. There were still plans for upgrade but it remained grounded in poor condition. After five years with American Jet the registration was cancelled on 05 February 1979. It was scheduled to be broken up at Van Nuys by the end of the year. Jack Conroy had been diagnosed with cancer late in 1978. By the fall of 1979 he was terminally ill. Allen Paulson, in respect for his dying friend, instructed the staff to delay any plans to scrap the B-377PG as long as Conroy lived.

Jack Conroy passed away on 05 December 1979. Few people are aware of his patriotism or his contributions to aviation and the space program which put America on the moon. The day after he passed away dismantling of the Pregnant Guppy began. Salvageable parts were removed before it was chopped up for scrap. However, this was not the end of the first guppy because the rear section 45 was salvaged for the next SGT-201 scheduled to be built.

The First B-377PG Lives On

One of the most unique stories of the guppy fleet came about because of a chain of unplanned circumstances and shortage of B-377 airframe components. Eighteen years after the first guppy was built UTA was contracted by Airbus to build additional turbine guppies in France from components supplied by the Aero Spacelines division of Unexcelled. The salvaged rear fuselage section 45 of the Pregnant Guppy was acquired from American Jet after it was broken up at Van Nuys. The assembly was purchased for the next aircraft being constructed for Airbus at Le Bourget. The section 45 was shipped to France to be used on the next SGT-201 c/n 004 built by UTA. There was no active intent to use parts from the first guppy on the last one built. The section 45 was acquired for use on the next aircraft in production. It was not until sometime later it became the last guppy. An Air Force C-97 and the section 45 from an ex–Northwest Stratocruiser were acquired by Airbus to build a fifth SGT-201. Eventually the decision was made not to build additional SGT-201 guppy aircraft in favor of the A300 Beluga. The parts previously acquired to build the fifth Super Guppy including a section 45 from Northwest N74603 were not used. As a result, the lower rear section from the B-377PG N1024V was on the last guppy built by default. As of 2017, the last SGT-201 remains in service with NASA as N941NA.

When Airbus negotiated for the

On 06 December 1979, the day after Jack Conroy passed away, work began to break up the Pregnant Guppy at Van Nuys. The scrap dealers showed little mercy chopping it up in short order. It was a sad end to a landmark aircraft that contributed to America's effort to put a man on the moon (courtesy Clay Lacy).

first two SGT-201s which were built at Santa Barbara the contract included a licensing agreement with options for up to four more to be constructed by UTA in France from parts supplied by ASI. Photographs of the salvage and crating of section 45 from the B-377PG at Van Nuys have yet to surface. The commonly circulated photo said to be section 45 from N126AJ (N1024V) is actually from ex–Northwest B-377 N74603 which was owned by ASI for many years. Conroy reversed the Northwest B-377 color scheme and had Aero Spacelines lettering painted on the nose. Sometime between September 1981 and June 1982 ASI fabricator Tom Roberts was dispatched to Tucson to remove the rear section for shipment to France for use on SGT-201 number five. He was reminded to use extreme caution in cutting along the pre-marked cut lines to insure critical structural components were not damaged. Roberts believed the section 45 of N74603 which he removed was actually for use on the last SGT-201 and not the one from N1024V (N126AJ), but there is no conclusive proof.[62]

The fuselage section from B-377 N74603 (c/n 15949) along with KC-97G, 53-280 (c/n 17062) was purchased by Airbus for construction of a fifth SGT-201, which was never built. The last two SGT-201s were built from USAF KC-97G airframes. The commercial B-377 rear section 45 is required to replace the 38-foot military C-97 tapered rear fuselage. The original C-97 sub-assemblies were built in San Diego by Ryan Aeronautics for Boeing. They are considerably heavier than the B-377 unit and structurally different because of strengthened framing to accommodate the refueling pod or clamshell doors with hydraulic operating system and ramps.

An invoice obtained from UTA in France indicates the Pregnant Guppy section 45 was purchased from Aero Spacelines on 05 September 1980 at a cost of $9,680 and was shipped on 09 September. Also shipped at the same time was a part listed as MLT-MG1025 Tool, Pallet Rail loaned to UTA until December 1982 when it would be returned to ASI. The parts cleared French customs on 31 October 1980.[63] The section 45 was intended for use on the next SGT-201 built because other guppies were planned. Additional parts and assemblies had been purchased to build two to four more. As it turned out, SGT-201-4(F) F-GEAI (N941NA) was the last built using the section from the first B-377PG creating a unique story of part of the first one flying on the last guppy built. It is a fitting tribute for an aircraft that may well have saved the United States' fledgling space program from delay, failure and international embarrassment.

INVOICE

AERO SPACELINES INC.
AIRPORT • 4958 SOUTH FAIRVIEW AVENUE
GOLETA, CALIFORNIA 93017
PHONE 805 967-4571

INVOICE DATE: 5 Sep 80
INVOICE NO.: 6535

PLEASE SHOW OUR INVOICE NUMBER ON YOUR REMITTANCE

SOLD TO:
Union de Transports Aeriens
Division des SErvices Comptables
DF/FI/FGF 2
B.P. 7 – 93350 Aeroport du Bourget
France
Atten: P. Vellay LBGIGUT
M. Reinold LBGYJUT

SHIP TO:
DOUANE MATERIEL
31 OCT. 80 C00035
UTA—DI-GE

CONTRACT/P.O. #	SHIPPER #	DATE SHIPPED	VIA	F.O.B.	TERMS
Y600911J		9 Sep 80	Schenkers	Goleta	By Check

ITEM	QTY SHIP'D	PART NUMBER DESCRIPTION	UNIT	UNIT PRICE	AMOUNT
1	1	PREGNANT GUPPY B377 Section 45	Ea.	9,680.00	9,680.00
		(Price is for packaging, labor and material for ocean shipment.)			
		Condition: As is			
2	1	MLT-MG1025 – 1A Tool, Pallet Rail	Ea.	N/C	N/C
		S/N 001			
		This item is on loan to UTA until December, 1982 at which time UTA will return it to ASI at UTA'S expense.			
		Condition: Serviceable		TOTAL	9,680.00
		FOR SUPER GUPPY PRODUCTION			

Aero Spacelines acquired the salvaged B-377 fuselage section 45 of N1024V from American Jet. It was then resold to Airbus for use on the next of eight planned SGT-201 aircraft being built by UTA for Airbus at Toulouse, France. The invoice shows the rear section arrived at UTA on 31 October 1980 (ASI, courtesy Tom Smothermon).

13

B-377SG Super Guppy

A Turboprop Guppy

Early in the space program NASA predicted the need for additional larger volumetric transports. As the first Aero Spacelines Pregnant Guppy was being built agency officials advised Jack Conroy the size of boosters would increase over time. Once operational, the demand for the B-377PG made it obvious a single aircraft could not meet the delivery schedule for all the components and assemblies being produced. Wernher von Braun and Jack Conroy discussed a larger turboprop conversion in September 1962 just days after the Pregnant Guppy first flew. If a larger aircraft was not built, the space program would have to rely on the slower sea-going barge for the larger launch vehicle components. NASA planners realized the first guppy was merely a short term fix. The space program would be in the same logistical situation before the moon mission was completed if larger aircraft were not available. Officials at MSFC repeated their concerns for a larger and more powerful aircraft with greater volume than Conroy's upcoming second design designated the Very Pregnant Guppy (Super Guppy). The NASA logistical team suggested a third aircraft could serve as backup. Conroy envisioned a fleet of outsize transports. He began creating designs for jet powered giant aircraft larger than the Super Guppy with the capacity to handle Saturn V S-II second stage before the Pregnant Guppy was completed.

Conroy pointed out each larger aircraft would need engines with at least 15 percent more horsepower than the previous. The 28-cylinder P&W R-4360s were already at their maximum output leaving turboprops as the only alternative. NASA was acutely aware of power limitations after several engine failures grounded the B-377PG en route to the cape. The very precise delivery schedule could not be disrupted if America was going to put a man on the moon. The requirement for Aero Spacelines to place spare R-4360 engines for the PG at intervals across the country to prevent future delays was only a stopgap measure.

Aero Spacelines operated on a tight budget making fuel stops at SAC bases where government was providing the fuel. The aircraft was under exclusive contract to NASA and could not be used for commercial charters to produce additional revenue. Conroy reasoned a second aircraft could back up the B-377PG and be utilized as a commercial transport when not moving space hardware. He was not yet aware that NASA had other ideas of having exclusive use of a second aircraft. Conroy had always entertained the possibility of more lucrative commercial charters and wanted to negotiate a contract with Douglas to transport DC-10 fuselage sections from Convair at San Diego to Long Beach and wing sections from Canada. In addition there were L-1011 wing sections which needed to be transported to Lockheed at Palmdale. In an effort to generate interest Conroy initiated a promotional campaign targeting aircraft manufacturers with promises the guppy fleet would grow to six aircraft.

The NASA logistical team had not originally taken Conroy and his cumbersome outsize transport seriously. However, the success of the aircraft prompted a feasibility study of much larger transports. NASA's director of the Manned Space Flight Center Development at Houston, Robert Freitag, re-reviewed Conroy's 1962 proposal of a larger aircraft with von Braun in 1964. He saw the need and advantages of a second aircraft and praised the success of the first guppy. NASA planners considered the possibility of a giant transporter up to 40 feet in diameter. Marshall Space Flight Center issued a request on 02 February 1964 for quotations on Large Booster Carrier Aircraft of both heavier and lighter than air design. On 18 October 1964 Conroy offered several suggestions which included a second B-377PG or a larger aircraft with a swing nose which could handle most NASA cargo leaving only the largest boosters to be transported by sea.

The B-377 Super Guppy became the first aircraft built with a swing nose which includes the cockpit. The nose loading concept first patented by Charles Babb in 1938 with both a side hinged and a flip up door design was only for nose loading with an elevated cockpit. In 1941 Curtiss-Wright contracted to the USAAC to build a wooden transport with a flip up nose door similar to the B-747F, ultimately, becoming the first drive-on transport. Only five were built before the order was cancelled. A number of other cargo aircraft have been built using the Babb design. The nose loading concept with elevated cockpit had been proposed on a number of previous designs which included the advanced Convair C-99 (not built), Douglas C-124, ATL-98 and AW 660. Conroy took the concept a step further by proposing a hinged forward nose door opening which included the cockpit.

Outsize Aircraft Proposals

ASI engineering submitted multiple proposals of mega aircraft up to 40 feet in diameter which could handle the larger S-II second stage of the Saturn V rocket. Conroy proposed a jet powered 30-foot Boeing 707 and 38 to 40-foot B-52 guppy with the giant fuselage above the wing and a Convair B-36 and two Saunders-Roe guppy designs, one with props and the other with jets. Freitag and von Braun were acutely aware a decision was needed immediately to begin construction of a larger aircraft. A second transporter was obviously needed to maintain the supply line of components to keep the moon mission program on schedule. The NASA logistics team reviewed the proposals for a Large Booster Carrier from major aircraft manufacturers, airframe modification companies and Aero Spacelines.

The Fairchild Stratos Corporation proposed the ten-engine M-534 outsize transporter based on a highly modified Convair B-36J with a guppy type fuselage at a cost of $11,500,000. It was supposedly capable of carrying the Saturn V second stage. (See the B-36 to XC-99 section for details). Convair Aircraft proposed an alternate design using a conventional four-engine transport with very long fixed gear with the rocket boosters slung under the bottom of the fuselage similar to the Soviet Myasishchev M-90 aircraft design. Dee Howard proposed a 40-foot diameter aircraft using DC-7 components and ten engines.

NASA needed larger diameter transporter up to 40 feet but the most appealing and cost effective solution was the proposed Aero Spacelines B-377VPG (Very Pregnant Guppy). It was a larger version of his successful Pregnant Guppy which was being developed by Conroy with private funding. A model of the B-377VPG was presented to the press with wing tanks and end plates on the horizontal stabilizer. NASA realized most of the other proposals would take too long to develop. ASI had already developed many engineering techniques when converting the B-377PG. It was a proven design already in service for NASA. Delivery time of a second transporter could be greatly reduced because it would be an upgraded larger version of the same type.

The Very Pregnant Guppy, soon to be renamed Super Guppy, was originally designed with a 24-foot, 6 inch fuselage. Von Braun requested an increase in diameter of a few more inches to accommodate larger rocket components which were already designed and being constructed. Conroy continued to stress it was imperative to obtain turboprop power before building a bigger aircraft. In June 1964 Hayvenhurst Van Nuys Corporation announced a renewed Aero Spacelines outsize transporter contract with NASA for another year in the amount of $1,444,000. This extended the two previous transport contracts which had been in effect over the last 12.5 months.

Super Guppy Construction

One month before work began to cut up ex–Pan Am B-377 Stratocruiser N1038V (c/n 15938) for the next conversion, ASI Board Chairman J.H. Olverholser held a press conference. He announced the construction of a larger B-377SG with a proposed payload increase from 33,000 to 38,000 pounds. The range was expected to be 1,800 miles compared to the 1,300 of the B-377PG. At the time, obtaining suitable turboprop engines was still in question. He stated the new guppy would be fitted with a pair of auxiliary Westinghouse J34 jet pods in place of the two external wing tanks. The jets would only be used for takeoff and climb-out and would be shut down after reaching cruise relying only on the four Pratt & Whitney R-4360s.[1] This would be the same engine configuration of the Military KC-97L tankers.

This is further indication of Jack Conroy's tenacity. He started building the Super Guppy promising von Braun it would be powered by turboprops and, yet, he did not have the engines or a way of obtaining them. It never occurred to him that he would not succeed. He often made plans and set unrealistic schedules which were impossible to achieve. At the last minute when failure was eminent he would switch to plan B which seemed to always cover the impending doom. He never really failed but, rather, would completely change direction and continue on to achieve his goal.

In November 1964 ex–Pan Am Stratocruiser N1038V was pulled on the parking ramp at Van Nuys. The aircraft was placed on jacks with the rear half on a sliding cradle. Construction began as workers cut away the unneeded upper fuselage sections from behind the cockpit to behind the wing. The rear half of the aircraft in the cradle was slid back to make room for the additional fuselage sections to be grafted in behind the wing. Conroy was still looking at the possibility of building the Super Guppy with reciprocating R-4360s augmented with jet pods. The B-377 wings with nacelles for piston engines were left in place. The Pan Am B-377cockpit was also left attached. Conroy had inquired about the Pratt & Whitney T-34 turboprops from the two Air Force YC-97J test aircraft but was flatly turned down. Any chance of getting them was still in question. Although no data has been located with specific plans, ASI engineers were considering using C-97 wings with the P & W R-4360 power plants. The wings were beefed up and would have been necessary in order to have hard-points for the jet pods. Conroy was indeed making it up as the project proceeded.

J.H. Olverholser further announced the new guppy would be certified under Civil Air Regulation Part 4B verses Part 8 used for the Pregnant Guppy. The idea was to obtain FAA certification for the Super Guppy verses the "public aircraft" classification of Pregnant Guppy. During this same conference, Olverholser stated they were considering retrofitting the J34 jet pods on the PG.[2] Much of this was overoptimistic hype but the press ate it up.

In reality jet pods on the B-377PG probably was not possible because it had B-377 wings without hard point mountings. To assure NASA needs were met and to speed the process MSFC sent aeronautical engineers along with a flight text engineer from Edwards Air Force Base to Van Nuys in the fall of 1964. The group was assigned to work with the ASI design team on the second generation guppy. Von Braun was putting pressure on the Air Force and Pratt & Whitney but the use of YC-97J had not been approved.

The first B-377 guppy was an operational success for NASA creating cash flow for ASI. However, it was not enough to fund the development of a new aircraft. Conroy was still experiencing financial problems. Work continued to move forward at Van Nuys to build a more powerful larger version swing nose super guppy but at an irregular pace. Actually, the completed aircraft was more YC-97J with parts from three other C-97s. They were combined with parts from Transocean B-377, N406Q and N408Q plus just enough of Pan Am B-377 N1038V to acquire a civilian registration. The fuselage would be 141 feet long, with a 25-foot diameter 30-foot constant section in the 94-foot, 6 inch cargo bay which produced a volume of 49,750 cubic feet.

Acquiring the Boeing YC-97J

Adapting turboprops to a B-377 wing would be costly and require additional engineering. Conroy wanted to use as many off the shelf components as possible to reduce cost. He was in effect cobbling together parts from multiple airframes and building the large diameter fuselages. By doing this he could greatly reduced engineering cost. He had learned the two USAF turboprop YC-97J test aircraft (52-2693 and 52-2762) were soon to be set for disposal and lobbied the Air Force via NASA to acquire them. The two KC-97G aircraft model 367-86-542 had been modified by Boeing in 1955 at a cost of $2,106,542 with the installation of P & W T34-P-5 5700 ESHP turboprops and Curtiss CT735S-B104 turbo-electric propellers with 15-foot blades. The tail, wings and control surfaces had also been beefed up which would be perfect for Conroy's giant fuselage. This engine propeller combination was an early version of the power plants used on the Douglas C-133. The production C-133 was also fitted with the Curtiss CT735 props with larger 18-foot blades. The C-133 aircraft ultimately suffered engine and propeller problems throughout the entire operational life.

Pan Am N1038V was cut behind the wing and the rear half slid back for grafting of lower fuselage sections from two other aircraft. The B-377 wings and nose were left in place because the YC-97J with T-34 turboprops had not yet been acquired. The possibility still existed that the Super Guppy would have piston engines with outboard J34 jet pods similar to the KC-97L (ASI, courtesy Tom Smothermon).

The life expectancy of the T-34-P7 engines on the YC-97J was originally set at 250 hours. As testing progressed it was eventually increased to 750 hours. Even with the extended engine life, the problems continued with seven propeller changes on the YC-97J between July and December 1955. It is a recognized fact modifying older airframes to fly at higher speeds than originally designed creates additional problems relative to higher stress. Despite the fact YC-97J aircraft had its own problems, Conroy wanted to use it because the wings were already engineered and modified with nacelles for turboprops. It also flew faster than the original C-97 from which it was upgraded.

To compensate for some of the

Air Forced C-97 AF 52-2693 is one of two aircraft modified in 1955 to YC-97J by installing 5,700 ESHP Pratt & Whitney T-34-P-5 turboprop engines. Jack Conroy lobbied the Air Force for the aircraft after he learned they were scheduled to be retired without results. Wernher von Braun persuaded the Air Force to reconsider and allow Conroy to acquire them (USAF).

problems experienced with the YC-97J, the original aileron ribs were reinforced and re-skinned with a heavier material. The configuration of the modified YC-97J nacelles and engine mountings relative to the original C-97G wing did create some vibration problems. The vibrations were manifested in cracked welds and increased deterioration of the tail pipes. These problems were never resolved during the YC-97J test program. It is not documented as to how it affected the Super Guppy although it was reported to have moderate vibration in the center wing.

The two KC-97Gs when modified to turboprop power plants were first designated as YC-137s. They were changed to YC-97J and the C-137 designation was reused for the VIP C-135 jet transports modified for the White House. The test program for the pair of YC-97Js was operated out of Norton Air Force Base, San Bernardino, California. This was close by to Conroy's operation and he had seen them while flying standard C-97s in the California Air Guard. The lighter turbines provide 63 percent more power increasing the max speed of the converted KC-97G from 375 to 417 mph. The lighter engines decreased the empty weight by over 9,000 pounds. The installation of the more powerful engines increased the payload from the original 40,000 pounds to approximately 60,000 pounds.

Boeing attempted to interest the Air Force in a production run of the lighter more powerful turbine version C-97 without success. Conroy's was told under no circumstances would the experimental T-34 engines be made available for a civilian aircraft. His goal was to obtain the two airframes and lease the engines. Acquiring more efficient and powerful turbine engines for the second generation guppy was the perfect solution eliminating the research and development required to retro-fit turboprops to the old Stratocruiser wing.

Conroy expressed his need to von Braun. After repeated pressure by NASA on the Air Force, Conroy managed to secure a commitment to acquire the YC-97J airframes but without the engines. At least the engine mountings, nacelles and fairing would not have to be engineered leaving the task of finding an existing turboprop engine which could be adapted. His friendship with von Braun combined with the success of the Pregnant Guppy was instrumental in NASA continuing to press Pratt & Whitney for the engines.

The Super Guppy was well under construction with Stratocruiser wings and nose when Conroy went to Tucson on 08 March 1965 to inspect YC-97J 52-2693 and make arrangements to break it up for shipment by rail to Van Nuys. Don Stratman was dispatched to oversee the operation and hire local mechanics to disassemble a single YC-97J for shipment to Van Nuys. Lee Mansdorf's younger brother Norman accompanied the team to observe and represent his brother's interest in the project. Stratman designed and constructed support frames for the inner wings in order to remove them without damage and transported to the railroad siding for loading. He used the outrigger struts from surplus B-47 aircraft to convert the support frames into dollies which would roll easily. After the wings were loaded on the rail cars the dollies were trucked back to Van Nuys. They were used to support the wings when they were rolled into position and attached to the new expanded center-wing.[3] Only wing parts and nacelles were salvaged from the second YC-97J.

In the spring of 1965 NASA administrators pressured the Air Force to approve leasing of the Pratt & Whitney T-34s to Aero Spacelines on the grounds it was in the national interest. The NASA Administrator wrote to the Air Force stating: "We definitely feel that it would be in the public interest and advantageous to the government if these engines were made available."[4]

The P & W T34-P-7 was a test engine installed on the C-97s for data collection for use on the C-133 and never meant for a civilian contractor. As a result of testing the engines on the YC-97, YC-121 and YC-124B, the C-133 was fitted with the 7,500 eshp P & W T-34-P-9W version. The Air Force and Pratt & Whitney reluctantly complied with NASAs request for the engines. ASI fabricator Don Stratman recounted the story in 2016 stating, "NASA was very impatient, wanting the aircraft completed as soon as possible. We were doing all we could but some of the logistics people at NASA were outright demanding. They were under extreme pressure and wanted the plane yesterday." Precious time was being consumed in delivery of rocket assemblies by sea which was preventing NASA from performing all test needed for the moon mission and Pratt & Whitney was slow to release the T-34 engines.

The airframe of N1038V was stretched by adding two forward lower fuselage sections 42 from Transocean B-377s (c/n 15944 and 15945) behind the wing totaling 300 inches (25 feet). After the YC-97J aircraft were acquired, Conroy decided to replace B-377 nose with the upgraded military C-97 unit because the engineer's panel and controls were set up for turboprops. The nose of Pan Am N1038V was cut around the circumference of the fuselage at the rear cockpit bulkhead and moved away. The nose of the YC-97J was cut behind the cockpit bulkhead on the upper half down to floor level and 70 inches back on the lower half of the fuselage and grafted ahead of the wing to in-

Top: **Aero Spacelines fabricator Don Stratman was dispatched to Tucson to oversee the disassembly of the YC-97J. Local mechanics were hired to prepare the components for shipment. The forward fuselage, including the cockpit, inner wings, and tail and wing fuel tanks, were loaded on rail cars for shipment back to Van Nuys on 13 April 1965 (ASI, courtesy Tom Smothermon).**

Bottom: **The Boeing B-377 lower deck forward compartment entry door was located on the right side of the aircraft. The crew could access the lower compartment from the cockpit by descending a three-step ladder to the equipment bay then a four-step ladder into the compartment. The drop down door could also be used for crew or maintenance technician entry (author's collection).**

crease the length. Super Guppy cockpit crew entry via the lower deck created an engineering problem because the B-377 and C-97 lower forward bulkhead is different. This had not been a problem on the Pregnant Guppy because it was built with all B-377 parts.

Because the commercial B-377s have a main deck passenger entry door, crew access was not through the lower deck like the C-97. The B-377 has a single hatch in the cockpit floor aft of the engineer's chair to access the lower equipment bay via a three step ladder. It would not have been practical to have a floor hatch outside the rear cockpit door because it would have been in the aisle of the passenger cabin. In an emergency situation the crew could enter the B-377 forward baggage compartment by opening the hatch in the cockpit floor, descending a three step ladder into the forward equipment bay mounted on top of the nose gear, then step back 24 inches through an opening in the lower deck bulkhead and down a second four step ladder into the lower cargo compartment.

The C-97 has a lower deck entry door on the left side of the fuselage and the B-377 is on the right. Because the C-97 has a solid lower deck bulkhead there are two main deck floor hatches. One hatch in the cockpit floor with as three step ladder to the equipment bay. A second entry hatch is outside the cockpit in the cargo bay floor with a seven step ladder to the lower compartment. The C-97 crew members entered through the lower left entry door and up the seven step ladder through the main deck cargo bay floor hatch which is hinged on the left.

The military YC-97J nose was substituted for the original Pan Am B-377 nose on the Super Guppy. When the

Top: **The entry door of the Air Force C-97 was on the lower left side of the aircraft behind the cockpit. The crew entered via a seven-step ladder and through a hatch in the floor behind the cockpit door in the cargo bay. The left side door was eliminated in favor of the B-377 right side configuration utilizing the separate hatch inside the cockpit for access to the forward equipment bay (courtesy Patrick Dennis).**

Bottom: **The Super Guppy forward compartment and split ladder access closely resemble the configuration of the commercial B-377 Stratocruiser. The two hydraulic stabilizer struts are deployed before the nose is opened for stability. The entry door steps are visible on the right in the closed position. On the original B-377 airframe they were 70 inches farther forward (Pierre Cogneville collection).**

C-97 nose was grafted to the Super Guppy there was no reason for two trap doors in the floor. The door in the cargo bay floor in the YC-97J nose outside the cockpit was deleted when the new floor was installed in the eight foot section aft of the cockpit bulkhead which was grafted to the Super Guppy. The lower nose bulkhead was modified for crew entry. The original C-97 seven-step ladder to the upper deck was removed. An access hole was cut into the C-97 lower deck forward bulkhead. It was modified with parts from the B-377 lower bulkhead to access the existing three step ladder in the equipment bay below the cockpit floor hatch. The B-377 lower ladder was a four step unit mounted on the lower forward bulkhead. The configuration closely resembles the original B-377 allowing the crew to enter the lower right door and ascend the split ladder to the cockpit. A pair of hydraulic jacks extended from the lower fuselage behind the nose break to stabilize the aircraft for loading. A retractable chain driven two-wheel dolly fits behind the nose gear wheel well to power the nose around 100 degrees for loading. The nose gear is partially retracted during the open and closing of the nose.

The Super Guppy also has a rear drop down entry door for access to the lower rear compartment behind the wing. This provides access to a small catwalk to the extreme rear of the aircraft. A ladder and hatch in the main deck floor provides access to the rear cargo compartment after cargo has been loaded. Most NASA cargo only had inches of clearance on each side once loaded leaving no access to the rear of the aircraft to inspect and secure the load. This rear entry provided the loadmaster access behind the cargo.

The fuselage is 30 feet, 10 inches longer than the B-377 base airframe and 14 feet longer than the B-377PG. The upper half of the lengthened fuselage was built up from B-377/C-97 floor level to create the expanded cargo hold. The Super Guppy cargo bay is 94 feet, 6 inches long with a 30-foot, 8-inch constant section which is 25 feet in diameter. Sixty-seven feet of the fuselage is equal or larger than the diameter of the Pregnant Guppy and 81 feet more than 17 feet in diameter. By comparison the constant section of the PG was 19 feet, 9 inches in diameter. The SG and PG are the only two guppies which retained the Stratocruiser lower fuselage with 8-foot, 9-inch wide floor.

The B-377SG was a considerable improvement over the Pregnant Guppy with higher gross weight payloads but it was also unpressurized. The crew was required to be on oxygen anytime it flew above 14,000 feet. The cumbersome oxygen mask made it a bit awkward and uncomfortable for the crew. In addition the SG air conditioning system did not work below 5,000 feet. Flight deck air conditioning problems did not start with the guppies. It was also a problem on the military C-97 especially the reconfigured HC-97to Air-Sea Rescue standards. As aforementioned the HC-97s often flew at low altitude on search and rescue missions. A series of up to six small auxiliary AC units were installed in the cockpit for crew comfort.

The SG airframe was increased from the 110-foot B-377 to 141 feet by inserting fuselage sections from multiple aircraft. The nose and 70 inches of the lower fuselage behind the cockpit came from the YC-97J. The lower fuselage section 42–43 containing the center-wing to the trailing edge was the original Pan Am Stratocruiser N1038V. Immediately behind the wing was a 14-foot, 8-inch section removed from the forward fuselage of Transocean B-377 N406Q c/n 15945. Immediately behind it was an 8-foot, 4-inch section removed from the forward section of Transocean B-377 N408Q c/n 15944. It connects to the original rear fuselage section 45, 46 and 47 of B-377 N1038V. Both of the forward sections of the two Transocean B-377 aircraft had the lower deck cargo door on the right side. The Super Guppy retained the B-377

By 03 May 1965 the new 23-foot extended center wing was in place. The 7.5-foot extension on each side placed the inboard propellers a sufficient distance from the fuselage. The new 156-foot span reduced wing loading and spread the main gear farther apart for better stability on the ground (ASI, courtesy Tom Smothermon).

forward door on the right for crew entry door although it is located 70 inches farther back from the original position. All C-97 aircraft had the lower deck door on the left side.

Actually, the completed Super Guppy is more YC-97J than Pan Am B-377 airframe with multiple sections of other B-377 aircraft spliced in. It is surprising the FAA allowed Aero Spacelines to retain the civilian registration N1038V. The fact it did eliminated the problems of getting the YC-97J re-certified as a civil aircraft. It would have required a limited certificate which is issued to ex-military aircraft used for such work as crop dusting. A special licensing under the certificate for the modifications would have been needed which would have been nearly impossible and very costly to acquire. It was much easier to obtain a certification as a modified B-377 under the original certificate.

To extend the wingspan, a completely new stronger 23-foot wide dry center-wing was fabricated of heavier gauge material adding a 7.5-foot extension on each side and deleting the center fuel tank. The new center-wing was designed and built to allow clearance of the props from the fuselage and place the main gear farther apart for better stability on the ground. The wider span configuration and dry center reduced the wing load by 20 percent. The center tanks were deleted for structural reasons to carry as much fuel as possible the farthest distance from the aircraft centerline which reduces wing deflection and stress on the wing root. As construction progressed, the YC-97J wings were fitted to the new extended center-wing.

The original Pregnant Guppy retained the B-377 wings with an area of 1,720 sq ft and a span of 141 feet, 3-inches. In contrast, the Super Guppy had the YC-97J wings with an area of 1,768 sq ft plus the extra area added from the new wider center-wing making the span 156 feet, 3-inches. As a comparison the later built Mini Guppy had the extended center-wing but retained the B-377 wings and R-4360 engines. The Mini Guppy Turbine with Allison 501 engines and the first two SGT-201 guppies were built with the new center-wing using lower time C-97 wings. The last two French built SGT-201s used KC-97G wings with newly constructed nacelles to adapt the Allison 501s to the wings.

The B-377 wings were replaced with the more up to date YC-97J (52-2693) units which have the nacelles for the P&W T-34-P7WA turboprop engines and hard points for auxiliary external fuel tanks. The rear third of the airframe section 44–45 was removed from Pan Am Stratocruiser N1038V, c/n 15938 which Mansdorf acquired after Boeing took it on trade for newer 707 equipment in 1960.

The new extended center-wing and larger diameter fuselage of the SG create considerably different airflow characteristics. The empennage was radically modified by beefing up horizontal stabilizer tail plane and adding 48-inch extension on each end. The heavier duty vertical fin of the YC-97J was increased for better stability by adding a 58-inch section to the bottom and a 48-inch section to the top then mounted on the B-377 rear empennage section 46 of the base airframe of ex–Pan Am B-377 N1038V.

The angle where the ballooned 25-foot diameter fuselage is joined to the original lower half is much more pronounced on the Super Guppy because it is expanded at original floor level. The indentation runs the full length of the fuselage at the bottom of the expanded top at floor level. By comparison the Pregnant Guppy expanded top is six feet smaller in diameter and grafted in approximately 24 inches above the fuselage floor level indentation.

The Pregnant Guppy suffered from buffeting during the retraction of the flaps. The problem persisted for years and

Looking aft inside the rear section the indentation of the fuselage is visible in the lower rear cargo compartment under the bottom edge of the upper deck floor. Access to the rear of the aircraft is along the catwalk, ladder and hatch in the rear cargo compartment floor. Once NASA cargo was loaded this was the only way the loadmaster could gain access behind the cargo to inspect and secure the load (courtesy Patrick Dennis).

was not corrected because it could not be taken out of service long enough for modification and testing. Anticipating a greater problem with the larger fuselage of the Super Guppy, aerodynamicist Irv Culver designed a large wedge shaped air deflector which was mounted on the fuselage at the wing root trailing edge during original construction.

The takeoff gross of the SG was increased to 170,000 pounds with a 40,000-pound cargo capacity. It had 49,790 cubic feet cargo space which is two-fifths larger than the Pregnant Guppy at 29,187 cubic feet. The B-377SG could accommodate the Apollo lunar excursion module, the S-IVB stage of Saturn launch vehicles or the instrument unit of the Saturn 1B and V. The S-IVB was 21.7 feet in diameter and 58 feet, 4 inches long, weighing 25,000 pounds, which was a perfect fit for the SG's 60-foot usable cargo compartment with 30-foot constant section.

The SG was certified as a "public aircraft" preventing it from being used for commercial charter work when not needed by NASA. Conroy was not pleased; however, it had to be done to get it certified quickly because of the urgent need for the aircraft. If Conroy intended to offer commercial outsize cargo service he would have to build additional guppies. It appears he was entertaining the idea of turbine power for the Mini Guppy conversion. Although he could not obtain the second YC-97J (52-2762) or the T-34 engines, the nacelles were removed and taken to Mojave and placed with the remaining Northwest B-377s.

Unexcelled Chemical

The Super Guppy was nearly 70 percent complete when Conroy ran out of money again. Von Braun was expecting the aircraft to be completed, which increased the pressure to find additional financing. The financial rescue and eventual complete control of Aero Spacelines by Unexcelled Chemical is interesting and somewhat puzzling when considering it was a company not directly related to the aerospace industry. Unexcelled Chemical was initially a small fireworks manufacturer which was combined with multiple other unrelated companies by a group of investors with rather curious intentions.

There is no hard evidence of individuals in control of Unexcelled Chemical stock being interested in Aero Spacelines or providing any investment prior to 1965. How Unexcelled acquired Aero Spacelines is somewhat cloudy. There is a strong possibility Beverly Hills attorney and stock manipulator Burt Kleiner became aware of Aero Spacelines earlier and passed the information to some of his Wall Street friends who were looking for emerging growth companies. Unexcelled's methods of issuing and exchanging stock to acquire other companies in an effort to become a conglomerate is curious. Aero Spacelines was and had the potential of remaining the only operator of ultra-large aircraft for NASA. Conroy's cash flow problems and efforts to find investors were openly known in Los Angeles. The circumstances were right for Unexcelled to step in and gain control of ASI after severe financial trouble placed the completion of the Super Guppy in jeopardy.

The history of Unexcelled Chemical is interesting. After New York Congressman Louis Gary Clemente failed to win re-election in 1952, he became a corporate officer of Ohio Bronze Corporation, Premier Chemical and Modene Paint in 1953–54. In 1958 he became the executive vice president of Unexcelled Chemical under President James W. Crosby, who was 30 years old. Crosby was a former investment banker and stock broker who spent eight years on Wall Street after graduating from Georgetown University. He was awarded the presidents' position at Unexcelled Chemical in 1958 by Washington construction magnet, financier and investor Gustav Ring. Crosby had assisted Ring in gaining control of the 75-year-old Unexcelled Chemical.[5]

Unexcelled was originally a small New Jersey chemical company which manufactured fireworks and insect repellent. Under Ring's direction it evolved from a fireworks manufacture into a chemical and electronics company which was expanding into unrelated industries. Crosby had enlisted his friend and stockbroker Richard C. Pistell to assist in putting together the deal for Ring to acquire Unexcelled Chemical.[6] Pistell, who was known in the investment community as "Pistol Pete," was an associate and confident of stock manipulator Robert Vesco.[7]

Crosby was looking to expand Unexcelled into other fields by acquiring other companies. He acquired the rights to manufacture the Skyjector machine for Unexcelled in 1960. He believed it would be the next wave in advertising by projecting pictures on the clouds. Crosby then tried to interest Gustav Ring into purchasing Mary Carter Paints, which was expanding. When Ring turned him down he went to his father John M. Crosby, who bought half of the paint company. This appears to have nothing to do with Aero Spacelines (ASI). However, the financial dealings and stock manipulations devious practices of certain individuals are related to financial problems of Aero Spacelines.

Richard Pistell was attempting to form Unexcelled into a conglomerate by finding growth companies to be absorbed. He would receive a finder's fee and participate in insider trading to reap huge profits. He was close to Los Angeles attorney Barry Sterling, who was the head of the investment banking division of Investor Overseas Services (IOS), an offshore fund management firm based in Geneva founded by Bernard Cornfeld. It was primarily a pyramid scheme investment system of a mutual fund investing in other mutual funds. Sterling's wife was a limited partner in the brokerage firm of

Kleiner, Bell & Company for three years. Burt Kleiner, who was the senior partner, did business with the IOS, steering small companies to them as takeover targets and collecting a finder's fee. This practice eventually brought him to put together a deal for Unexcelled to acquire Aero Spacelines.[8] Kleiner also attempted to takeover Studebaker, Armour & Company, MGM and most interestingly Pan American Airways using stock manipulation techniques.

When a company in need of capital issues new shares the supply of shares on the market is increased which can cause the price of existing shares to go down. During a takeover battle the strength of the company may be jeopardized. To avoid SEC approval a substantial amount of unregistered or letter stock is placed with a fund manager. Funds are raised in this private sale while the market price of the stock is unaffected. In the unregistered sale the fund manager can buy a large block of stock at a prearranged price below market value. When the company's quote price moves up the shares are then registered and sold at a profit.[9]

Unexcelled growth was dependent on an expansion program into unrelated businesses. On 07 March 1962 Unexcelled Chemical made the unusual purchase of Twin Fair Department Stores of Buffalo, New York. The company which operated six discount department stores in western New York was established on 22 March 1956 by Anthony Ragusa, John J. Bona, Louis Battaglia and John J. Nasca.[10]

After the acquisition, Unexcelled vice president Harold A. Egan Jr. was named president of Twin Fair Department Stores. He announced it would also be expanding into the grocery business, and the two founders Nasca and Ragusa would be retained by Unexcelled as advisors. Vaughn Read moved into Egan's position of vice president at Unexcelled. Egan had been a CPA with Price Waterhouse and joined Unexcelled in 1958. The corporate takeover of Twin Fair eventually played a role in Unexcelled Chemical gaining control of Aero Spacelines.

Unexcelled Chemical had branched into munitions manufacturing. Although highly speculative and unsubstantiated it has been suggested some military munitions manufactured by Unexcelled were being shipped to the Army at Redstone Arsenal in Huntsville, Alabama, which also is the location of Marshall Space Flight Center. Wall Street deal maker Richard Pistell was looking for "emerging growth companies" to exploit as takeover targets. Kleiner, who was based in Los Angeles, had been watching Aero Spacelines as a potential growth investment. He was aware of ASI cash flow problems causing a delay in completing the Super Guppy for NASA. He directed Pistell's attention to Aero Spacelines as a candidate for cash infusion by way of a stock manipulation scheme. This appeared to be a way for Unexcelled to capitalize on the huge sums of money being spent on the space program. Whatever the situation when Unexcelled tendered a bail out offer for Aero Spacelines, Jack Conroy saw it as a financial rescue.

On 11 February 1965 Unexcelled Chemical filed with the SEC to register 27,500 outstanding shares of common stock held by James M. Crosby. He stated the shares were to be offered from time to time on the American Stock Exchange at a price not to exceeded $28 per share. Unexcelled at the time was primarily engaged in the leasing and sale of electronic weighing systems (meat scales for meat packers and butcher shops) and the operation of department and grocery stores. The company had 460,200 outstanding common shares. Crosby owned 62,386 shares. Five months later Unexcelled would acquire Aero Spacelines under Crosby's stewardship.[11]

The stage was set for Aero Spacelines to become a wholly owned subsidiary of Unexcelled Chemical by 27 July 1965. Vaughn Read succeeded Crosby as president of Unexcelled Corporation in late 1965 after the acquisition of Aero Spacelines was complete. It was unique for Unexcelled Chemical to be interested in Aero Spacelines because the only aviation holding it controlled was American Airmotive Corporation in Miami. The company was in the business of converting surplus Stearman aircraft to NA-75 agricultural crop sprayers. The other major holding of Unexcelled included Twin Fair discount department stores in Buffalo and a subsequent discount food chain; Attalla Pipe and Foundry which made iron castings in Attalla, Alabama; and The Cashin Company, which manufactured and leased meat slicing and packing machines.

Crosby's dealings prior to the Unexcelled acquisition brought attention to some of his business practices. Huntington Hartford, who was the heir to the Atlantic and Pacific Tea Company (A&P) fortune, embarked on a project to acquire Hog Island in the Bahamas and turn it into a resort, which became Paradise Island. The resort was an economic failure because it could not produce enough revenue to recover the cost of development. It was apparent the only alternative which could produce enough revenue was gambling. Unfortunately for him gambling was controlled by Wallace Groves, a Wall Street financier who went to the Bahamas after being released from prison for mail fraud and stock manipulation. He became associated with Mary Carter Paints in the acquisition of Paradise Island. Mary Carter Paints was a legitimate business which had been turned into a shell company. It was used to acquire the resort through stock manipulation. In order to acquire gambling to show a profit it was forced to give four-ninths of casino revenue to Bahamas Amusement Limited, which was controlled by Wallace Groves. He was a close associate and had many business dealings with Meyer Lansky.[12]

Mary Carter paints which started out as a small of New Jersey paint store grew to 69 retail locations by 1959. After

James Crosby could not interest Gustav Ring in a takeover, he presented the deal to his father, John Crosby, an attorney who once worked in the Woodrow Wilson administration. John Crosby was a partner in the firm identified as Crosby-Miller. Under that firm he was able to acquire 50 percent of Schaefer Manufacturing of Wisconsin. It was a foundry and metal working company established in 1908. In order to purchase Mary Carter Paints Crosby-Miller authorized a large increase in shares of Schaefer, which was half-owned by Crosby's father. Richard Pistell, who helped Crosby junior acquire Unexcelled Chemical for Gustav Ring, brought in outside investors. They included former New York governor and presidential candidate Thomas E. Dewey and his neighbor, famous radio broadcaster Lowell Thomas. Pistel also encouraged them to purchase shares of a third investment company named R.R. Williams. The assets of the new Crosby-Miller stock issue were used to acquire 80 percent Mary Carter Paints and to forward their investment activity in the Bahamas buying Paradise Island under Crosby-Miller.[13]

The activity was embroiled in controversy and became the target of an IRS investigation. Pistell also tried to merge shares of R.R. Williams Company into Unexcelled Chemical, which had increased in value via stock manipulation. This diluted the value of the Williams company stock. Eventually Pistell was sanctioned by the Securities and Exchange Commission for multiple violations. He went on to become further involved with Robert Vesco.[14] Mary Carter Paints was used as a shell company to acquire Paradise Island. James Crosby changed the name of his Paradise Island operation to Resorts International in 1967. Resorts International acquired Chalks International Airlines in 1974. Chalks purchased 13 HU-16 Albatross G-111 aircraft commercial conversions to operate between Miami and the Bahamas. They were only operated for a few years before being retired to the desert. Chalks then opted for the Grumman Turbo Mallard aircraft.

Tracking the development of Unexcelled into a conglomerate which eventually acquired complete control of Aero Spacelines is extremely confusing. The trail is riddled with the greed of questionable individuals and stock manipulations. In a very curious twist freelance rogue Chauncey Marvin Holt who was a friend of Meyer Lansky and associate of crime operatives Dominic Bartone and Louis Babe Triscaro in the Havana/Castro C-74 Globemaster deal (see Chapter 5 for the Castro connection) was a frequent processor of unregistered and suspicious stock through Kleiner, Bell & Company. Holt played both sides working for Lansky and doing black bag work with the CIA and FBI. He stated in 1967 he was still on good terms with Bartone and Triscaro. On the other side Holt claimed to be a business associate of Robert Prescott of Flying Tigers Airline. Prescott was close friends with John King founder of King Resources, which invested in Paradise Island Bahamas. King Resources was absorbed by the firm of International Overseas Services (IOS), operated by Bernie Cornfeld and subsequently controlled by Robert Vesco. IOS was instrumental in making Unexcelled stock worthless. There are several unique connections here also. Chauncey Holt restored art for John King, who made a fortune early in IOS. Burt Kleiner's brokerage firm was assigned Unexcelled as his conglomerate to manipulate by Bernie Cornfeld.[15]

Aero Spacelines Desperate for Capital

When Aero Spacelines was originally formed Jack Conroy was so desperate to raise money to build the first guppy aircraft he did not maintain the majority of the stock leaving him in the position of not having total control of the company. The success of the first guppy did not go unnoticed. The NASA contract was generating revenue and the space program budget was growing. Von Braun and his team visited Van Nuys to inspect the progress on a regular basis. A second larger aircraft with an exclusive contract had the potential of generating considerable profit because America was determined to reach the moon. After the moon the next step was presumed to be Mars. If the space program continued past the moon landing there appeared to be no limits to need for outsize logistical transports.

The B-377 Very Pregnant Guppy (Super Guppy) project began with the expectation it would be considerably more complex and costly than producing the first one. Up to this point there had only been a few investors and Conroy had struggled to acquire them. He had limited funds but it never occurred to him the project would not be finished. He was a super salesman with admirable intentions, but he continually operated on the financial edge. By early 1965 Hayvenhurst Van Nuys Corporation (holding company for ASI) was insolvent. The company credit was exhausted and the Super Guppy production was in jeopardy. Lack of capital brought construction to a crawl. NASA was pushing for the aircraft to be completed. Conroy was frantically seeking investors to provide financing to complete the aircraft. The situation was perfect for corporate raiders with capital.

In a desperate move to interest investors Conroy put on a spectacular presentation for the press, bankers and invited guest on 21 April 1965. The majority of the super guppy enlarged fuselage had been completed. The new 23-foot centerwing had been installed and the inner wings with the nacelles. The old Pan Am B-377 cockpit nose had been removed and placed alongside. The main section of airframe with cockpit of the YC-97J fuselage, which had arrived by rail from Tucson, was placed inside the Pregnant Guppy. Conroy conducted a presentation with a large model of the Super Guppy and the Saturn V rocket in a hangar. He had all the guests and press entertained when in typical Conroy showman fashion the Pregnant Guppy taxied up on the ramp. Here was the operational B-377PG across the ramp from the obviously larger under construction Super Guppy. Conroy invited the guests out on the ramp as he took a megaphone and told everyone what they were about to see. He explained the capacity of the Pregnant Guppy and how it could transport rockets and another aircraft inside. He stated the aircraft inside was being delivered to be used to complete the larger Super Guppy. The crowd applauded when he stated there was no limit to the size Aero Spacelines could build an aircraft and these were just the first two of a proposed fleet. After the B-377PG was shut down, the tail was detached and rolled away to reveal the forward half of the YC-97J fuselage inside. The YC-97 fuselage slowly emerged on to a loader as the press took photographs. It was placed in front of the partially completed Super Guppy along with the large model. Invited guest and the press took turns having their picture taken with the model which had been placed in front of the partially completed Super Guppy. The YC-97J forward section and cockpit which they had just seen unloaded was in the background.

It was a show which would have impressed P. T. Barnum. Jack Conroy's genius had the crowed fascinated. Research did not uncover the list of invited guest and potential investors. However, the presentation struck a chord with Burt Kleiner, who was

Top: **Jack Conroy was desperate for cash. In an effort to keep the project going he invited potential investors, bankers and the press to a spectacular presentation on 21 April 1965. He warmed the crowd with a speech describing the giant aircraft while unveiling a model of the Super Guppy and the Saturn V rocket, which is visible over his right shoulder (ASI, courtesy Tom Smothermon).**

Bottom: **Conroy had the B-377PG taxi on the ramp while he was unveiling the Super Guppy model. The tail of the B-377PG was rolled away, revealing the forward section of the YC-97J inside. Conroy took up a bullhorn and described the unloading procedure to the crowd. The C-97 fuselage was placed near the partially completed Super Guppy, and guests took turns having their photos taken (ASI, courtesy Tom Smothermon).**

looking for emerging growth companies which could be acquired by the growing Unexcelled conglomerate. In short order Unexcelled was offering Conroy the needed capital to complete the Super Guppy while maneuvering into the position of gaining control of Aero Spacelines.

NASA had been closely monitoring the progress of the build. Von Braun never doubted Conroy's ability to get the B-377SG completed and certified for service. He had endorsed the guppy program from the first time the project was proposed. Von Braun was convinced the larger guppy was necessary after the original Pregnant Guppy proved to be airworthy. His enthusiasm was reinforced when he visited Van Nuys on 14 May 1965. He and his team inspected every aspect of the build as Conroy described the progress and verified the construction of the Super Guppy was on schedule. The Pan Am B-377 nose had been removed and the YC-97 nose was in place but had not been attached. The Air Force had only provided the minimum to Conroy. He was able to salvage some wing parts and the nacelles from the second YC-97J (52-2762) but the Air Force donated the fuselage to fire rescue for practice where it was eventually burned.

The NASA team was favorably impressed with the progress on the Super Guppy setting in motion preliminary contract negotiations. Von Braun and Conroy posed for the press in front of the partially completed aircraft on 14 May. As a result of the visit, NASA offered Conroy a

Top: **Von Braun requested the fuselage diameter be increased a few inches to accommodate larger boosters. He made multiple visits to Van Nuys to inspect the progress. Conroy and von Braun discussed details of the progress with engineers inside the Super Guppy while work was being performed in May 1965 (ASI, courtesy Tom Smothermon).**

Bottom, left: **On 14 May 1965 Von Braun and Conroy pose for the press in front of the Super Guppy. Von Braun was very pleased with the progress but his team was very impatient. They inspected the build in detail. Precious time was being consumed with component logistics in the race to get to the moon by the end of the decade (ASI, courtesy Tom Smothermon).**

Bottom, right: **From left, ASI director of engineering Bob Lillibridge, Wernher von Braun and Jack Conroy on the Van Nuys ramp on 27 July 1965. Von Braun was concerned because NASA was in desperate need of the aircraft and Unexcelled Chemical was now in control of Aero Spacelines (courtesy Angelee M. Conroy).**

renegotiable oversize air transportation contract, NAS 8-15476. The contract was specifically for transportation and did not contain any provisions for aircraft manufacturing. Conroy was left to produce aircraft independently with private funds. Conroy believed the renewable service contract could be use to encourage future investors. The contract included provisions stating Aero Spacelines would perform instrument calibration, maintenance and engine overhaul on the aircraft. The B-377SG was designed and being built to recommended dimensions provided by NASA. In order to transport the S-IV-B stage booster the diameter had been increased by a few inches. The contract with NASA was very clear. The agency would have exclusive use and control of the aircraft during the duration of the contract. NASA had little choice they needed the transport and no other contractor possessed such and aircraft.

Unexcelled Acquires Aero Spacelines

Unexpectedly James Crosby of Unexcelled Chemical expressed an interest in providing the needed capital to complete the Super Guppy. The infusion of capital came at the most crucial time but would ultimately lead to Conroy losing control of Aero Spacelines. As mentioned it was curious that a chemical company which owned grocery stores, department stores, a foundry and leased meat packing machines would be interested in an aircraft construction and operation company. Conroy needed the cash and had little choice. Unexcelled announced on 22 July 1965 it had negotiated an agreement effective 30 July 1965 to purchase controlling interest in Aero Spacelines. The agreement represented an exchange of shares of Unexcelled stock for the transfer of all of the business and assets of Hayvenhurst Van Nuys (DBA Aero Spacelines) making it a wholly owned subsidiary of Unexcelled Chemical. Unexcelled president John M. Crosby stated "this affords Unexcelled the opportunity to enter the air freight field with an all new fuselage design." He was obviously referring to the new 13-foot wide floor fuselage design for the planned B-377MG which Conroy was planning for commercial cargo service.

Giving Unexcelled Chemical majority control of Aero Spacelines produced the cash Conroy urgently needed to complete the aircraft, but it prevented him from having the absolute word on operation of the company, which gave NASA officials concern. It was openly known Conroy wanted to enter the commercial outsize transport business. Von Braun and his team returned to Van Nuys again on 26 July 1965 on the pretense of inspecting the progress of the aircraft. Considerable progress had been made since his visit in May. The fuselage was skinned and partially painted. Several of the engines were in place. Von Braun witnessed the installation of the modified vertical fin on 27 July.

There was the underling need for NASA to protect its interest and make sure Unexcelled would maintain a commitment to the space program. Marshall Space Flight Center needed the aircraft and had a good relationship with Jack Conroy. There was a fear he was no longer in control and the new owners may want to get the aircraft type certified and divert it to commercial venture. NASA needed to insure it had control and exclusive use of the two guppy aircraft.

The Unexcelled acquisition was not as clean as it appeared. It has been suggested Los Angeles brokerage firm Kleiner, Bell & Company devised a rather convoluted reorganization plan to raise the needed capital to acquire the company and provide funds to complete the Super Guppy. In theory a company with a high quoted stock price can take over another with a lower price stock. The company will purchase a controlling portion of the smaller company by offering its own more expensive stock as payment. This maintains or increases the conglomerate's share price by appearing as growth, creating a house of cards situation.

Henry Brown Murphy, who was on the board of Unexcelled subsidiary Twin Fair (grocery stores), Trenton Trust and National State Bank, approached the management of Twin Fair Corporation of New York with a proposal. Twin Fair was a most unlikely choice to acquire a specialty aircraft manufacturer. The proposed plan stipulated Twin Fair, owned by Unexcelled Chemical, would acquire the assets of Conroy's Hayvenhurst Van Nuys Corporation, which was the holding company of Aero Spacelines. It effectively gave Unexcelled Chemical control of the entire Aero Spacelines operation with minimum investment. Hayvenhurst only had one valuable asset, which was the lucrative NASA contract. The contract was on a year to year basis, but with there being no other operator of volumetric guppy aircraft, it was assured it would be renewed. Stock was issued to raise capital for a buyout. Then Aero Spacelines was provided with a loan from a New York bank guaranteed by Twin Fair for $950,000 which provided the capital to complete the B-377SG. The price paid for the company is not known as it was based on number of factors related to values of stock exchanged and assets. However, the transaction prompted an IRS investigation because the price was alleged to exceed the assets of the company by $2,185,340.

By 12 August 1965 the infusion of capital via Unexcelled put Super Guppy construction back on track. Unexcelled via Twin Fair acquired the worthless stock of Hayvenhurst, the assets of Aero Spacelines and the NASA contract in exchange for an undisclosed cash amount and shares of Twin Fair retail discount and grocery stores voting common stock. As a result, Unexcelled as owner of Twin Fair gained controlling interest of Aero Spacelines.

It was not immediately realized but it soon became apparent Unexcelled management did not subscribe to Conroy's sometimes cavalier methods of getting the job done and worry about the paperwork later. It was only a matter of time before a clash of ideas and management practices would cause Conroy to leave Aero Spacelines. As a condition of the takeover and financial bailout, a covenant was set forth stating if /when Jack M. Conroy should separate from Aero Spacelines anytime thereafter he could not construct and/or operate anywhere in the world any heavier than air aircraft having a diameter in excess of 13 feet and usable cargo capacity of more than 16,000 cubic feet.[16]

Certification and Potential Disaster

Completion of the B-377SG went into high gear at Van Nuys after the acquisition of Aero Spacelines by Unexcelled Chemical. It had taken a total of ten months to complete. The Super Guppy was debuted to the public in typical Conroy fashion on 31 August 1965. The aircraft was towed to the ramp in front of Johnny's Skytrails restaurant and bar. This was the local watering hole where the guppy design was first hatched by Conroy and his pilot friends. Many design and engineering problems had been worked out there over afternoon sessions and drinks. Conroy informed the press and invited a long list of guest. Television coverage brought more curious onlookers.

The crew for the first flight exemplified Conroy's efforts to hire the most qualified pilots available. In command would be Jack Pedesky, an ex–Navy pilot who had flown for Pan Am with over 18,000 hours and 1,500 in the Pregnant Guppy. The co-pilot was Air Force Lt. Colonel P.G. Smith with 11,500 hours. He was a former Air Force instructor pilot on the YC-97J from which the Super Guppy was built. The flight engineer was Galen Hull, who had 3,000 hours on the YC-97J while serving in the Air Force. The systems engineer was Ercel Oliver, who had been the maintenance foreman on the YC-97J at Norton Air Force Base. The experience of flying the YC-97J before conversion made them uniquely qualified to evaluate the aircraft after conversion. The B-377SG made the first flight on 31 August 1965 from Van Nuys airport to Mojave.

Multiple flight tests were required before the FAA would approve the aircraft for NASA service. There is no indication in any of the flight test records located of ASI, FAA or NASA pilots performing a test where a stall was actually executed. Stall tests were conducted up to buffet, but it appears no one was prepared to take it any further. The indications are since it was being operated as a public aircraft with multiple restrictions, the minimum to get it certified was good enough. In order to be certified as a commercial aircraft, a full stall exercise would have been required.

In typical Conroy tradition the Super Guppy was towed in front of Johnny's Skytrails Restaurant for inspection by the press and a large crowd. This was the same bar where the guppy idea went from a sketch on cocktail napkins to a working concept, which became Aero Spacelines (Dottie Furman, courtesy Angelee M. Conroy).

During flight testing it was determined the Super Guppy could be flown to a point where it would pitch up as it approached stall. It appears ASI had unknowingly created the first lifting body aircraft long before NASA or any studies had been done to exploit the concept. In this attitude the guppy fuselage would still be lifting after the tail and outer wing had given up. Aero Spacelines test pilot J.K. Campbell stated, "Once it started to pitch up, pushing forward on the yoke did nothing. The only way to recover was to roll it over on a wing which required a lot of altitude. We consulted with Douglas and installed a DC-9 type Stall Prevention System using an

Left: **The flight crew selected for the first flight was extremely qualified. From left, Jack Pedesky, pilot; Ercel Oliver, systems engineer; Colonel P. G. Smith, co-pilot; and Galen G. Hull, flight engineer. All had flown the YC-97J during the Air Force test program and knew the characteristics of the Pratt & Whitney T-34 engines (Dottie Furman, courtesy Angelee M. Conroy).**

Right: **From the first time the turboprop B-377 Super Guppy lifted off from Van Nuys on 31 August 1965 the crew realized it handled differently. Photographers were stationed at multiple intervals along the entire runway to capture the rollout and liftoff. It had the tendency of the main gear lifting off before the nose. Pilots described it as trying to drive a giant wheelbarrow down the runway (Dottie Furman, courtesy Angelee M. Conroy).**

angle-of-attack computer and a loud warning horn."[17]

In September 1965, after only a month of testing, the B-377SG was flown across the country on a reliability demonstration flight for NASA. Flights were conducted to the Manned Space Flight Center at Houston, Marshall Space Flight Center in Huntsville, Dulles International at Washington, D.C., and Andrews Air Force Base. As test continued it was noted the aircraft had some peculiar characteristics on takeoff. For some unknown and never determined reason the B-377SG, N1038V, did not rotate on takeoff as other normal aircraft. When parked on the ramp it was obvious the aircraft sits at a nose up angle with the original C-97 nose gear. As it rolls down the runway the main gear lifts off

Right: **In September 1965 the B-377SG was flown across the country for a reliability demonstration for NASA. The frontal area above the cockpit was originally built with smaller structural rings without reinforcement similar to the Pregnant Guppy, which had not experienced structural problems. Also visible is the partially retracted nose gear and the portable strut with electric motor, which was used to power the nose open (NASA S-65-45073).**

first, placing it in an attitude parallel to the runway with the nose gear still on the ground. Because of the nose up attitude the aircraft actually tries to fly before it reaches a safe rotation speed. To maintain control the pilot must keep the nose wheel on the ground until sufficient speed is obtained to rotate. NASA pilot Frank Marlow said, "It feels like you are driving a giant wheelbarrow down the runway."[18]

No other guppy including the predecessor B-377PG had this quirk. The second generation SGT-201 aircraft were fitted with the B-707 nose gear, which was shorter producing a different attitude with the nose lower when on the ground. Interestingly, the commercial B-377 Stratocruiser suffered from a similar handling characteristic. The main gear would lift off before the aircraft was ready to rotate while the nose gear was still on the ground. This prompted the CAA to require the installation of leading edge spoilers on the wings between the fuselage and the inboard nacelle. The spoilers of the B-377s were eventually removed but the civil B-377s had a tendency to land on the nose wheel throughout the operational life.

After a successful demonstration tour the Super Guppy was returned to California for further testing and certification. A velocity dive test was scheduled for 25 September 1965 to complete the airworthiness certification. The aircraft takeoff and climb out was normal. After reaching altitude the crew leveled off and prepared to conduct a series of test. All seemed normal as the aircraft cruised over the Mojave Desert at 10,000 feet. Colonel P.G. Smith was in command. Smith later became vice president of marketing for Conroy Aircraft. The copilot was Larry Engle with engineers Alex Analavage and second engineer John Kinzer. Also in the cockpit was systems engineer Ercel "Ollie" Oliver and instrumentation engineer Sandy Friezner, who was in the compartment below the flight deck.

The dive started as the guppy was nosed over and began to accelerate. As they passed through 275 mph there was a slight shutter followed by an explosive bang above the cockpit. The aircraft shuttered again and began to shake violently with increased intensity. Both captain and copilot reached for the controls to throttle back. The cockpit door had blown off hitting Oliver and fracturing his leg. There was bright sunshine on the flight deck floor coming through the top inspection window from a giant hole in the roof behind the rear bulkhead. It would have normally been a darkened cargo bay. The unreinforced upper

After passing 275 mph there was a loud explosion and the upper forward fuselage collapsed. The crew managed to maintain control of the aircraft despite violent shaking. After a harrowing 17 minutes Colonel P.G. Smith was able to land safely at Edwards AFB with a 23-foot hole in the fuselage above the cockpit (Dottie Furman, courtesy Angelee M. Conroy).

fuselage structure above the cockpit had collapsed. Col. Smith keyed his mic, calling, "Mayday, Mayday."

Aero Spacelines had struggled for cash just to get the aircraft completed. Even after being acquired by Unexcelled and obtaining a $950,000 loan to complete the aircraft the company was cash-strapped. The money was just not available to obtain and install tanks in the cargo bay which could be filled with water for ballast during flight testing. As a cost cutting measure, instead of using water tanks for ballast sacks of borate were borrowed from the nearby mining operation at Boron, California, near Edwards Air Force Base. Thirty thousand pounds of sacked borate had been loaded in the cargo bay to duplicate maximum gross takeoff weight. Flying debris ripped open the bags, creating a dust storm inside the aircraft. There was a gaping 23-foot hole above the cockpit. There was dust in the cockpit from the blown-off door limiting visibility. The forward motion was pressurizing the cargo compartment, causing uncontrollable buffeting and vibration. As power was reduced, the speed dropped to 150 mph. The buffeting increased as the aircraft approached stall. As the power was increased to prevent stall they were able hold it at 175 mph, however, the aircraft was coming apart.[19]

The crew was very experienced with considerable time in the aircraft but a catastrophic airframe failure is a challenge even to the most qualified crew and has a high probability of fatal consequences. Colonel Smith had flown forty different types of aircraft since World War II and Larry Engle had over 20,000 hours experience but this was an in-flight emergency like no other. Smith called for the crew to prepare to bail out through the emergency floor hatch. After a short exchange with the crew, he reconsidered. The antennas had been moved to the bottom of the aircraft. In all probability the slip stream would drag them along bottom and they would become entangled, causing serious injury. They were trapped but rather than fall 8,000 feet in the crippled aircraft Smith decided to take a chance of landing if it stayed together. The aircraft was becoming a pressurized large flying canister which was violently shaking with debris flying everywhere. Pieces of stringers were flying like spears and penetrating other parts of the airframe structure.

Fortunately Colonel Smith had the professional forethought to bring aboard six helmets which he had taken some joking about before departure. They were quickly passed out as the situation was reaching a critical point. The pressure in the cargo bay became so great the access doors and hatches in the rear of the fuselage suddenly blew out. There was an instantaneous clearing of the dust as it vented out the rear. The violent shaking continued but the aircraft was slightly more controllable. It appeared they might be able to maintain control long enough to attempt a landing if more of the fuselage did not collapse or sections of the skin peel back and separate.

Smith called George Air Force Base at Victorville, California, for a chase plane to inspect the damage. Victorville acknowledged and scramble two fighters. At the same time the tower at Edwards Air Force Base where they had departed volunteered a DC-9 which was taking off on a flight test. Smith responded, "We will take the DC-9," which arrived in a few minutes, taking a position off the right wing. The DC-9 crew surveyed the damage and made a sweep of the tail to inspect for damage. The tail was intact but there was a lot of metal peeled back and flapping on the upper forward fuselage above the cockpit.

While Colonel Smith still had moderate control of the aircraft the decision was made to attempt a landing at Edwards Air Force Base. Smith was able to turn the guppy toward Edwards as it made a slow shallow descent at 175 mph. The aircraft was still shaking violently at 4,000 feet about 10 miles out. Maintaining 175 mph, the approach was flat with no flaps because it was not possible to predict how the aircraft would react. Smith did not call for the landing gear until they crossed the threshold at only 260 feet off the ground. The gear deployed and locked as the aircraft remained stable. The landing was surprisingly smooth and they rolled out to a dead stop. It had only been 17 minutes since the forward top had collapsed but the crew had lost track of time and felt it was much longer.[20]

Jack Conroy arrived on the scene in a short time to inspect the damage. This was quite a setback for Aero Spacelines and could affect certification and impact NASA component delivery. Engineers from ASI immediately inspected the extent of damage. All heavy damage was confined to the upper forward fuselage. There did not appear to be any structural damage to the rest of the enlarged airframe although there was considerable minor damage caused by flying debris. Aero Spacelines engineers believed the high frontal area could be beefed up and successfully repaired to accept the increased pressure. A complete structural re-design with increased reinforcement of the upper superstructure was completed in just five weeks at Edwards.[21]

The collapsed front incident brought additional reliability questions from FAA inspectors. The Super Guppy was designed with an opening nose which swings left past 100 degrees for easy loading. Guide pins align the nose to the fuselage with automatic latching devices secured by six 1.25 inch and three 1.75 inch bolts. The upper section of the nose above the cockpit was originally built with smaller structural ribs which collapsed under pressure during flight testing. After reviewing the damage and calculating the amount of pressure it took to collapse the framework, engineers came up with a reinforced set of ribs for the upper fuselage which could withstand considerably greater pressure. Consequently, when the upper fuselage was rebuilt, the structure above the cockpit was redesigned with considerably heavier material.

The severity of the damage is reflected in the amount of aircraft structure, which was torn away. The aircraft became a flying wind tunnel. The large section hanging down was flapping and shedding debris. The crew demonstrated amazing skill and airmanship in being able to maintain control of an aircraft with a third of the front missing and still land safely. They were extremely fortunate to survive. The 30,000 pounds of sacked borate used for ballast is in the foreground (Dottie Furman, courtesy Angelee M. Conroy).

The modifications and repair were completed in record time but there was still doubt regarding the structural integrity of the frontal fuselage. The overhead catwalk was not in place when the upper fuselage caved in during the flight test. After rebuilding the upper fuselage, a mobile scaffold was built and mounted on a track which extends approximately one third the length of the cargo compartment. The unit moves back on rollers to allow technicians to secure the load and inspect tie downs and roof latches. During flight the scaffolding is retracted forward into the swing nose and secured just behind the cockpit.

It was reasoned the expanded fuselage produced previously unknown forces not fully understood. No explanations were offered as to why the Pregnant Guppy did not suffer the same fate with the same lighter structural front. It is possible but not conclusive the Super Guppy, being larger in diameter, suffered from an exponential stress increases because of the larger diameter.

When the Super Guppy was being built the Air Force resisted supplying the experimental Pratt & Whitney T-34-P7 engines from the outset. NASA influence and pressure assisted ASI in negotiating two contracts with the Air Force. One contract was for the loan of the engines from the YC-97J and another for the loan of spare parts, engines and propeller hubs from government C-133 surplus. The Air Force set a value on the parts at $1,424,476 with specific language that they were to be maintained and returned when the contract was completed. This was just one of the ways the government sought to control the exclusive operation of the guppies for NASA and prevent Conroy from using the aircraft for any commercial venture. As time progressed the supply of spare engines increased and the Air Force seemed less concerned.

Unlike the Douglas C-133, the Aero Spacelines B-377SG experienced fewer problems with the P & W T-34 engines. There were a number of propeller malfunctions and failures with the Curtiss CT735S-B104 turbo-electric propellers with 15-foot blades mounted on the SG. The failures were minor compared to those experienced by the Air Force with the C-133. The Aero Spacelines problem was manageable if the propellers had not been in such short supply. Aero Spacelines only had eight spares which were timed at 1,100 hours between overhaul. The props were a shortened version of the earlier 18-foot blade CT735S-B319 used on the C-133 which were at the maximum limits of technology of the time (see C-133 section).

The prop assembly used on the B-377SG weighed about 1,200 pounds. One benefit of the Curtiss props was the blades could be removed individually from the hub. This proved to be a benefit on several occasions when there was a failure. A new hub was flown in and the blades transferred from the old unit. The aircraft did not experience the same extreme center-wing vibration problems created by the engine-propeller combination on the C-133. The Super Guppy did have a high frequency vibration in the center-wing of the aircraft at cruise which would give you a foot massage. However, it was not as severe as the vibration on the Douglas C-133 and never cited as contributing to any failures.[22]

Engineers designed a reinforced compound structure with larger ribs, angular supports and thicker panels. The new design could withstand considerably greater pressure, which was originally not believed to exist. Interestingly, the original Pregnant Guppy did not encounter this problem (author's collection).

Limited cash flow and lack of any government funding reduced the amount of wind tunnel testing done prior to construction of the Super Guppy. Only the minimum amount of data had been obtained on performance, stability and controllability. A total of 96 hours of flight testing was conducted which is between one-fourth and one-sixth of what is normally done on an aircraft to evaluate a modified airframe for commercial use. By the time most of the wind tunnel tests were completed the aircraft was well under construction. The FAA found a number of structural deficiencies relative to items required for commercial aircraft. Because of the severe need to put the aircraft in service, Aero Spacelines agreed to impose special operating restrictions as a "public aircraft" rather than correct the deficiencies and NASA concurred.

NASA needed the aircraft to transport larger boosters and the deadline to the moon was fast approaching. Confident with the success of the B-377PG predecessor, Marshall Space Flight Center chief of Project Logistics John Goodrum continued to press the FAA for the aircraft to be approved in early 1966 to move critical cargoes even if on a limited basis. By 01 March the FAA still had not flown the Super Guppy. Marshall Space Flight Center officials continued to press for FAA approval, pointing out the national importance of keeping the space program on schedule. NASA requested the FAA allow the Super Guppy to be placed in operational status because it was needed for immediate service. The Saturn Instrument module built at IBM facilities at Huntsville needed to be transported to Huntington Beach for testing. The FAA yielded to NASA granting permission for the flight. Within a week the SG was in Huntsville. The Instrument Unit was transported to Douglas in California on 10 March 1966 without incident. After inspection the module was mated with the S-IVB stage for testing.

Although NASA had not been given an FAA exemption, the SG transported the instrument unit to California. It went without incident, but further test were still required to certify the aircraft and verify the structural redesign and repairs were sufficient. The flight test program for N1038V ended on 25 March 1966. Test pilot Joe Walker flew the final Super Guppy acceptance flight for NASA. Tragically Walker lost his life a short time later on 08 June 1966 when his F-104 chase plane collided in mid-air with the XB-70 while on a photo test flight.[23] The Air Force was conducting a series of flights to photograph the five supersonic aircraft in inventory when it occurred. Ironically Clay Lacy was flying the Learjet camera plane for the Air Force that day.

The success of the SG test flights prompted NASA's decision to revise the transportation mode of the S-IV-B stages from seagoing to air. The aircraft had been originally scheduled to be FAA certified by the end of 1965. It was delayed until spring 1966 because of the incident of the forward front fuselage cave in the previous September. Repairs were not completed until November 1965 and the flight tests had to continue to obtain certification.

Unresolved propeller stress problems and questions about the performance, reliability and safety of the aircraft were exempted by the FAA under pressure from NASA. The FAA was forced to compromise because the "national" need for the aircraft was paramount. There was no alternative to transport hardware for the moon mission. The testing was finally completed and the aircraft certified on 30 March 1966. It entered contract service on 06 April transporting a booster stage to the Cape. Within a few weeks it delivered the Saturn instrument unit and made another visit to MSFC with the S-IV-B test stage. The FAA had been granting special waivers at the request of NASA for each flight up until then. The flight manual operating instructions were approved on 21 June 1966 over three months after it had been in full service.[24]

The pair of Aero Spacelines guppies resolved many NASA transport issues for outsize hardware. Congress and the Government Accounting Office were questioning the cost of logistics. NASA was questioning the cost of hardware logistics by 1966. Budget planners projected future transport needs would eventually cost up to one third of the launch vehicle budget.[25] Budget restrictions imposed by Congress brought into question the logistical cost of space hardware delivery and the future of the space program. Although transport costs were in question, Conroy still believed the political pressure to explore space would overcome any budget restrictions.

Aero Spacelines Expands to Santa Barbara

Prior to Unexcelled Chemical acquiring control of Aero Spacelines in 1965, Jack Conroy considered moving the operation to San Diego. He was seeking larger indoor production facilities and more favorable rates for office space. If he expected to build and sell commercial versions, construction would have to be moved inside and engineering brought in house as opposed to contracting to On Mark Engineering.

Conroy and his wife, Milbrey, looked at housing in the San Diego area. She suggested they consider Santa Barbara. It had always been an area she liked and it had a more favorable climate. Conroy agreed and discussed his plans with Santa Barbara city officials. The mayor offered him a very attractive package to bring new industry and jobs to the city.

The Aero Spacelines operation was moved from Van Nuys to the Santa Barbara airport in February 1966 after Unexcelled gained control and before construction started on the Mini Guppy.[26]

New Guppy Designs

The success of the two operational guppies and move to Santa Barbara prompted Conroy to propose new design ideas. By the beginning of September 1966 he was negotiating the lease of a B-52 to begin building a giant guppy with a fuselage diameter of approximately 38 feet. He had envisioned a jet powered aircraft before the Super Guppy was completed. The preliminary design work for an enlarged B-52 conversion was given the appropriate name of the Colossal Guppy. The name is not to be confused with two CL-44 Conroy projects which used the similar name Colossus. Conroy liked the term "colossal," which he frequently used to describe larger proposed models with diameters from 25 to 40 feet. The new

Repairs were completed by November 1965. The amount of new aircraft skin shows the extent of damage and yet it still could fly. The gap in the nose hinges has been taped over in preparation for a repainting of the repaired area. The engines are being serviced and inspected for any damage from flying debris not previously identified. Pilot Ercel Oliver is in front of the nose and Los Angeles television newsman Clete Roberts is on the right (Dottie Furman, courtesy Angelee Conroy).

"heavy duty" design was planned with 12 engines. He proposed extending the center-wing and mounting three of the B-52 double engine pods on each side. It was presented in brochures as being capable of transporting all stages of the Saturn rocket for the Apollo program with optimistic captions.

> Aero Spacelines' objective is to anticipate industry's requirements by developing new concepts whereby even larger outsize designs can be moved by air. Preliminary studies have been completed on the modification of a B-52, incorporating a new 40-foot diameter fuselage with a proposed payload of approximately 200,000 pounds.
>
> The concept has great versatility.... It will be used for the movement of the second stage of the giant Saturn Rocket.[27]

Conroy began negotiating a lease for a B-52 in September 1966. He proposed a fuselage diameter of 38 to 40 feet with a 200,000-pound payload. A new center wing would increase the wingspan for an additional paired engine pod would be added on each side inboard of the existing location for a total of 12 engines (ASI/Ettridge, courtesy Tom Smothermon).

Conroy previously reviewed a 707 design and even considered a 747 guppy with a diameter greater than 40 feet. The irony is years later in 2003 Boeing did indeed build a B-747 guppy. It was 235-feet long with a 27-foot wide, 65,000-cubic-foot cargo hold. The success of Conroy and Mansdorf's 1962 first outsize fuselage Pregnant Guppy design created an entire new class of aircraft which survives today.

By the end of 1966 Conroy planned to build larger outsize transports based on the B-52 and 707 but was constrained by the Unexcelled board and lack of capital. Construction of the Mini Guppy began in December but funds were tight. How Canadian financier and entrepreneur Allen Manus was introduced to Conroy is unknown. Manus, who was a resident of Palm Beach, Florida, expressed an interest in investing in Aero Spacelines and offered some financial alternatives. Conroy and his wife had been invited by NASA to view the launch of Apollo I in 1967. Manus suggested they stay at his home in Palm Beach while in Florida. Tragically Apollo I burned on the launch pad during a live test on 27 January, taking the lives of Gus Grissom, Ed White and Roger Chaffee. After the disaster Manus expressed an interest in seeing the burned capsule up close. Conroy arranged for them to see it, which impressed Manus. Consequently Manus made several trips to Santa Barbara to visit with Conroy and discuss acquiring a loan.[28]

It is doubtful if Conroy was aware of the past dealings by Manus in the Bahamas. Several years before Manus purchased the Lucayan Beach Hotel in Freeport from Wallace Groves (see Chapter 13). Groves was a rather questionable character who had acquired the hotel in a forced sale from Louis Chesler. The Lucayan was a financial disaster. It was one of the costliest hotels built in the world at the time on a per room basis of $25,000. Gambling was the only possible way it could ever gross enough to cover the cost and Groves had the only gambling permit in the Bahamas. He sold the hotel to Manus at a $1,000,000 loss. Manus secured a loan from Aurora Leasing Corporation and Atlantic Acceptance Corporation for $2,500,000 to purchase the Lucayan. The Monte Carlo casino, which was paying 30 percent of the profits to Meyer Lansky, was not part of the deal. The owners paid $600,000 rent annually to the hotel for the casino. However, the casino owners considered the payment as a subsidy to keep the hotel open.[29]

In June 1965 Atlantic Acceptance collapsed, defaulting on $104,000,000 which caused an international financial scandal. The Lucayan went into receivership and the shares were purchased by Montreal Trust. Allen Manus also sold his interest to Montreal Trust in 1965. It is quite interesting that Montreal Trust was the agent and registrar for the IOS Fund of Funds.[30] Even more curious was the fact Los Angeles broker Burt Kleiner of Kleiner, Bell had steered the stock transfer deal for Unexcelled to acquire control of Aero Spacelines through Twin Fair. Kleiner also conducted business with IOS and ran unregistered stock through his firm. "Kleiner knew and did business with almost everyone.... Until Kleiner, Bell went bust in 1970 it popped up in almost every deal the conglomerate wove ... taking a finder's fee and shifting stock."[31] In 1965 Wallace Groves, who sold Manus the Lucayan hotel, was contacted by the CIA to work on a project in Cuba. It is not clear if Manus participated in or arranged a financial deal to build the Mini Guppy. Six months after meeting with Manus, Conroy was squeezed out of Aero Spacelines.

Guppy Flight Training

Both the B-377PG and B-377SG continued transporting NASA components to Huntsville and the Cape from the new base at Santa Barbara. In the spring of 1966 the Super Guppy transported the Surveyor II from Hughes Aircraft at Los Angeles to the Cape. The Surveyor II was scheduled to be the second unmanned Lunar Lander. It was launched on 20 September aboard an Atlas—Centaur rocket. After a successful launch NASA lost contact with the spacecraft on 22 September.

The SG had taken up exclusive service for NASA without any serious problems. At the time of the first flight ASI only had ten qualified crew members, six pilots and four flight engineers. With two guppies in service, the company was confronted with training an additional six pilots and four flight engineers. A new flight training program was phased in between August 1965 and November 1966. All flight training had to be conducted in the aircraft because no ground flight simulator for a guppy type aircraft existed. Pilot training was a very expensive proposition because all 20 crew members had to be qualified on the Super Guppy and ten of those qualified on the older Pregnant Guppy. The B-377SG flew 224,588 miles for NASA in 1966. When considering it was not in service the full year this is quite an achievement because it was almost half of the total 464,965 miles flown by both aircraft.[32]

Guppy operational cost reported for 1966 totaled $2,491,614. The contract guaranteed 18,000 miles per month at an average rate of $12.92 per mile. Additional charges of $3.15 loaded and $3.05 empty was paid for each mile which exceeded the guaranteed minimum. NASA was very pleased with this arrangement because the more miles flown the lower unit cost to the government. The result was 464,965 miles flown by both guppies compensated at $3,624,614 placing the NASA overall cost at $7.63 per mile.[33] The reported profit for Aero Spacelines was $1,132,940 for the year 1966. President of Unexcelled Roy C. Schoenhaar announced the guppy operation revenue for the first six months of 1966 was $14.5 million, which was a 50 percent increase over the previous year. The Unexcelled acquisition of Aero Spacelines and revenue from the NASA transport operations obviously contributed more to the bottom line than the Twin Fair department and grocery stores division.

The turnaround time of the Super Guppy with the swing nose feature was considerably shorter than the removable tail predecessor. The more primitive and smaller B-377PG was a success, however, assembly of the tail cradle and removal of the rear half of the fuselage took longer to load. A special cargo scissors type loader known as the Cargo Lift Trailer (CLT) was developed by NASA at Marshall for loading the S-IV into the Pregnant Guppy. A modified and improved model was built which could handle the larger S-IV-B with ease when loading the Super Guppy. The new CLT was also used as a ground transporter to move the oversize rocket assemblies for short distances at MSFC.

Air transport of the Saturn boosters also reduced the cost of preparing the units for shipment. One of the noticeable differences of transporting the units by air was the improved lightweight covers which fit over the ends of the booster units. For sea travel a costly heavy full shroud was designed and installed to protect against salt air. Air transport also resulted in additional cost savings because the lightweight covers for each unit could be reused on each successive shipment.

The two guppies were equipped with an environmental instrument unit to measure humidity, pressure and temperature. The booster propellant tanks were fitted with static desiccators for environmental control. The improved loading methods for the Super Guppy was superior to seagoing transport but not without incident. On 27 December 1967 a nose door incident almost proved fatal.

Shortly after takeoff from Mather Air Force Base near Sacramento with booster S-IVB-503N on board, the crew heard loud popping noises. The flight engineer went to investigate and found a latch which holds the nose closed was disconnected and there were multiple latch failures causing a gap between the nose and fuselage joint. The crew declared an emergency and returned to the field without incident. The latches were inspected and secured and the guppy departed for the second time en route to Kennedy Space Center. The potentially fatal malfunction prompted an immediate review and redesign of the swing nose hinge latch mechanism.

At the end of 1967, the Pregnant and Super Guppy had grossed $11,591,633 in NASA contracts. Conroy built the two aircraft for under $3 million, making it easy to understand why he envisioned a large potential market proposing a number of other guppies.[34]

Jack Conroy was aware of a congressional concern of growing space program budgets. America was committed to put a man on the moon. However, President Kennedy was gone and the Vietnam War was escalating. President Lyndon Johnson remained committed to landing on the moon but had little interest in post–Apollo commitments. The NASA budget was held at current levels and targeted by Johnson for a $500,000,000 reduction in 1967.[35] Conroy began to consider the possibility the NASA contract may not last if the space program got in trouble. He concluded once America landed on the moon the space program could be cancelled or drastically cut back. Conroy's fear became reality when the federal budget crisis in 1967 dealt NASA a blow with major budget cuts.

A larger Super Guppy with six turboprop engines was proposed as the next step during the 1965 construction. The design was seen as a commercial transport for aircraft subassemblies and oddly wears the SG registration of N1038V. Donald Malcolm and Tex Johnston reintroduced the design in February 1968 as part of a proposal to Boeing for a transport capable of moving B-747 subassemblies (Doug Ettridge, author's collection).

Conroy had envisioned a market for multiple commercial guppies. He believed transporting airframe components for Boeing and other manufacturers was the future. Ultimately he was correct; however, Boeing did not realize it until 25 years after his death. Aviation industry planners believed Conroy was overly optimistic in pursuing Boeing to purchase outsize transports. In reality Boeing officials led him to believe they were seriously considering outsize aircraft to transport components for AOG charters and future aircraft production. At the time all focus was on the B-2707 Super Sonic Transport (SST). Boeing stated the SST will have larger subcontract programs requiring transport of components from around the country. In a letter to Conroy Boeing stated:

> We are not encouraged the available transportation will be able to satisfy our future requirements. The present need for support to the B-747 program exist because of sales to many foreign carriers. The shipment of control surfaces, body sections and other items for grounded aircraft (AOG) is very real. Management in this division continues to show interest in and outsized airplane. We are willing to discuss some kind of a "use contract" which would commit The Boeing Company to use such an airplane if one were available.[36]

The letter certainly indicates Boeing's interest and anticipated need for outsize transports. Aero Spacelines officials were thinking in terms of transporting components for all Boeing production. Boeing was placing all emphasis on the SST. Unexcelled president Roy Schoenhaar announced a $5,000,000 commitment to Aero Spacelines to build four more SGT-201 aircraft. However, production would not begin until a commitment is received from Boeing. He further stated the new guppy could transport the components for a B-747 fuselage in three trips instead of the current method of sixty assemblies by rail.

Jack Conroy received very encouraging letters from Fairchild Hiller, Northrop-Norair and Martin stating and anticipated need for volumetric aircraft. Fairchild's R.J. Calhoun wrote to Conroy stating it would be supplying fuselage sections for the SST and would require outsize transport from Long Island, New York and Seattle: "We are interested in and anticipate participation in other programs which require movement by super guppy aircraft." He stated the movement would commence in late 1968 continuing into the 1970s. Likewise John Reid of Northrop wrote, "it is our belief that the public interest requires immediate go-ahead in design and construction of the super guppy." Referring to the assumed production of the SST, he stated, "necessity makes such a cargo carrying plane as the Super Guppy absolutely mandatory." Martin-Marietta traffic manager J.F. Kraft wrote, "In our opinion availability of guppy type aircraft is urgently needed ... certification of this type aircraft for commercial use would open new areas of development and propel some programs which currently are faced with transportation restrictions."[37]

Boeing press releases stated the production of the SST prototype would begin in 1967. One hundred and fifteen were ordered by 25 airlines. The SST was an extremely optimistic project when considering it was three times longer than the Concorde and projected to carry 250 passengers. Ultimately the SST was cancelled on 20 May 1971 when the

U.S. Senate rejected funding. The cancellation combined with the downturn in the airline industry drastically affected the ability of Aero Spacelines to sell the next generation SGT-201 guppy aircraft.

Unexcelled officials had recognized the commercial outsize transport potential in 1965 after gaining control of Aero Spacelines, but not in the same way as Jack Conroy. The company continued to operate with plans to build an improved version of Conroy's guppy design using the Boeing C-97 as a base. The B-377PG and SG were still under exclusive contract to NASA, restricting them from commercial cargo contracts because of certification restrictions imposed by the FAA.

Prior to 1965 Conroy had envisioned an even larger six-engine B-377 guppy powered by Allison 501 turboprops. It was introduced in advertising brochures produced to promote the Santa Barbara operation. It was proposed as an upgraded version of the Super Guppy for the commercial market with a 60,000 pound payload. The design was revised and reintroduced by Donald Malcolm to Boeing as part of the ASI proposal to transport B-747 subassemblies in 1968. The idea never came to reality at Aero Spacelines, but Conroy pursed it again in 1971 after forming Conroy Aviation. Multiple designs for a six-engine guppy were considered using several different airframes as the base but none were ever built.

Unexcelled continued to pursue Conroy's ideas in a campaign to expand the role of guppy aircraft into commercial transport of boats, jet engines, helicopters, oil drilling equipment and airframe components for aircraft manufacturers. Planned updated turboprop versions of the Mini Guppy and Super Guppy became the focus of Aero Spacelines. The success of the first two guppies in NASA service and the prospects of the third Mini Guppy as a commercial transportation proved the viability of outsize volumetric aircraft. Unexcelled president Roy Schoenhaar expressed the belief the company could further capitalize on the success of the guppies by transporting airframe components. Officials at Unexcelled sought to exploit the close relationship the company had with noted Boeing test pilot Tex Johnston. His influence could be instrumental in gaining Boeing contracts.

In 1968 the Super Guppy transported the environmental chamber in preparation of the manned Apollo module launch. It also transported the Apollo telescope mount and hardware for an experimental nuclear rocket. In an effort to generate more revenue, Unexcelled made and aggressive move to amend the contract to allow the B-377PG and SG to transport supplies for the Department of Defense (DOD) and other government agencies. This is exactly what NASA officials feared when Unexcelled gained control of Aero Spacelines in 1965. NASA stated the following: "The existing launch schedule for the next 12–18 months requires rapid reaction to changing requirements, demanding immediate availability of the guppies. If the guppies are not under the proximate control of NASA our response time may be seriously jeopardized. This could delay priority cargoes 2–3 days with a resultant impact on launchings and added cost."[38]

If the guppies had not been under the immediate control of NASA they may have been used on commercial work if certified delaying response time. In addition one of the guppies could suffer a mechanical failure of even loss of the aircraft which could jeopardize the launch schedule. Even if NASA had agreed to commercial charters the FAA would not have allowed it. Some exceptions were made with conditions after NASA landed on the moon in July 1969. The Super Guppy was allowed to transport DC-10 fuselage sections for Douglas as a test program. The B-377PG continued as an exclusive transport of critical space hardware for NASA into the 1970s. The B-377SG continued in NASA service transporting components for the Skylab workshop after the Pregnant Guppy was sold off.

Both transports were certified as public aircraft on a very restricted operating certificate for limited NASA service. Before Conroy created the guppy the only transportation available to NASA was a fleet of sea-going barges which had to pass through the Panama Canal. The guppies cut the delivery time from weeks to hours. Wernher von Braun wrote a letter to Conroy stating, "We could not have made it to the moon by 1969 without it."[39]

Tex Johnston and TIFS

Conroy's relation with test pilot Tex Johnston began early. Johnston was named as technical advisor when Aero Spacelines was formed in 1962. Although not planned he would eventually replace Jack Conroy as president of ASI in 1968. Johnston brought Tex Johnston Inc., which was formed in October 1967 for the development of the Total in-Flight-Simulator (TIFS), to Unexcelled. The board appointed Johnston to chairman of TIFS Inc. and William C. Lawrence of Twin Fair (grocery and department stores) was named president. The Total In-Flight Simulator was being developed by Cornell Aeronautical Labs and needed a contractor with the capability of building it. Unexcelled was moving to split the Aero Spacelines heavy airframe modification away from the guppy flight operations division. W.L. Traylor, who was a pilot on the B-377MG around the world flight, became TIFS director of flight development.

Johnston had formed his own company to take on the TIFS project. It was one of the reasons he made the move from Boeing to Unexcelled. Johnston became involved with the TIFS program because it was theorized the cost of pilot training would be reduced. It was too costly and risky to type-rate pilots in the Concorde, B-747, DC-10 and L-1011. He was totally committed to developing a flying simulator as the answer to pilot training and convinced the board at Unexcelled to take on the project. The TIFS program consisted of an extreme modification of a Convair C-131H which had been converted to CV-580 standards. The C-131 fuselage was shortened forward of the wing then extended forward of the nose with a second B-707 type cockpit added ahead and below the standard flight deck. The unique configuration created a very strange aircraft.

The TIFS Convair under construction at Santa Barbara. The shortened Convair 580 has a B-707 forward cockpit grafted ahead of the standard cockpit. It was designed for hands-on flight training at lower cost. The introduction of computer-operated in-flight simulators soon made it obsolete (ASI, courtesy Tom Smothermon).

Johnston officially left Boeing in 1968, bringing his company to the Unexcelled conglomerate and taking up a corporate position at Aero Spacelines vacated by Jack Conroy's departure. Tex Johnston Inc. became TIFS Inc. Unexcelled Chemical issued 26,300 shares of common stock to finance the TIFS enterprise. TIFS Incorporated filed with the SEC on 29 July 1969 to sell 435,000 shares of common stock. The buyers of TIFS stock acquired 19.1 percent of the company while Unexcelled retained 80.1 percent. The $3.4 million from stock sales was used to retire bank loans for TIFS development cost and the balance of $2.58 million used to complete the aircraft.

The dynamics of Tex Johnston's advisor relationship with Aero Spacelines began in 1962. However, there is no mention of him as an advisor or holding any corporate position after the first ASI papers were filed. How long he remained with the early company in the advisory role is sketchy. After Unexcelled gained control in 1965 there is no mention of a relationship with Johnston. It is not clear what if anything was quietly offered by Unexcelled relative to a position at Aero Spacelines to gain influence with Boeing or obtain the TIFS program and a guppy contract to transport B-747 subassemblies. There had been friction between Conroy and the Unexcelled board for some time.

Tex Johnston resigned as Boeing's vice president of the space division on 04 November 1967. This was only five days before the first launch of the Saturn V. The company stated Johnston had done an excellent job on the Boeing portion of the Apollo-Saturn program. Unexcelled issued a press release on 08 November 1967 stating Tex Johnston would become president of Aero Spacelines effective January 1968.[40] It has been speculated Boeing may have wanted Johnston to take the position at Unexcelled to develop an in-flight simulator (TIFS) and to have access to developments at Aero Spacelines. Both could benefit Boeing at some future date. Digital computer technology, which was in its infancy in 1968, could not possibly have been considered a threat during development of the Total In-Flight Simulator. The developing digital age soon made the live aircraft in-flight concept obsolete. The advance of computer technology allowed for ground simulators which could duplicate any condition without burning any fuel or leaving the ground, replacing airborne training forever.

Unexcelled announced there would be a major expansion of the guppy fleet and in-flight simulator training for Boeing 747 airline crews would soon be done at Santa Barbara with the acquisition of the Tex Johnston Incorporated and the TIFS program. Aero Spacelines announced the heavy airframe modification division would be expanded with the building of three new aircraft in 1968 and another three to follow when they were completed. Although Conroy had resigned from Aero Spacelines under tense conditions, the company stated Conroy would possibly take over the manufacture of the guppies as a separate operation. It was not disclosed but assumed ASI would contract construction to him in a similar manner as On Mark was contracted in the building of the first two guppies. This would free ASI from the restrictive provisions and cost of guppy certification by contracting it to Conroy. The company would be able to sell the aircraft to other operators without the expense of obtaining certification.[41] The Unexcelled plan was speculative. Consideration was given to separating the Aero Spacelines flight operations from the airframe modification and aircraft construction operation. It appears to have been a cost cutting measure. The development and certification cost would be done by another company at a fixed contract cost. Whatever the plan was it never progressed past preliminary talks.

Tex Johnston announced the appointment of Donald F. Malcolm as executive vice president of Aero Spacelines on 27 November 1968. He stated Malcolm had experience in worldwide marketing as the senior project supervisor at Boeing. Malcolm and Johnston were associates while employed at Boeing. A plan was announced to expand the Aero Spacelines fleet of outsize volumetric aircraft. As part of the program, Unexcelled president Roy Schoenhaar announced in January 1969 the appointment of William C. Lawrence as president and chief operating officer of both Aero Spacelines and Tex Johnston Incorporated.

The success of the swing nose design of the Super Guppy N1038V prompted Unexcelled to review production of a more profitable commercial outsize transport. Aero Spacelines wasted no time in exploiting Jack Conroy's proposed expansion into commercial transport of airframe components and engines for the major aircraft manufacturers. The Mini Guppy was used to transport L-1011 wings from Avco in Nashville to Lockheed at Palmdale. Conroy had dreamed of commercial contracts and the possibility of building even bigger guppy types shortly after forming the company in 1962.

Tex Johnston and Donald Malcolm began using their influence with Boeing on a possible contract to move B-747 components from Boeing-Wichita and other contractors in Dallas, Los Angeles, Farmingdale and Phoenix to Everett, Washington. Boeing traffic managers were shown one of Conroy's larger six-engine guppy designs. Johnston and Malcolm projected the guppies could reduce Boeing's cost of the current rail delivery program, saving $7.5 million on the production of 400 commercial aircraft. Northrop was building B-747 fuselage sections and shipping them in 39 pieces to Seattle for assembly.

The proposed six SGT-201 commercial versions could transport the same parts in three assemblies. Malcolm held a press conference in February 1968 stating a proposal had been made for the transport airframe sections of the new B-757 as well as B-747 components and he expected a decision from Boeing any day. He further stated Aero Spacelines would be proposing a contract to transport Pratt & Whitney JT9D engines from Hartford, Connecticut, to Everett, Washington, and they were negotiating with Rolls-Royce for engine transport utilizing the new Mini Guppy.[42] Jack Conroy had formed a new company and was also bidding for the Rolls-Royce contract with his proposed Conroy 103 CL-44 guppy. The Rolls-Royce contract was eventually awarded to Saturn Airways in an arrangement to purchase Lockheed C-130 aircraft.

In retrospect the Aero Spacelines proposal to Boeing and press releases stating the new guppies would be transporting B-747 assemblies were overly optimistic. The proposed new six-engine guppy with 60,000 pound payload only existed in older artist renderings commissioned by Conroy. It was a Jack Conroy design which appeared in Aero Spacelines brochures prior to the Mini Guppy flying in 1967. Aero Spacelines had been successful and gained experience in building the Pregnant, Super and Mini Guppies using Stratocruiser B-377 and C-97 airframes. Years earlier Conroy had put fourth ideas of using Boeing 707s, a twelve-engine B-52 and even the Saunders-Roe SR.45 Princes sea plane with a 40-foot diameter fuselage. All of these proposals required considerable capital for development and even more time-consuming engineering.

The B-377 and C-97 was the only suitable low cost airframe which could have been completed in a reasonable amount of time to secure a transport contract with Boeing. The majority of the engineering and design work had been done and paid for with the previous aircraft. A cargo contract with Boeing to move enough components to maintain production would require a fleet of guppies. It takes multiple surplus airframes to supply enough parts to build just one guppy. Boeing seemed more aware of the limited amount Stratocruiser airframes left in existence than Donald Malcolm. Boeing planners looked at the number of airframes available and concluded there were not enough surplus B-377 parts to build the number of aircraft needed.

Boeing never committed to accepting bids from Aero Spacelines or anyone else for air transport of the B-747 components. Years later in the development of the Boeing 787 Dreamliner, it became evident to Boeing planners a super transporter was needed. Conroy had predicted the need in the 1960s for extremely large guppy designs to transport airframe components which he referred to as colossal guppies. His predictions were correct but they had fallen on deaf ears.

The GAO Questions Aero Spacelines Profit

The two original guppy conversions transported NASA space program hardware for nearly thirty years without any damage. This is a remarkable record when considering they were built from 1940s technology and still flying on a B-29 wing design. In addition the base Stratocruiser was a problematic commercial aircraft and ASI took many shortcuts to produce the volumetric giants. The aircraft were underpowered and demanded constant hands on attention. Conroy tried to overcome the problem with the second guppy B-377SG with experimental turboprop engines. The looming threat of mechanical failures dictated a list of Air Force bases where spare engines and other parts were stationed. The bases also served as diversion fuel stations in the event storms and headwinds caused excessive fuel consumption.

The program worked quite well for both Aero Spacelines and NASA. Marshall Space Flight Center paid Aero Spacelines $21.9 million between May 1963 and July 1971 under seven contracts to airlift outsize cargo for a total of 2.2 million miles. The GAO questioned the cost but NASA countered by stating the Pregnant Guppy could lift a 29,000 pound payload, 19.5 feet in diameter and 30 feet long and the Super Guppy could lift 39,000 pounds up to 24.5 feet in diameter and 60 feet long. This could only be accomplished by Aero Spacelines because no other aircraft existed capable of lifting the Saturn stages.

A number of audits and law suits were filed by the GAO alleging Aero Spacelines had charged for maintenance performed by the government which allowed the company to make too much profit. The courts stated the GAO overstepped its authority by attempting to dictate the amount of profit a contractor could make. Ultimately the courts ruled in favor of Aero Spacelines, stating no profit comparison could be made because there was no alternative airlift.[43]

Super Guppy N1038V was developed with limited private funds in a very short time frame from worn out airlines and experimental YC-97J components. The urgency to get it completed for NASA service resulted in multiple FAA required tests for certification of a new commercial aircraft not being done. Circumventing the process by making the B-377PG and SG a "public aircraft" for a specific government agency accomplished the mission. However, making it exempt from many FAA and CAB regulations created a condition of planned obsolescence from the outset. It prevented any future commercial work after the NASA contracts expired because it could not meet the requirements for a commercial aircraft.

14

B-377MG Mini Guppy

An Outsize Commercial Swingtail

The B-377MG came about because of Jack Conroy's belief a need existed for an outsize commercial transport. He maintained a theory of private industry being confronted with the same problems as NASA when transporting large components and materials. Water had been the only mode of transport for extremely large items since before the Roman Empire. In modern times many outsize machines could not be transported intact from the point of manufacture to site of operation. Until volumetric aircraft became a reality, large items were built and tested then disassembled and shipped in pieces and reassembled on site. Once the manufacturing industry realized the need, the guppies were considered a reliable mode of transport. Conroy anticipated multiple medium size volumetric aircraft could be built. The primitive process of building the first two Aero Spacelines guppies outside at Van Nuys and later a third at Santa Barbara could not continue if the anticipated orders for multiple aircraft were to be built simultaneously.

The construction of both the B-377PG and SG had been contracted to On Mark Engineering in Van Nuys, California, because Aero Spacelines did not have facilities, engineering staff or craftsmen. On Mark Engineering, which specialized in converting Douglas A-26 aircraft to Counter Invaders for Air Force and CIA operations in Southeast Asia, had an excellent design and engineering staff. Conroy managed to lure a number of highly qualified engineers and technicians to come to Aero Spacelines to work at the new company facilities, which were being built in Santa Barbara.

The B-377MG (Mini Guppy) was based on Pan Am Stratocruiser N1037V c/n 15937 but with a completely new 13-foot wide lower fuselage, making it the first wide-body guppy. It was the first Aero Spacelines guppy transport built in house and the first designed solely to capture the commercial outsize market. It was the only ASI conversion built with a swing tail. When Aero Spacelines first relocated to Santa Barbara, the new hangar facilities were not yet complete. Conroy acquired an old parachute hangar for offices and shops but there were no covered facilities available large enough for guppy construction. He did not let the lack of facilities spoil his plans for a new type of commercial outsize transport. The previous two were built outside with no facilities and were successful. Constructing a third guppy outside was not considered an obstacle, just an inconvenience.

The construction for the B-377MG was set up on an old military base hardstand area across Hollister Avenue adjacent to the Santa Barbara airport. As work progressed a temporary shed was built over the aircraft. The corporate offices and engineering were set up in multiple small buildings and a hangar on airport property. The hangar would eventually become the headquarters for Conroy Aviation after his dust-up and departure from Aero Spacelines. Conroy was not aware at the time that the Mini Guppy would be the last B-377 conversion built under his direction.

It was during the move from Van Nuys to Santa Barbara when Jack Conroy first encountered Frank Clark. *Golden Coast News* had dispatched Clark to cover the airport expansion and the company which would create jobs in the area by building a new type of aircraft. The two men would become friends and work on many projects together. Clark had been interested in aircraft since he was a child because his father had been a sheet metal fabricator at Lockheed. Once engineering for the Mini Guppy started, Conroy convinced Clark to come to work at Aero Spacelines in the technical publishing department. He would oversee Conroy's ideas and act as engineering liaison with former Messerschmitt engineer Moritz Asam. Conroy had convinced Asam to leave On Mark and move to Santa Barbara to become head engineer at Aero Spacelines. The three men were very compatible, developing a long term working relationship. Asam was Conroy's right hand man when it came to engineering. Clark's ability to visualize concepts spurned many new ideas and projects which he and Conroy pursued.

Before beginning construction of the third generation mini guppy, market studies were conducted to determine if there was a commercial need for a scaled down volumetric aircraft. Conroy believed the commercial market was untapped because up until now there were no other outsize transports other than the two he had built which were under exclusive contract to NASA. The Mini Guppy with a swing tail was promoted as the answer for outsize commercial cargo market demands. The swing tail concept already had a proven record of efficient operation with the Canadair CL-44D, which was introduced in 1960. The swing tail is much more practical and efficient in a commercial operation where turn-around time is an issue. Because the break is behind the wing, the number of disconnects are less than a swing nose configuration. It was presented as the logical transport for aircraft sub-assemblies, helicopters, jet engines, oil-drilling equipment and military hardware for friendly foreign governments and as a car-ferry.

Optimistic projections were to build at least 10 to 12 of the type which would be completed and in service by 1972. The new design generated considerable interest from Boeing, Douglas, Fairchild Hiller, Northrop, Martin and the Department of Defense. NASA's request for a guppy backup prompted the decision to proceed with production. Conroy was even more optimistic when the petroleum exploration industry expressed interest in the possibility of guppies being used to transport drilling equipment.

The oil exploration industry was very interested in straight in loading outsize aircraft which were capable of transporting drilling and exploration equipment to remote operations in the Middle East. Shell Oil reviewed several aircraft including the front-loading British modified Aviation Traders DC-4/ATL-98 Carvair. Originally designed for the car-ferry market, removing the rear cabin made it an excellent cargo-liner with an 80-foot constant section cargo hold. Shell Oil and Shell Nigeria were two producers which found the ATL-98 suitable, leasing it for Middle East and African oil field supply work. Eventually the Mini Guppy was chartered to transport components for an oil refinery in Iran.

Amazingly Conroy was looking at every aspect of the transport market, including car-ferries. No one could have imagined he was interested in the British car-ferry market and yet, he proposed a guppy type to either compete with the ATL-98 Carvair or as a replacement. No evidence has been obtained indicating Jack Conroy and Freddie Laker exchanged ideas or compared aircraft construction operations either formally or informally. It is quite evident Conroy was aware in the early 1960s of Laker's Aviation Traders ATL-98 conversion of DC-4s. There are no confirmed meetings of the two men prior to 1967. There is a high probability Conroy met Laker and possibly Mike Keegan at the 1967 Paris Air Show where the Mini Guppy was debuted. The ATL-98 Carvair guppy conversion for British United Air Ferries car-ferrying operation obviously stirred Conroy's imagination. He proposed a B-377 Mini Guppy car-ferry aircraft which was depicted in the 1966 Aero Spacelines sales brochure with the caption "especially configured for travel in Europe." The swing tail car-ferry Mini Guppy featured a double deck for automobiles. The passenger cabin occupied the stretched lower rear deck where the bar-lounge area was on the commercial Boeing B-377 Stratocruisers plus a smaller forward lower cabin.

We can only speculate if it was designed as a potential replacement for Laker's ATL-98 Carvair or as a competing design. The titles on the artist rendering are "Auto-Air Ferry" using Laker's Air Ferry trade name. It is possible Conroy was toying with selling it as a competing design to the Aviation Traders proposed turboprop DC-7 Carvair conversion. A number of carriers opted not to purchase the DC-4 Carvair, waiting for the Dart turboprop powered Carvair 7, which was

Conroy proposed a car-ferry version of the Mini Guppy that could have been either a replacement for or competitor to Freddie Laker's Aviation Traders ATL-98 Carvair, which had a passenger cabin in the rear. The Conroy version had both front and rear lower deck passenger cabins and double car decks (Doug Ettridge, courtesy Tom Smothermon).

never produced. This left the potential of a market for a competing design. He and Freddie Laker may have been reviewing niche commercial guppy markets at the same time.

The Mini Guppy car-ferry artist concept is shown with Pratt & Whitney R-4360 reciprocating engines, which were gas guzzlers. They would not have been economically practical for car-ferry operations. However, the Aero Spacelines brochure presented Mini Guppy specifications for both P & W piston powered engines and Allison turboprops. It is known that Conroy was considering every possible market for a commercial guppy.

There is an interesting twist to the British Air Ferries ATL-98 Carvair operation. At the end of the 1971 summer car-ferry season British Air Ferries' parent company Air Holdings announced the car-ferry operation and the remaining Carvair fleet had been sold to Transmeridian CL-44 operator Mike Keegan. Jack Conroy was no longer with Aero Spacelines and was building a CL-44 Guppy known as the Conroy 103. After it was completed Keegan acquired the 103 CL-44 guppy in a lease purchase agreement from Conroy. Once again this leads to the speculation of Conroy, Laker and Keegan being professionally acquainted early on.

Mini Guppy Construction

The selection of a specific airframe to modify for the new commercial B-377 Mini Guppy was based on no more than which B-377 aircraft was easiest to get in position at Santa Barbara and less on low time or components in good condition. Aero Spacelines did not single out individual airframes, making N1037V a random selection. It was originally delivered to Pan Am on 18 September 1949. The carrier flew it for eleven years until Boeing took it in on trade in 1960 for newer Boeing 707 equipment. It was acquired in a group of B-377s by aircraft broker Lee Mansdorf before being transferred to Aero Spacelines in 1963 for the guppy program. Optimism ran high on what was believed to be the first of a fleet of commercial guppy transports. In reality only two of the smaller diameter guppy aircraft would ever be built.

The Mini Guppy was fitted with the original Pratt & Whitney R-4360 engines even after the previous B-377 Super Guppy conversion was built with turboprop power. The original B-377 Pregnant Guppy was known to be underpowered, establishing the need for all subsequent conversions to be fitted with turboprops. Conroy did not have the funds to acquire new commercial Allison 501 engines or fabricate nacelles for them. It should be pointed out the Super Guppy predecessor was originally designed with piston power. The only reason it was fitted with turboprops is von Braun had pressure put on Pratt & Whitney to release the engines. There were only two YC-97J aircraft built. Conroy got the first airframe for the Super Guppy because NASA put pressure on the Air Force, stating it was for the national good. The Air Force would not consider relinquishing the other airframe for a commercial venture. Conroy had the forethought to salvage the nacelles from the second YC-97J as spares in anticipation of building another turboprop guppy. The nacelles were transported to Mojave and stored with the remaining Stratocruisers. The idea was to use the nacelles on the Mini Guppy, however, Pratt & Whitney had only released the T-34 engines to ASI under government pressure with the stipulation they could only be used for NASA work. Once Conroy realized he could not acquire the engines for a commercial guppy, he was forced to retain the piston engines, making the MG an obsolete step forward. It appears he was willing to adapt in order to exploit the commercial market. Contingency plans were even put forth to install auxiliary jet pods on it, however, the B-377 wings did not have the mounting hard points. The Mini and Pregnant Guppy are the only two built using B-377 wings.

T-34 nacelles from the second YC-97J were salvaged and taken to Mojave. They were placed with the Northwest B-377s, which had been moved there for storage until broken up for parts. Conroy was planning on powering the next Mini Guppy with turboprops if he could secure the engines. Lack of capital forced him retain the old piston engines (Don Falenczykoweski, author's collection).

In 1965 Pierre Jorelle, representing Sud Aviation, which had been involved in developing the Concorde since 1962, contacted Jack Conroy. Jorelle expressed interest in the Super Guppy as a possible transport for Concorde airframe parts. He pointed out the B-377SG in the current configuration was not practical because the floor was too narrow to transport the Concorde wing box. It has been speculated but not conclusively proven that the Mini Guppy was originally designed with the 8-foot Stratocruiser lower fuselage and floor. Jorelle's visit prompted ASI to redesign the MG with the 13-foot floor and new slab sided fuselage in an effort to obtain a contract with the consortium for multiple conversions. All subsequent conversions were built with the wider lower fuselage utilizing lower time C-97 cockpit nose, main gear, empennage and wings with Allison 501 turboprop engines.

The fuselage of N1037V was cut in sections, leaving only the cockpit area, a narrow strip of lower belly pan and the empennage. Once the upper sections of fuselage were removed it was lengthened with pan sections from other B-377s. The new wider fuselage could then be built up from the existing ribs visible in the foreground (Dottie Furman, courtesy Angelee M. Conroy).

Parts were acquired from two other B-377 airframes for the Mini conversion. Some parts were obtained from Stratocruiser N31227 c/n 15967 Mainliner Hana Maui, which was delivered to United Airlines on 15 December 1949. United operated it for exactly five years before phasing it out for more economical Douglas DC-6 and DC-7 equipment. It was acquired by BOAC on 16 December 1954 as G-ANTZ. It was either leased or leased to own to Ghana Airways for several years. In July 1958 it was operated by Nigerian Airways for a short time before reverting back to BOAC. Transocean acquired the Stratocruiser from BOAC in 1959. It was first registered as N106Q then re-registered as N411Q but remained in BOAC livery. Neither registration was applied to the aircraft. Lee Mansdorf acquired the group of B-377s Transocean purchased from BOAC but never put into service. Aero Spacelines broke it up for belly pan construction parts for the Mini Guppy.

Conversion work began in December 1966 after only three months of design and stress calculations. The wings, center-wing and vertical stabilizer were removed, leaving only the B-377 fuselage. The fuselage was cut around the circumference behind the cockpit leaving only a small strip of belly pan of section 42 and the forward lower entry door steps hanging down. The complete section 43 and center-wing section was cut away and removed to make way for the new extended center-wing. Sections 44 and 45 were cut away, leaving only a 6-foot wide section of belly pan all the way to section 46 and the horizontal stabilizer. Approximately 22 feet of additional 6-foot wide strips of section 42 forward belly pans cut from the other B-377s were spliced together to increase the overall length from 110 feet to just under 133 feet.

The B-377 narrow 8-foot wide cargo floor of the previous two conversions limited the transporting of heavy cargo with the concentrated weight in the upper half of the item. The narrow floor required items to be mounted on a fixture, placing the vertical center of gravity too high. A wider floor would allow the cargo to be loaded closer to the floor, reducing the vertical center or gravity. The idea of having the fuselage walls extend up at an angle from the belly pan as opposed to curving back to form a figure eight fuselage increased cargo floor space. The result was an upside down teardrop fuselage cross section instead of the B-377 figure-eight. The all new scratch built fuselage was lighter than the beefy Stratocruiser lower half designed to contain lower lounges or transport cargo.

A 13-foot-wide slab side deep-V lower fuselage was designed with the outsize ballooned top beginning at floor level. The new wider center-wing was identical to the one designed for the Super Guppy. It placed the wing root outside the new wider fuselage floor. The new deep-V fuselage design was a vast improvement over the previous B-377PG and SG conversions. The Mini Guppy used only the nose, tail and wings of the Pan Am B-377 N1037V c/n 15937 base airframe. The

new second generation fuselage design was ultimately used on all subsequent guppy conversions.

The deep-V fuselage was built up using the strip of bottom fuselage belly as more of a gauge for spacing the new ribs (rings) and bulkheads. The lower slab sides were built up at an angle from the ribs of the old belly strip and tied into the new 13-foot wide floor cross beams. The ribs continued up to form the new cargo bay, which measured 18 feet at the widest point. The fuselage was 132 feet, 10 inches long, which was 22 feet, 6 inches longer than the original B-377 Stratocruiser and almost six feet longer than the original B-377 Pregnant Guppy. It had a volume of 23,615 square feet. It was essentially a new deep-V fuselage with the B-377 nose and empennage fitted on each end with the original wings mounted on a completely new wider, heavier constructed center-wing.

The cockpit of N1037V was stripped completely including all panels and the center console. The engines were zero timed and virtually all components and systems were removed and overhauled or made new, including all controls, plumbing and wiring. This had not been done on the previous two guppy conversions. The all new fuselage was constructed using only a six-foot-wide strip of sheet metal belly from Pan Am Stratocruiser N1037V combined with strips from other B-377s. Just enough B-377 parts were used on each end to retain the registration.

The new 23-foot center-wing increased the B-377MG wingspan to 156 feet, 3 inches. The increased width between the engines and main gear spacing from the airframe centerline reduced the wing loading by 20 percent. The completed aircraft had a combination of rivet

Top: **The upper rear fuselage has been cut away leaving the lower half of section 45 and the complete section 46 with the horizontal stabilizer. The rear sections have been moved back 22.5 feet. Belly pans strips from two other B-377s were spliced in behind the center wing to extend the fuselage length to 132 feet, 10 inches (Dottie Furman, courtesy Angelee M. Conroy).**

Bottom: **Looking rearward from the center wing, the dark area is all that is left of the belly pans of other B-377s that were spliced together to extend the aircraft length. The new fuselage ribs were grafted to the trimmed ends and angle up to meet a new 13-foot wide floor. The swing-tail break will be installed at the end of the dark area. The bulkhead of section 45 is visible in the rear section (ASI, courtesy Tom Smothermon).**

types which created additional drag. The skin of the new enlarged upper fuselage was overlapped and attached with standard or round head rivets while the Boeing B-377/C-97 cockpit nose unit and tail section had the original flush rivets. The same technique used on the original Pregnant Guppy was employed to cut around the circumference of the rear fuselage to form the separation joint for the swing tail. The mating rings were considerably smaller and somewhat lighter than those used on the Pregnant Guppy with removable tail.

The fuselage interior was 15.5 feet high from the cargo floor, 18 feet, 2 inches wide at the widest point and 13 feet wide at the floor. The cargo bay was 91 feet, 6 inches long with a constant section of 75 feet, 3 inches, which is the longest constant section of any of the guppies built. There was slightly over 23 feet of usable cargo space behind the break in the tapering swing tail. The total fuselage length was 132 feet, 10 inches. The tail was slightly more than 38 feet tall.

The new smaller diameter fuselage featured a swing tail with dual hinges on the right side behind the wing. Seventeen double-action hydraulic jack pins locked the tail and fuselage together for flight. To steady the aircraft before unlocking a support strut or "tail stand" was manually lowered under the rear of the fuselage at the hinged break. The swing tail section had a hydraulic actuated retractable tandem-wheel adjustable strut which extended from a small gear well in the tail section. The wheels were located below the center of gravity of the tail section to reduce stress on the hinges as it swung 90 degrees.

Top: **A temporary shelter was built over the aircraft. The rear tapered section is considerably shorter on the Mini Guppy, with all the ribs and stringers in place before the skin is attached. The dorsal fin has been repositioned at an angle. The area between it and the vertical fin was filled and skinned. The number four engine is already in place as technicians connect the lines and cables (Dottie Furman, courtesy Angelee M. Conroy).**

Bottom: **The construction shed was removed from around the aircraft. The nearly complete Mini was towed across Hollister Avenue in the middle of the night. Several fences and obstructions had to be moved to get the aircraft on the active airport ramp, where engine run-ups could be done. The pressure was on for first flight in order to take it to the Paris Air Show (Dottie Furman, courtesy Angelee M. Conroy).**

The new B-377MG fuselage design was considered to have a more balanced profile because it was not as extreme as the two previous conversions. It was long and straight with a much longer constant section. Marshall Space Flight Center was still very concerned with maintaining delivery schedules of space components to Huntsville for testing. NASA chief of project logistics John Goodrum contacted Jack Conroy in February 1967 complimenting the performance of the Pregnant and Super Guppy. He expressed interest in the Mini Guppy, stating Conroy's plan for a third aircraft as a backup for the other two aircraft was very desirable.[1]

The construction of the Mini Guppy was nearly complete in only five and a half months. Almost all the work was done in an area across the street from the Santa Barbara airport, which was a former military air base. During construction a temporary shelter and scaffolding was built up over the aircraft as work progressed to provide access for the workers and some protection from the wind and sun.

The MG was partially painted before the temporary shelter was removed from around the aircraft. The engines were mounted and several of the props hung when the decision was made to tow it across Hollister Avenue to the airport in the middle of the night. The destination was the ramp area in front of the old hangar where Aero Spacelines had temporary offices. The engines could be run-up there and the final work was completed outside on the ramp before the first test flight. The hangar still stands today. It was imperative to complete the aircraft in time to be flown to the Paris Air Show because the show is only held in odd years. Otherwise Conroy would have to wait two years to present the Mini Guppy to the aviation world. He was very anxious to generate interest in the commercial arena. Revenue cargo had been booked for transport to Paris for the show.

Only seven months after the first blueprints were drawn the B-377MG flew from Santa Barbara on 24 May 1967 at 13:30. The aircraft had an empty weight of 85,000 pounds, maximum payload of 32,000 pounds with a takeoff gross of 142,000 pounds and was projected to cruise at 250 mph. The crew consisted of ex–Navy pilot Captain Larry Engle as chief pilot, former FAA pilot Robert Tripp as co-pilot and Travis Hodges, who had retired after 23 years in the Air Force as a flight engineer. Also on board was systems engineer William Hickey who designed and installed the completely new instrument panel. Conroy held a news conference, charming the press with a demonstration show of the latest guppy capabilities. It was christened Spirit of Santa Barbara by Jack Conroy's oldest daughter, Barbara. While the press was being entertained, mechanics were on the other side of the aircraft frantically trying to seal a fuel leak.

The first flight was conducted without any airflow directing fairings. During the test flight, the crew reported sluggish response from the elevator and buffeting in several configurations. The improved lighter weight fuselage with wider floor was a step forward in guppy design. However, the long uniform fuselage with lower slab-side allowed the air stream to wash up the sides above the inboard trailing edges, which interrupted the flow across the horizontal control surface, causing buffeting at high angles of attack.

The Pregnant and Super Guppy which utilized the old C-97 figure eight fuselage had experienced similar problems especially when the flaps were deployed, causing buffeting. It was originally believed the depression the length of the figure eight fuselage seam at floor level would help control and direct the airflow on the Pregnant Guppy. However, buffeting was experienced during flight test each time the flaps were retracted. Wing root airflow fillets were also installed on the Super Guppy soon after it first flew and years later added to the Pregnant Guppy. The Mini Guppy and all subsequent guppies experienced buffeting until a proper fillet could be designed and installed.

A wedge shaped airflow fillet was not added to the B-377MG until after it returned from the Paris Air Show. At least two different designs were installed on the MG to alleviate the problem without the desired results. Finally a very large wedge fairing was constructed above the wing trailing edge to control this problem and direct the air stream over the control surface of the tail and improve handling.[2]

First Flight and Paris Air Show

During construction, Conroy had announced Aero Spacelines had applied to the CAB for authority to operate the new B-377MG commercial guppy in worldwide outsize cargo service. The request was limited to cargo too large to be transported by DC-6, DC-7, L-1049, 707, DC-8 and CL-44 aircraft. The same announcement to the press also stated after the Mini Guppy was completed construction would start on a six-engine Turbine Super Guppy capable of transporting B-747 nose sections, which were being built for Boeing at Wichita. It is interesting that Conroy was negotiating with Boeing to transport B-747 assemblies before Tex Johnston and Donald Malcolm ever came to Aero Spacelines.

The B-377MG was originally planned to fly on or before 18 May 1967 in order operate enough test flights to be ready for a trip to the Paris Air Show. The six day delay could have jeopardized the debut in Paris but in typical Conroy fashion nothing was impossible. After the first flight on the 24th the aircraft was moved to Mojave to make preparations for a planned flight to North Philadelphia Airport en route to Paris. Conroy departed Santa Barbara for Philadelphia to await the arrival of the Mini Guppy, where it was scheduled to pick up revenue cargo for the Paris Air Show.

The cargo consisted of the XB-1 Sky Lounge Transporter, or People Pod, built by the Budd Company. The 23-passenger Sky Lounge was designed for passenger ground use and could be slung under a Sikorsky S-64 Skycrane helicopter. The Budd Company was looking to expand into other areas of transportation hardware after building railroad cars since the 1930s. The mobile transporter concept was briefly tested in California as a proposed way of reducing automobile traffic. As a result of the study the City of Los Angeles agreed to pay a portion of the cost of transporting the Sky Lounge Pod from Philadelphia to Paris.[3]

Plans to display the Budd People Pod at the Paris Air Show were under way far in advance of the completion of the Mini Guppy. The project was announced by Budd Company spokesperson Ray G. Shaffer on 06 December 1966. He detailed a mass transportation study project which was being conducted jointly by the federal Housing and Urban Development Department and the City of Los Angeles. The project was designed to reduce traffic to and from airports in major cities by having passengers board a mobile lounge at locations some distance from the airport. The lounge would be transported by S-46 Skycrane helicopters to Los Angeles International over congested city streets. The irony is the People Pod never caught on but years later the Mini Guppy was purchased by Erickson, a fire suppression contractor, to transport S-46 Skycrane helicopters.

The estimated cost of the 19-day round-trip B-377MG flight to Paris for debut and display at the air show was estimated at $39,938. To recover a portion of the cost of flying the aircraft to Paris for its debut, Conroy came up

Top: **The B-377MG flew from Santa Barbara on 24 May 1967 commanded by Larry Engle. The increased constant section of the cargo bay is obvious by the uniform lines of the fuselage. There is no airflow direction fairing at the rear wing root, which caused sluggish elevator response and buffeting in certain configurations (Dottie Furman, courtesy Angelee M. Conroy).**

Bottom: **Jack Conroy in his "lucky" flight suit after the "People Pod," or Budd Skylounge, has been loaded on the Mini Guppy at Philadelphia for the flight to the Paris Air Show. Conroy always took an active role in getting the job done and was not afraid to get his hands dirty (Dottie Furman, courtesy Angelee M. Conroy).**

with the ingenious idea of transporting cargo to promote the versatility of the aircraft. He negotiated with the Budd Company to transport the People Pod to Paris (if it was completed in time). To seal the deal he offered to split the cost of the flight with the Budd Company, Sikorsky Aircraft and the City of Los Angeles Department of Transportation. The city agreed to pay $4,500 on one half the cost of the flight to Paris. Officials with the city believed the success of the Sky Lounge People Pod would generate a considerable number of skilled jobs in California and reduce automobile airport traffic. Aero Spacelines was left with the balance $15,460 to complete one half of the expense. The other half ($19,960) of the total cost was split between the Budd Company and Sikorsky Helicopters. Conroy believed additional jobs would also be created if he could interest the European aerospace industry in purchasing guppy aircraft for airframe component transport.[4]

Conroy also had plans to present the Mini Guppy as a competing or car-ferry replacement aircraft to British and European airlines around the Mediterranean. The British ATL-98 had only been in service for four years and holiday travel business was good. He theorized because it was produced from surplus World War II era Douglas DC-4s and only held five cars it was obsolete. A turbine powered B-377MG could increase capacity and reduce cost. He produced drawings and plans of an upgraded Mini Guppy turboprop version which could compete with the Aviation Traders proposed DC-7 Carvair conversion.

Two days after the first flight the B-377MG arrived at North Philadelphia Airport (PNE), where the People Pod was loaded. The Mini ferried to Philadelphia International Airport (PHL) for fuel and to take advantage of the longer runway which would allow operation at max ATOG. William Ostrander replaced Robert Tripp as the co-pilot for the leg to Paris. Reserve Captain John Pinney, second engineer Howie Chalupsky and navigator N.E. Andross also came onboard to assist with the long overwater flight. Conroy took a commercial flight to Paris to prepare and promote the arrival of the Mini Guppy. He planned a large press briefing in Paris to demonstrate the ability of the B-377MG to transport not only outsize but very long cargo. The unloading of the People Pod allowed him the perfect demonstration of the volumetric swing-tail transport.

The guppy made a rather circuitous flight from Philadelphia via Syracuse, Massena, Prince Edward Island and Stephenville, which gave the crew the opportunity to review the aircraft performance in light to moderate turbulence. The smaller diameter fuselage had a lower vertical center of gravity than the previous guppies which was an advantage by making it potentially less top heavy. The elevator was slower to react because of the lack of fuselage airflow fairings; however, it was longer and heavier than the standard B-377 Stratocruiser, which made it more stable with less pitch and yaw at cruise. The Mini was fitted with dual RMI, VOR, HIS and DME, which is sufficient for flying within the United States. However, it was not properly equipped for flying over water with only an APN-9 Loran and an A-14 hand-held sextant.

Despite the navigational equipment shortcomings the flight continued without incident on to Gander then another six and a half hours to Santa Maria, Azores. After only two hours on the ground at Santa Maria for refueling it departed for the last leg to Le Bourget Airport, Paris, arriving on 27 May. Conroy was already courting European aviation manufacturers in an effort to solicit orders for next generation Mini and Super Guppy Turbine conversions for airframe component transport. At the end of the Paris Air Show the B-377MG was flown to Toulouse to the Sud (Aerospatiale) facility for a demonstration review before the French, German and British representatives who were meeting to negotiate the forming of the Airbus consortium. After the demonstration the B-377MG returned by the same route to Santa Barbara via Philadelphia. The new guppy had performed well, without any serious incidents.[5]

Conroy was optimistic his demonstration had touched a chord with the Airbus group which would produce orders for multiple turbine guppy aircraft to transport airframe components. Airbus was not officially formed until sometime after. Two years later on 29 May 1969 at the Le Bourget air show the French, German and British politicians announced they had signed an agreement to create an aircraft manufacturing group which eventually became Airbus. The company did not originally see the need for outsize volumetric transports. Conroy foresaw the need in 1967; however, it would be November 1971 before Airbus acquired its first SGT-201 Super Guppy.

The French demonstrated their appreciation for Conroy's contribution to aviation and honored his achievements during the 1967 Paris visit. The Council Municipal de Paris presented him with the Medal of Greatness for the greatest contribution to aerospace development in the last two years. The award was based on his creation of the B-377SG, which saved the space program from falling behind schedule. Upon his return to Santa Barbara, Conroy praised the performance of the Mini Guppy and made an optimistic announcement to the press. He stated Aero Spacelines would be entering into a lease or purchase contract with a French company for guppy aircraft. The announcement gave the impression that the side trip to Toulouse for a demonstration to Sud-British Aircraft Corporation (Aerospatiale) officials for subassembly transport was fruitful. No orders were received and the two companies were not merged into Aerospatiale until 1970. Aerospatiale later became part of EADS, the parent of Airbus. European aviation officials were adequately impressed with the concept of an outsize component transport but it appears Conroy was overoptimistic on the timeframe.

The French pointed out that the interior of the Mini was too small for the transport of airframe sections. It was capable of carrying wing assemblies for Airbus next generation commercial aircraft production. Interest in a larger version of the concept was quite high. It was not until four years later that Airbus acted on the need for an outsize transporter for airframe subassemblies. Conroy had long left Aero Spacelines but the development of the larger next generation Super Guppy Turbine SGT-201 prompted Airbus to reconsider the effectiveness of an air transporter system. The EADS group determined it would be more cost effective to purchase an existing aircraft off the shelf from Aero Spacelines rather than attempt to develop a new volumetric aircraft in house.

Conroy Resigns

By July 1967 Unexcelled Chemical Corporation, a publicly traded company, was in complete control of Aero Spacelines. Jack Conroy was well received in Paris with the debut of the Mini Guppy. He believed orders were forthcoming and released to the press his plans for an Allison 501 turbine powered Mini Guppy. He also reintroduced his design for a six-engine commercial Super Guppy and a 12 engine B-52 Colossal Guppy with a 40-foot diameter fuselage and 300,000 pound payload. After all the hype and interest the Paris trip produced nothing. Aero Spacelines was experiencing financial difficulties with the only source of regular revenue coming from the two guppies under contract to NASA. The Mini Guppy still had not been certified for ad hoc commercial charters. Plans were in place to build the SGT-101 turboprop Mini Guppy but there was no firm interest. Conroy became disenchanted with the lack of aircraft sales and slow acceptance of commercial guppies. He believed there was a market for commercial outsize transports especially in aerospace but the concept still had not adequately impressed aircraft manufacturers.

Conroy had lost control of Aero Spacelines in July 1965 when Unexcelled Chemical acquired a majority position. It can be argued he had no choice but to give up control because he needed to raise capital to complete the Super Guppy and there were no other options. Aero Spacelines was always on the financial edge because investors were skeptical of a broad need for volumetric transports.

The Unexcelled board was not always pleased with Conroy's maverick style of operating Aero Spacelines. The friction was often overlooked because the board members gambled on him complying with their decisions and marketing plans. In 1967 the flight operations division of Aero Spacelines was embroiled in an attempt by the Teamsters to unionize the pilots. Conroy despised unions dictating pilot standards but he no longer held a majority position in the company and did not have the authority make decisions regarding the situation. He and Clay Lacy along with several other pilots were attempting to regain control of the company. Conroy decided he and a small group of pilots who were close friends would fly all the NASA charters to break the attempt to unionize. The Unexcelled board wanted the problem resolved with the Teamsters and a settlement of the pilots' union contract.

This is curious and not surprising when considering Unexcelled stock was being controlled and manipulated by brokers Kleiner, Bell. Although circumstantial, the indications are there was behind-the-scenes influence. Chauncey Holt, who was a free-lance operative for the crime syndicate, CIA, FBI or anyone who needed his services, openly admitted in his biography he was friends with Detroit crime boss Peter Licavoli. Teamster leader Jimmy Hoffa and Licavoli had been associates since 1935.

Holt stated, "We were processing securities through Kleiner, Bell, the brokerage firm that was later suspended from the New York Stock Exchange for flagrant violations of insider trading practices.... Many stock swindlers were able to convert stolen securities, issued in street name by using Kleiner, Bell's omnibus account with MBT in New York."[6]

Holt's statement leaves the impression he was laundering stocks for crime bosses affiliated with the Teamsters through the broker who was controlling Unexcelled stock. The union organizing situation could have been influenced by a few well placed phone calls.

The Unexcelled board was not satisfied with Conroy's efforts to circumvent the union. While Conroy and some of his crew were on a fishing trip, the Unexcelled board of directors without consulting with him signed a contract with the Teamsters. When Conroy returned and found out he was outraged. Senior management at Unexcelled maintained they were in majority control and could do what they wanted. In reality, there was nothing he could do because he no longer controlled the company. This was a costly miscalculation by the Unexcelled board. Their reaction was that he would live with their decision although he threatened to resign.

Without a doubt, Jack Conroy was rather creative with his accounting and insisted on doing things his way; however, the board failed to realize Jack Conroy was also the soul of Aero Spacelines. He built the company and was the inspiration who created the phenomena of guppy transports. His public relations, vision and showmanship was a key asset in building the aircraft and securing contracts.

Conroy became so disenchanted with decisions made without his approval that he tendered his resignation in August 1967. He walked out, selling his minority interest in the company and giving Unexcelled Corporation complete control. It has been speculated he did not think they would accept it, however, there had been friction with the parent company for some time and it was accepted.

Unexcelled announced to the press he resigned to engage in other business enterprises. Jack Conroy was Aero Spacelines. He built the company from an idea. He made a great contribution to the space program by reducing the delivery of components by months to hours, which kept NASA logistics on schedule. Von Braun and Goodrum acknowledged his contributions. He received awards for his achievements while demonstrating the Mini Guppy at the Paris Air Show. His dust-up with Unexcelled management was a power play on both sides. If Conroy had stayed until the Mini Guppy was certified, he would have been forced to endure the humiliation of no longer controlling the direction of the company he built.

Unfortunately his projection of a market for a fleet of commercial guppy transports was 20 years ahead of the need. Aero Spacelines guppy design never progressed after Conroy's departure. The Super Guppy Turbine 201s were built; however, they were only improvements of Conroy's original B-377 design. He had explored outsize volumetric designs using multiple airframes. He moved on to create the Conroy 103 CL-44 guppy and many other projects. If he had remained at Aero Spacelines or if he could have acquired sufficient financial backing for Conroy Aviation, one can only dream of what aircraft he would have created.

Aero Spacelines Reorganizes

Conroy's departure coincided with Unexcelled filing with the SEC on 26 September 1967 to register $10,000,000 in convertible subordinated debentures, due in 1982. They were offered for sale to the public by Kleiner, Bell & Company of New York. The company stated it engaged in leasing and sales of meat processing machines; operation of department stores; conversion and operation of aircraft for outsize cargo; manufacture of iron castings; and the overhaul and maintenance of aircraft and aircraft engines. The company further stated $1,800,000 of the stock sales would be used to repay debt and defray cost of planned new guppy aircraft built by subsidiary Aero Spacelines. The balance would be used to stock a new department store under construction. It was further stated Unexcelled had 1,411,774 shares of common stock outstanding and the management owned 15.05 percent.[7]

Unexcelled president Roy Schoenhaar announced Tex Johnston would succeed Jack Conroy by taking over as Aero Spacelines president on 01 January 1968. In the interim, vice president Frank Higgins would fill in as acting president. It appears the Unexcelled board wanted Johnston for his name and reputation in order to sell aircraft and expand into other projects. The company was desperate because there were no pending orders for guppy conversions. Johnston left his position as director of Boeing's 5,000-man Cape Kennedy operation. He had been with Boeing since 1949 as a test pilot and was in charge of B-47, B-52 and 707 testing. Johnston had flown America's first swept wing airplane, first jet bomber, first rocket engine aircraft and first jet airliner, which he rolled during a test flight. Now he was going to a small aircraft company which had built three guppy conversions to date and had no pending orders.

In 1961 when Aero Spacelines was formed Johnston along with Herman Salmon were listed as advisors. Salmon and Conroy developed a close friendship which lasted until their deaths only six months apart. Johnston and Conroy were not known to be close. On occasion for undisclosed reasons there had been reported distance between them. Jack Conroy left behind many loyal employees and friends at Aero Spacelines. Both Moritz Asam and Frank Clark, who he brought in after relocating in Santa Barbara, stayed with Aero Spacelines. They continued to give "Jack" an inside eye of the certification of the Mini Guppy and other planned projects.

Certification of the B-377MG

The task of getting the Mini Guppy certified immediately became the responsibility of Tex Johnston. Unlike the previous guppies which were built using every shortcut possible, the Mini Guppy had a new fuselage and every part was overhauled, new or zero timed. However, it proved to be a more challenging task than expected. It had been fairly easy for Conroy to cut corners with the Pregnant and Super Guppy because NASA was putting pressure on the FAA to approve the aircraft. The B-377MG was built for the commercial market, making it impossible to be certified as a "public aircraft." Indications are the FAA was determined to insure a guppy for commercial use complied with regulations.

One of the requirements was the installation of a fire suppression system for the cargo compartment. The interior structure of ribs, formers, stringers and longerons were exposed with no liners on previous guppies. As a commercial transport, the FAA demanded some form of fireproofing in the cargo bay. ASI engineers pleaded the size of the aircraft presented problems in meeting the requirement. The installation of a heavy fireproof liner would add weight, increasing the aircraft ZFW and reducing payload.

Conroy had relied on NASA to pressure the FAA to bypass regulations. He was a master of having his shops fabricate parts then working out the details and creating the engineering drawings later. Conroy was no longer at Aero Spacelines and NASA could not intervene. Savvy engineers fell back on Conroy style ingenuity to circumvent the problem. The engineering department was challenged to find a solution and still comply with the FAA requirement. Malcolm Miles, who was considered to be the wizard of the electric shop, proposed the mounting of a pyroelectric vidicon (PEV) camera above the cargo bay. This was basically a video camera used to detect forest fires. It had a sensor tube which could detect the yellow flame even with thick smoke. It was connected to a closed circuit television monitor, which was mounted above and to the left of the engineer's panel in the cockpit. The system was installed for testing but failed miserably. Apparently it worked well on large forest fires but could not detect a small flame in an enclosed area.

It was back to the drawing board because the FAA was standing firm on the requirement for a fire suppression system in a commercial aircraft. After considerable negotiations, the FAA agreed to either a thin layer of fire retardant liner throughout the cargo compartment or an observer seated in a permanently installed chair at the rear of the cargo area. This presented another problem when the plane was loaded because the additional crew member was trapped in the rear of the aircraft with no access forward. An intercom was installed along with a porta-potty. To the amazement of ASI engineering, the FAA accepted the plan for an extra crew position as a suitable fire prevention measure.[8] Once the aircraft was in service the extra crew became less of an issue and eventually forgotten.

After the Mini Guppy was operational but still not certified, Aero Spacelines moved from the temporary facilities at Santa Barbara to the newly built hangars. The new B-377MG was chartered for a special mission with international political overtones. The government of the Shah of Iran was involved in a major expansion of that country's oil refining capacity. The B-377MG departed Santa Barbara on 24 August 1967 en route to Minneapolis, Minnesota. The next day Unexcelled president Roy Schoenhaar announced the FAA had granted an exemption for the Mini Guppy to operate a special revenue cargo flight around the world. The crew included Aero Spacelines chief pilot Larry Engle, reserve captain W. L. "Bill" Traylor, co-pilot Bill Ostrander, flight engineers Travis Hodges and Howard Chalupsky, navigator Dick Gadfield and flight mechanics Roland Hipley and George Conway.

The charter for the Fluor Corporation with assistance from Lufthansa Airlines uplifted cargo from Minneapolis–St.Paul to Tehran. The exemption was granted for the not yet certified B-377MG extreme transport because it was the only aircraft in public service capable of uplifting the outsize cargo. The flight required approval of the Department of Commerce and the State Department. The B-377MG arrived at Minneapolis to uplift a five section 160-foot construction boom built by American Hoist and Derrick Company. The specially constructed crane was built for use in the building of an oil refinery at Tehran.

The refinery, which was estimated to process 85,000 barrels of oil a day, was being built by a German-American consortium for the National Iranian Oil Company. The partners in the project included Fluor International, NIOC Tehran Refinery, Thyssenrohr International, Stahlunion and Mannesmann Export. The guppy departed Minneapolis with the sections of the 7.5 ton crane staging through Newfoundland, Iceland, France, Greece and Turkey en route to Tehran.

After offloading the crane in Tehran, the guppy ferried to Munich to upload a rocket booster stage for delivery to the launch range at Woomera, Australia. The rocket, which was a modified British Blue Streak, was delivered for testing by the European Launcher Development Organisation (ELDO) as part of a three stage system to launch satellites.

The cargo flight included stops in Pakistan, Ceylon, Singapore, Jakarta and Darwin before reaching Woomera on 12 September 1967. The return flight across the Pacific staged through Fiji, Samoa and Hawaii before arriving back at home base in Santa Barbara. Co-captain L.W. Traylor stated the aircraft performed well with only minimum maintenance problems. He said from the cockpit it felt like a Stratocruiser

In August 1967, before the B-377MG was certified, the FAA granted a special exemption for the aircraft to transport a five-section, 160-foot construction crane for the Fluor Corporation to Tehran, Iran, for the construction of a refinery. The flight continued on around the world, transporting a rocket to Australia and returning across the Pacific to Santa Barbara (Dottie Furman, courtesy Angelee M. Conroy).

because you do not realize what you have behind you. It did not rotate like a normal aircraft. On landing it had a tendency to roll over as it slows down. The operation was considered a successful demonstration of the reliability and uplift capability of the first Mini Guppy. The Aero Spacelines corporate office was optimistic more guppies would be built. Vice president Donald Malcolm announced the proposal made to Boeing to transport B-747 assemblies would be accepted.

In February 1968 the Mini Guppy operated an FAA test flight at all up gross weight over the Pacific off the coast of Santa Barbara. After completing the test N1037V made several runs up and down the coast to dump ballast before returning to the field at Goleta. Emergency authorities and the Coast Guard began to receive multiple phone calls from the public reporting a large aircraft crashed into the ocean. The aircraft had been fitted with bladder tanks filled with 2,000 gallons (16,680 pounds) of water for the test. The crew flew low over the ocean to jettison the water to reduce weight before landing. The water spray created an illusion of smoke, prompting citizens on the beach to alert authorities a large aircraft trailing smoke had burst into flames and crashed. The Coast Guard along with a Navy helicopter, four fire trucks and the California Highway Patrol were dispatched to the area only to witness the Mini Guppy making a slow approach and safe landing at Goleta, Santa Barbara airport.[9]

Trouble Begins at Unexcelled

In December 1968 Morton Sterling, who was the brother of Los Angeles attorney Barry Sterling, purchased 5,000 shares at $34 each of the conglomerate known as Unexcelled. Sterling was a close associate of Burt Kleiner and was the head of investor banking division of Bernie Cornfeld's Investor Overseas Services (IOS) Fund based in Geneva. Bruce Rozet, chairman of Commonwealth United, which was another conglomerate in which Los Angeles broker Burt Kleiner was involved, also purchased 5,000 shares. Morton Sterling's wife died in January 1969 and half of the holdings passed to her estate. The Sterling brothers and Burt Kleiner became the trustees of her estate.[10] That same month Unexcelled reorganized, appointing Tex Johnston to chairman of the board, W.C. Lawrence, president and COO of ASI and Tex Johnston Incorporated, Roy Schoenhaar to president of Unexcelled and Vaughn Read as executive vice president of Unexcelled.

On 07 February 1969, with the assistance of Kleiner and Sterling, Unexcelled formed a subsidiary in Curacao as Unexcelled International N.V. On 05 March Investors Bank of Luxembourg agreed to manage the private sale of debentures of Unexcelled N.V. totaling $10,000,000, carrying an interest rate of 7 percent. Barry Sterling, as president of IOS, was in charge of Investors Bank. In May and June 1969 the IOS purchased over $3,000,000 worth in two transactions. Unexcelled was to be Burt Kleiner's conglomerate to manipulate by working inside trades. The conglomerate consisted primarily of Twin Fair discount grocery and department stores, a company which manufactured meat processing and packing machinery and Aero Spacelines.[11] Unexcelled had once been linked to Mary Carter Paints, which became Resorts International. The development was started by Huntington Hartford, who was a friend of Unexcelled former president James Crosby. It appears Kleiner had delusions that he could parlay Unexcelled holdings into a big payoff. In just six months ASI was in trouble. President Lyndon Johnson began cutting the NASA budget in 1967. After the United States landed on the moon in 1969 he had no interest in the space program, which caused massive cutbacks in the NASA budget by 1970.

Unexcelled reevaluated its slipping financial position in January 1970. The decreasing cargo demand at NASA prompted Unexcelled to write off $13,000,000 in losses at Aero Spacelines. On 12 May the new MGT-101 Turboprop Mini, which was predicted to revolutionize commercial air cargo, crashed during a test flight. The aircraft was insured for approximately $5,000,000 but $3,000,000 of the payout was used for debt obligation and accounts payable. By September Unexcelled N.V. could not make interest payments. The debentures of Unexcelled International N.V. became worthless. On 03 September the American Stock Exchange halted trading of Unexcelled stock. Chase Manhattan Bank as trustee demanded $350,000 in past-due interest on the $10,000,000 in Eurobonds of Unexcelled International N.V. A week earlier Unexcelled attempted to buy the bonds back at 10 cents on the dollar. Roy Schoenhaar stated Unexcelled had a 30 day grace period. He presented a plan where investors would get ten cents on the dollar plus a non-negotiable note of $50 for each $1,000 in bonds payable in five years. He further stated the note could be payable the following April when Aero Spacelines would sign a contract for the production and certification of a new guppy aircraft. Chase Bank announced it would resign as trustee as soon as Unexcelled found a replacement.[12] Kleiner, Bell & Company was suspended from trading by the New York and American stock exchanges on 17 March 1970. Burt Kleiner was charged by the SEC for manipulating stock and insider trading of multiple companies. Unexcelled suffered drastic cutbacks but survived to produce two SGT-201 aircraft and the parts for two more before fading away.

B-377MG Operational Service

Despite all the testing, effort and influence Tex Johnston brought to Aero Spacelines, the FAA did not type certify the

B-377MG (N1037V) until March 1969. The certification of it as a commercial cargo guppy paved the way to the development of the next generation Mini Guppy Turbine (MGT-101) completed in 1970. It was tragically lost in a crash during testing (see Chapter 15).

The Mini Guppy N1037V transported hardware for NASA and the DOD on an irregular basis but not under exclusive contract, leaving it available for the commercial charters. The Army utilized it to transport helicopters. In 1972 NASA chartered it to transport the Pioneer 10 Spacecraft to Cape Kennedy. Unexcelled embarked on a commercial program with the Mini transporting components for aircraft manufacturers Boeing, Douglas and Lockheed with limited success.

A number of unique commercial charters were flown throughout the world. Goodyear contracted to Aero Spacelines to transport the Blimp Europa from Akron, Ohio, to Cardington, Bedfordshire, England, in 1972. Helicopters were transported to Nigeria to support the Biafra food lift. A 707 nose was uplifted from Seattle to Damascus to repair an aircraft which was bombed by terrorists and a DC-8 fuselage was moved from Mexico City to Burbank for a Universal Studios movie set. The Mini was even used on several occasions to transport yachts and large commercial boats.

The demand for cargo work was increasing when one of the Mini Guppy charters came very close to ending in disaster. The aircraft was chartered to transport six swift boats to the Congo. They were cut in sections in order to make them easier to load. The planned routing staged through the Azores. Approximately halfway across the Atlantic the generator overheated on the number one engine. As a precautionary measure, the engine was shut down. It was a necessary precaution because in the 1950s the Pratt & Whitney R-4360 engines suffered a number of failures resulting in crashes which were traced to engine driven generators overheating and bursting into flames. The aircraft approached Santa Maria, Azores, with the number one engine shut down. It was standard procedure on landing to pull back the middle throttles on the inboard engines two and three. Most pilots conducted the procedure by feel by placing their hand on the throttles and pulling back the two levers in the center. Upon landing the pilot placed his hand on the throttles and skipped the first lever pulling back the next two. He overlooked the number one lever which was out of position because it had been shut down and was against the peg. Engines three and four were pulled back in error. Power was still on number two engine causing the aircraft to swerve to the right. The crew was able to correct and maintain control of the aircraft, but not before both nose wheel tires were shredded.[13]

The MG suffered another close call while operating a cargo flight to Brazil. The guppy was chartered from New Jersey to São Paulo, Brazil, to transport components for a petrochemical plant. The flight proved to be quite costly when an engine failed en route, forcing a diversion to Caracas, Venezuela. A replacement engine was obtained locally from an old parts supply and changed out. The engine was run up and all checks were done. Departing Caracas as the aircraft climbed out at takeoff power the freshly replaced engine suffered a catastrophic failure and loss of oil. It was fitted with the engine oil activated Hamilton Standard propellers. The crew was unable to feather the wind-milling propeller because of loss of oil to the propeller hub. It was the same Hamilton Standard Hydromatic 24260 hubs which were problematic on the Pan Am Stratocruisers, resulting in multiple failures and aircraft losses in commercial service.

The Hamilton props are actuated by oil from a supply line to the engine oil tank. The tank is supposed to have a 2.5 gallon reserve in a standpipe, preventing the tank from being pumped dry, which insures a constant supply of oil to the props. However, the runaway prop would not feather and the quantity gauge indicated the tank was empty. Engineer Bob Harridge stated he frantically transferred oil from the central tank to the engine tank, getting enough to finally get the prop feathered. The aircraft returned to the field safely to await delivery of another engine.[14]

Subsequent inspection found there was no standpipe in the oil tank. The Mini Guppy wings were from a Pan Am B-377 which should have the tanks with standpipes. It was reasoned the tank had been changed out sometime in the past using one from an ex–United or Sila B-377 which did not have a standpipe. Those carriers ordered the Curtiss Electric C644S-BS02 prop hubs with an internal oil supply which did not require the reserve standpipe.[15]

After several days a second engine was located and flown in and installed. The aircraft arrived in São Paulo almost a week late. The president of Brazil was on hand to inspect the B-377MG. He expressed an interest in this unique giant aircraft. Aero Spacelines marketing reasoned this to be an opportunity to make a sales pitch to a foreign government which could purchase multiple aircraft and bring in long term contracts.

The president of Brazil requested a demonstration flight to Brasilia on the pretense his government was interested. The flight was more about internal politics than buying aircraft. The people of Brazil had never seen an aircraft of this size. The press could be expected to run stories of the president flying in the largest aircraft in the world. While returning from Brasilia the guppy suffered a third engine failure, which gave the president an excuse not to be interested in the aircraft.

NASA Reduces Demand

Although N1037V was built as a commercial outsize transport and operated ad hoc charters, it was frequently flown with the B-377PG /SG transporting NASA cargo. The three logged over three million miles supporting NASA's Gemini, Apollo and Skylab programs. The planes transported components for 90 percent of the moon mission program. They moved 11 of the 13 major Saturn Apollo components. The list includes the command, lunar and service modules; five engines for the first stage; the lunar module adapter; the S-IVB stage; and an instrument unit. After Apollo the cargo ranged from gigantic subassemblies to very delicate and crucial components for the manned space station.

The designing and building the guppy aircraft was an incredible achievement but the success was its own worst enemy. After Apollo and Skylab were completed interest in space waned with public outcries over excessive government spending. President Lyndon Johnson was never really committed to NASA. He had continued the program because of the promise of President Kennedy. Johnson lost interest in the space program after America put a man on the moon. Billions had been spent on the Vietnam War and there were public outcries of government uncontrolled spending. The expense and lack of interest in space exploration generated investigations into the immense cost, forcing Congress to cut funding. With the reduction in the space program NASA had very little work for three guppies, causing Aero Spacelines to seek commercial transport contracts.

There was one major problem. Both the Pregnant and Super Guppy were certified as "public aircraft," leaving only the Mini Guppy for commercial charters. Special waivers were issued for the Super Guppy to transport DC-10 airframe components but the Pregnant Guppy was restricted to only NASA and DOD service. With fewer calls for NASA and DOD outsize lift, the development of a second generation guppy for commercial service was the only avenue for Aero Spacelines and the possible sale of more aircraft. Unexcelled could not justify the cost of owning the Pregnant, Super and Mini Guppy for an occasional government charter. Mini Guppy N1037V and Pregnant Guppy N1024V were sold to Clay Lacy's former business partner Allen Paulson's American Jet Industries in 1974. The Mini was retitled American Jet and re-registered N422AJ. The Pregnant Guppy had all lettering removed and was re-registered N126AJ but remained grounded pending a planned overhaul.

American Jet operated limited service with the Mini Guppy for six years before selling it to Aero Union in July of 1980. The registration was changed to N422AU but not until October 1981. After eight years with Aero Union supporting fire suppression operations and ad hoc charters it was acquired by Erickson Air Crane in 1988. Once again it was used to support aerial fire suppression and the transporting Sikorsky S-64 Skycrane helicopters which were also used for firefighting.

In 1992 the Mini Guppy became the only Aero Spacelines built guppy to appear in a major movie. It was used in the early scenes of the sci-fi adventure *Universal Soldier*. This same year Erickson obtained the license from Sikorsky to manufacture the S-64 helicopter. The Mini Guppy remained active transporting heavy equipment and the Sikorsky S-64s helicopters until 1995. It had seen considerable abuse and was quite in need of major overhaul. It was withdrawn from service and placed in flyable storage. The only B-377MG made the last flight to Oregon in 1997, where it was placed on permanent display at the Tillamook Air Museum. Unfortunately sitting in the moist salt air for many years has caused severe corrosion.

In 2013 Erickson Aero Tanker reviewed plans to move the Mini Guppy to Madras, Oregon, where the museum is being relocated. The wing spars were found to have extensive corrosion, which would require major maintenance at exorbitant cost. In addition a hole had been cut in the rear lower belly for a set of wooden steps to allow access by museum visitors. Several attempts have been made to bring it back to airworthy status for one ferry flight to a new location at Madras. It has been suggested wing replacement may be a possible solution to the corrosion problems for the one flight. However, there are no B-377 wings in existence and it is not clear if any C-97 wings are left. NASA had a set which were acquired for possibly overhauling the Super Guppy N940NS and upgrading to Allison 501 turboprops. It is not known if they are still in storage. It is generally believed the C-97G wings, if available, are not interchangeable but it has not been conclusively determined. The B-377MG remained at Tillamook in 2016 with an uncertain future.

Super Guppy in Commercial Service

The Super Guppy N1038V joined the Mini Guppy N1037V in commercial charter service in 1970 on a temporary basis. Despite the collapse of parent company international division Unexcelled N.V. stock, the domestic Unexcelled parent of Aero Spacelines managed to survive. In preparation for the planned production of more volumetric SGT aircraft Aero Spacelines began increasing their employee base in mid–1969 from 75 to 150. The buildup for production of the planned Mini Guppy Turbine (MGT-101) and Super Guppy Turbine (SGT-201) and aircraft overhaul business expanded the number of ASI employees to 600 by 1970 with an annual payroll of $5,000,000.[16]

The Super Guppy was built in the 1960s when NASA

needed a means of transporting the individual components and stages of the Gemini and Saturn rocket from the manufacturers to the assembly and testing facilities. As previously covered the FAA and CAB prohibited using the PG and SG for commercial operation. After considerable lobbying, the CAB partially revised the earlier decision. On 02 January 1968 the CAB granted a five year authority for Aero Spacelines to operate worldwide charter service with the Super Guppy N1038V based on certain upgrades and restrictions effective in 01 March. It has been suggested Tex Johnston and Donald Malcolm were instrumental in obtaining authority because of their prior affiliation with Boeing. The aircraft would have to be certified for commercial service to transport B-747 subassemblies. If Boeing had let a contract the Super Guppy would have been allowed to fill in until the next generation commercial SGT-201 was constructed.

The Super Guppy had been certified as a public aircraft for an exclusive NASA contract. After being granted special permission for commercial charters, on 09 January 1970 the Super Guppy began transporting 22-foot diameter Douglas DC-10 fuselage sections from San Diego to Long Beach (ASI, courtesy Tom Smothermon).

A contingent contract to transport DC-10 fuselages from Convair at San Diego for Douglas Aircraft was negotiated by 1969. The Super Guppy Turbine 201 was nearing completion at Santa Barbara and would fly in August 1970. Until testing was completed on the SGT-201 and FAA certification was received, it would not be available to begin service. Aero Spacelines and Douglas suggested a plan to the FAA in order to circumvent the "public aircraft" certification. Aero Spacelines requested an exemption to operate the original Supper Guppy N1038V on a temporary basis until the new SGT-201 N211AS was certified. Douglas also enjoyed a good relationship with the FAA. Being a major contractor to NASA provided needed leverage to work out an agreement in 1970. The Super Guppy as long as certain conditions were met would operate on a temporary basis transporting DC-10 fuselage sections from Convair at San Diego and wings from Douglas Aircraft in Toronto to Long Beach. The details are not clear, but the FAA granted the special exemption. It was a most unusual situation. The many shortcuts taken when it was built had not been corrected yet it was allowed to operate commercial charters with certain restrictions.

One of the conditions laid down by the CAB prohibited the Super Guppy transporting anything which would fit in a standard commercial cargo aircraft. To circumvent the restriction on several occasions, frames and pipe fixtures were welded to items to make them too large to fit through the door of a standard cargo aircraft.[17] The flights transporting fuselage sections were publicly stated to be primarily to demonstrate the viability of the guppy to aircraft manufacturers and generate interest in the next generation SGT-201 guppy. After the SGT-201 was registered as N211AS and certified as a commercial aircraft it transported two 22-foot diameter DC-10 fuselage sections per day from San Diego.

Aero Spacelines announced on 31 December 1973 it would cease flight operations as soon as existing contracts were satisfied but would continue in building guppy aircraft. The Apollo program was coming to an end, prompting NASA to review newer and larger transportation systems for the proposed Space Shuttle. Most of the hardware was so large not even the Super Guppy could accommodate it. NASA would purchase the first of two surplus B-747s for conversion to a Space Shuttle Carrier Aircraft (SCA) in 1974. Studies were being conducted to transport outsize hardware inside a pod which could be mounted in a similar manner as the Shuttle on top of the B-747. The Apollo program lasted until 1975 but had been severely affected by budget cutbacks. In 1978 Aero Spacelines announced it would no longer operate the Super Guppy. Without the Super Guppy NASA logistical planners predicted future problems transporting outsize components.

15

MGT-101 Mini Guppy Turbine

Turbine Guppy 101 Construction

The limited success of the piston powered B-377 Mini Guppy combined with the belief that there was a market for a more powerful improved version prompted the design of a second generation swing-nose turbine powered Mini Guppy model B-377MGT-101. The slab-side lower fuselage with 13-foot-wide cargo floor was carried forward from the swing tail piston powered B-377MG. Aero Spacelines presented the improved MGT-101 as the standard for a new series of outsize aircraft built to meet the demands of an expected commercial market. It was the prototype of the next generation turbine outsize transports and not a one-of-a-kind like the Pregnant, Super and Mini Guppy. In reality it was more of a transition aircraft with a composite of previous guppy features. It had the slab-side lower fuselage with 13-foot-wide floor of the MG; the center-wing and swing nose of the SG with production Allison 501 turbines and a 707 nose landing gear. The construction was the same method of stringing six-foot-wide strips of B-377 and C-97 belly pans together to create spacing for the ribs and formers of the new lighter fuselage.

The next generation MGT-101 under construction in the left side of the new hangars constructed for Aero Spacelines at Santa Barbara. The new slab-side fuselage is built up from sections of B-377 belly pans in darker color at the bottom. ASI presented it as the new standard in cargo transports, expecting to build a fleet of up to 10 aircraft (ASI, courtesy Tom Smothermon).

The MGT-101 interior was 15 feet, 6 inches high from the cargo floor and 18 feet, 2 inches wide at the midpoint and 91 feet, 6 inches long. The 132-foot fuselage had a constant section of 73 feet, 2 inches which was slightly less than the B-377MG because of the swing nose. Switching to the swing nose allowed for the utilization of the tapered rear section for a total usable cargo hold length of 99 feet. The previous swing nose B-377SG was built on a hurry-up schedule for NASA, which left little time for advanced engineering. Because the MGT-101 was a commercial venture, the ASI engineering department had taken the time to improve on previous designs and refine the complexity of disconnecting cables and lines for the swing nose. These improvements made it commercially feasible because it could be loaded and unloaded in minimum time without requiring a team of technicians to reconnect multiple lines and torque locking bolts.

The vertical stabilizer was extended and measures

the same height as the Mini and Pregnant Guppy with the round top. Although taller, it appears the same as the original commercial B-377.

The horizontal control surface was extended 40 inches on each end with squared tips. This was done to increase handling by allowing the airflow from around the widened fuselage to engage the elevator. To compensate for the increased maximum gross weight a stronger Boeing 707 nose gear was adapted. The 707 gear was turned 180 degrees and mounted approximately 30 inches to the rear of the original C-97 mounting. Additional engineering was required to move the mounting point rearward and reinforce the area to properly transmit the stress load to the airframe. The B-707 nose gear was shorter, which allowed the nose to sit lower. This eliminated the problem experienced with the Super Guppy of the main gear lifting off before the nose on takeoff.

The nacelles were formed to match the Allison 501 engines cowlings. The MGT-101 had a constant fuselage with a swing nose sharing the same round top vertical fin as the Mini Guppy. Except for the dorsal fairing the tail profile was similar to the B-377PG, which can be seen parked on the ramp behind it, giving a good comparison of the tail and engines (AAHS Archives).

Shortly after construction began on the MGT-101 the first SGT-201 was started in the adjacent bay of the hangar. There were no firm orders for either aircraft. Aero Spacelines expected orders, optimistically planning for production of six to ten MGT-101s. ASI negotiated tentative long term agreements with Convair, Douglas, Lockheed and Avco to transport DC-10 and L-1011 TriStar components relative to the new aircraft being built and certified. The manufacturers were led to believe a fleet of new volumetric transports was in production. Unfortunately, the marginal financial position of Aero Spacelines combined with unforeseen circumstances resulted in only one MGT-101 being built. The single example N111AS is the shortest lived of all the guppy conversions, first flying on 13 March 1970.

Turbine Engines

The improved design features of the MGT-101 carried through on all subsequent SGT-201 models. The 5,700 horsepower P & W T-34-P7 turboprop engines used on the Super Guppy were not available, thus prompting the selection of 4,680 hp Allison T-56-501-D22C turboprop engines used on the Lockheed P-3A Orion and Lockheed C-130. The engines were a proven unit, readily available, with a cowl design which could be adapted to the C-97 wing, making them an excellent choice.

All previous guppies, both piston and turbine powered, utilized existing engine mountings and nacelles. The building of the second generation turbine guppies took the lower slab side fuselage design of the first piston powered Mini Guppy with extended center-wing and mated it to C-97 wing, cockpit and tail components along with the B-377 fuselage section 45. Aero Spacelines opted to purchase fully outfitted QEC Allison T-56-501engine packages from General Motors. Because of previous stock manipulation scandals and a history of cash flow problems and marginal credit rating, General Motors required Unexcelled to put up the Aero Spacelines factory facilities as collateral for the engines. In the event the aircraft were not sold or the company went into default General Motors was covered.

The Allison 501 is not a simple bolt-on upgrade to the C-97G wing. The Lockheed P-3 cowlings adapted fairly easy to the existing C-97 lower nacelles and landing gear bay housing but required the fabrication of panels on top of the wing. Use of the C-130 cowling was not possible because the engines were slung under the high wing without a provision to retract the gear below the number two and three nacelles.

Mounting the Allison 501 engines to the existing C-97 firewall presented an engineering problem. The turbines were not as wide as the old P&W R-4360 engines. Without knowing how many aircraft were going to be produced it was cost-prohibitive to purchase expensive metal forming machines to fabricate a small number of filler panels for the cowlings.

ASI engineers selected 17.7 PH stainless steel stock as the material for the nacelles because it had excellent fatigue properties and good corrosion resistance.

In order to form the nacelle filler panels, a buck was fabricated out of fiberglass about two inches thick, which matched the required contour of the fairing. Picture a large box three feet on each side with a rounded top of the proper contour of the nacelle with an open bottom. The buck, or box, was turned upside down and filled with pea gravel mixed with resin. After the resin set it created a very heavy and solid rounded stretch form block. Each sheet of 301stainless steel was then pulled down and stretched over the buck using large hydraulic cylinders to form the filler skin panels between the engines and the firewall.[1]

The propeller blades used on the Lockheed C-130 were selected because they were engineered for high loads, whereas the Lockheed P-3 props were built for speed. The new engines increased low altitude airspeed to about 200 knots and limited upper altitude speeds to 185 knots. The P-3 cowlings, cuffs and spinners, C-130 props blades and Allison T-56-501 engines made the perfect combination. They proved to be efficient and reliable enough to be used on all four subsequent SGT-201 conversions.

The MGT-101 was the first swing nose guppy with a single hinge which opened past 90 degrees. The more sloping upper fuselage above the cockpit and slab sides with the 13-foot-wide floor produced a different geometry than previous guppy conversions. The change in fuselage contour prevented a double hinge configuration as seen on the B-377SG with swing nose. The slab side fuselage of the MGT-101 was built with the same methods of the B-377MG utilizing a thin strip of belly pan from multiple B-377s and C-97s as a gauge for spacing the attachment of the ribs and formers.

The pressurized C-97 cockpit was grafted to the forward fuselage swing nose section. A cross section of the fuselage was the shape of an upside-down teardrop as opposed to the original B-377 figure eight design. A highly modified C-97 empennage with extended horizontal tail plane and B-377 section 45 lower section was grafted to the rear of the deep-V fuselage. The interior constant section of the fuselage was slightly shorter but the empennage was 2 feet, 8 inches longer than the Mini Guppy, which is mostly accounted for by the tail-plane modifications. The new center-wing increased the wingspan to 156 feet, 3 inches, which is the same as all other guppies built except the first B-377PG.

The Allison 501 powered MGT-101 was the first conversion built at the new Aero Spacelines Santa Barbara hangars. Construction was well under way before the first SGT-201 registered N211AS was started in the adjacent bay. Without any orders or letters of intent, Aero Spacelines made the decision to build the larger SGT-201 on speculation as a demo aircraft. These are the only B-377 ASI Conroy Class guppies constructed at the same time. All others were built as single projects. The MGT-101 first flew on 13 March 1970, commanded by Aero Spacelines vice president Van H. Shepard.

The geometry of the new fuselage design allowed for a single ball-joint hinge for the swing nose opening. Twin forward and single rear hydraulic jacks stabilized the fuselage. The nose was powered open by a small electric motor on the bottom of a two-wheel removable strut forward of the fuselage break (ASI, courtesy Tom Smothermon).

Aero Spacelines maintained the Conroy tradition of only engaging the most qualified pilots to perform the certification flight test. Van Shepard was an outstanding test pilot who had experienced many challenging in-flight situations and malfunctions. His experience in large aircraft included flying in the North American XB-70 program with noted test pilot Al White (lone survivor of the crash of the second XB-70). Shepard's ability to deal with adverse flight conditions had been demonstrated in March 1966 while at the controls of XB-70 on the 37th test flight. The giant aircraft suffered a hydraulic failure on both a primary and secondary systems. The main landing gear did not fully deploy, leaving the right main gear not in line with the direction of travel. The left deployed with the bogie standing on end with the rear wheels above the front. On landing it veered sharply to the right. Shepard was able to apply just enough power to the number six engine to keep it from cartwheeling. After rolling three miles on the partially deployed landing gear it had turned 110 degrees. The aircraft was saved, limiting damage only to the main landing gear.

The Mini Guppy Turbine N111AS was tragically lost on 12 May 1970 at Edwards Air Force Base during certification test at all up weight. The entire crew perished when the left wing contacted the ground during an engine out test, causing the aircraft to cartwheel into the ground (USAF AFFTC/HO).

Fatal Flight of N111AS

The optimism of building a fleet of commercial turbine Mini Guppy transports was short-lived. The first in a planned series of up to ten conversions was lost during FAA certification testing the morning of 12 May 1970 at Edwards Air Force Base. The test flight of N111AS was being conducted under the command of Aero Spacelines vice president Van Shepard. He had logged 6,827 hours in multiple aircraft as an instrument rated test pilot and logged 34 hours in the guppy aircraft. On the sixth takeoff at maximum takeoff weight during test flight number twelve the plan was to cut number one engine at 112 knots to simulate a three-engine takeoff with a catastrophic in-flight failure.

At approximately 08:30 the wind was from 200 degrees at ten knots as N111AS rolled out on runway 22. The rudder boost was on and the right rudder force was normal throughout the roll-out and rotation. The number one engine was pulled back a few seconds earlier than plan at about 109 knots and the aircraft rotated normally at 114 knots. One second after rotation the rudder rapidly changed direction from right to left, requiring the pilot to exert considerable force on the pedal. Within a few seconds the aircraft rolled to the left and began to slip as the wingtip contacted the ground, changing heading to approximately 245 degrees. It then cartwheeled to a heading of 020 degrees as the nose slammed into the ground. The forward section burst into flames with the intense fire burning the cockpit and fuselage back to the centerwing. Tragically, all on board perished.

It had only been eight weeks since N111AS was flown for the first time. The Air Force identified those lost as pilot and vice president of Aero Spacelines Van H. Shepard, 45, of Santa Barbara; co-pilot and Aero Spacelines chief test pilot Harold Hansen, 44, of Goleta; flight engineer Travis B. Hodges, 44, of Santa Barbara; and Convair Aircraft test engineer Warren Walker of San Diego.[2] Shepard was also president of Tex Johnston Incorporated and a former North American F-100 production test pilot. Hodges had served 23 years as an Air Force flight engineer and had crewed the piston powered B-377 Mini Guppy on the round-the-world flight.

Post Crash Investigation (NTSB-LAX70SL071)

Complete data was obtained from the flight recorder and instrument panel test camera film. The data included airspeed and aileron position, angle of attack, degree of side slip, elevator deflection and stick force, engine rpm, g load factor and rudder position.

A representative from Hamilton Standard inspected the propeller blades at the scene. The shims were found to be normal, indicating the number one propeller was feathered and all others were in normal power position. The engines were transported to Aero Spacelines in Santa Barbara for disassembly and inspection. Engine numbers 2, 3 and 4 were found to have FOD damage caused by aluminum debris ingested during the crash. Sand was found in the compressor sections, indicating they were functionally normally before impact. Number one engine did not show any FOD damage or sand, indicating it was shut down.[3]

Investigators examined the control systems for any evidence of rudder failure. The rudder bell crank arm and rudder boost control link assembly were found to be broken. Both assemblies were removed for fracture analysis by an NTSB metallurgist. They were determined to have typical breaks caused by overload bending. No determination was made as to point in time the fractures occurred.

The pilot's rudder pedal assembly was destroyed and not recovered. The co-pilot's rudder pedal assembly was recovered and appeared to be functioning properly. The rudder cable system showed severe impact and fire damage. The right rudder cable quick disconnect was found unlatched and disconnected with no evidence of any safety-wire. All other cable disconnects were inspected and found to be latched and safety-wired. It could not be determined if the right rudder cable quick disconnect was safety-wired or not prior to the crash. The cable and latch assembly was removed and forwarded to the NTSB. It was determined the forward latch handle portion was severely burned but the attaching clevis was not. Although this gave the appearance of possibly being disconnected prior to the crash, no conclusion was made.[4]

The rudder boost system was removed and taken to Santa Barbara for testing on sister aircraft B-377SGT N211AS, which was not yet complete and would not fly for two more months. The first test on 03 June 1970 was inconclusive because hydraulic pressure was not available. A second test was conducted on the same aircraft on 01 July 1970 with hydraulic pressure and the rudder boost system operating. The test produced similar data to what was on the flight recorder and test data film recovered from the crash site. Hydro-Aire of Burbank, California, enlisted a Boeing expert to conduct a functional test of the rudder boost. All parts were disassembled after the test. The inspectors concluded there was no evidence of any malfunctions.

A review of maintenance records and log book entries reveal multiple write-ups of previous rudder incidents. On the second test flight on 18 March 1970 when the gust lock was released the rudder moved unexpectedly left, hitting the rudder stop quite hard. The crew was unable to activate the rudder boost and began takeoff with it inoperative. At 50 knots the aircraft required increased right steering and began leaving the centerline of the runway. Increased right rudder force would not move the rudder, prompting the crew to abort takeoff. An inspection of the system by maintenance personnel found no malfunctions. The inoperative rudder boost was corrected by adjusting the gust lock switch. Investigators speculated it was possible the safety wire on the quick disconnect cables may have snagged on adjacent cables or framing. As a preventative measure the quick-disconnect cable latches were taped over to prevent any binding or snagging of the safety wire between them.[5]

On test flight number three of 19 March 1970, while attempting an exercise similar to the day of the crash, the pilot reported when the number one engine was reduced to flight idle at V1 there were no problems with VMCG. The VMCG is the minimum speed while on the ground at which directional control can be maintained, using only aerodynamic controls, with one engine inoperative. The pilot reported he had trouble maintaining directional control after 121 knots.[6]

On fourth test flight of 20 March 1970 the pilot reported while passing through 120 knots with the rudder boost on the amount of rudder available varied between 9 and 21 degrees. Maintenance personnel inspected the system and recorded the corrective action as "bled excessive air from system, operation checks Ok."[7]

On a test flight of 25 March 1970 the pilot reported there was no rudder difficulty with the number one engine at flight idle at V1 and shortly after takeoff with rudder boost off control was normal with 20 to 23 degrees of travel available. However, during the majority of the flight only 9 to 10 degrees of rudder deflection was available. He reported with constant full

View of a typical control cable system disconnects, which runs beneath the floor. They are in the latched and safety wired position. The rudder cable quick disconnect of the MGT-101 was found unlatched with no evidence of any safety-wire (ASI, courtesy Tom Smothermon).

pedal, rudder position was erratic above 10 degrees part of the time.[8]

Conclusion

Post-crash review of test data recovered from the wreckage was analyzed and graphed by test engineers. All possibilities of cable binding, hydraulic failure, rudder boost failure and release of cable quick disconnect were considered. It was concluded at times the pilot's rudder pedals were jammed in the near neutral position while the co-pilot was applying right pedal input. Further review of the data indicates the jammed condition temporarily subsided with both pilot and co-pilot applying right rudder and backing off. The data possibly indicated as the pedals returned to the position where the jam previously occurred they appeared to bind again and remain jammed until impact.[9]

Assuming calibrations of rudder force vs. position done on 07 May were correct, the co-pilot's rudder pedal force was logical and consistent with binding or jamming of the pilot's pedals. Of all the data reviewed, only a binding pilot's rudder pedal fit the data recovered and charted. It was suggested an unlatched cable disconnect would have prevented any possibility of further rudder movement and it would not produce "created" left force on the rudder. There was some rudder movement up to impact possibly discounting an unlatched cable. In all tests conducted and analyzed only the binding of the pilot's pedals match the data recorded.[10]

FAA inspector James Bugbee, who claims to have flown flight test for certification on the Pregnant, Jumbo (Super), Mini and Turbine (MGT-201), stated:

> I was called to assist in the investigation and in my opinion the indications are the rudder cable disconnected. In the Aero Spacelines hangar in Santa Barbara we were able to simulate a sequence of events on a similar guppy. The guppy fuselage nose area is hinged to allow cargo insertion. The cable actuated controls are connected at the fuselage split line by a toggle and hook arrangement each time the nose is opened and closed. Each cable connection is safety wired in place but the rudder connection was found in the wreckage not to be safety wired and disconnected. When the pilot applied full right rudder at engine out test the hook hasp caught on another control cable and disconnected the rudder. As a precaution on all subsequent aircraft the control cables were run in a separate channel in a cable connection box and safety wired. Inspection of the box was put on the flight engineers before start check list.[11]

The final report presumes it crashed because of "rudder power-boost reversal." The exact cause has never been conclusively determined. Aero Spacelines electrician Bob Harridge observed the takeoff and crash. He stated the rudder was in full right position all the way to impact. If the cable had been disconnected it would not have remained in this position. Harridge had breakfast with the crew prior to the flight. He had flown with the crew many times and knew them well. He stated, "They were highly qualified. Van Shepard was my friend, it was tragic but it was pilot error. I was there and saw it all. They took off at maximum takeoff weight, downwind and pulled back Number one engine at maximum weight."[12]

The aircraft had been built on speculation. There had been no serious interest or preorders. The corporate facilities were used as collateral for the engines. The crash of the MGT-101 combined with the construction cost of the SGT-201 dealt a severe setback to Aero Spacelines. NASA budget investigations and cutbacks in the space program had reduced the revenue stream from outsize cargo transport. Unexcelled gambled the future of Aero Spacelines on building the next generation MGT-101 and SGT-201 with the belief there was a market for volumetric commercial transports. Despite expectations there were no customers and the crash of N111AS left Aero Spacelines without a demonstration model.

ASI was already in a very precarious financial position because of Unexcelled stock devaluation. General Motors had secured a lien on the Aero Spacelines facilities as collateral for the engines and demanded payment. All future production was in jeopardy. The company had declared a major loss in 1969 and was without cash. The focus turned to promoting the larger SGT-201. Hopes for producing a fleet of smaller MGT-101 aircraft for the commercial market appeared to be forgotten. Tex Johnston stayed on after the crash but the pink slips went out to many employees because there were no orders, no work and no cash flow. One of those dismissed was Jack Conroy's loyal friend Frank Clark. Ever since Conroy's dust-up with Unexcelled, Clark had been his eyes on the inside at Aero Spacelines. Clark went directly across the field to Conroy's new company, Conroy Aircraft, as his personal technical advisor working with P.G. Smith. Conroy was beginning the Conroy 103 CL-44 guppy project and needed qualified people. It was a year later before Airbus purchased the first SGT-201 N211AS from Aero Spacelines which had been built on speculation.

16

SGT-201 Super Guppy Turbine

The Next Generation of Super Guppy

The Super Guppy advanced version turboprop SGT-201 was announced in the last quarter of 1968, over a year after Jack Conroy's departure from Aero Spacelines. It was actually proposed as the next generation commercial large guppy over two years before. Conroy believed there was a market for a commercial turboprop guppy which could be produced in numbers prior to the flight of his first B-377SG N1038V. His original proposal for the next generation outsize transport was a six-engine Super Guppy Turbine with a 60,000 pound payload. The Aero Spacelines 1966 promotional brochure introducing the piston powered Mini Guppy B-377MG N1037V also featured the proposed six-engine super guppy with Allison 501 turboprops. The copy presented it as the commercial super guppy:

> This version of the Super Guppy is being introduced to fulfill a requirement in the commercial market for an outsize airlift capability greater than the Mini Guppy. Engineering improvements and operational projections for this model providing for greater size, payload and performance have been completed.[1]

Interestingly the early six-engine design was carried forward in promotional material although the design was flawed. The original B-377 Stratocruiser and B-377PG conversion have a main gear stance of 28.5 feet apart. The development of the new 23-foot center-wing for the second conversion B-377SG placed the positing of the main gear stance at 43.5 feet, which was a definite advantage. The relocation provided better handling and weight distribution on the ground. The proposed six-engine guppy design was quite different from the four-engine production SGT-201 with the main gear only 31 feet apart. The engines and main gear were positioned inboard on the new 35-foot center-wing. The nose gear was moved back behind the fuselage break closer to the main gear, decreasing stability on the ground. This design places the main landing gear 12 feet closer together than the B-377SG. All of the guppies built except for the original B-377PG have the 23-foot center-wing which extends 7 feet outside the fuselage on each side with the inboard engines and gear mounted on the wing. This advanced design with 35-foot center-wing would increase the span from the already extended 156 feet to 168 feet. The inboard prop arc is curiously close to the fuselage. The most unique feature of the advanced design SGT is the location of the nose gear. It is positioned behind the break in the fuselage requiring the nose to swing on the two hinges (production SGT-201 had a single hinge). A supplemental strut would fit into the nose similar to the production SGT conversion with a small pair of electric powered wheels to power the nose open.

Conroy was considering the Allison 501 turbines early as an alternative because of the problems encountered in acquiring the P & W T-34 engines for the B-377SG. The objective was to obtain a turbine which would easily adapt to the B-377 wing. The easiest and least costly choice was the YC-97J, which already had the nacelles. It was obvious an alternative turbine would eventually have to be found if additional guppy aircraft were to be built because the T-34 engines were in short supply. It is possible the six-engine design was created because the Allison 501s were lower horsepower than the P & W T-34s. The original B-377SG with lower Stratocruiser fuselage was considerably heavier than the later scratch built lighter deep-V slab side unit of the MGT and SGT. By the time the Super Guppy Turbine 201s went into production the six-engine design was changed to four Allison 501s.

Aero Spacelines projected a production run of ten to twelve next generation SGT-201 transports. The projection was based on an assumed need derived from overoptimistic predictions of forthcoming orders and contracts to transport airframe components for aircraft manufacturers. Boeing was presented with a purchase proposal for the advanced version to transport assemblies between Kansas and Renton/Seattle. After careful consideration, Boeing officials determined it would take a fleet of SGT-201s greater than the existing inventory of B-377/C-97 airframes and parts available to build them. There were no possible alternative aircraft available. After considering the proposal Boeing rejected purchasing the SGT-201s to transport airframe assemblies.

Aero Spacelines continued with an aggressive marketing campaign without obtaining the Boeing order or any strong interest from other U.S. aircraft manufacturers. Construction of the first SGT-201 N211AS (later F-BTGV) was built on speculation as a demo aircraft for the commercial market. Senior Aero Spacelines officials continued to seek a buyer for oversize commercial aircraft, a market Conroy had predicted existed. Conroy's optimism can possibly be attributed to French interest, which began in 1962 with the Anglo-French Aerospatiale/BAC consortium inquires. Sales and market planners were convinced a commercial version could also serve as a transport for future NASA charters. Unfortunately, NASA demand was drastically reduced from the pre-moon mission needs.

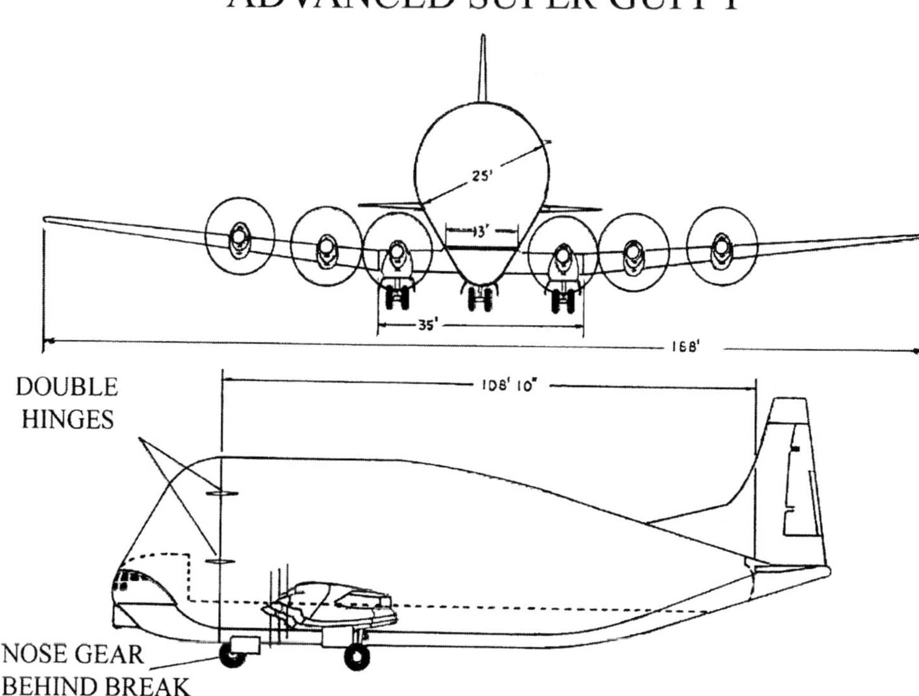

The proposed six-engine design reverts to a narrow positioning of the main gear. Unlike other Super Guppy conversion, the main gear is mounted on the center-wing, which is approximately 35 feet wide. In addition the nose gear has been repositioned behind the fuselage break, which reduces the handing stability on the ground (ASI, courtesy Tom Smothermon).

SGT-201 Construction

The conversion of N211AS began with the disassembly of a low time KC-97K airframe with 5,550 hours. The airframe selected, 52-2625 c/n 16656, was originally a KC-97G which had been converted to K standards. Construction began in the hangar bay next to the MGT-101. A deep-V fuselage was built up in the same manner as the previous B-377MG and MGT-101 with the 13-foot wide-floor using multiple strips of the lower fuselage belly pan from donor airframes. The new slab side fuselage was fabricated to accept the 23-foot extended center-wing. The nose gear, like the MGT-101, was taken from the 707, turned 180 degrees and moved approximately 30 inches rearward. Relocating the nose gear lowered the front of the aircraft leveling the cargo floor. This made it easier for the flight engineer, who doubles as the loadmaster and directs all loading.

To stabilize the aircraft for loading jacks were fitted to the lower fuse-

The first SGT-201 takes shape adjacent to the MGT-101 at Santa Barbara. The sections of belly pans and B-377 section 45 have been trimmed and mounted in the cradles, forming the bottom of the aircraft. They serve as the base for the build-up of the new lighter fuselage. These are the only two ASI aircraft built at the same time (courtesy Fred Weir).

lage, one aft of the wing and two forward. These two forward support jacks were located behind the fuselage break and were lowered before the nose is opened. The nose was secured by sixteen hydraulic locks which were disconnected around the fuselage. The flight controls were disengaged when opening. The nose wheel was cocked to the left. An access panel on the lower right side of the nose at the break opened to allow access to the electric dolly mount. A small strut fitted with two electric driven wheels extended down from the nose. It supported the nose section as it was driven open to 110 degrees. The electric tow dolly was activated by a hand-held control mounted on a long steering arm as the engineer walked along beside the moving nose.

The lower half of the original R-4360 engine nacelle was retained on the C-97 wings the same as the MGT-101. The wings were mated to the new extended center-wing which increased the span to 156 feet, 8 inches. The original main gear was retained along with the lower C-97 bay and doors. The new wider center-wing provided better weight distribution and handing on the ground. The upper nacelles were designed to match the Allison 501 engines and upper cowlings from the Lockheed P-3 Orion. The nacelle filler panels were formed by using the stretch block method developed for the MGT with sheets of 301-302 stainless steel stock. The propeller hub was a four blade 54H60 Hamilton Standard with blades from the C-130 Hercules and P-3 Orion cuffs and spinners. The forward area of the P-3 cowlings was modified to meet civil certification requirements. The C-97K cockpit nose section was pressurized. The tail planes were extended to compensate for the airflow over the larger diameter fuselage. In reality the SGT-201 was a new fuselage with pressurized C-97 cockpit nose grafted to one end and highly modified C-97 /B-377 combination empennage grafted to the other. All four SGT 201 guppies were built using C-97 wings.

De-Icing System

While modifying the wings engineers encountered a problem with the wing heaters. Bleed-air from the Allison turbines could be used to heat the inboard wing but considerable engineering would be required to fabricate a system using bleed-air for the outboard wing. To achieve the desired heating and obtain the most cost savings the decision was made to use the existing wing heater system.

Electric shop foreman Bill Hickey was given the task of retaining and adapting the C-97 avgas fired heaters located in the outboard nacelles. This was not expected to be easy because the new engines were turboprops which burned JP-4 turbine fuel. An initial proposal of dividing one of the fuel tanks in two sections to provide a supply of 115/145 avgas was considered. After careful review this appeared to be a disaster in the making. Ultimately, the decision was made to convert the heaters to burn JP-4 jet fuel.

Adapting the old system to the wing after modifications were made for turbines was probably not the best idea. No prints existed as to how to make this work. Aero Spacelines electrician Fred Weir was assigned the task of reviewing and consolidating old Pan Am B-377 wiring diagrams and producing a new single print which would allow the use of the old wing heaters. Weir completed the drawings but left the company before the units were overhauled, rewired and installed. When he returned three years later he was lauded for the detailed prints he had produced and told the heaters worked the first time they were fired up.[2] The system could have blown up and burned the wings off the airplane. Although Weir created exceptional wiring diagrams, after the aircraft entered service the old heaters had a tendency to malfunction.

The cargo hold of the SGT-201 had a maximum interior width of 25 feet, 1 inch with a height of 25 feet, 6 inches. The overall cargo compartment length is 111 feet, 6 inches with a constant section of 32 feet. The floor had a 125-pound per running inch limit and fitted with a rail system for ease in loading. The rails accommodated three 24-foot pallets of cargo leaving 39.5 feet for additional bulk cargo. The total volume of the cargo compartment was 49,750 cubic feet with a usable volume of 39,000 cubic feet. This was slightly different from the original Super Guppy which had a constant section of 30 feet, 8 inches and cargo compartment length of 94 feet, 6 inches. The SGT-201 had a maximum payload of 52,500 pounds with a range of 564 miles at 25,000 feet. The range was increased to 2,000 miles with a 16,000 pound payload.

Aero Spacelines anticipated worldwide commercial sales of the SGT-201, which would require it to be operated in year-round conditions. Because of planned worldwide service the aircraft had to be capable of operating in cold weather and extreme conditions. In theory, engineers believed icing could build up above the cockpit on the upper frontal fuselage, which could be a problem. A de-icing system was developed and installed on the large upper frontal area of N211AS. Tests were conducted which proved ice would not build up or cling to the surface on the section above the cockpit, rendering the de-icing system worthless.

It was theorized the shape of the front of the aircraft created an air boundary area preventing icing from ever forming. The reasoning behind the air boundary theory was questionable when it was noted the upper frontal area caved in on Super Guppy N1038V during testing. The incident proved the entire front was a wetted area and there was no air boundary area as suggested. The reasons for lack of ice buildup on the frontal fuselage above the cockpit was never conclusively determined and remains a mystery. The development of the de-icing system was a costly miscalculation. The expense could have been avoided if an ice buildup test had been conducted before building the system instead of theorizing it existed.

Airframe Parts Shortages

All guppies including the SGT-201, regardless whether they were based on B-377 or C-97, require the section 45 from the commercial B-377 Stratocruiser. Section 45 has a three dimensional curve which is fairly complex to construct. It was a series of 12 circular ribs with each a smaller diameter than the preceding. They tapered on the bottom with the rear face a smaller diameter than the front face. It is not that it could not be reproduced, it was just not cost-effective. The lower rear tapered section 45 from the C-97 military fuselage was not usable for conversions because it was structurally different and much heavier. It had semi-circle upper ribs which attached to a side beam. The area between the beams had either the clamshell doors or in-flight refueling pod. Obviously, this presented a parts shortage because only a limited amount of surplus B-377 rear sections were available. In fact, only 55 B-377 airframes plus the prototype were ever built. The supply was further reduced because ten crashed while in commercial service, BOAC broke up two at Heathrow, RANSA scrapped eleven at Miami and one was ferried to Quito, Ecuador, fate unknown. An additional five of the Pan Am B-377s escaped being broken up for the guppy program because they were purchased by the Israeli Air Force in 1962. This left a total of only 27 B-377 airframes with section 45 available.

The commercial B-377 Stratocruiser section 45 is much lighter weight, with circular ribs that each become progressively smaller to form the rear taper of the fuselage. This area of the B-377 would have been fitted with floor panels for baggage storage. The cargo compartment door is on the left looking rearward (courtesy Patrick Dennis).

Israeli Connection

Israel opted to recondition the B-377s for military use after the U.S. embargoed the sale of Lockheed C-130s. The five B-377s (N1025V, N1030V, N90946, N90947 and N90948) were flown to Tel Aviv, where they were overhauled by Israeli Aircraft industries. The cargo floor was reinforced and two had the B-377 tail sections replaced with C-97 components with clamshell doors. The disposition of the B-377 sections 45 is unknown and they were most likely scrapped. Two airframes, N1025V (4X-FOF) and N1030V (4X-FOH), were converted to swing tail similar to the CL-44 conversions. The swing tail was secured with 12 latches and opened to 92 degrees. After completion they were re-registered 4X-FPW and 4X-FPV respectively.

A pair of original Stratocruisers remained in service with the Israeli Air Force into the 1970s. The former Pan Am N1030V (4X-FPV) was de-certified in 1972 and Pan Am N1025V (4X-FPW) became the last active B-377 until de-certified in December of 1975. The pair retained the B-377 section 45 after modification to swing tail; however, it is not known if this fuselage section was structurally changed enough to make the rear unusable. Pan Am B-377 N90948 c/n 15964 and another aircraft were converted to VIP seating for the Israeli military and retained the B-377 rear section.

Aero Spacelines acquired 27 Stratocruisers in the 1960s via Lee Mansdorf for airframe and parts inventory used in the guppy program. The company accumulated a large stock of P & W R-4360 engines removed from the Transocean B-377s. The two piston powered guppies went through a lot of engines because they experienced considerable failures and timed out at 600 hours. A total of 18,697 Pratt & Whitney R-4360 engines were built. They were fitted to the

The military C-97 section 45 is considerably heavier, with beams and extra structure along each side of the lower section. This aircraft has the retractable ramps removed. The clamshell doors can be opened in flight for airdrops. This system was used on the HC-97 for air-sea rescue drops (courtesy Patrick Dennis).

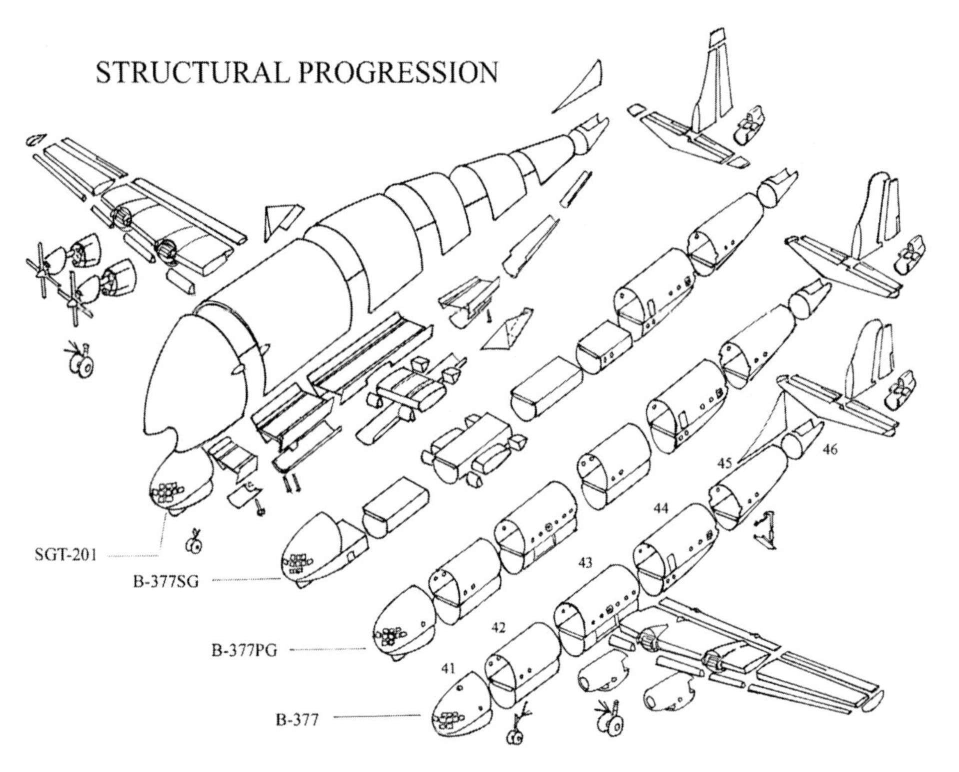

A progressive breakdown of multiple B-377 and C-97 fuselage sections needed to construct the PG, SG and SGT gives an indication of how many additional sections were required. At least 35 more sections in addition to cockpits were needed to build the other five guppies not identified here (author's collection).

Boeing B-50, B-377, C-97; Douglas C-74 and C-124: and multiple other military aircraft. Many of these aircraft were still in service into the 1970s but eventually left a seemingly inexhaustible supply of engines from multiple sources. A number of Aero Spacelines spares were never used because all future guppies were planned with turboprops.

In addition to the B-377s the Israeli military also acquired five KC-97G aircraft without engines which were stored at Davis-Monthan Air Force Base in Tucson. The Israeli military contracted to Aero Spacelines to mount engines from Air Force inventory to make the C-97s airworthy. Jack Conroy's longtime friend and ASI master fabricator Don Stratman was dispatched to Tucson in May 1967 to oversee the work on KC-97s 52-2683 and 52-2612 and assist the Israeli maintenance crew. He was impressed with the efficiency of the team but was shocked one morning in June 1967 when they did not arrive for work. He learned from Air Force personnel the Israelis suddenly left in the middle of the night. They had returned home to fight in the war against the United Arab Republic (Egypt and Syria). The war ended in only six days. In about ten days the Israeli crew returned and resumed work on the KC-97s.[3]

After the aircraft were made airworthy a former Pan Am flight crew was contracted to fly the aircraft to Israel via the U.S. East Coast. While the aircraft were parked overnight the embassy staff filled the aircraft with household appliances purchased for friends and family in Israel. The staff did not have any concept of weight and balance or overloading, and the cargo almost created an international incident. Much of the cargo had to be unloaded and left behind in order to take on the maximum fuel load required for the ferry flight and stay under Max ATOG.[4]

Airframe Salvage Miscalculation

Conroy acquired 27 B-377s Stratocruisers from Lee Mansdorf. If there had been more available it is most likely they would have been transferred to Aero Spacelines. When the Pregnant Guppy was built in 1962 it was not anticipated the section 45 assemblies would be so critical and eventually in short supply. Fuselage sections were cut from many donor B-377s to stretch the airframe of the Pregnant and Super Guppy without consideration for future needs. In addition, a small strip of fuselage belly was cut from multiple B-377 airframes and combined with C-97 belly sections to form the bottom of the second generation slab side fuselage fabricated for the B-377MG, MGT-101 and first two SGT-201s. Only one section 45–46 was needed per aircraft, leaving the unused airframe sections to be broken up for salvage. Don Stratman was instructed to break up multiple Transocean aircraft parked at Oakland without regard to salvaging the section 45 fuselage assemblies. Two of the Transocean aircraft were saved and flown by Clay Lacy from Oakland to Mojave where most of the B-377 inventory owned by Mansdorf and used on the guppies was parked.[5]

Airbus Reconsiders Guppy

EADS (Airbus) representative Felix Kracht, who became head of Airbus production, visited Santa Barbara in April 1969 to review the proposed SGT-201. As early as 1965, Pierre Jorelle, representing French aviation manufacturer Sud, had contacted Conroy with a request for information on guppies for airframe transport for Concorde components. The European aircraft manufacturing consortium showed cautious interest again in guppy transports beginning with Conroy's demonstration of the smaller B-377MG to Aerospatiale representatives in 1966. The specifics of the 1969 Santa Barbara visit were confidential. It is believed to have been part of the planning process to evaluate a means of transport for Concorde airframe assemblies. Kracht was aware of the poor financial condition of Aero Spacelines and questioned whether the company would be able to complete the SGT-201 certification process. The Airbus decision to eventually purchase SGT-201 guppies came very slowly.

It is unclear why Aero Spacelines engineering decided to move another B-377 to Santa Barbara prior to one of the Airbus logistics team visits. It has been speculated the sales team believed an order was forthcoming from the Anglo-French group and an abundance of airframes would impress Aerospatiale BAC representatives and speed up the order process for conversions.

ASI engineering rushed to get a B-377 to Santa Barbara by contracting a group of Air National Guard mechanics to go to Mojave and prepare ex–Pan Am Stratocruiser N90942 c/n 15958 for the ferry flight. There was no plan at the time to use this airframe for construction of a designated conversion. Some vague explanations have been offered as to why it was being moved. Aero Spacelines was using the wings and nose from a C-97G for the first SGT-201s but some parts (lower belly and section 45) could be used if needed.

The mechanics at Mojave were able to get the ex–Pan Am B-377 N90942 systems operating. They were powering up the engines in an attempt to get the aircraft out of the soft earth where the tires had sunk in the ground. Whoever was at the controls throttled up the R-4360s to rock the aircraft underpower in an effort to break the tires lose from the ground. The nose and right main broke free; however, the left main remained in a depression, causing the aircraft to pivot 180 degrees into ex–Transocean (BOAC) B-377 N402Q c/n 15974 which was parked next to it. The number one prop chopped up the side of the N402Q cockpit. The

The remains of Pan Am N90942, which pivoted 180 degrees into Transocean B-377 N402Q at Mojave. The imprint of the wing of N402Q can be seen in the nose of the Pan Am aircraft. Both B-377s were written off. The engines from N90942 were used for spares on the PG. Section 45 is believed to have been used on SGT-201-1, N211AS (AAHS archives).

nose of N90942 was damaged where it impacted the right wing and engine of N402Q.[6]

Fortunately, no one was injured but both aircraft were damaged beyond repair. An ex–Northwest aircraft N74601 c/n 15947 was hastily made airworthy and substituted for the flight to Santa Barbara. This incident severely damaged two more usable airframes of the already limited number of B-377s.[7] The rear fuselage section 45 of Transocean N402Q c/n 15974 was eventually salvaged by Aero Spacelines for use by Union de Transports Aeriens (UTA) to build SGT 201-3, F-GDSG, at Le Bourget-Paris. It is believed but not conclusively verified that section 45 of Pan Am N90942 was used on the first SGT-201-1, N211AS.

Top: **Chase plane observes SGT test flight from Santa Barbara. Aircraft is still in bare metal with only the area around the nose painted. The large airfoils behind the wing, which were necessary to direct airflow to the tail, are clearly visible.**

Bottom: **Unexcelled CEO and Twin Fair president Harold Egan in front of N211AS. Tufts have been placed on the right side of the fuselage and tail to determine airflow characteristics. A technician is placing tufts on the left side along the hinges before a series of test flights. ASI encountered serious delays in getting it certified (both photographs, ASI, courtesy Tom Smothermon).**

The new generation SGT 201-1 flew from Santa Barbara on 24 August 1970 with U.S. registration N211AS. The crew reported the guppy handled very well with the Allison 501 engines. The aircraft easily rotated (Vr) at 110 kts (126.5 mph) increasing to 150 kts (173 mph) and continued to climb out at 1,500 ft/min at 185 kts (212 mph). The Allison 501s are capable of carrying the aircraft along at much higher speeds but the maximum cruise speed is limited by the size of the fuselage. In addition, the maximum cruise speed (Vmo) is reduced by 4 kts for every 1,000 feet flown above 14,000 feet, limiting the speed to 185 knots at 22,000 feet.

Aero Spacelines began an extensive promotional program for the SGT-201 with expectations that multiple orders for the second generation super guppy would be generated. Unexcelled, Twin Fair and Aero Spacelines corporate officers were very optimistic contracts would soon be negotiated with major aircraft manufacturers and suppliers for airframe component transport. Unexcelled CEO and Twin Fair president Harold Egan as well as ASI President Kirk Irwin along with other corporate officers and financial backers were photographed in front of the new guppy.

The improved SGT was placed in an extensive test program. Tufts were placed along the fuselage and vertical fin to determine airflow characteristics. ASI was spending a considerable amount of assets which it did not have to obtain certification, which was progressing slowly. Airbus had shown a strong interest but was concerned if ASI could get it certified. The large upper forward fuselage was constructed at a slightly different angle than the B-377SG. The aircraft was experiencing stall problems in certain configurations because the horizontal control surface was not getting sufficient airflow. After a year of trials and testing, it was finally certified by the FAA as normal commercial aircraft on 26 August 1971.

The first SGT-201 had been constructed using lower time military C-97G cockpit nose, wings and empennage combined with the deep-V fuselage. Although the new fuselage contained only strips of belly pan and section 45 from the commercial B-377, it was certified as a civilian aircraft. The only B-377 piece was the section 45. It was a new fuselage and center-wing with C-97 parts grafted to each end.

By comparison, obtaining a civilian registration for the original Super Guppy N1038V with P&W T-34 turbine engines had been as difficult and even then it never received a type certification. It was built from an experimental military YC-97J which had been modified in 1954 from a C-97. Jack Conroy built the first Super Guppy from considerably more civilian B-377 parts combined with anYC-97J to make it easier to get a civilian registration. It was considered a modified B-377. Even then the FAA would only certify it with a restricted classification as a "public aircraft." Then the FAA filed a suit against Aero Spacelines claiming it was a civil aircraft and NASA did not have the authority to contract as such because it was non-military. The FAA certification process was different on the SGT-201; however, it was still quite rigid.

The first SGT-201 N211AS was planned on speculation that the tentative transport contracts with American aircraft manufacturers would materialize. The second generation turbine guppy was far superior to the original Pratt & Whitney turboprop Super Guppy. The expanded pressurized cockpit had seating for four passengers. The improved air conditioning system worked below 5,000 feet and on the ground plus the swing nose had hydraulic latches. Despite improvements, repeated efforts to interest American aircraft manufacturers failed. In addition, the crash of the MGT-101 at Edwards in May 1970 dealt the company a devastating financial blow. The construction cost of the SGT-201 combined with the crash of the N111AS placed Aero Spacelines near bankruptcy.

The aerospace industry was moving at a speed far greater than the Aero Spacelines design and production. The first guppies demonstrated an ability to move airframe components but it was never taken seriously because there were only three aircraft operational and two of them were in exclusive NASA service. The B-377SG temporarily transported DC-10 fuselages for Douglas but the B-377MG was the only one which could fly commercial charters. Aero Spacelines was betting on aircraft orders coming from American suppliers of the Boeing 2707 SST production. In retrospect one can see the Boeing 2707 was never feasible. It was too expensive and too large of a supersonic transport with the technology available at the time. The hype of future passenger travel being predominately supersonic was promoted and believed. Letters from Boeing and multiple subcontractors encouraged Aero Spacelines by stating a next generation guppy would be needed. Congress cancelled funding for the SST on 20 May 1971, which ended the program.

EADS (Airbus) officials conducted preliminary contract talks with ASI in 1969. They proposed an offer to purchase a SGT-201 relative to ASI being able to finish it and receive certification. Airbus monitored the certification progress for a year with concerns the FAA would not certify it. After a year it was finally certified, prompting EADS to act on their proposals. Airbus Industries offered to purchase the unsold SGT-201 at a price of $32 million.[8] Unexcelled jumped at the opportunity to sell the aircraft and recover some of the expense of building and certifying the aircraft. The sale would help bring some financial stability to ASI.

The agreement was for the single SGT-201 with options for additional aircraft. An interest had been expressed in purchasing the original Super Guppy B-377PG N1038V if it was modified and upgraded to the same configuration of the new SGT-201 (pressurized cockpit, Allison 501 engines, 707 nose gear, etc.) in order to be certified as a commercial transport. ASI officials explained it was not possible. Jack Conroy had long since left Aero Spacelines. The sales pitch he made at Toulouse in 1967 for a commercial Super Guppy was finally bearing fruit but not with American aircraft builders.

As a condition of purchase, Airbus required Aero Spacelines to agree to a backup clause providing another aircraft in the event the SGT-201 was out of service. The contract stated the SGT-201 could be out of service for routine maintenance but any prolonged grounding would require Aero Spacelines to provide a backup plane. The agreement presented a problem from the outset. There were no backup aircraft available. The B-377SG N1038V was not available because it was built specifically for NASA contracts and could not be certified as a commercial aircraft. It would require major modifications to be a substitute aircraft. Special exemptions had been issued for it to transport DC-10 fuselage sections in America. No such exemptions could be granted for work in France. The Mini Guppy N1037V was certified for commercial service and could fill in; unfortunately, it was not of a diameter which was suitable for all the work Airbus required.

EADS—Airbus SGT-201

The first SGT-201-1 N211AS was certified by French authorities while still in California. The registration F-BTGV was assigned on 14 September 1971. It arrived in Paris on 28

September for trials and training. Flight operations were contracted to UTA subsidiary Aeromaritime based at Le Bourget Airport in Paris. Aeromaritime was the largest independent airline in France at the time. It was created in 1966 to compete in the cargo charter market. F-BTGV made the first paying flight on 02 November and was placed into regular service as the primary transport for A300 wings. Almost immediately industry jokesters stated every Airbus A300 first flew on Boeing wings.

After a short time in service it became obvious F-BTGV (formerly N211AS) could not handle the work load. To maintain Airbus production schedules it was required to fly a demanding schedule transporting A300 sub-assemblies. The single SGT-201 was initially purchased to fill the volumetric transport need to move A300 components from locations around Europe to Toulouse. To build a single A300 aircraft F-BTGV was required to make eight flights, first picking up wing boxes, two at a time, from British Aerospace at Manchester and moving them to Bremen. They were then fitted with flaps and slats flown in from Amsterdam. Each completed A300 wing was then flown from Bremen to Toulouse for final assembly. While this method was far easier and more efficient than transporting the parts and assemblies by surface, it was putting considerable hours on the single SGT-201 aircraft.[9]

The transport flights were increased to include Concorde components along with A300 wing assemblies. The SGT-201 was utilized to the point of maintenance problems having an effect on reliability. EADS officials were concerned about the logistics of the supply system for the future production of the A320, A330 and A340. The guppy transport system would not be possible with only one aircraft. The new models would require subassemblies from a supply network all over Europe. The amount of components needed to maintain production created a serious need for additional volumetric lift.

Aero Spacelines Builds Second SGT-201

After the delivery of the first SGT-201-1 to France, Airbus was in continuous contact with Aero Spacelines regarding reliability issues. ASI concluded the aircraft was overutilized. Airbus was not in agreement, pointing out the hours flown relative to reliability was not overutilization. ASI made the decision to construct a second SGT-201, aircraft N212AS (later F-BPPA), on speculation. If Airbus did exercise an option the aircraft would be in progress, reducing delivery time. It was designed with long-range capability as a demonstration commercial plane for the anticipated U.S. market. The construction of the deep-V fuselage used strips of belly pan from multiple unidentified B-377 and C-97s which were combined with parts from two KC-97s. The nose, empennage section 46–47 and horizontal stabilizer were taken from KC-97G (C-97K) 53-0163 c/n 16945 which had only 4,521 hours. The wings and vertical stabilizer were taken from KC-97G 53-0121 c/n 16903 with 4,230 hours. The required section 45 came from a still unidentified B-377.

The first SGT-201, N211AS, was purchased by Airbus to transport A300 components. It arrived at Le Bourget airport in the fall of 1971. After certification it entered service on 02 November as F-BTGV. Airbus soon realized one guppy could not fulfill the transport schedule of both A300 and Concorde subassemblies (courtesy Richard Vandervord).

The second SGT-201 first flew on 24 August 1972. The Aero Spacelines assumption that Airbus would exercise an option for a second aircraft did not materialize. Aero Spacelines' bankruptcy was almost a certainty. In an attempt to regain financial stability, Unexcelled management reviewed the possibilities of selling the Pregnant, Super and Mini Guppy along with the latest version SGT 201-2 N212AS. The Pregnant and Super Guppy were in less demand with NASA. Funding for the space program had been drastically reduced. There was a move in Congress for the Air Force to transport all NASA cargo. Both the guppies had served NASA well but were in need of major maintenance.

Planners at Airbus reviewed Aero Spacelines financial problems from two perspectives. A backup aircraft was now available. However, if ASI went out of business the new owner of the SGT-201 might not honor the alternate aircraft agreement in the backup clause of the original contract. The second SGT-201 (N212AS), which would be the last built at Santa Barbara, was available. Airbus logistic planners realized the situation could become critical if the existing guppies were sold off.

The remains of Transocean N402Q, which was damaged in a collision with Pan Am B-377 N90942. The rear section 45 was cut away in 1973 and sold to Airbus as a spare with SGT N212AS. The rear section was eventually used in 1981 by UTA to construct the third SGT-201 F-GDSG in France. The remains of N402Q were eventually scrapped at Mojave (AAHS archives).

After a year of ad hoc charter work and nonproductive demonstration flights with N212AS, Unexcelled was ready to divest itself from the guppy business. In August 1973, a year after first flying, Airbus offered to purchase the second SGT 201-2 (N212AS). Aero Spacelines had been seeking a U.S. buyer but eagerly accepted the French offer. Fearing the financial problems would lead to the demise of the company, Airbus negotiated the guppy purchase agreement to include a spare Allison 501 QEC engine and a B-377 section 45. It was reasoned if the section 45 was damaged on either SGT-201 due to a tail strike or other incident the aircraft would be grounded and possibly not repairable. The second SGT-201 (N212AS) departed Santa Barbara for France with the miscellaneous parts, the spare engine plus 950-pounds of technical documents and blueprints. The agreed extra section 45 was not on board because it had not yet been removed from Transocean B-377 N402Q which was derelict at Mojave. It was later shipped by sea.

After arriving in France, N212AS was certified and reregistered as F-BPPA. It entered component transport service for Airbus in September 1973. Airbus logistics managers predicted the two SGT-201s would soon be overworked and additional guppies would be required. It has been speculated the extra B-377 section 45 was not bought as a spare but for a plan to produce additional guppies in the event Aero Spacelines became insolvent. This theory appears to have been verified when Airbus came back to Aero Spacelines with another offer.

Airbus proposed for a second time purchasing the original Super Guppy B-377SG N1038V as a spare backup. The purchase offer came with a long list of requirements for modifications and upgrades to bring it up to the same standards of the two SGT-201s. The cockpit had to be pressurized and the P&W YT34-P-5 5,229 hp turbine engines replaced with the 4,680 hp Allison 501D22C turboprop engines to match the two SGT-201 guppies already in Airbus service.

The entire proposal was pure folly. Airbus officials were previously informed the Pratt & Whitney T-34 engines could not be swapped out for the Allison 501s. The B-377SG was built using the heavier C-97 figure eight lower fuselage. The SGT-201 had a scratch built lighter slab side fuselage which reduced SGT-201 overall weight and drag. The double lobe fuselage of the Super Guppy weighed approximately 15,000 pounds more. The figure eight fuselage also had considerably more aerodynamic drag requiring the more powerful engines to pull it through the air. It was borderline underpowered even with the higher horsepower T-34 engines. Aero Spacelines engineers pointed out it would not be possible to re-engine the Super Guppy. Conroy had only been able to acquire the engines with the help of NASA. The T-34 test engines were on loan from the Air Force and could not be sold with the aircraft, leaving no alternative. Additionally, there were a limited number of spare props available.

The newer SGT-201 model also had the 13-foot-wide floor which Airbus required. The fuselage profile was an upside down teardrop, making it easier to load cargo. Furthermore, the 8-foot wide cargo floor of the original Super Guppy could not be modified to the 13-foot floor of the new SGT-201 airframes. It would require the building of an all new slab side fuselage with new 23-foot center-wing and mounting the upper half of the Super Guppy on top, which was cost prohibitive. Many shortcuts were taken in building the Super Guppy, resulting in FAA certification restrictions.

In reality, Airbus was asking for a new guppy using the cockpit wings and empennage of the Super Guppy. After it was pointed out rebuilding the guppy was not possible, the proposal was withdrawn. Aero Spacelines management still believed Airbus would place an order for a third SGT-201 guppy after it was explained the original Super Guppy could not be modified. However, Airbus did not demonstrate any interest and there were no further negotiations. Without orders Aero Spacelines was not in the financial position to produce another aircraft on speculation.

NASA projects were drastically cut back reducing the call for guppy charters. There was still an occasional need by NASA for a volumetric transport but not enough to warrant the cost of maintaining the Super Guppy. Unexcelled's international counterpart Unexcelled N.V. Corporation and subsidiary Aero Spacelines were experiencing litigation on multiple fronts combined with financial difficulties. In an effort to reduce expenses, Aero Spacelines announced on 31 December 1973 the company would cease flight operations at the end of current NASA contracts. The company would divest itself of guppy flight operations to concentrate on repair and modification work.

The Pregnant and Mini Guppy were sold to American Jet in 1974. The remaining Super Guppy continued contract flight operations for five more years with ASI. Military aircraft were being developed for transport of outsize hardware. Without Aero Spacelines for outsize cargo, NASA was confronted with the possibility there would not be another outsize transport. The Air Force was bringing online new cargo planes. There were still logistical questions of the military aircraft being able to handle NASA's outsize needs. The Air Force operational costs were as much as ten times greater than previous Aero Spacelines cost per flight.

It appeared Aero Spacelines was out of the guppy manufacturing business entirely. Tex Johnston stepped down as chairman in 1975 to pursue other interest. He took a position at Stanley Aviation developing personnel escape systems (ejection seats). He was replaced at Aero Spacelines by W. C. "Bill" Lawrence, who had been president of Twin Fair department stores and grocery chain, which was owned by Unexcelled. Twin Fair stock and assets were used for the stock swap takeover of Aero Spacelines years earlier. Lawrence also became president of the Total In-flight Simulator System (TIFS) sub-company which Johnston brought to Unexcelled. Lawrence remained as president and COO of Aero Spacelines until 1979. He was replaced by Kirk Irwin, who began as an engineer with Aero Spacelines before the development of MGT-101.

Unexcelled began seeking a U.S. buyer in 1972 who could maintain the NASA contract after informing EADS (Airbus) the B-377SG could not be converted to SGT-201 standards. They were still seeking a buyer in 1974 after the Pregnant and Mini Guppy were sold to American Jet. The situation was quite unique because the Super Guppy was still the only aircraft in America in 1978 capable of transporting outsize cargo. There were no cargo operators interested in purchasing the B-377SG because it was restricted to only transporting cargo for NASA and the DOD. If it had been certified as a commercial transport there may have been interested cargo carriers. The cost of modifying it to meet FAA certification for commercial cargo was cost-prohibitive, making it unsellable.

NASA transportation officials concluded once Aero Spacelines disposed of the Super Guppy, the space agency could have a logistics problem. The transport situation became critical at NASA by the end of 1978. The Air Force C-5 component transport which was intended to replace the B-377SG was costing up to $1,000,000 per flight. With the agency confronted with unsustainable transport cost and Unexcelled divesting itself of all guppy flight operations, NASA was faced with no choice but to consider purchasing the original Super Guppy (N1038V) for in-house operation.

Super Guppy Acquired by NASA

In February 1979, only months after Airbus came back to ASI to negotiate a contract for additional SGT-201 aircraft to transport subassemblies in France, NASA announced a purchase agreement for the original B-377 Super Guppy N1038V. The decision to acquire the aircraft was necessary to maintain control of transport cost. Aero Spacelines had previously announced it would no longer be in the business of guppy air operations. Without an outsize transport the space program would be totally dependent on the military for cargo logistics. NASA officials acquired the B-377 Super Guppy at an agreed amount of $2,940,000, which included spares, support equipment and training.[10]

The actual purchase of the Super Guppy was not from Aero Spacelines as would be expected but from Unexcelled sub-company Twin Fair Corporation, a retail department store chain. It seems bizarre NASA would purchase a specialty aircraft from a grocery/department store chain which belonged to a chemical company. A check dated 29 March 1979 in the amount of $2,583,537 paid to the order of Twin Fair Incorporated was drawn on the Nassau Bay National Bank of Houston, Texas. The check noted "Escrow Agreement between Twin Fair Inc., National Aeronautics and Space Administration and Nassau Bay National Bank of Clear Lake." The Super Guppy was delivered to NASA and registered as

N940NS. The Aero Spacelines division of Unexcelled Corporation continued to operate as an FAA approved airframe overhaul station.

There could be no other buyers for the B-377SG because of the operation restrictions. As mentioned an offer by Airbus to purchase was rejected by Aero Spacelines for multiple reasons. It would have been impossible for Airbus to operate it without major overhaul and upgrades. Technically it was restricted to government service only, even though some commercial charters were operated for aircraft manufacturers. It could never be certified as a commercial aircraft for Airbus. Selling it to NASA was the only alternative.

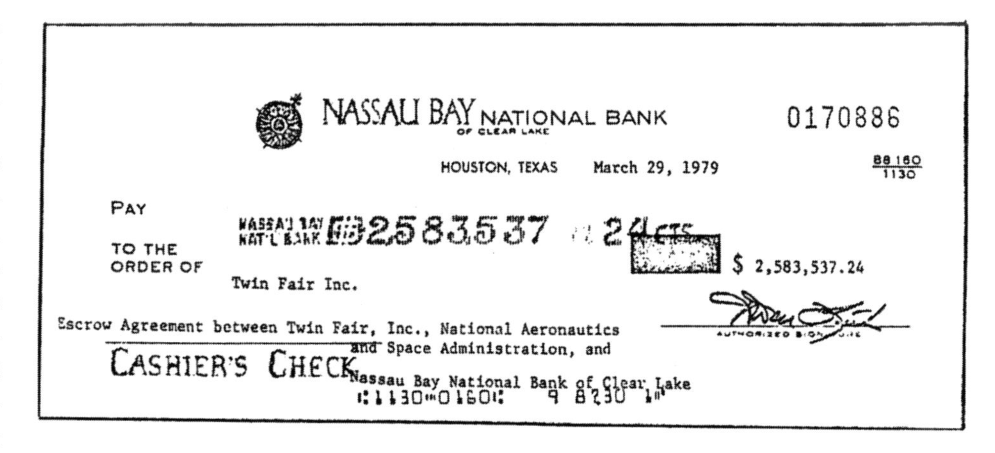
NASSAU BAY NATIONAL BANK
OF CLEAR LAKE
0170886
HOUSTON, TEXAS March 29, 1979
88-160/1130
PAY
NASSAU BAY NAT'L BANK $2583537 24CTS
TO THE ORDER OF
Twin Fair Inc.
$ 2,583,537.24
Escrow Agreement between Twin Fair, Inc., National Aeronautics and Space Administration, and Nassau Bay National Bank of Clear Lake
AUTHORIZED SIGNATURE
CASHIER'S CHECK

NASA announced an agreement in February 1979 to acquire Super Guppy N1038V at a cost of $2,940,000. The transaction is unique when considering NASA purchased an aircraft from the Twin Fair department and grocery store chain. A check for the balance less deposit was issued on 29 March 1979 (NASA, courtesy Charles Gillespie).

The NASA purchase agreement was signed on 29 March 1979. The space agency assigned Frank Marlow as project pilot and Charles J. Gillespie as project flight engineer. Prior to the transfer the NASA pilots flew with Aero Spacelines flight crews as observers. Multiple familiarization flights were conducted for NASA to become familiar with cockpit operations and duties. The B-377SG was flown to Hayes International at Birmingham, Alabama, for a thorough structural inspection as a condition of purchase. Any defects found were repaired prior to delivery. Once N1038V was transferred to NASA, Aero Spacelines shut down the flight operations division.

The aircraft ferried from Birmingham to Ellington Field, Texas, on 13 July 1979, where it was received by NASA. The crew consisted of Aero Spacelines pilots Paul Heyn, Dick Peters, flight engineer Bob d'Agostini and maintenance superintendent Tony Sacchi assisted by NASA pilot Frank Marlow and flight engineer Charles Gillespie. The Aero Spacelines crew conducted multiple training flights with the NASA crew consisting of Frank Marlow, Al Manson and Charles Gillespie. The three were the only NASA flight crew trained by Aero Spacelines pilots. They trained all subsequent NASA crews.[11]

Aero Spacelines flight engineer Bob d'Agostini was a guppy veteran from the beginning. He was the flight engineer with Jack Conroy and Clay Lacy on the first Pregnant Guppy flight at Van Nuys in September 1962. The trio had flown C-97 Stratocruisers together while serving in the 146th Transport Wing of the California Air Guard. D'Agostini was also the engineer on the B-377PG demonstration flight with Wernher von Braun at Marshall Space Flight Center and he was the engineer on the last Aero Spacelines guppy flight.

The B-377SG was assigned to Johnson Space Center (JSC) with an annual NASA operating budget of $300,000. Aircraft use was shared with the Department of Defense. All military and other government agencies were then required to reimburse NASA for use of the aircraft. NASA contract NAS9-18075 was negotiated to Northrop Worldwide Aircraft Services for maintenance and services. Northrop assigned Milo Kohl as the onsite B-377SG crew chief.

After Aero Spacelines sold the remaining B-377PG guppy transport to American Jet, several aviation industry news magazines reported in October 1979 Aero Spacelines was in liquidation. The story was quickly countered by Kirk Irwin, who stated the company was quite strong financially and specializing in aircraft structural modifications, painting and avionics installation. He further stated Aero Spacelines was bidding on contracts to provide painting and interiors for all BAE 46 and Gulfstream 31 aircraft sold in the United States. Irwin noted Aero Spacelines had contracts to produce airframe components for 727 and 747 aircraft.[12]

Irwin stated the company would be reviving guppy manufacturing because Aero Spacelines had been the only builder of outsize volumetric transports. He also speculated Super Guppies N211AS and N212AS, which were built in 1970 and 1972, were likely the last 25-foot diameter outsize transports the company would ever build. He stated there was renewed interest for the smaller version MGT-101, which was being planned for production. His prediction was partially correct because the two SGT-201s were the last two 25-foot diameter transports built by Aero Spacelines in the United States.

Market evaluations were conducted for an 18-foot diameter commercial guppy similar to the MGT-101. Irwin continued to promote the idea of producing the smaller MGTs for the commercial market. He further stated the smaller turbine guppies would be based on Boeing KC-97L airframe components. The Air Force had retired 50 tankers which were in storage at Tucson. They would provide more than enough parts to fill the demand he believed existed.[13]

Airbus Requests More SGT-201 Aircraft

In 1977 Airbus assigned Erin Braun to determine if Aero Spacelines was still in business and if so, did they still produce the SGT-201 aircraft. When NASA made the inquiry to Aero Spacelines in 1978 regarding the possibility of purchasing the B-377SG, Airbus unexpectedly showed renewed interest in an additional SGT-201. The Aero Spacelines sales department was quite surprised when Airbus came back to Unexcelled with a request for two more SGT-201 conversions. Felix Kracht invited Kirk Irwin to Toulouse to review the possibility of ASI tooling up to produce additional SGT-201 aircraft. Airbus had experienced significantly more commercial orders which increased the delivery schedule of airframe subassemblies. The operation of the two guppies had reached an unsustainable utilization schedule, creating a need for more volumetric lift. Airbus was advised guppy production at Aero Spacelines was shut down and the company was no longer in the position to build additional conversions. Many of the craftsmen, engineers and technicians had long ago been furloughed and moved on to other companies.

Kirk Irwin's optimism and confidence in the financial stability of Aero Spacelines was stoked by the fact the Super Guppy had been sold to NASA. Airbus was pressing for a contract to purchase additional SGT-201 aircraft with provisions for up to six more. However, Unexcelled initially rejected the Airbus request, reiterating it could not tool up for the production of one or two aircraft.

After lengthy consideration, Airbus Industries made an alternate proposal. The European aircraft consortium offered to pay Aero Spacelines a $2.6 million licensing fee to build two SGT-201 aircraft in France with a free option for a third (fifth SGT-201). The offer included a negotiation clause for the possibility of additional conversions six through eight at a later date. The contract stipulated construction would be subcontracted to UTA (Union de Transport Aeriens) Industries at Le Bourget Airport in Paris. In return, UTA would sub-contract back to Aero Spacelines at Santa Barbara about half the component fabrication work which would then be shipped to France for assembly. ASI had previously held itself out to the aerospace industry as an airframe component manufacturing business. In an effort to save the failing company, an agreement was signed on 17 December 1979 with the provision UTA would purchase the five remaining KC-97G and K models aircraft in Air Force storage. ASI was instructed to salvage the KC-97s for conversion components.[14]

Aero Spacelines fabricator Tom Roberts was dispatched to Tucson along with shop foreman Roland Hipley. They were instructed to put a crew together to salvage the

The incredible size of the SGT is obvious as fabrication of airframe sections take shape at Santa Barbara after Airbus negotiated a contract to build additional SGT-201s in France. The new center-wing has been mated to the midfuselage section primary ribs. As large as this subassembly was, it was transported to Le Bourget, Paris, by the second SGT-201 F-BPPA (ASI, courtesy Tom Smothermon).

needed C-97 sections at Davis Monthan for shipment back to Santa Barbara. Hipley had been with ASI since 1962. He had worked on the construction of every guppy since he cut the old B-377 fuselage from inside the Pregnant Guppy. Four of the C-97s were broken up for parts to be shipped back to Santa Barbara. Two cockpit noses, wings and empennage which were not immediately needed were stored as spares for additional conversions expected from options in the contract. A fifth, C-97G 53-0280 c/n 17062, was scheduled to be made airworthy and flown to France for storage.

The Santa Barbara operation began setting up to fabricate components for the two SGT-201 guppies which would be built by UTA in Toulouse, France. The airframe subassemblies for SG-201-3 (F-GDSG) were to be completed on a rapid schedule for shipment by air to Le Bourget. The second set of components was scheduled for shipment by sea at a later date.

Aero Spacelines built up the pressurized C-97G cockpit nose unit by attaching a large structural V to the modified rear bulkhead. The V framework mates to the newly built fuselage with 13-foot-wide floor which was being fabricated in France. A cone shaped framework was built up above the flight deck eyebrow windows on the top of the C-97 nose section. The frame provided an attachment point for the stringers and skin of the high frontal fuselage above the cockpit. Airbus set up the flight deck with Sud Aviation Caravelle flight instrument gauges along with old style American engine gauges for the Allison 501s turbines.[15]

Aero Spacelines did not have to search for the required B-377 section 45 for the first French built guppy

Top: **The nose of KC-97G 52-0828, last flown by the Mississippi Air National Guard, was removed at Davis-Monthan for shipment back to Santa Barbara. It is one of four KC-97Gs acquired by Aero Spacelines under contract with Airbus for airframe components. This nose was used on F-GEAI, which was the last SGT-201 built. It was eventually acquired by NASA and was still in service in 2017 (courtesy Tom Roberts).**

Bottom: **The overhauled and modified KC-97 nose being mounted on a shipping frame. A ladder would be added to the lower rear bulkhead for crew entry up through the hole and to a hatch in the cockpit floor. The cockpit door to the cargo deck is in the center upper half. The large V frames gives some idea of how much the original fuselage has been expanded (courtesy Fred Weir).**

SGT-201-3(F). The section 45 from damaged Transocean B-377 N402Q c/n 15974 was part of the spares inventory Airbus acquired with the purchase of the second SGT-201-2 N212AS in 1973. The B-377 rear belly section was already on hand at Toulouse. Most of the parts, nose, wings and empennage came from KC-97F 53-0215 c/n 16997, which had 9,286.5 hours. The section 44 came from another C-97G 52-2766 c/n 16797.

A new extended center-wing was fabricated at Santa Barbara with wide V upper framework which mated to the contour of the new fuselage. Seven lower belly pan ribs were fabricated and attached to the bottom of the center-wing. Interestingly ASI was able to fabricate these ribs to attach to the bottom of the center-wing and yet airframe section 45 for the rear of the aircraft was considered too costly and impractical to duplicate because of the three dimensional curves. The sub-assemblies along with C-97G wings and main landing gear were prepared for shipment to UTA. Boeing 707 nose gear components were acquired and modified to fit the C-97 nose in the same manner as first two Santa Barbara SGT-201s.

Aero Spacelines also purchased a C-97G 52-2766 c/n 16797 (N22766) which was owned by the Foundation for Airborne Relief (FAR). Tom Roberts was dispatched to Long Beach to break up the C-97 and transport the usable parts to Santa Barbara for overhaul and modification before shipping to UTA in Paris. The wings and most of the belly were found to be severely corroded and unusable. The section 44 of 52-2766 along with seven sections of belly pan from C-97 c/n 53-0215 c/n 16997 were salvaged. Aero Spacelines invoiced the seven sections of C-97 belly pan from C-97G 53-0215 at $30.17 each and the section 44 from C-97G 52-2766 $2109.80. They were shipped to UTA on 16 November 1979, arriving at Toulouse on 03 December.[16]

UTA Industries subcontracted much of the fabricating and modifying other components at Le Bourget while awaiting delivery of the QEC engines, props, modified cockpit nose, access doors, center-wing, wings and tail components from California. Fabrication of the lower fuselage section forward and aft of the center-wing was contracted to REVIMA, a subsidiary of UTA located at Caudebec en Caux, which is about 124 miles (200 km) from Le Bourget. The company was primarily an overhaul contractor for landing gear components.

No jigs were ever produced for the rib and former spacing, prompting REVIMA to use the same Aero Spacelines method of connecting sections of belly pans to form the bottom. Fabricators virtually started from scratch on building the deep-V fuselage section. The Santa Barbara Built B-377MG, MGT-101 and two SGT-201s were built using six-foot-wide strips of multiple B-377/C-97 fuselage belly sections to form the lower pan rib spacing.

The construction method of the lower portion of the deep-V fuselage differed from the two previous Santa Barbara SGT-201s. In those aircraft the ribs were connected to the matching part in the salvaged belly pan and the slab-side ribs extended up to slightly above floor level where the complete upper arch fuselage top ribs were attached. The REVIMA built lower fuselage has the ribs extending up approximately seven feet above the cargo floor level for attachment of the arch ribs which form the top. The lower portion was skinned up to the top rib attachment point. The large wedge shaped wing root airflow control fairings were fabricated and attached to sides of the aft lower fuselage half at REVIMA. Once completed the lower fuselage sections fabricated by REVIMA were moved by truck to UTA at Le Bourget. The newly fabricated lower fuselage modules were

The rear section of F-GDSG was fabricated by UTA subsidiary REVIMA at Caudebec-en-Caux and transported to Paris for final assembly. The French-built fuselage sections were built in several sections with ribs up to seven feet above the floor then tied together. Aero Spacelines guppies were built in place with all the belly pan sections spliced together before the sides were built up with frame rings attached at floor level. The darker colored C-97 belly pans are visible at the bottom of the assembly (courtesy Tom Roberts).

grafted to the center-wing with primary ribs which had arrived from Santa Barbara. The extended forward former matches up with the V frame on the aft cockpit bulkhead which was fabricated by Aero Spacelines.

The second SGT-201-2 N212AS (F-BPPA) which was purchased by Airbus in 1973 was originally built as a demonstrator for the commercial market. It is the only SGT-201 fitted with an inertial navigation system giving it worldwide capability. Airbus made additional modifications for the long-haul flights to retrieve the SGT-201 assemblies from California. Three trips were made by F-BPPA between Paris and Santa Barbara to pick up the components for SGT 201-3 (F-GDSG).

The first round trip was 27 October to 03 November 1980 when it uplifted the nose section and new extended center-wing box. The second flight was around 01 August 1981 staging at Le Bourget, Keflavik, Gander, Indianapolis and Santa Barbara to pick up the left wing with nacelles, flaps, horizontal stabilizer and two QEC 501 engines with props. The engines were loaded first toward the rear of the aircraft because they require less ceiling clearance. It left Le Bourget again on 15 August for the third trip to Santa Barbara via the same routing. The other two QEC engines with props, right wing and vertical stabilizer were loaded departing on 17 August and arriving back at Le Bourget on 22 August. After all the components were delivered the aircraft was completed at Le Bourget. The first French built SGT-201-3 first flew with registration F-WDSG on 01 July 1982. After testing and certification it was reregistered as F-GDSG before entering service. To indicate the construction location the first UTA built guppy was designated as Model B-377 SGT-201-3(-F). The new designation was set up because EADS (Airbus) seriously planned the production of additional SGTs as defined in the 1982 licensing agreement.

Frank Robinson and Aero Spacelines president Kirk Irwin monitor the loading of the center wing section and cockpit nose aboard F-BPPA on 30 October 1970 for transport to Paris. The nose has cone-shaped framing above the windows for attachment of the expanded top. The flight arrived back at Le Bourget on 03 November. Construction commenced on F-GDSG once the fuselage sections were delivered by truck from REVIMA (ASI, courtesy Tom Smothermon).

At the time the SGT-201-3 was being built it was assumed the supply of B-377 airframes with section 45 intact was exhausted. Airbus had wisely acquired the rear section from B-377 N402Q as a spare when N212AS was purchased in 1973. That section 45 was used on SGT-201-3(F).

The first B-377PG was built at a cost of approximately $1,200,000 by Jack Conroy in 1962. Twenty years later the cost of building the SGT-201 was set at $32,000,000. The presumed shortage of additional rear sections was communicated to Airbus when parts were being gathered and modified for SGT-201-4. The expense of producing one from scratch was still considered prohibitive.

REVIMA was already producing the entire constant section of the lower fuselage using strips of C-97 belly pans. The question is then raised as to why a tapered section 45, which is only about 16 feet long, could not be reproduced. It is true that modifying the lower half of a C-97 section 45 to B-377 standards would require considerable engineering and fabrication. It was cost prohibitive for Aero Spacelines when considering Jack Conroy built the first Super Guppy for approximately $1,900,000. However, when compared to the $32,000,000 being spent by Airbus to essentially take existing parts from multiple airframes and do the same thing how much expense was too prohibitive?

When the components for the last SGT-201 were being gathered, the question arose as to locating additional rear section 45 units if Airbus exercised their option for more aircraft. It was assumed there were no more B-377 airframes available to rob the section 45–46 intact. They were in short supply; however, in 1978 there were at least three and possibly more existing. The ex–Northwest Stratocruiser N74603 c/n 15949 owned by Aero Spacelines was complete and parked at Bobs Air Park in Tucson. The ex–BOAC (G-AKGL) Transocean N404Q (N9600H) c/n 15978 remained at Mojave. It was never used and was photographed intact on 01 December 1989. The Pregnant Guppy which was registered N126AJ after being sold to American Jet in 1974 remained intact through 1979. The registration was cancelled on 05 February 1979; however, it was still complete. The process of breaking it up at Van Nuys did not begin until after the death of Jack Conroy on 05 December 1979.

These three airframes were intact with the rear sections, one at Mojave, one at Tucson and one at Van Nuys. It is quite possible at least two more B-377 rear sections were in existence. Stratocruiser c/n 15940 went to Quito and was seized by the Ecuadorian government and placed in storage for years. The cost of shipment to France may have been considered cost-prohibitive. It is not known if the tail section of an Israeli B-377 c/n 15964 formerly N90948 was available; however, it was converted to VIP seating and was not modified to rear cargo loading. It remained intact through 2009 as an officer's club in Israel. It appears Aero Spacelines procurement did not conduct in-depth research opting to take the easiest, lowest cost alternative.

The Last Turbine SGT-201 Super Guppy

Aero Spacelines began modifying the C-97 subassemblies for the fourth SGT-201 in 1980, prior to being acquired by Tracor. Parts from KC-97 52-0828 c/n 16522 with 8,090 hours were modified at Santa Barbara. Aero Spacelines acquired the section 45 off the Pregnant Guppy from American Jet at Van Nuys. Tom Roberts was dispatched to remove the section. It is a mystery why the rear section of ex–Northwest B-377 N74603, which was then at Mojave, was not used because it was owned ASI.

The salvaged section 45 from the Pregnant Guppy was brought back to Santa Barbara and prepared for shipment to UTA at Le Bourget. It was invoiced on 09 September 1980 for $9,680 and shipped to France by boat with special tools on loan until December 1982 (see section on Pregnant Guppy). It arrived at UTA on 31 October 1980. This was two years before the majority of the modified parts from C-97 52-0828 used to build SGT 201-4 (F-GEAI) were shipped. This gives some clarity to why Tom Roberts always thought the PG section 45 was used on SGT-201-3, the first built in France.

The fact of the rear section of the first Aero Spacelines guppy being used on the last one built was not intentional but purely coincidental. When Airbus negotiated a contract with Aero Spacelines to supply components to build additional SGTs in France, the agreement included an option to build one more at no licensing cost and possibly as many as four additional. If more had been built as planned, F-GEAI would have been the fourth of six or as many as eight SGT-201s. Consequently, claims of parts from the first guppy being purposely used on the last are pure aviation folklore.

Aero Spacelines' Tom Roberts supervises the loading of the nose for SGT-201-4 at Goleta Beach in September 1981. It was transported to Long Beach and transferred to a cargo ship for the voyage to France. The other components were not crated and shipped until the following spring. The components were intended for use on the next French built guppy F-GEAI. It was not known at the time this would be the last SGT-201 (courtesy Tom Roberts).

Tracor Acquires Aero Spacelines

Tracor Corporation of Texas acquired Aero Spacelines from Unexcelled subsidiary Twin Fair for $9 million in 1981. The Aero Spacelines name was retired after 20 years of being the premier builder of ultra-large guppy transports. The company was reformed as Tracor Aviation but retained the ASI management team headed by Kirk Irwin. The first project under the new owner was to complete the contract inherited from ASI to modify KC-97G components for UTA construction of the fourth SGT-201.

The majority of parts for the final SGT-201 (F-GEAI) modified at Santa Barbara were transported by ship to France in four boxes and overland by truck to Le Bourget. A fifth box containing the outer wings and tail components was flown by the Mini-Guppy N1037V.

As the components modifications were completed the parts were crated and shipped at multiple times. The overhauled KC-97 nose and framing parts used to attach the nose to the new fuselage was crated and loaded on a boat by crane at Goleta Beach in September 1981. It was transported to the port of Long Beach for transfer to a larger ship. A second large crate containing the center-wing was trucked to Long Beach in October to join the first crate. The final two very large crates measuring 42.5 feet long, 19.5 feet high and 9.75 feet contained the inner wings and nacelles standing on their sides. They were loaded by crane on a barge at Goleta Beach on 21 April 1982 then transported to Long Beach where they were transferred to the cargo ship *Lafayette*.[17] They arrived at Le Havre, France, on 14 May 1982, then were barged up the Seine as far as Gennevilliers. There they were moved by road to Le Bourget. The final shipment containing tail components, outer wings, entry doors were secured on large pallet. The Mini Guppy was then called in to transport the pallet to Le Bourget probably in March 1982.

Union de Transport Aeriens projected a completion date of April 1983.

The overhauled wings with newly fabricated nacelles for Allison 501 turboprops were mounted on pallets at Tracor-Santa Barbara for shipment to UTA in Paris. The crates were more than 43 feet long and 19 feet high. They were barged to Long Beach, then transferred to a cargo ship to France (courtesy Tom Roberts).

On 05 November 1982 F-BPPA was being loaded with a section of A300 fuselage at Finkenwerder (Hamburg, Germany) when a sudden gust of wind caused the aircraft to shudder. The hinge snapped and the open front section was twisted over on the nose. The nose was lifted back into place with cranes. Repairs were conducted from 11 to 16 November. After five more days of testing it was returned to service (DGAC Ministry of Aviation, courtesy Pierre Cogneville).

The SGT-201-4 first flew on 21 June 1983 as F-WEAI. After testing and certification it was reregistered F-GEAI before beginning service at Toulouse. It joined the other UTA built guppy F-GDSG plus American built F-BTGV (N211AS) and F-BPPA (N212AS). The four aircraft were operated by Aeromaritime under contract to EADS (Airbus) which began with the first one purchased in1971.

In 1989 Aeromaritime was absorbed into Air France. Airbus Industries took over the operation of the four guppies, forming a Transport Directorate bringing the operation in house under the sub-company Airbus Inter Transport known as Skylink. The livery was changed and major maintenance was transferred from UTA to SOGERMA, a subsidiary of Aerospatiale at Toulouse and sub-company under the EADS consortium. In 2013 the EADS name was changed to Airbus Group.

The four Turbine guppies maintained a heavy schedule transporting Airbus A300/ A310/ A319/ A320/ A321/ A330 and A340 wings. The wings were built in Wales and trucked to Manchester to be airlifted out. The narrow-body wings were flown to Hamburg-Finkenwerder and the wide-body to Bremen-Lemwerder, where control surfaces and systems were added. After completion they were uplifted again to Toulouse for final aircraft assembly. The SGTs also transported A300/A310/A320 fuselage sections to Toulouse.

The daily transport of sub-assemblies between fabrication plants in Europe and final assembly plant in France is a logistical nightmare. Weather is always a determining factor in daily operations. The size of the guppy fuselage dictates strict limits on crosswind landings. Because of the frequency of operation each aircraft is inspected every 150 hours in addition to scheduled annual checks. The fleet flies on average 62 hours per week or about 16 hours per aircraft putting at least one aircraft down for inspection just under every nine weeks.

Damage to Airbus Guppies

The heavy flying schedule combined with winds and mechanical problems increases the possibility of serious incidents. Three of the Airbus guppies suffered serious damage while in service. The first guppy sold to Airbus F-BTGV SGT 201-1 experienced the most incidents:

F-BTGV nose gear collapse on landing, 21 April 21 1972 in Finkenwerder (Hamburg)

F-BTGV nose gear collapse on takeoff, 22 March 1974 in Lemwerder (Bremen)

F-BTGV nose ripped off during opening and number two prop damaged, 29 March 1977 in Getafe (Madrid)

The second SGT-201, F-BPPA, was severely damaged at Finkenwerder (Hamburg). The final checks were being completed to secure the cargo after the loading of an Airbus A300 fuselage section. At approximately 11 a.m. on 05 November 1981 the winds were moderate. Suddenly an unexpected extreme gust of wind caught the nose. The open nose shuddered and there was a loud cracking noise. The first officer was sitting in the cockpit doing pre-flight checks when he felt the nose start to twist. The forward portion of the fuselage hinge broke in two places, causing the entire cockpit section to rotate forward over the nose gear, turning straight down and crushing the radome and causing major damage to the nose section. Only the control cables between the forward section and fuselage prevented the nose from rolling on its top. The first officer was shaken but able to escape uninjured.

After engineers secured the nose two cranes were brought in to slowly lift the forward section back to an upright position above the nose gear. It was rotated to the closed

The first French-built Airbus SGT-201 F-GDSG was en route from Hamburg to Toulouse in October 1993 when headwinds forced a diversion to Lyon, France. Upon landing the nose gear collapsed. There was no fire and damage was confined to the nose belly and gear doors. The aircraft was repaired and returned to service (courtesy Jacques Lienard-France).

position and bolted shut to maintain structural integrity then towed to a hangar for repairs. Between 11 and 16 November UTA engineers and maintenance personnel repaired the damage. After five more days of testing the electrical circuits it was once again declared airworthy. It departed for Toulouse to deliver the A300 fuselage section. The heavy flying schedule was severely impacted by each incident that put a guppy out of service for unscheduled repairs. As orders for new Airbus aircraft increased more guppies were needed to maintain production schedule.

Tom Roberts was dispatched to Tucson in the winter of 1981–82 to salvage the sections 45–46 from ex–Northwest Stratocruiser N74603. Roberts was given specific instructions on how to cut the section out to maintain the structural integrity. It was one of the last remaining B-377s know to exist. Airbus purchased the parts for a fifth planned French built SGT-201 (author's collection, courtesy John Wegg).

The third SGT-201-3(F) guppy F-GDSG, which was the first built in France, experienced an incident when the nose gear collapse during landing in October 1993 at Satolas Airport (Lyon, France). This was the last reported serious incident with the SGT-201s while in Airbus service. Only the last SGT-201-4(F) built, F-GEAI (formerly F-WEAI and later NASA N941NA), has no reported incidents of serious damage while in service with Airbus.

Airbus Plans to Build More SGT-201s

Airbus anticipated a need for additional guppies prior to the 1982 construction of the fourth SGT-201 F-GEAI. Acquiring parts for SGT number five began sometime after September 1981 Tracor (ASI) dispatched Tom Roberts to Tucson to remove the section 45 from ex–Northwest B-377 Stratocruiser N74603 for use on the next guppy. Jack Conroy had reversed the colors of the Northwest paint scheme and had Aero Spacelines lettering placed on the nose. It had been used for support work and testing. The Northwest paint scheme and square windows makes it quite easy to identify. Airbus purchased it for component salvage for use on a fifth SGT-201 as stated in the licensing agreement with Tracor (ASI) for additional guppies. The salvaged section 45 of N74603 was planned for use with KC-97G, 53-0280 c/n 17062, which had received U.S registration N49548. The KC-97 was part of the lot purchased by Airbus for conversion. It was flown to Le Bourget-Paris and placed in storage awaiting conversion. The photos of section 45 that are widely circulated and presented as being from the Pregnant Guppy are actually from ex–Northwest B-377 N74603 which was parked at Tucson as late as September 1981. It clearly has square windows and is painted orange. The Pregnant Guppy N126AJ (formerly N1024V) parked at Van Nuys was painted white and built from Pan Am Stratocruiser N1024V, which had round windows.

As late as 1988 Airbus reviewed the possibility of producing SGT five and six. In a letter dated 02 September 1988 Airbus requested a quote from UTA to build two additional SGT-201 aircraft. The letter was also copied to Tracor director Kirk Irwin. The letter stated the purchase of more guppy aircraft would be based on the quote and the decision would be linked to the cost of technical improvements of the existing SGT-201 fleet. Also affecting the decision would be the availability of spares required for operation of the Super Guppies for the next 20 years. The letter referenced the previous work sharing agreement between UTA and Tracor (ASI) which produced F-GDSG and F-GEAI. Airbus requested a quote for one additional guppy; two additional guppies; or one additional Super Guppy plus the delivery of refurbished and retrofitted inboard and outboard wings including nacelles for a further Super Guppy or for replacement.[18]

The request further stated Airbus would provide the stored C-97 (N49548) and B-377 section 45 from B-377 N74603, which was previously acquired. Additional sections would have to be located or newly produced for a second (number 6) SGT-201 aircraft. It was also stated the license to produce SGT 201-5 was authorized by parent company Tracor on a no-charge basis in the 1982 agreement. Airbus needed confirmation this would also apply to SGT 201-6.

The proposal was to start construction by July 1989 and SGT 201-5 be in service by mid–1992.[19]

Tracor responded by stating the company was no longer capable of guppy component modification. The proposal for Tracor Aviation to supply components for the construction of additional B-377 guppies five and six never progressed past the request for cost analysis.

After the acquisition of Aero Spacelines in 1981, Tracor Aviation expanded, becoming an FAA certified overhaul station at Santa Barbara. The company operated 24 hours a day seven days a week, ultimately employing over 1,000 people. After the SGT-201 component assemblies were completed for guppy conversions three and four the company withdrew from all heavy airframe modification.

After completing the inherited Aero Spacelines contract with Airbus for SGT components Tracor pursued more lucrative government contracts. The company was successful in winning a $5,900,000 Navy missile contract in 1985 and a $19,100,000 contract with the Warner Robins Air Force Base Air Logistics Systems for a countermeasures systems. The health of the company looked even better in 1986 when Tracor successfully bid on a five year $17,200,000 contract to train U.S. Air Force personnel stationed in Japan and Korea.

In 1987 Tracor was acquired by Westmark Systems for $694,000,000. Despite the buyout and numerous government contracts Westmark slipped into heavy debt and bankruptcy by 1989. The Santa Barbara operation was in serious financial trouble because of the heavily leveraged acquisition. As part of a restructure the Austin, Texas–based Tracor was split into three independent companies. The Santa Barbara (Goleta) aviation technology division filed for chapter 11 bankruptcy in January 1991. Lucas Aerospace, a subsidiary of the $4,300,000,000 Lucas Industries, acquired the Tracor Goleta, California, aircraft modification operation for $27,000,000. Tracor employed over 600 workers and had a gross income of $72,000,000 in 1989.[20] By October 1993 Lucas Aviation announced the closing of the Goleta/Santa Barbara operation.

The 1988 Airbus request for additional assemblies came at a time when Tracor Santa Barbara was experiencing financial problems. Airbus requested a quote on additional guppies because it was becoming apparent the aircraft were nearing the end of the expected service life. All but one of the fleet had experienced multiple damage incidents including nose gear collapse and damage by wind while unloading. To reduce turnaround time and avoid wind damage, in 1989 a special hangar was constructed at Toulouse for inside unloading.

Logistical problems had increased with the aging aircraft. The volumetric giants were constructed from parts of nearly 50-year-old airframes. Airbus feared additional airframe maintenance and lack of spare parts could jeopardize delivery of new Airbus products. Even if UTA and Tracor Aviation could have worked out a schedule for producing SGT-201 five and six it would only have been a stopgap measure. Airbus soon concluded the B-377 based guppy was not large enough to meet future needs. Although there may have been several B-377s scattered around the world, for planning purposes there were none available and only so many aging C-97 airframes left. Airbus still had in storage the section 45 removed from B-377 N74603 in 1982. Any additional section 45 rear fuselage sections would have to be located or fabricated. Airbus planners were also weighing the possibility of an all new jet powered guppy. In order to be cost effective it had to be based on an existing type because it would be built in very limited numbers for in house work. The age of the SGT-201 guppy fleet combined with the need for reliable larger component transport eventually led to the development of the Airbus A-300-600ST Beluga.

In 1988 Airbus requested a quote from Tracor for additional SGT-201 guppies five and six. The KC-97 N49548 was still in storage at Toulouse along with the sections 45–46 from ex–Northwest B-377 N74603, which was acquired in 1982. The planned construction was to begin by July 1989 (courtesy Tom Roberts).

No additional B-377 based guppy aircraft were ever built, consequently, the rear section 45 of Aero Spacelines (ex–Northwest) Stratocruiser N74603 was never used. The KC-97 c/n 53-0280 N49548 which Airbus purchased in 1979 had been on display at the Musee de l'Air at Le Bourget Airport, where it remained until 1988. It was then flown to Nancy-Essey for display in a museum. Subsequently the museum became insolvent and bankrupt. The aircraft fell into disrepair and was broken up in 2005. The tail was preserved for a time at Zruc Park for children to play in but has now been scrapped.

Airbus announced in 1990 it would produce a new volumetric transport designated A300-600ST. In time it would be known as the Beluga. In March that year the four Skylink Aero Spacelines designed SGT-201s were reported to have the following total time hours: F -BTGV 16,160 hours, F-BPPA 14,240 hours, F-GDSG 13,610 hours and F-GEAI 11,850 hours. The total hours included time as KC-97s. The next generation volumetric transporter first flew on 13 September 1994 with proving flights which continued until July 1995.

After the Airbus A-300-600ST Beluga became operational the B-377/C-97 based SGT-201 guppies were declared redundant. The first two were retired in 1996 after 25 and 23 years respectively. The last Airbus SGT-201 flight occurred on 24 October 1997 when SGT-201-3 F-GDSG operated a Toulouse—Bristol-Bremen—Hamburg route, and it has been placed on permanent display in Hamburg. As a tribute to the SGT-201 fleet, the first three have been displayed in three countries where Airbus parts are manufactured; F-BTGV at the British Aviation Heritage Museum at Bruntingthorpe, England; F-BPPA at Airbus Toulouse has been restored and is now on display with the nose door open in the Aeroscopia museum; and F-GDSG is on display at Deutsche Airbus facility at Finkenwerder, Germany.

The four SGTs were operated for 26 years transporting over 100,000 tons of aircraft parts for Airbus. European aircraft production was dependent on the SGT-201 fleet for airframe components. Early on the consortium relied on the concept of building aircraft parts in multiple locations throughout Europe then transporting them to Toulouse, France, for final assembly. The method was first utilized to build the A300 and Concorde. It is ironic the Boeing B-377 based SGT-201 guppy was acquired to transport A300 parts and it was replaced by the A300 based Beluga. It can be reasoned the guppies contributed to their own demise by transporting A300 assemblies.

NASA Guppy Operation

When NASA acquired the original B-377SG N1038V (N940NS) in 1979 Airbus was flying the first two SGT-201s in Europe. The space agency continued to operate N940NS as Airbus acquired sub-assemblies from ASI to build two more SGT-201s in 1982–83. NASA purchased the original B-377 Super Guppy because it was the only outsize transporter available and could be operated at far lower cost than any large Air Force transport. Not only was it utilized for space hardware but for multiple aircraft recovery missions for NASA, the Air Force, Navy, Army and defense contractors. It was used to transport damaged aircraft and partial airframes which were being upgraded to later models. The DOD reimbursed NASA for the cost of flights operated on behalf of branches of the military.

Shortly after NASA acquired the Super Guppy in 1979, the DOD wasted no time in requesting a flight for aircraft recovery. A Wyoming Air National Guard C-130B s/n 59-1533 had been damaged by a tornado at the Cheyenne air base. The aircraft was flipped over, causing considerable damage. The wings and tail were removed in order for it to be shipped to a depot for overhaul. NASA 940 was dispatched to uplift the complete fuselage, weighing 25,830, pounds to Hayes International at Birmingham on 09 July 1980. The C-130 was completely rebuilt and returned to service. It was eventually sold to the Tunisian Air Force.

The DOD requested NASA 940 again on 02 November 1983. It was dispatched to Dallas to uplift a Rockwell B-1 A1F bomber midfuselage section weighing 36,082 pounds from NAS Dallas, Texas, to Palmdale, California. The B-1 had a considerable amount of electronics and classified systems, prompting the decision to transport it by air rather than overland. There were security concerns and fears of damage from excessive vibration if moved by truck. The flight, in typical guppy fashion, did not go exactly as planned. The aircraft departed Dallas but because of the weight and excessive fuel burn landed at Dyess Air Force Base in Abilene. After departing Dyess it diverted to El Paso with a prop failure. After two days to change the prop it departed to Palmdale without further incident. The Air Force had heavy security meet and guard the aircraft at every stop.[21]

In early 1986 an Air Force A7D was damaged at Howard Air Base in Panama when it landed with the gear up. NASA was contracted to recover the aircraft with the guppy. On 28 January 1986 NASA 940 departed Houston for the first leg of a rather circuitous route to Panama. The first stop was Nassau for fuel. Upon arrival the local ground crew told the captain, "Your space shuttle just blew up." The crew contacted Houston and was advised the Challenger had exploded and to stand by for instructions. After nearly a day of waiting they were instructed to proceed on to Puerto Rico, where they remained over night.

The next day N940 departed for El Libertador Air Base at Maracay, Venezuela. The guppy experienced an in-flight propeller failure en route landing in at Maracay on three engines. The crew waited two more days for the prop to arrive then another day for installation and checks. The guppy was on the sixth day of the mission as the crew made the final approach to Howard Air Base. As the aircraft turned on final one of the props on the left went into reverse, damaging the engine. The guppy made another three engine landing. It took two more days for another engine and prop to be delivered and several more days to install and check it out. The A7 aircraft was finally loaded and the guppy departed Panama for a return to Houston by the same route. NASA 940 arrived back at Ellington Air Force Base on 11 February 1986. The round-trip flight had taken 14 days.[22]

Near Disaster Fuel Spill

NASA dispatched the Super Guppy to Los Alamitos in February 1987 to recover a damaged T-38 aircraft which had been struck by lightning. The mission to recover a damaged NASA T-38 began like all the others, as a routine NASA cargo flight. A special 16,000 pound mounting fixture designed to fit the eight-foot-wide cargo floor had been built to secure aircraft for transport. The guppy had routinely transported multiple aircraft in the past and this was not considered to be any different.

The Northrop T-38 being recovered had been flown by NASA Aircraft Ops Division Instructor Robert Rivers with astronaut Brewster Shaw as second pilot. Shaw, who had flown two Shuttle missions as a pilot and the other as commander, was offered a ride to California by Rivers in the T-38. Shaw was consulting with the Rockwell Space Division at Downey on safety enhancements for the Space Shuttle. Rivers gave his account of the T-38 lightning strike incident:

> We were flying NASA T-38 N914NA in February of 1987 to Los Alamitos. I was in front, Brewster in back, scattered clouds at about 2,500 feet. We were struck by lightning over the Pacific about 8 miles offshore. Seventeen lightning strikes occurred in an hour that day with only scattered clouds. A bolt came through the cockpit going through my left hand which was on the throttles leaving a burn mark and it exited out the right horizontal stabilizer. It went through the dorsal fuel tanks which in a manufacturing error had never been properly grounded. I turned immediately to Los Alamitos as I got a Fire Light right away along with a huge explosion which shook the plane. The main fuel tank had exploded but we did not know it then. I shut down the left engine with secondary confirmations of fire. Crossing the coast, the right engine Fire Light came on. I told Brewster he could get out, but I was going to try to land it. He said he was sticking with it. I could not maintain altitude as the right engine was maxed somewhere around 85 percent and we had tremendous drag from the blown out fuel tank and trailing fuel lines and debris. Two minutes from landing Brewster told me flames were coming through the rear bulkhead. I made a no flap landing and barely made the under run 500 feet short of the runway having to pull up to clear a line of Australian pines on the airport boundary. I shut the right engine down on touchdown, stopped, and saw a huge fireball on the left side of the plane. Brewster was out first, and my ox mask tangled up on my way over the side causing me to land flat on my back knocking the wind out of me. I was on the ground and could not move. Brewster ran back, pulled me up, and we ran as fast as I could still with the wind out of me. Brewster was a real pro to come back for me. That's my story and I'm sticking to it.[23]

NASA sent a QA representative and Northrop dispatched a maintenance specialist to prepare the T-38 for transport by the Super Guppy back to Ellington Air Force Base in Houston. Maintenance had prepped the aircraft and supposedly the fuel tanks had been drained and purged in preparation for the recovery flight. The next day the T-38 was loaded into NASA 940. With all secured, it departed Los Alamitos for El Paso and Houston. The five man cockpit crew consisted pilots Frank Marlow and Ace Beall with flight engineers Ken Sanders, Henry Marshall and Lee Lawson. The standard procedure after takeoff is for one of the engineers to go back and check the load and verify it is secure. Flight engineers are rotated on the panel and Marshall was taking his turn for this flight. Sanders was sitting in one of the adjacent seats monitoring

NASA T-38 N914NA flown by NASA flight instructor Robert Rivers and astronaut Brewster Shaw after being struck by lightning on approach to Los Alamitos in February 1987. They were able to land safely on one engine with a fire in the dorsal fuel tank. The Super Guppy was dispatched to recover the aircraft back to Ellington Air Force Base, Texas (NASA, courtesy Larry J. Glenn).

the panel with Marshall. After they leveled off it was his duty to verify the load was secure. Before he got up to check, one of the two loadmasters ran into the cockpit shouting, "Fuel spill!" Sanders went with the loadmaster to check. He ran back into the cockpit shouting, "Fuel is running out of the T-38 and puddling in the cargo bay floor! It is running into the center drive bay!" Frank Marlow declared an emergency calling March Air Force Base at Riverside, California, which has a 13,000-foot runway. Sanders said, "Don't touch the flaps. We could get a spark!" The flaps are electric and if deployed there was the possibility of a spark, which could blow up the aircraft. Sanders pulled the circuit breaker to cut the power to the flaps.[24]

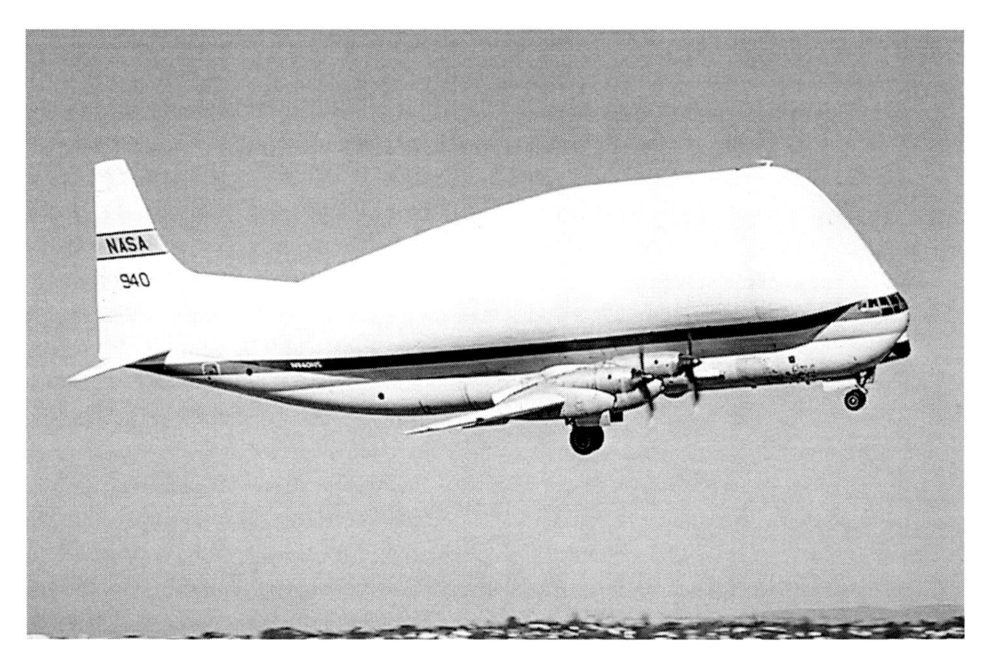

NASA 940 departed Los Alamitos with the T-38 on board. As the guppy leveled off, fuel began spilling from the T-38 into the cargo compartment. Pilot Frank Marlow declared an emergency. All nonessential circuit breakers were pulled to prevent a spark as they turned on a heading to March Air Force Base at Riverside, California (NASA, courtesy Larry J. Glenn).

Marlow and Beall rolled the guppy to the left and turned on a heading for March Air Force Base. Crash trucks were scrambled as the NASA Guppy made a straight-in smooth approach with no flaps. It landed and rolled to a stop off the runway. Crash-rescue crews followed until it stopped. They approached the aircraft as the engines were shut down and the crew evacuated. The aircraft was vented and the crew was not allowed near it. After several hours it was cleared for the flight engineer and one pilot to taxi it to the ramp area. Sanders said, "I can't tell you how many gallons of fuel were in the T-38 but it was flowing over the side of the T-38 aircraft and puddling in the floor in a large amount."[25]

The T-38 was removed and the fuel tanks were emptied and purged. All the bays and compartments effected were cleaned and secured. After several days the T-38 was reloaded. The return flight to Ellington Air Force Base was operated without incident. The T-38 was rebuilt and returned to service. It holds the distinction of being the only T-38 to be transported on the both NASA guppies. It was removed from service in 2012. On 12 April it was transported on NASA SGT-201 N941NA from El Paso to Davis-Monthan for storage.

NASA B-377SG Super Guppy Readiness

Eight years after NASA acquired the original B-377SG (N940NS) the Space Station Freedom Transportation Working Group (SSFTWG) conducted a study known as the "Space Station Program Transportation Implementation Plan." The group issued a report on 12 June 1987 stating Super Guppy readiness was questionable and no longer considered a viable transport option. The study conducted by Boeing for NASA stated, "We believe the Lockheed C-5 SCM currently under modification will be superior and meets the needs of the space station program." A second study dated 18 March 1988 conducted by the Moonspace Corporation for the Space Station Oversized Element Transportation Study reached the same conclusion of the 1987 study. The study declared, "Recently modified C-5M SCM could be cost effective." The study further stated the newly acquired NASA B-747 SCA (space shuttle transport) aircraft and/or a modified (Clark) DC-10 could be used for transporting space station hardware provided a transport canister were developed.

The group recommended the production of a canister in which cargo could be loaded and transported in a controlled environment. The canister could be placed on the top of the B-747 transport aircraft in a piggyback fashion or inside the proposed Clark swing nose DC-10. This was conditional on a canister being developed. Beech Aircraft studied the lightweight canister concept and estimated it would cost $11,000,000 to produce. The canister could also be used in the B-377 Super Guppy if it was refurbished and brought up to readiness standards and eliminating environmental problems which may be encountered with the older aircraft.[26]

The DC-10 referenced in the report is a swing nose conversion of a standard DC-10-30 designed and proposed by Jack Conroy's long time friend Frank Clark. The conversion would have been a joint project between Clark Aircraft Corporation, TRW and NASA. The main cabin floor would be removed back to the center-wing, providing an 18-foot diameter cargo hold creating a non-guppy appearing outsize transport. Clark also proposed a second stretched DC-10 advanced swing nose conversion. It consisted of a DC-10-30 with a 17-foot fuselage section added forward of the wing and 10-foot section aft of the wing with the cabin floor removed ahead of the wing. The plan essentially produced a jet guppy type with an 18-foot diameter cargo hold equivalent to the Mini Guppy without having to expand the fuselage.

The findings of these studies apparently only focused on the need for transportation and not the construction and operational cost. Although Super Guppy N940NS was in need of major upgrades, the actual cost per cargo mission was just over $100,000. The modified Lockheed C-5 SCM mission cost estimates were approximately $1,000,000 per cargo haul. Even if the Clark DC-10 had been completed the cost would have also been cost prohibitive just in fuel burn alone. NASA Super Guppy N940 fuel burn is a little over 4,500 pounds (663 gallons) per hour and the DC-10 is nearly four times greater at over 15,000 pounds (2,238 gallons) per hour at cruise after the first hour.

The use of the B-377SG by NASA had diminished significantly from 1986 to 1989. It had been flown continually by Aero Spacelines and NASA for 24 years without major upgrades. In 1989 a critical evaluation determined overhaul and considerable upgrades were needed. The upgrades would be mandatory if it were to be used to transport support components for the Space Station Freedom Program, the mission to build a permanently manned orbiting space station. The report estimated $6,000,000 would be required to obtain and install new engines on the B-377SG. A total budget of $28,000,000 would be required to bring it up to readiness standards suitable for the Space Station support operation.[27]

This was unquestionably cost prohibitive when considering the aircraft only cost Jack Conroy $1,900,000 to build in 1965. The cost to produce the latest SGT-201 aircraft operated by Airbus was $32,000,000. It was quite obvious it would have been more cost effective to purchase a new aircraft if Tracor (formerly ASI) were still producing them. Interestingly, this was at the same time Airbus was querying Tracor for quotes on the production of a fifth and sixth SGT-201 aircraft. If circumstances had been different it is possible there would have been more guppies produced for not only Airbus but also NASA.

Super Guppy N940NS flew 65 flights for NASA and 26 for the DOD between February 1979 and 30 April 1989. Most of them were before the end of 1985. Only 25 flights were operated between 01 October 1985 and 30 April 1989 and only six of those were for NASA. Actually from September 1986 to April 1989 all flights were for the DOD. Marshall Space Flight Center discussed the use of the B-377SG on 01 March 1989 for two flights to transport space telescope hardware. The space station transportation group replied to Marshall stating the Lockheed C-5SCM was the only option.

NASA requested the SSFTWG review the rising cost of the C-5SCM modification proposal and to reconsider the B-377SG in comparing cost. The review concluded the Super Guppy airframe life can be extended indefinitely provided it is properly maintained. However, the guppy is unpressurized, the turbo engines require replacement and there are limited spares, making it obsolete.

Furthermore, there was no overhaul capability for the existing Curtiss CT735S props and no surplus T56-A-14 engines were available for use on the guppy. (The report is puzzling because Allison T56 engines were not previously considered an option. They are lower horsepower than the P & W T-34, which was necessary for the heavier B-377SG.) The estimated

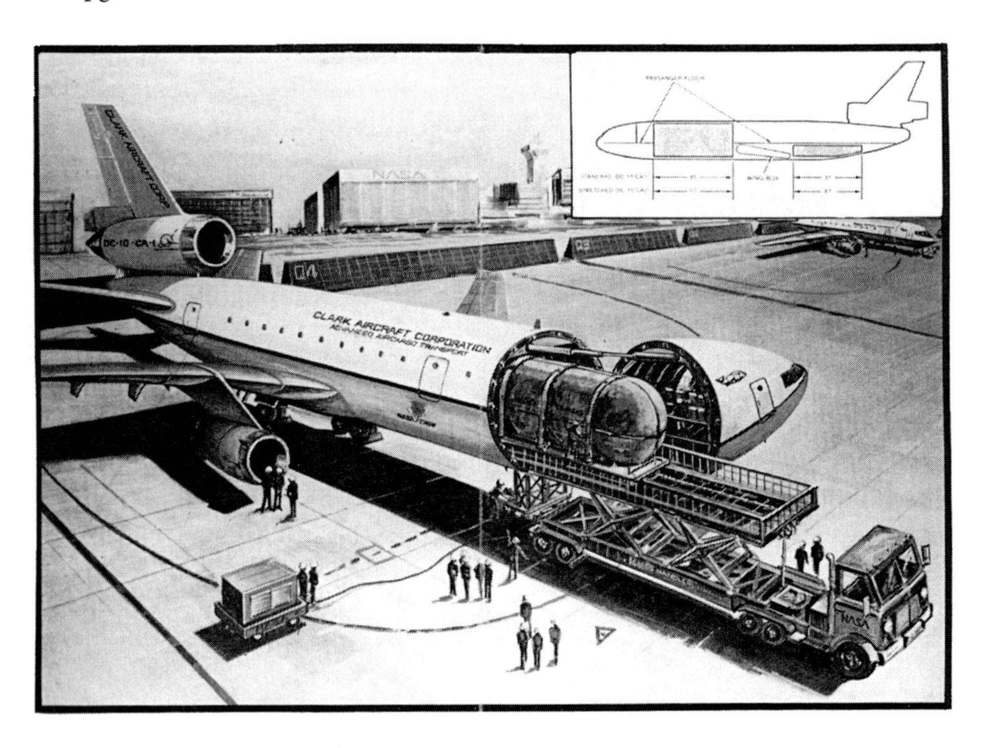

The Clark swing nose DC-10-30 was proposed as an alternative transport for space station components. The aircraft would have the main cabin floor removed ahead of the wing to produce a cargo compartment the diameter of the Mini Guppy. The DC-10 was to be followed up with a stretched second version by extending the fuselage 27 feet (Clark Aerospace Group).

cost for five new Allison T56-A-14 replacement engines was $8,000,000. An additional $3,300,000 to $4,500,000 would have been needed for spares, props and ancillary equipment. The group estimated the cost to keep the B-377SG operational at $28,000,000 ($9 million for airframe modifications, $8 million for engines and $11 million for the transport canister).[28]

The advisory group concluded the cost comparison to the (Clark) DC-10 proposal could not be addressed because no DC-10 swing nose conversion existed. The Office of Space Station (OSS) concluded the B-377SG was no longer needed and funds for upgrading should be deleted from the budget, including the $6,000,000 previously appropriated for engine replacement. It was recognized the transport canister was needed regardless of mode of transportation and should be built. The report issued on 29 September 1989 recommended the director of aircraft management officer take action if necessary to dispose of the B-377SG Super Guppy.[29]

The B-377 Super Guppy transported NASA hardware on a continuous basis beginning in 1965. NASA acquired it from Aero Spacelines in 1979. Reliability became an issue because it was a one-of-a-kind aircraft built from another one-of-a-kind aircraft (YC-97J). The base B-377 airframe design was expanded from Boeing B-29 wartime engineering. It was built with 1940s technology using parts of Pan Am Stratocruiser N1038V which first rolled out on 22 July 1949 and one of the two YC-97J turbine engine test aircraft. The engines and props were experimental, taken from a test aircraft. After years of service spare parts were in very short supply. The situation became critical after the supply of props dwindled to only three spares, forcing the grounding of the aircraft.[30]

Top: **In 1991 N940NS was transferred to AMARC in flyable condition and stored at Davis-Monthan Air Force Base in Tucson. It sustained damage in 1999 when the right wing was hit by a forklift, damaging an aileron. In 2014 the GSA made the decision it would never fly again. It remained under NASA control until 2015 when it was transferred to the Pima Air Museum (courtesy Patrick Dennis).**

Bottom: **One of only two Lockheed C-5 SCM transports, which was specially modified for NASA airlift support. The rear upper passenger compartment was removed to increase the cargo compartment ceiling height behind the wing. The cargo doors were modified to full clamshell configuration for loading of oversize cargo (USAF, SSgt Kelly Goonan).**

The Aero Spacelines Pregnant and Super Guppy flew more than three million miles together for sixteen years supporting the Gemini, Apollo and Skylab programs. The original Super Guppy N1038V continued to transport critical space hardware after the Pregnant Guppy N1024V was decommissioned. The primary mission was transporting components for the space program. The Super Guppy racked up millions more miles, flying for a total of 25 years.

In addition to NASA work and because there was no other aircraft with extreme outsize capability, the DOD engaged it for multiple aircraft recovery missions. The giant transport remained in service until 1989 when reliability and lack of spare parts became an issue. It was flown to Hayes International at Birmingham for depot maintenance in late 1989. The overhaul facility verified it was no longer viable for transport of critical space hardware. After routine maintenance work it returned to NASA. The agency was confronted with the decision of grounding the aircraft or budgeting for serious upgrades. It remained in limbo for nearly 18 months.

Reliability reports stated NASA had 23 spare 6,500 horsepower Pratt & Whitney T-34 engines obtained from C-133 spares storage. The engine situation could support the operation for some time. However, the lack of spare props was critical. A propeller overhaul vendor was located but the initial investment for service was over $1,000,000. The cost was considered too prohibitive by NASA AOD division chief Joe Algranti. After the cost for upgrades and modifications were reviewed the decision was made to withdraw Super Guppy N940NS from service.

The severe propeller shortage left NASA with no alternative. The Super Guppy was flown to Davis-Monthan Air Force Base, Tucson, Arizona and transferred to Aerospace Maintenance and Regeneration Center (AMARC) on 09 July 1991. While a decision was being considered on whether it should remain in NASA aircraft inventory, it was placed in flyable storage at a cost of $3,000 per year. A pair of spare KC-97 wings were acquired and stored with the guppy. The idea of returning it to service at a future date refitted with the Allison T-56 engines was still considered an option. The T-56 is the military version of the commercial Allison 501 power plant.

Aero Spacelines had refused to re-engine the Super Guppy with Allison 501 engines for Airbus years before because they lacked sufficient power for the heavier airframe. The aircraft was thousands of pounds heavier because the more beefy B-377 lower fuselage half. The figure eight fuselage with enlarged top was skinned with dome rivets. The rivets were also used on other modified portions which contributed to considerably more drag. The Allison powered second generation slab side SGT-201 aircraft was considerably lighter. NASA administrator Vice Admiral Richard H. Truly pledged a supply of six Navy Allison T-56-14 engines with six Hamilton Standard 54H-60 propellers for the re-engine project.[31]

Why NASA was not aware of the previous study advising Airbus of the need for more powerful engines is a mystery. If the agency had been advised there would be no reason to consider a plan to modify the spare wings to accept the less powerful Allison T-56 engines. The plan was eventually dropped with no formal reason given.

Consequently N940NS was sealed and placed on display at the Pima Air and Space Museum located just south of Davis-Monthan Air Force Base in Tucson. Although it was still considered available to fly again the Air Force cocooned and mothballed it for storage. In 1999 the right wing was hit by a forklift, causing serious damage and destroying the right aileron. The aileron was removed but no repairs were made. In 2014 the GSA made the decision N940NS would never fly again. It remained under NASA control until 2015 when it was officially transferred by the GSA to Pima Air Museum.

The Pregnant and Super Guppy played a major role in transporting critical hardware which kept the space program on schedule. Man would have eventually gotten to the moon; however, it most likely would not have been by the end of the decade as promised by President Kennedy. In reflection John Goodrum, head of MSFC's logistics said, "The payoff of the guppy operations was exceptional.... It would be too strong to say the guppy operations saved the Saturn program, but without the availability of the unique planes, NASA might have been forced to scrub some of the schedule launches and might have incurred horrendous cost in money and time."[32] Wernher von Braun is quoted as saying, "The guppy was the single most important piece of equipment to put man on the moon in the decade of the 1960s."[33]

NASA Acquires the Last SGT-201

Plans were being made for the International Space Station (ISS) launch in 1998. NASA had not relied on the Super Guppy since 1986 although it had flown missions for the DOD until 1989. For over ten years NASA relied on the Air Force for transport of critical space hardware. The reality of a critical logistical problem existed without a purpose built outsize transport to move space station components across the country. The completion of the two Air Force Reserve Lockheed C-5s modified at Kelly Air Force Base for DOD and NASA took far longer than expected at a considerable cost overrun. After completion it became apparent the cost to move a single space station assembly across the country with the C-5 SCM (space cargo modified) could exceed a $1,000,000.

The C-5 maintenance cost had been as high as 75 hours

for every one hour of flying. Many improvements and enhanced maintenance procedures had reduced it to 46 hours for every one hour of flight. However, it was still the highest operational cost of any cargo transport in Air Force inventory. This exceeded previous guppy operational cost by a considerable margin and was far too prohibitive. Without an operational guppy NASA was forced to rely on the C-5SCM and other military aircraft for component transport for six years.

The space agency was still in need of a volumetric transport. However, Aero Spacelines and successor Tracor had long disappeared, leaving no alternative because there were no manufacturers of outsize guppy type aircraft. The only serviceable guppy aircraft which existed were the C-97 based SGT-201s, which were transporting airframe components for Airbus. The outsize Antonov AN-225 created for the Soviet space program first flew in 1988. It was created to transport outsize items such as Buran which was the equivalent to the American Space Shuttle. The AN-225 was more costly to operate than the C-5 SCM, making it impractical and politically impossible to charter. The Soviet Union did not collapse until the end of 1991 paving the way for better cooperation between space agencies and development of the International Space Station. The AN-225 eventually found its way into commercial charter service.

International Space Agency

Within the framework of the International Space Station program a barter system was set up among the participating space agencies and NASA. The objective was to provide a way for agencies to exchange goods and services among themselves, eliminating the need for budget appropriations and financial transactions. The introduction of the next generation Airbus A300-600ST Beluga guppy brought the retirement of the EADS (Airbus) SGT-201 Super Guppies. This provided the opportunity for NASA to acquire a much needed outsize transport. Johnson Space Center director George W. S. Abbey instructed AOD Bob Naughton to find a way to procure a guppy from the French.[34]

The reality of the Super Guppy being still the most viable cost effective transporter for NASA since 1965 is quite a compliment to its creator Jack Conroy. The last Super Guppy SGT-201 built in 1983 had relatively low hours and was the only SGT which had not been involved in a serious damage incident. However, the large expense of an aircraft purchase was not in the NASA budget and would require congressional appropriation of funds. In addition Airbus is not an approved vendor to NASA, making it difficult to obtain funds for an aircraft purchase.

The European Space Agency (ESA), a NASA space program counterpart, is closely connected through contracts to subsidiary companies of the European Aeronautic Defense and Space Company (EADS) which is the parent of Airbus. The European Space Agency did not have a full time need for a surplus SGT-201. If a critical need arose for outsize heavy lift the case could be made to Airbus to provide a next generation A300-600ST Beluga.

The American space program did have a critical need, prompting negotiations, with help from ESA, for one of the retiring Airbus SGT-201 aircraft. Craig Stencil at the Johnson Space Center brokered a deal with the European Space Agency and Airbus. NASA would acquire one SGT-201 with all ground support equipment, one spare engine, two spare propellers and all spare parts inventory from all four Airbus guppies.[35]

In order to circumvent a cash transaction and the restriction from buying directly from Airbus, ESA negotiated the purchase of the last SGT-201-4(F) F-GEAI from EADS (Airbus). The European Space Agency (ESA) would then trade the SGT-201 with support equipment and spares to NASA in exchange for shuttle services consisting of a total of 450 kg payload distributed on multiple shuttle flights. ESA made one fixed price payment to Airbus for SGT-201-4(F). NASA then traded shuttle service to the International Space Station without making cash payments to Airbus. The ESA/NASA Super Guppy Transporter Barter Contract agreement was signed on 15 August 1997.

NASA sent an air crew to Toulouse in September and October to train on the SGT-201. The group, made up of John Breintebach, Larry Glenn, Frank Marlow, David Finney and Arthur "Ace" Beall, were qualified on multiple NASA aircraft including the retired B-377SG N940NS.[36] The SGT-201-4-(F), F-GEAI, was transferred to NASA in October 1997 after NASA crews spent four weeks in France for flight training. The aircraft was assigned to NASA as N941NA, arriving at Ellington Field near Houston on 23 October 1997. The SGT-201 is FAA type certified giving it the ability to transport commercial cargo. However, it operates as a public aircraft in NASA service.

Crew training on the SGT-201 with the Allison 501 engines proved to be more than just becoming familiarized with aircraft. The previous NASA guppy N940NS was fitted with the higher horsepower P & W T-34 engines. The guppy size and shape contribute to handling characteristics and limits which make for interesting nontraditional flying. Airspeed restrictions on NASA 941 were set at .433 mach because of the incident of upper front cave-in during testing of the original B-377SG N1038V in December 1965. The maximum indicated airspeed was set at 210–215 knots at 14,000 feet and below.

The SGT-201-4's (N941NA) max Turbine Entry Temperature (TIT) was 1,010 C. In order to establish an operational standard the NASA crews set the TIT limit at 930 degrees C at cruise, which produces airspeed between 205 and 210 knots. The curved structure of the guppy has the tendency to cause the airflow to go supersonic above .433 mach, which is a good reason to monitor the speed. A clucker alarm is set to warn when approaching an overspeed condition. Above 14,000 feet the cruise is reduced to 205 knots or about 240 mph true airspeed. Because the SGT-201 is more aerodynamic than its predecessor B-377SG it operates very well under most conditions with the lower horsepower Allison engines.[37] This is not to say that it does not struggle on hot days with heavy payloads but it is aerodynamically more efficient.

NASA crews found the last SGT-201 to be far superior to the original B-377SG Super Guppy. However, it is still old school technology. None of the SGT-201 controls are boosted with hydraulic assist. All surfaces including the ailerons, elevator and rudder are manually activated. The trim is manually cranked in for all surfaces from the center pedestal. There is no autopilot, not even a partial autopilot. The propellers are a unique hybrid system using Hamilton Standard 54H60-123 blades combined with components from the Lockheed C-130 sync system and Lockheed P-3units. Power is not related to shaft horsepower as in the Lockheed P-3 Orion (Electra). Instead it is related to torque like a Lockheed Hercules C-130. The engine instruments are basically a C-130 panel of analog gauges. The crew members who had flown C-130s found it to be an easy transition.[38]

As stated the SGT-201 is more aerodynamic than previous guppies; however, takeoffs from high altitude fields on hot days or from short runways put the aircraft at its power limits. To provide extra boost on takeoff a water methanol anti-detonate injection (ADI) is used for an extra boost of power. The system has four 56-gallon fiberglass tanks mounted behind each engine in the aft section of the nacelle. During takeoff when the throttles hit 70 degrees of travel on the quadrant, the ADI pumps kick in, giving an additional 2,000 pounds more torque on each engine, which makes a big difference. On average about 58 gallons of water methanol is used during two minutes of max power setting on takeoff. The engines produce 20,972 inch/pounds of torque and are quite capable of inlet temperatures (TITs) greater than the 930 degrees set by NASA. To insure the aircraft is operated within safe limits, a TIT overheat control system was installed as a precaution to prevent over temping the engines under these extreme max takeoff conditions.[39]

The possible need to operate flights with heavy loads at high altitude on hot days emphasizes why NASA's 1989 plan to re-engine the first B-377 Super Guppy N940NS with the lower horsepower Allison T-56 engines was ill-planned. The original B-377/C-97 Super Guppy struggled under these conditions even with the higher horsepower P & W T-34 engines because it was heavier with considerably more drag than the SGT-201.

The cockpit of the last SGT-201 is far superior to the original super guppy. Although turbine powered, there are still similarities to the original Stratocruiser. The engineer's panel is similar to the Lockheed C-130 unit with analog gauges. There is no autopilot and the controls are not hydraulic assisted. Even the trim is manually cranked in from the center pedestal. It is a hands-on aircraft, which requires constant attention (NASA, courtesy Bob Burns collection).

NASA SGT-201 Cargo Mission

After nearly eight years without an operational guppy, NASA took delivery of SGT-201-4(F) in October 1997. It was assigned the registration of N941NA with the call sign of NASA 941. Initially crew training was the primary focus before the expected routine of supporting the International Space Station (ISSP), which was the reason for acquiring the French built guppy. Additionally it would be used to support other NASA projects and aircraft recovery missions. Contrary to what was expected, the first mission of NASA 941 was not for the ISSP but transporting the rear section of a Boeing 747 for the SOFIA airborne telescope project. The telescope was a NASA joint project with the German Aerospace Center (DLR) to put an infrared telescope above the moisture in earth's atmosphere which blocks infrared wavelengths.

The fuselage section 46 of a retired United Airlines Boeing 747SP, N141UA, was needed to build a full-size mockup of the rear mounted airborne telescope. The tail section of the retired aircraft would be used as a mockup to engineer and fabricate bulkheads and mounts for the SOFIA telescope. The operational mounting would then be installed in the aft section of another operational B-747SP acquired by NASA which had also been retired by United Air Lines and stored since 1995. The completed SOFIA B-747SP was given the call sign NASA B-747 and expected to transport the largest airborne observatory in the world.

The first mission of NASA 941 was not transporting space hardware. It was dispatched to Ardmore, Oklahoma, on 04 May 1998 to uplift the 24,450 pound rear fuselage section from a retired United B-747SP. NASA required the airframe to engineer and construct the mountings and bulkhead for the SOFIA airborne telescope, which was installed in another B-747SP (NASA, courtesy Larry J. Glenn).

The SGT-201 arrived at the aircraft bone yard in Ardmore, Oklahoma, on 04 May 1998 to uplift the B-747 rear section for delivery to L-3 Communications (Rayethon) at Waco, Texas. The fuselage section had been cut away from behind the wing and the empennage removed behind the rear pressure bulkhead. It was mounted on a special designed cargo loading fixture which will slide into the NASA 941. Transporting of the section had presented NASA with a serious problem because it was not practical to move it overland and there were no military cargo aircraft capable of handling it. The section weighing 24,450 pounds is 50 feet long and 21 feet in diameter making it a close fit even for the Super Guppy-201. Without the guppy the project would have been in serious jeopardy. The first NASA payload flight for the guppy was just the beginning of many missions uplifting complete aircraft and partial airframes. NASA 941 made 22 flights between June 1998 and March of 2000 transporting components for the International Space Station.

Critical Weight and Balance

Weight and balance are factors which must be considered on all aircraft in order for them to operate efficiently and safely. Commercial airliners are relatively easy for the load planner to calculate the center of gravity (CG) because passenger weight is distributed across the horizontal plane of the airframe. Luggage, cargo and mail are distributed on the lower deck in the fore and aft cargo compartments to make it fly level. Fuel loads are set for each tank to maintain the attitude of the aircraft as it is consumed and weight is burned off. The majority of commercial aircraft are relatively easy to balance but there are some exceptions like the Boeing 747SP. It is difficult to load plan because it is relatively short and carries a high fuel load for great distance. It even carries fuel in the horizontal tail plane. Although more difficult to plan because of a critical CG the weight and balance challenges are relative to only the horizontal plane.

The Super Guppy aircraft have some handling characteristics which are unusual compared to other transport aircraft because of the height of the fuselage relative to the cargo floor. Commercial airliners like the Boeing 747 have a large diameter fuselage. However, the main deck floor is usually at midlevel with concentrated cargo weight below the floor on the lower deck. This more evenly distributes weight both above and below the wing. The cargo bay of the outsize fuselage SGT is above the main floor to accommodate extremely large items which can be mounted in fixtures. This can result in considerable weight concentrated high above the cargo floor and wing. Many of the guppy cargo loads can have considerably more weight concentrated above the midpoint of the cargo bay. For this reason the aircraft performance charts are predicated on both the horizontal center of gravity, which is standard on all aircraft and the vertical CG on the guppy. The loadmaster and flight engineer are tasked with making the complex calculations for both CG limits.

The height and weight of the extreme outsize payloads must fall within certain parameters and values to be in the safe zone. Excessive weight high above the wing can make the aircraft top heavy. Picture a bowling ball in the top of a box. When picked up it tends to flip over because too much weight is above the center line. The same applies to an aircraft. Too much weight in the tail makes it fly nose up. Too much weight above the wing when banking makes it want to roll over.

When N941NA was acquired through the European Space Agency, NASA crews spent several months at Toulouse for flight training with Airbus. The crews were instructed on vertical CG calculations and given the charts with specified weights up to the certified 162 inch high weight limit. The 162 inch height is the limit at which an unequal amount of weight can affect aircraft stability and handling. More easily explained, the aircraft cannot have more weight above 162 inches (13.5 feet) than below.[40]

The French had conducted tests up to the 175 inch level but the trainers refused to provide all the high vertical CG data to NASA flight crews. They had concluded it was not practical to fly the aircraft in a top heavy configuration because it could produce an extremely unsafe condition. The French attitude was if the data was not supplied then it would not be flown above 162-inch certified operating limits. The reasoning was if the pilot made a left turn at the maximum bank angle of 45 degrees with concentrated weight at 162 inches it would take 90 seconds for the aircraft to come back to wings-level attitude. If the number one engine flamed out or suffered any malfunction it would constitute a nonrecoverable situation. The Airbus trainers did not want to be held accountable in the event of an incident caused by extreme payloads at NASA.[41]

Aero Spacelines pilots on the Pregnant and first Super Guppy encountered this top heavy condition when transporting NASA cargo in the 1960s. The guppy was so revolutionary in 1962 these were conditions which had not been previously experienced. Years later several Aero Spacelines pilots acknowledged they inadvertently encountered some scary situations when operating at high weights. There were a few technicians who refused to be on the guppy with certain high cargo loads. The B-377PG and SG did have a heavier bottom because of the old B-377 lower fuselage. The SGT-201 was built with a much lighter new lower fuselage which made the vertical CG top heavy situation more critical.

NASA needed additional data outside the 162-inch envelope which Airbus would not provide. NASA was confronted with the problem of some space station cargo planned for transport having considerable weight in the upper half and considered top-heavy. NASA flight management decided to conduct in-house tests to determine the maximum limits of operation. The data was needed or they would be unable to transport some of these extreme payloads because there would be no tables to calculate the operational limits. In September 1998 a series of yaw and velocity dive tests with NASA 941 almost ended in disaster. The flight test identified as phugoid (aerodynamic yawing roll test with high vertical CG payloads) tests a series of aircraft behaviors. It covers the proper trimming of the aircraft to maintain speed stability and prevent porpoising and recovery from turns at high bank angles with maximum payload and high vertical CG.[42]

The guppy fuel tanks are large rubber bladder type bags which are fitted in cavities of the wing. The SGT-201 is equipped with two 28-volt boost pumps per tank in the same configuration of the B-29, C-97 and KC-97. The pumps in all the tanks are hard mounted to the bottom of the tank without a sump, surge box or hopper tank around the pumps to prevent starvation in extreme attitude. Boeing standard requirements were for all pumps to be on during flight which was adhered to by Airbus crews. NASA changed the French checklist after nearly a year of operation to using only the four forward boost pumps in cruise or level flight to reduce wear on running all eight. Up until this point no problems were encountered with the single pump operation after reaching cruise.

On the first day of the phugoid test the aircraft had over 40,000 pounds of dead weight to simulate maximum load conditions. Frank Marlow was in the left seat and Larry Glenn was at the engineer's panel. The aircraft had about 12,000 pounds of fuel in the wing tanks when the test began. The SGT-201 also has a center fuel tank with 8,000 pound fuel capacity and two pumps but it was not in use. After conducting several attitude tests, fuel was burned off to approximately 10,000 pounds.[43]

Pilot Marlow was still conducting the phugoid test with hard rudder. Engineer Glenn began to get intermittent low fuel pressure lights at certain attitude because the fuel in the tanks was forced away from the pumps, creating a starving condition. Glenn advised the pilots he was turning on all eight pumps. He also suggested they should consider terminating the test and continue on the next day with a greater quantity of fuel to which the pilots agreed. Upon return the alternate crew was briefed on what they had encountered with low fuel. It was advised, conversely to NASA prescribed flight procedure, all boost pumps should be running on any further phugoid high CG testing.[44]

On the following morning the crew assignments for engineers were changed. Marlow was in the left seat and Dave Finney in the right; however, Glenn had been replaced by Ken Sanders on the panel. There were also two flight test engineers on board. The aircraft was at altitude and they had conducted multiple yaw slips and slides at higher angles. All fuel selectors valves were on tank to engine. The test had begun with a high fuel load. Only one pump per tank was on as standard NASA practice. Sanders was watching the fuel because they were within a few hundred pounds before setting the selectors on cross-feed to insure uniform distribution of fuel as it is burned off.[45]

As the next yaw test started Marlow kicked hard right rudder and the right wing tucked to the rear. The left wing was very high as the aircraft banked. Suddenly engines number three and four rolled back 70 percent to start limiting power. The propeller RPMs were constant but there was no power. Sanders realized the props were controlling the engines and feared they would go into feather. The engines would not respond and come back up. The aircraft weighed over 140,000 pounds and they were in a sliding descent with no power on the right side.[46]

Boeing/Bell contracted to NASA for the SGT to transport V-22 Osprey airframes from Philadelphia and Amarillo for the installation of the drive system and rotors. Airframes due for overhaul were transported on the return trip to reduce the expense of the guppy flying empty (NASA, courtesy Bob Burns collection).

Marlow called a "Mayday," which was heard at Ellington Air Force Base. Sanders later stated, "We didn't think we would make it back." The other NASA flight crews at Ellington at flight ops turned all ears to the radio transmission. There was real concern they were not going to regain control of the aircraft and could possibly go into a slide or roll over. They were descending rapidly in a right wing down attitude when Sanders suggested to Marlow they shut the two starved engines down one at a time and try to restart. Marlow agreed and Sanders shut down number three. He was able to restart and bring it back up to power. Sanders repeated the procedure with number four and it responded with no problems. As soon as the aircraft was straight and level and no longer losing altitude, they climbed as high as it would at maximum weight and turned back to Ellington Field. The aircraft landed without incident and was impounded.[47]

As a result of the incident and data obtained from the test, new limits were placed on degree of turn when transporting cargo with a high vertical CG. To insure the engine fuel starving situation did not occur again the flight engineer procedures reverted back to the 1950s Boeing standard of flying with all boost pumps on.

NASA N941NA Aircraft Transporter

On 11 July 2000 NASA 941 transported the X38, V-131 test vehicle from Ellington Air Force Base, Texas, to Edwards Air Force Base, California. Two days later the guppy mission took a greater role of transporting aircraft for NASA and the DOD by uplifting a Northrop T-38 from Moffett Field to El Paso. Two more T-38s were transported the next day from El Paso to Ellington. This was the beginning of more than 91 flights not associated with the International Space Station mission. The last guppy cargo flight for International Space Station was on 16 December 2002.

Beginning in 2002 Boeing/Bell contracted to NASA to transport new V-22 Osprey airframes with the tail mounted from Philadelphia and Amarillo. The fuselage is built in Philadelphia and the drive system rotors are assembled in Fort Worth, Texas, with final assembly in Amarillo. In addition fuselages in need of overhaul are flown to Philadelphia for modifications then back to Texas. This arrangement gave the guppy a round-trip payload on half the missions. A total of 15 were transported from Amarillo for overhaul and 31 new Ospreys were moved back from Philadelphia.

The aircraft recovery flights which were becoming more numerous included a Navy SH-3 helicopter and F-14 plus a total of 28 Northrop T-38s for NASA, the Air Force and Navy. A special pallet fixture was designed to mount and secure aircraft for easy loading on the SGT. The pallet allowed for the loading of two T-38 aircraft at the same time.

In 2003 two Navy T-38s from the test pilot training school at NAS PAX River were severely damaged by hail. The pair made an emergency landing at Seymour Johnson Air Force Base in Goldsboro, North Carolina, with hundreds of holes in the nose, wings and empennage. The Navy contacted NASA at Ellington with a request for the SGT to recover the aircraft. The guppy was dispatched to Goldsboro on a three day mission moving the aircraft to Randolph Air Force Base for repair. Many other T-38s were transported from Moffett Field, El Paso, Titusville, Ellington, Biloxi, Davis-Monthan, and Edwards. A total of 72 aircraft were uplifted between 1998 and 2016 for NASA and the DOD.

Deicing and Heater System

NASA N941NA is the last SGT-201 built and is more refined than earlier models. However, it still suffers from problems which date back to the first B-377/YC-97 Super Guppy built in 1965. The SGT-201 wing heater system first designed for the B-377SG was considered a nightmare by the flight engineers. The inboard wing leading edge uses engine bleed air for anti-ice, however, outboard of number two and three utilizes the old B-377 1940s era combustion heaters.

The system has three heaters mounted in a cluster behind number one and four engines. Just above the heaters is the wing fuel tank. Warm air is forced outboard through ducts in the leading edge of the wing to the wingtips where it exits overboard. There was always a fear a thermistor could go bad allowing the heaters to keep running and cause major damage or burn the wing off. An incident occurred at Ellington Field in 2002 while engineer Larry Glenn and crew chief Tom Gordon were doing a ground system check. The heater system in the right wing malfunctioned. Glenn and Gordon attempted an emergency shutdown but the system ran away when the heater control valve behind number four engine inside the "skate doors" stuck in the open position. The fire department was alerted of possible fire in the wing. As they arrived Glenn and Gordon were able to cut the fuel supply. NASA technicians reviewed the incident and concluded the only option was to deactivate and remove the combustion heater system. NASA could not take the chance on losing the only airworthy guppy to a wing fire. Because of the incident there currently is no anti-ice for the outboard wing. Bleed air is used for the inboard wing and for the leading edges of the vertical and horizontal stabilizers.[48]

NASA estimates acquiring the last SGT-201 for in-house operations saved over $40,000,000 in space component shipping cost. Without it the agency would have relied exclusively on the Air Force C-5 SCM aircraft at approximately $1,000,000 per mission. The Super Guppy Turbine was able to make 26 payload flights with components for the International Space Station for around $3,000,000.

NASA acquired the last SGT-201 F-GEAI from ESA through a barter agreement. The European Space Agency purchased it from Airbus. It was then traded by ESA to NASA via the space agency barter agreement for a 450 kg payload on multiple space shuttle flights. A NASA flight crew was sent to Toulouse to train on the guppy before delivery as N941NA (NASA, courtesy Bob Burns collection).

On 14 April 2005, after eight years of service, N941NA was flown to Dryden for some much needed maintenance and upgrades. The landing gear was replaced and retraction checks conducted. It was returned to service on 28 April after two weeks maintenance. In the summer of 2006 it spent three months at Tinker Air Force Base, Oklahoma, where it was completely overhauled, repainted and buffed. It took 3.5 weeks to polish the skin of the upper half of the fuselage to a mirror finish which burned up six air operated buffing machines.

The last SGT-201is far advanced over Jack Conroy's 1962 original B-377PG but is still a dinosaur by today's standards. The A-300-600ST Beluga and B-747LCF Dreamlifter were built with up-to-date technology. The SGT-201-4-(F) is for practical comparison an old style post–World War II manual aircraft with simple hydraulics and no autopilot. It has 1960s Sud Caravelle flight gauges which are more modern than the old 1940s Stratocruiser but still old school with the hybrid C-130 engineer's panel. There is no comparison to the all-glass cockpit of the Airbus A300-600ST Beluga which replaced it.

NASA 941 is the most advanced of the Jack Conroy ASI designed guppy series yet it still requires constant monitoring and flies out of trim most of the time. A crew of nine which includes two flight engineers and three mechanics is needed, proving it is a handful to maintain and operate. NASA keeps it in outstanding condition but flying has been reduced to only six to 10 flights a year which can be a week in duration. In April 2012 DynCorp International was awarded a contract to maintain the guppy along with other NASA aircraft at facilities in El Paso.

NASA N941NA is the final B-377/C-97 based SGT-201. At the end of World War II aircraft designers could not have envisioned anything like the Super Guppy. Postwar development gave us the XC-99 and C-124, which were technically guppies. Boeing designed the C-97 as a large military transport with a sub-model B-377 commercial version. The B-377 Stratocruiser was a commercial failure but the guppy series developed from it has defied the odds of longevity for transport aircraft. In 2017, after 55 years of guppies transporting space hardware, the lone Super Guppy survivor is still in NASA service. This is an impressive achievement when considering it is nothing more than a 1970 upscale version of Lee Mansdorf's and Jack Conroy's first 1961 Aero Spacelines design of a modified 1940s airliner. The 1940s vintage lower aft fuselage on N941NA came from the third Boeing B-377 Stratocruiser N1024V built in 1948 via the Pregnant Guppy. The wings, tail, cockpit nose and main gear are from a 1952 KC-97 (52-0828) which were overhauled and modified by Aero Spacelines (Tracor) in California. They were combined with parts from multiple donor airframes to fabricate a fuselage which was assembled by UTA in France in 1983. The aircraft is a stunning tribute to the quality, strength and structural integrity of American wartime aircraft engineering which began with the B-29, evolved into the Boeing B-377 Stratocruiser then given a second life by Aero Spacelines and UTA Industries. The record and longevity stands for itself proving the aircraft is truly a "Super" Guppy.

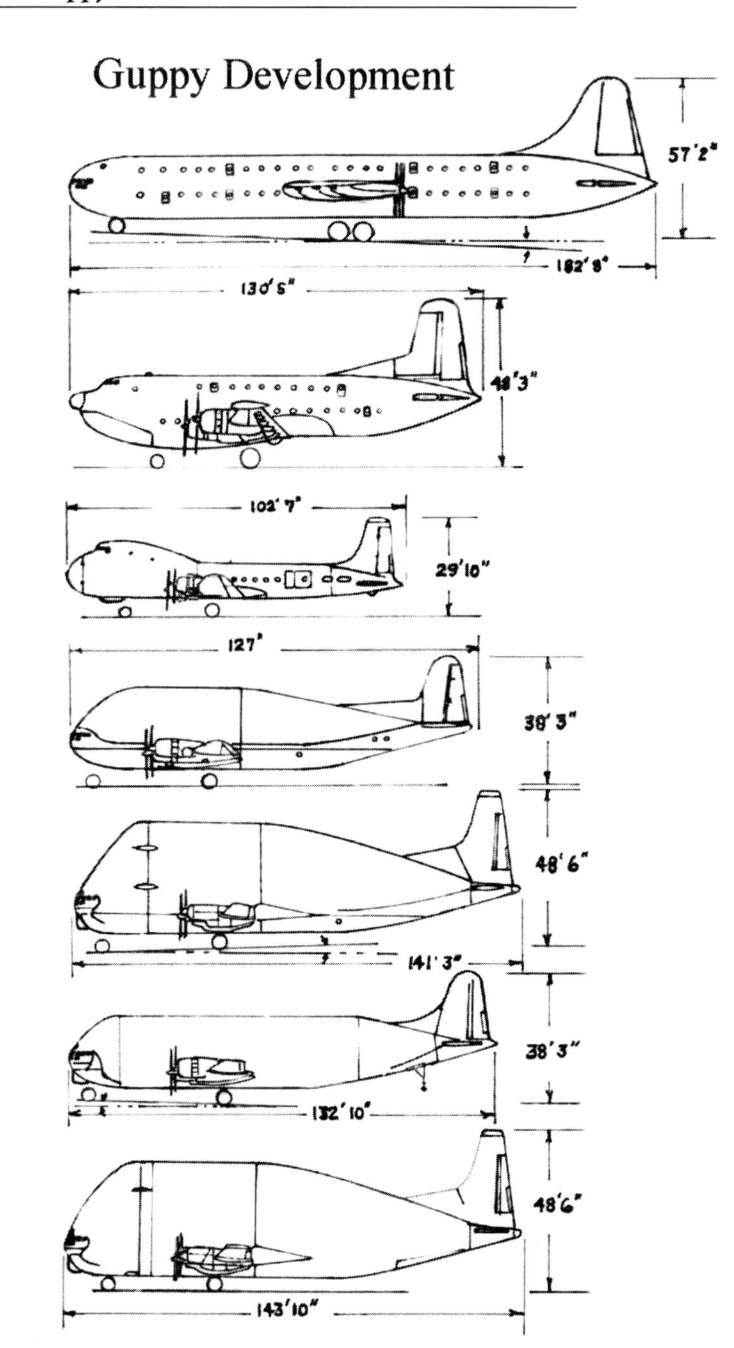

A comparison of postwar transports and Jack Conroy's outsize airplane design are some indication as to why he was not taken seriously. In 1962 the fuselage diameter was considered relative to wingspan and length. The B-377 Pregnant Guppy defied conventional thinking by maintaining the standard B-377 wing while lengthening and expanding the fuselage diameter. Subsequent models were even larger (author's collection).

Epilogue

It is not clear who conceived the concept of expanding the fuselage of an existing airframe to create an outsized volumetric aircraft, or when. In the 1930s giant aircraft were considered science fiction. The need for large military transport aircraft became evident in the 1940s. Boeing introduced the C-97 transport in 1944, followed with the commercial sub-model B-377. The C-97 had clamshell doors in the aft belly and a ramp system but was limited by fuselage diameter. The 1948 Douglas C-124 Globemaster was the first major design change with straight in loading. The two-story fuselage with folding upper deck, drive on/off capability and self-contained ramp system was a major step forward. The outsize C-124 and Consolidated XC-99 were considered to be at the maximum allowable limits of fuselage diameter. They were volumetric but never officially identified as guppy class aircraft although technically they were. Both came about by expanding the fuselage of existing designs.

Turboprop and jet powered transports were developed in the 1950s but fuselage diameter remained in the same range. Conventional design was revolutionized in 1962 when Aero Spacelines modified a Boeing B-377 Stratocruiser to create the first extreme diameter heavy lift volumetric transport. The strange appearance was mocked and jokingly said to look like a "guppy." The name soon became the official identity of all extreme diameter aircraft.

Aircraft broker Lee Mansdorf toyed with the idea of modifying a B-377 airframe to transport NASA hardware about 1960 while seeking a buyer for the fleet of surplus Stratocruisers he had acquired at bargain prices. When John M. "Jack" Conroy approached him to buy several aircraft to start a VIP airline the process was set in motion. Mansdorf realized Conroy had the drive and conceptual genius required to take the challenge and create a new class of high volume heavy lift transports.

Traditional aerodynamicist believed the high frontal area of the expanded fuselage made it too unstable to safely fly. The few military aircraft with expanded fuselages which came before were small in comparison. Aircraft manufacturers were still thinking in terms of limited fuselage diameter relative to wingspan. Even passenger airliners were single aisle with the belief very narrow needle like supersonic aircraft would be the future of passenger travel. Seven years before the first Boeing 747 flew Conroy set the standard for wide-body aircraft with the Conroy Class B-377PG. His adaptation of Mansdorf's idea of a volumetric aircraft got the attention of engineering departments of every major aircraft manufacturer. He set a new standard and proved it could be done. His influence is seen in all wide-body expanded fuselage aircraft both cargo and twin-aisle passenger jets which have followed.

Conroy is possibly the most underrecognized genius in American aviation. His contributions place him in the very small club of aviation pioneers. Many consider his achievements equal to the Wright brothers, Charles Lindbergh,

"Smiling" Jack Conroy squeezed an incredible life into only 58 years. His life and achievements read like an action movie script. He only built a handful of aircraft but his designs forever changed cargo aircraft. He is not universally recognized for his achievements and yet he is in the company of William Boeing, Donald Douglas and Jack Northrop (courtesy Angelee M. Conroy).

Donald Douglas, William Boeing and Jack Northrop. His circle of friends included the most achieved test pilots and airmen. He enjoyed a long friendship with Clay Lacy, who has logged more hours than any pilot in history. Lacy said, "Jack got me into airplanes I would have never gotten to fly, he was my best friend."

Conroy's life could have been an action movie script. He squeezed an incredible life into only 58 years. He became an actor who appeared in nine movies and learned how to fly early in life. He was working at Pearl Harbor on 07 December 1941 and that prompted him to join the Army. He was assigned to the USAAF and became a B-17 pilot. He was shot down over Germany on his nineteenth combat mission, becoming a POW. After the war he was a pilot for five different non-scheduled air carriers while working as a swimming pool salesman in Los Angeles. In 1955 he set a cross-country speed record in a National Guard F-86 by flying Los Angeles—New York—Los Angeles in 12 hours by refueling in-flight three times each way. He did it a second time in 1966 in a Lear Jet accompanied by his friend Lacy. The boldness of his personality and love for his children is demonstrated in the fact he took three of them along on the second record setting flight. After creating a turboprop DC-3 in 1969 he took his family along on a world demonstration tour with famed test pilot Herman "Fish" Salmon. This was followed by creating the Tri-Turbo Three by mounting a third turbine in the nose.

Aero Spacelines was formed for the express purpose of building an outsize transport aircraft. Conroy's passion for doing what others believed was impossible captured the imagination of NASA director Wernher von Braun. The two men developed a respect for each other possibly because they dreamed big. Von Braun acknowledged, if not for Jack Conroy's guppy airplanes America would not have been on the moon by the end of the decade.

Jack Conroy was the ultimate promoter and showman. He was always underfinanced, operating on the financial edge. However, he possessed the ability to convince you of anything. He set impossible goals for himself and his employees. At the moment failure was imminent he could change direction, refusing to accept the inevitable. Kirk Irwin, who succeeded him at Aero Spacelines, said, "Jack Conroy was a friend to whom I would not loan a dime," joking about Conroy's reputation for big ideas without consideration of finances. "He believed in the concepts considered impossible. We need more men like him." British aviator and entrepreneur Kevin Keegan (Transmeridian) described him as an eccentric American with incredible imagination.

When Jack Conroy left Aero Spacelines in 1967 he was not finished building unique aircraft and proposing new designs. His aviation adventures took a new direction creating an expanded fuselage CL-44 based guppy. He experimented with converting DC-3s, SA-16s and C-119s to turboprops. His most intriguing design proposal was the Virtus, a twin fuselage B-52 monstrosity with a 450-foot wingspan to transport the Space Shuttle Orbiter. Boeing and Lockheed countered with a twin fuselage 747 and C-5 designs. Conroy proposed the piggyback Space Shuttle Transporter which became the NASA B-747SCA.

A derivation of Conroy's Virtus lives on today. A twin fuselage high-tech monster named the Stratolaunch is being developed. The aircraft, using Boeing 747 components, is being built at Mojave where Conroy stored the B-377s used by Aero Spacelines and conducted many test flights. The evolution of outsize aircraft will be continued in another volume of post–Super Guppy and hybrid giants.

Jack Conroy's youngest daughter Angelee treasured the many illustrations, letters, photos and files which were left in her care. She spent many years researching and adding to the collection which we were privileged to have access to for this tome. On the morning of 5 December 2017 a raging wildfire swept through Ventura, California, destroying Angelee's home and hundreds of others. Ninety percent of all the original Aero Spacelines, Turbo-Three Corporation and Specialized Aircraft Corporation documents as well as personal letters and photos were destroyed. The aviation world has lost a part of history which will never be restored. Much of the story told in these pages came from those lost files. Coincidentally Jack Conroy passed away 38 years ago to the day on 5 December 1979.

Appendices

A. *Aircraft Size Comparison*

Type	*Span*	*Length*	*No Engines*	*Gross Wt (lb)*	*Fuselage Size*
Airbus (B377) SGT-201(F)	156 ft 8 in	143 ft 10 in	4	341,713	25 ft dia (upper)
ASI B377PG (Conroy)	141 ft 3 in	127 ft 0 in	4	141,000 (133,000)	19.5 ft dia (upper)
ASI B-377SG (Conroy)	156 ft 3 in	141 ft 3 in	4	175,000	25 ft dia (upper)
ASI B377MG (Conroy)	156 ft 3 in	132 ft 10 in	4	142,000	18 ft × 15.5 ft oval
ASI B377 MGT-101	156 ft 3 in	136 ft 6 in	4	180,000	18 ft × 15.5 ft oval
ASI B377 SGT-201	156 ft 3 in	143 ft 10 in	4	170,000	25.5 ft dia
ATL-98 (DC-4) Carvair	117 ft 6 in	102 ft 7 in	4	73,800	8.6 ft dia
Boeing B-29	141 ft 3 in	99 ft	4	105,000 (140,000 post war)	9 ft dia
Boeing B-50A	141 ft 3 in	105 ft 1 in	4	168,708 (J 179,500)	9 ft dia
Boeing B-377	141 ft 3 in	110 ft 4 in	4	135,000 (later 148,000)	11.75 ft (upper)
Boeing C-97(G)	141 ft 3 in	110 ft 4 in	4	175,000	11 ft × 9 ft upper oval
Boeing YC-97J	141 ft 3 in	110 ft 4 in	4	175,000	11 ft × 9 ft upper oval
Boeing 747-100	195 ft 8 in	231 ft 4 in	4	710,000 (later 735,000)	21 ft × 24 ft oval
Boeing 747LCF	211 ft 5 in	235 ft 2 in	4	803,000	27.5 ft × 32.6 ft oval
Convair YB-60	206 ft 5 in	175 ft 2 in	8	410,000	12.5 ft dia
Convair XC-99	230 ft	182 ft 6 in	6	265,000	14.3 ft × 20.5 ft oval
Convair B-36	230 ft	162 ft 0 in	6 (later 10)	410,000	12.5 ft dia
Douglas DC-4 /C-54	117 ft 6 in	93 ft 10 in	4	73,000	8.6 ft dia
Douglas C-74 (415A)	173 ft 3 in	124 ft 1.5 in	4	165,000	13.2 ft dia
Douglas C-124C	174 ft 1.5 in	130 ft 5 in	4	185,000 (208,000†)	11.6 × 11.3 ft useable
Douglas C-133B	179 ft 7.75 in	157 ft 6.5 in	4	286,000 (max 300,000)	12 ft dia
Douglas DC-10-30	165 ft 4 in	180 ft 8 in	3	580,000	19.75 ft dia
Lockheed C-5SCM	222 ft 9 in	247 ft 1 in	4	769,000 to 840,000†	19 ft × 13.5 ft*
Saunders-Roe SR.45	210 ft (219 w/floats)	148 ft	10	330,000 (max 345,000)	n/a

*useable oval (dimensions larger). † wartime overload.

B. *Aircraft Volume and Payload Comparison*

Aircraft	*First Flight*	*Volume*	*Max Payload*
ATL-98 Carvair (Guppy)	21 Jun 1961	4,350 cu ft.†	18,000 lbs
B-377PG Pregnant Guppy	19 Sep 1962	29,187 cu ft.	33,000 lbs
B-377SG Super Guppy	31 Aug 1965	49,750 cu ft.	45,000 lbs
B-377MG Mini Guppy	24 May 1967	25,192 cu ft. (23,615 useable)	40,000 lbs
B-377MGT-101 Mini Guppy	13 Mar 1970	25,800 cu ft.	64,000 lbs
B-377SGT-201 Super Guppy	24 Aug 1970	51,280 cu ft. (39,000 useable)	52,500 lbs
Boeing B-29	21 Sep 1942	Bomb Bay n/a	20,000 lbs
Boeing B-50A	25 June 1947	Bomb Bay n/a	20,000 lbs
Boeing C-97	09 Nov 1944	6,140 cu ft. (upper deck)	35,000 lbs
Boeing 377	08 July 1947	4,857 cu ft.	30,000 lbs
Boeing 747-100	09 Feb 1969	23,519 cu ft. (both decks)	108,000 lbs

Aircraft	*First Flight*	*Volume*	*Max Payload*
Boeing 747-100 SCA	1976‡	External Shuttle Cargo	240,303 lbs
Convair B-36	08 Aug 1946	12,300 cu ft. (bomb bay)	72,000 lbs
Convair YB-60	18 Apr 1952	12,300 cu ft. (bomb bay)	72,000 lbs
Convair XC-99 (Guppy)	23 Nov 1947	16,117 cu ft.	117,000 lbs
Douglas C-54/DC-4	14 Feb 1942	3,020 cu ft. (useable) 18,000 (war)	22,000 lbs*
Douglas C-74	05 Dec 1945	6,800 cu ft.	72,160 lbs
Douglas C-124 (Guppy)	27 Nov 1949	10,000 cu ft.	50,000 to 70,000 lbs
Douglas C-133	23 Apr 1956	13,000 cu ft.	80,000 lbs
Douglas DC-10	29 Aug 1970	(-30F series) 17,096 cu ft.	161,000 lbs
Lockheed C-5	30 Jun 1968	34,734 cu ft.	240,000 (war) 291,000 lbs*

*maximum wartime overload. †4,630 w/raised ceiling; 3,428 with 22-seat passenger cabin. ‡active aircraft modified in 1976.

C. Convair B-36 and Sub-Model Aircraft Statistics

Stats:	*B-36*	*XC-99*	*YB-60*
Gross Wt.	278,000 lbs.	265,000 lbs.	300,000 lbs.
Take-off Gross	357,500 lbs.	322,000 lbs.	410,000 lbs.
Span	230 ft.	230 ft.	206 ft.
Length	163 ft.	182 ft. 8 in.	171 ft.
Height	46 ft. 7 in.	57 ft. 6 in.	50 ft.
Max Speed	300 mph.	300 mph.	335 mph.
Ceiling	45,000 ft.	30,000 ft.	44,650 ft.
Take-off	5,000 ft.	5,000 ft.	6,710 ft.
Range	10,000 miles	8,100 miles	8,077 miles
Fuel	21,116 gal.	21,070 gal.	41,462 gal.
Wing Area	4,770 sq. ft.	4770 sq. ft.	5,239 sq. ft.
Payload	72,000 lbs.	101,000 lbs.	72,752 lbs.

D. Douglas C-124 Globemaster II Production Blocks

A total of 449 C-124 aircraft were built: One YC-124 converted from C-74 (w/four blade Curtiss Electric props), One YC-124A, 204 C-124As, One YC-124B (w/ T-34 turboprops) and 242 C-124Cs. The 448 production aircraft were produced in nine construction blocks. Aircraft 52-1090 and a 10th block 53-053 to 53-105 were cancelled. Airframes from s/n 49-245 forward were converted to "C" model standards. Aircraft 51-177 tested with carburetor scoops on top of engine cowling.

Production Blocks

Block 1: 28 Airframes s/n 49-232 to 49-259
Block 2: 36 Airframes s/n 50-083 to 50-118
Block 3: 14 Airframes s/n 50-1255 to 50-1268
Block 4: 111 Airframes s/n 51-073 to 51-182
Block 5: 15 Airframes s/n 51-5173 to51-5187
Block 6: 26 Airframes s/n 51-5188 to 51-5213
Block 7: 14 Airframes s/n 51-7272 to 51-7285
Block 8: 150 Airframes s/n 52-939 to 52-1089
Block 9: 52 Airframes s/n 53-001 to 53-052
Block 10: 52 (Cancelled) 53-053 to 53-105

Multiple changes were made to the forward panel and control pedestal but not at the same time as the engineer's panel or other airframe upgrades.

Forward Panel changes: 48-795 to 49-259
50-83 to 51-157
51-158 to 53-052

Control Pedestal changes: 48-795 to 50-94
50-95 to 51-107
51-108 to 51-5187
51-5188 to 51-7285
52-939 to 52-1085
52-1086 to 53-052

E. Boeing B-29, B-50, C-97, B-377 and ASI Guppy Production

Operator	Model	Name	Type	Designation	Built
U.S. Air Force	345	Superfortress	Bomber	XB-29	3
	345	"	"	YB-29	14
	345	"	"	B-29	2537
	345	"	"	B-29A	1122
	345	"	"	B-29B	311
				Total	3970
U.S. Air Force	345-2-1	Superfortress	Bomber	B-50A	79
	345-2-1	"	"	B-50B	45
	345-9-6	"	"	B-50D	222
	345-31-26	"	"	TB-50H	24
				Total	370
U.S. Air Force	367-1-1	Stratofreighter	Transport	XC-97	3
	367-5-5	"	"	YC-97	6
	367-4-6	"	"	YC-97A	3
	367-4-7	"	"	YC-97B	1
	367-4-19	"	"	C-97A	50
	367-4-29	"	"	C-97C	14
	367-4-29	Stratotanker	Tanker	KC-97E	60
	367-76-29	"	"	KC-97F	159
	367-76-66	"	"	KC-97G	592
				Total	888
Commercial Airlines	377-10-19	Stratocruiser	Prototype	B-377	1
	377-10-26	"	Pan Am	B-377	20
	377-10-28	"	Sila (SAS) to (BOAC)	B-377	4
	377-10-29	"	American Overseas	B-377	8
	377-10-30	"	Northwest Orient	B-377	10
	377-10-32	"	BOAC	B-377	6
	377-10-34	"	United Air Lines	B-377	7
				Total	56
Aero Spacelines	377-10-26	Stratocruiser	Pregnant Guppy	B-377PG	1
	367-76-29	YC-97J	Super Guppy	B-377SG	1
	377-10-26	Stratocruiser	Mini Guppy	B-377MG	1
	367-76-29	KC-97G Mini Guppy Turbine	B-377MGT-101	1	
	367-76-29	"	Super Guppy Turbine	B-377SGT-201	2
Airbus Industries	367-76-29	"	Super Guppy Turbine	B-377SGT-201(F)	2
				Total	8

F. Aero Spacelines B-377s and C-97 Conversion Airframes

Aero Spacelines acquired 27 B-377 airframes via Lee Mansdorf; six Pan Am, eight Northwest and 13 Transocean. At least six of those (#15944 65*, 66, 67, 68, 76) were previously purchased from BOAC by Transocean but never entered service. They are believed to have remained in BOAC colors with lettering removed while stored at Oakland. *(#15965 was purchased from Mansdorf by Lloyd Dorsett then transferred to Aero Spacelines.)

Three Guppy conversions from B-377s: Pregnant Guppy N1024V c/n #15924, Super Guppy N1038V/N940NS c/n #15938 (primarily YC-97J using nose, wings and control surfaces) and Mini Guppy N1037V/N422AU c/n #15937. The MGT-101 was built primarily with C-97 parts and B-377 lower belly. The lack of B-377 airframes caused the last four SGT-201 guppies to be built using all C-97G cockpit nose section, wings and empennage with B-377 sections 45–46 however they are identified as B-377SGT-201 aircraft.

B-377 Airframes Acquired by Aero Spacelines

c/n 15924: NX1024V ff. 17 Oct 48—Boeing test acft.—N1024V, Pan Am, 12 June 49—Boeing 1961 (35,769 hrs)—Lee Mansdorf, Aero Spacelines 1961 (35,769 tt. hrs) for "Pregnant Guppy" conversion, with P&W R-4360, combined with parts from c/n 15976,

N407Q—ff. 19 Sept 1962, damaged when struck by taxing 707 wing 10 Sep 1963—N126AJ, American Jet 1970s—Broken up for spares Van Nuys 1980—Section 45 salvaged reused second time on SGT-201-4, F-GEAI.

c/n 15927: N1027V ff. 12 Feb 49—Pan Am 02 March 49 "Clipper Friendship"—G-ANUM, BOAC, 03 Sep 1954 (11,722 hrs)—West African Airways, March 1958—N1027V/N401Q Babb Company to Transocean 1958 (tt. 19,247 hrs)—Sold at auction by Airline Equipment Company to Lee Mansdorf 01 Sept 60—Aero Spacelines for spares (tt. 20,119 hrs). Broken up at Oakland.

c/n 15937: N1037V ff. 11 Aug 49—Pan Am 08 Sept 49 "Clipper Fleetwings"—Boeing 1960—Lee Mansdorf—Aero Spacelines 1963, for conversion to "Mini Guppy" with P&W R-4360—ff. 24 May 1967—Combined with parts from c/n 15947 (Northwest N74601) and c/n 15967 (United Air Lines "Mainliner Hana Maui" (later G-ANTZ, N411Q)—B-377MG N422AJ to American Jet—N422AU, Aero Union Oct 81—Erikson Air Crane—Preserved at Tillamook, Oregon.

c/n 15938: N1038V ff. 20 Aug 49—Pan Am, 29 Sept 49 "Clipper Constitution" then "Clipper Hotspur"—Stored NY sometime after 1954—Boeing 1960—Purchased by Lee Mansdorf—Aero Spacelines, conversion to "Super Guppy," ff. 31 August 1965 with T34-PWA 5,700 hp turboprops. Used parts from YC-97J c/n 16724, B-377 c/n 15944 N408Q (N103Q, OY-DFY) and B-377 c/n 15945 N406Q (N101Q, LN-LAF)—Purchased by NASA on 29 March 1979 as N940NS—AMARC flyable storage 09 July 1991—Official transfer to Pima Air Museum 2015.

c/n 15942: N1042V, ff. 05 Dec 49—Pan Am, 30 Dec 49—Damaged 05 April 1950, left main gear collapsed and ground loop on landing at SFO. Repaired—Stored Miami 1960—Boeing 25 Jan 61 (tt. 32,406)—Lee Mansdorf 1961—Aero Spacelines, Broken up for spares Mojave.

c/n 15944: OY-DFY, SILA—G-ALSB, ff. 16 Sept 49—BOAC, "Champion," 24 Oct 49—Babb Company to Transocean N103Q/N408Q, 03 Feb 59; registration not applied to aircraft. (tt. 19,303)—Sold at auction by Airline Equipment Company to Lee Mansdorf, 01 Sep 60—Aero Spacelines for spares 1963. Broken up Oakland, parts used for "Super Guppy."

c/n 15945 LN-LAF, SILA—G-ALSC, ff. 28 Oct 49—BOAC "Centaurus," 02 Dec 49—The Babb Company to Transocean N101Q/N406Q, 08 Jan 59 (tt. 19,242)—Sold at auction by Airline Equipment Company to Lee Mansdorf, 01 Sept 60, (tt. 20,853)—Aero Spacelines for spares 1963. Broken up at Oakland, parts used for "Super Guppy.

c/n 15946: SE-BDR, SILA—G-ALSD, ff. 09 Nov 49—BOAC "Cassiopeia," 16 Dec 49—The Babb Company to Transocean N85Q/N403Q, 07 Sept 58 (tt. 18,669)—Sold at auction by Airline Equipment Company to Lee Mansdorf, 01 Sept 60 (tt. 21,428)—Aero Spacelines for spares 1963, Broken up Mojave.

c/n 15947: N74601, ff. 18 March 49—Northwest, 29 July 49—Damaged 05 August 1955 over ran runway Chicago/O'hare, props failed to reverse on landing. Repaired—Lockheed 1959—Lee Mansdorf 1960—Aero Spacelines 1963 stored Mojave. Broken up Santa Barbara 1967. Parts used on "Mini Guppy" N1037V.

c/n 15948: N74602, ff. 28 March 49—Northwest, 22 June 49—Lockheed 1959 (tt. 30,860)—Lee Mansdorf, 1960—Aero Spacelines 1963. Broken up for spares Mojave.

c/n 15949: N74603, ff. 09 April 49—Northwest, 10 July 49—Damaged by fire during engine run-up at YSP 28 July 57—Lockheed 1959 (tt. 24,844)—Lee Mansdorf, 1960—Aero Spacelines 1963, repainted in reverse Northwest paint scheme and flown for promotion and test purposes. Stored Mojave—Moved to Tucson stored. Section 45-46 removed early 1982 and shipped to France for SGT-201-5. Balance of airframe Broken up Tucson.

c/n 15950: N74604, ff. 20 July 49—Northwest, 11 August 49 inaugurated coast to coast service—Lockheed October 1959 (tt. 29,763)—Lee Mansdorf, 1960—Aero Spacelines 1963. Stored Mojave, broken up for spares.

c/n 15951: N74605, ff. 10 August 49—Northwest, 28 August 49—Lockheed 29 December 59 (tt. 30,841)—Lee Mansdorf, 1960—Aero Spacelines 1963. Broken up for spares Mojave.

c/n 15952: N74606, ff. 22 August 40—Northwest, 13 Sept 49—Lockheed 1959 (tt. 30,497)—Lee Mansdorf 1960—Aero Spacelines 1963. Converted to Freighter with forward cargo door. Not used on any Guppy. Stored Mojave and broken up.

c/n 15955: N74609, ff. 05 Nov 49—Northwest, 18 Nov 49—Damaged 12 Nov 57 when Number 2 cowl flap separated and struck tailplane.—Lockheed 29 Dec 59 (tt. 31,005)—Lee Mansdorf, 1960—Aero Spacelines 1963. Broken up for spares Mojave.

c/n 15956: N74610, ff. 08 Dec 49—Northwest 21 Dec 49—Lockheed 29 Dec 59 (tt. 30,866)—Lee Mansdorf, 1960—Aero Spacelines 1963. Broken up for spares Mojave.

c/n 15958: N90942, ff. 08 July 49—American Overseas 30 July 49—Pan Am, 25 Sept 50—Stored Miami 1960—Boeing 25 Jan 61 (tt. 33,191)—Lee Mansdorf, 1960—Aero Spacelines 1963 stored Mojave—Ground collision with N402Q at Mojave during run-up before ferry to SBA—Written off and Broken up for spares for Guppy program at Mojave. Section 45 believed used on N211AS.

c/n 15965: N31225, ff. 08 July 49—United Airlines, 28 Sept 49 "Mainliner Hawaii"—BOAC, G-ANTX, (8,979 hrs), "Cleopatra," 29 Dec 54—NAT 81 seat tourist class—Ghana Airways 1959—Babb Company to Transocean, N107Q/N412Q, 25 July 59 registration not applied to aircraft (tt. 18,355). Registration never applied to aircraft.—Sold at auction by Airline Equipment Company to Lee Mansdorf, 01 Sept 60—Lloyd Dorsett, June 1963—Trsf Aero Spacelines—Broken up Oakland 1964.

c/n 15966: N31266, ff. 12 Oct 49—United Airlines, 28 Oct 49 "Mainliner Kauai"—BOAC, G-ANTY, (9,161 hrs), "Coriolanus," Oct 1954–18 March 59 shed prop on take-off a Accra Ghana, returned safely—Babb Company to Transocean, N108Q/N413Q, 27 July 59, registration not applied to aircraft—Sold at auction by Airline Equipment Company to Lee Mansdorf, 01 Sept 60—Aero Spacelines 1963—Broken up Mojave.

c/n 15967: N31227, ff. 21 Nov 49—United Airlines, 15 Dec 49 "Mainliner Hana Maui"—BOAC, G-ANTZ, (9,063 hrs), "Cordelia," Dec 54—Ghana Airways, July 58—Nigerian Airways—Babb Company to Transocean, N106Q/N411Q, 10 April 59, registration not applied to aircraft, (tt. 18,605).—Sold at auction by Airline Equipment Company to Lee Mansdorf, 01 Sept 60—Aero Spacelines 1963—Broken up Oakland, parts used on "Mini Guppy" N1037V.

c/n 15968: N31228, ff, 18 Nov 49—United Airlines 03 Dec 49 "Mainliner Waipahu"—BOAC, G-ANUA, (9,289 hrs) BOAC, "Cameronian," 29 Dec 54—Nigerian Airways 1959—Babb Company to Transocean, N109Q/N414Q, Transocean, 01 Aug 59, registration never applied to aircraft, (tt. 19,104).—Sold at auction by Airline Equipment Company to Lee Mansdorf, 01 Sept 60—Aero Spacelines, 1963. Broken up Oakland.

c/n 15974: G-AKGH, ff. 01 Sept 49—BOAC, "Caledonia," 16 Nov 49—Babb Company to Transocean, N137A/N402Q, 02 Aug 58—Sold at auction by Airline Equipment Company to Lee Mansdorf, 01 Sep 60 (tt. 21,588)—Aero Spacelines 1963, ground collision with Pan Am N90942 c/n 15958 at Mojave April 69.—Written off and broken up at Mojave for spares, rear fuselage section 45 used on SGT 201-3, F-GDSG.

c/n15975: G-AKGI, ff. 05 Dec 49—BOAC, "Caribou," 10 Jan 50—West African Airways 01 May 57—Babb Company to Transocean, N100Q/N405Q, 06 Jan 59 (tt. 19,091)—Sold at auction by Airline Equipment Company to Lee Mansdorf, 01 Sep 60 (tt. 21,072)—Aero Spacelines 1963, Broken up for spares at Oakland.

c/n 15976: G-AKGJ, ff. 29 Dec 49—BOAC, "Cambria," 07 Feb 50—Babb Company to Transocean, N102Q/N407Q, 17 Sept 59, registration not applied to aircraft, (tt. 18,985).—Sold at auction by Airline Equipment Company to Lee Mansdorf, 01 Sep 60—Aero Spacelines 1962, parts used on N1024V c/n 15924 "Pregnant Guppy." Broken up Oakland.

c/n 15977: G-AKGK, ff. 17 Jan 50—BOAC "Canopus" 17 Feb 50—Ghana Airways 1958—Babb Company to Transocean, N104Q/N409Q, 03 March 59 (tt. 20,341)—Sold at auction by Airline Equipment Company to Lee Mansdorf 01 Sep 60—Aero Spacelines 1963, Broken at Oakland up for spares.

c/n 15978: G-AKGL, ff. 25 Jan 50—BOAC, "Cabot," 09 March 50—Prestwick, Scotland, 24 April 1951 at 08:27, Landed heavy on nose. Nose gear torn off, lower fuselage damaged back to wing trailing edge, all four props damaged. Rebuilt at Prestwick—Babb Company to Transocean, N86Q/N404Q, 13 Sept 58 (tt. 21,175)—Sold at auction by Airline Equipment Company to Lee Mansdorf, 01 Sep 60—Re-registered H9600H to Aero Spacelines. Stored Mojave, Broken up for spares.

c/n 15979: G-AKGM, ff. 01 Feb 50—BOAC, "Castor," 24 March 50—Damaged in-flight, west bound over North Atlantic, loss of prop on Number 4 then suffered engine fire 25 Dec 57. Emergency landing Sydney, Nova Scotia. Repaired and returned to service—Babb Company to Transocean N105Q/N410Q, 10 March 59—Sold at auction by Airline Equipment Company to Lee Mansdorf, 01 Sept 60 (tt. 21,175)—Aero Spacelines 1963, Broken up Oakland, March 1968.

C-97 Airframes Recorded as Acquired by Aero Spacelines

c/n 16522: 52-0828 KC-97G-21-B0, ff. 26 June 53—to MASDC as CH544 20 Feb 75 (tt. 8,090)—Sold to Aero Spacelines, section 44 used on SGT-201-4, F-GEAI.

c/n 16903: 53-0121 KC-97G—wings and vertical fin used on SGT-201-2, N212AS (F-BPPA).

c/n 16656: 52-2625 KC-97G-24-BO—converted to C-97K—MASDC 27 April 66 (tt. 5,550)—Sold to Aero Spacelines parts used for SGT-201-1 N211AS (F-BTGV).

c/n 16724: 52-2693 KC-97G-26-BO—converted to YC-97J, USAF Turbo-Stratocruiser—to MASDC as CH626 03 Aug 74—power plants, wing, nose and forward fuselage used for first turbine B-377 "Super Guppy" N1038V.

c/n 16797: 52-2766 KC-97G-28-BO—converted to C-97G—converted to KC-97L—to MASDC 29 April 70—Foundation For Airborne Relief N22766—Broken up Long Beach, section 44 and seven sections of belly pan used on SGT-201-3, F-GDSG along with section 45 of B-377 N402Q.

c/n 16945: 53-0163 KC-97G converted to C-97K—MASDC 29 Mar 66—Sold to Aero Spacelines, nose, section 46–47, horizontal stabilizer used on SGT-201-2, N212AS (F-BPPA).

c/n 16997: 53-0215 KC-97G converted to C-97—197th MAS Arizona ANG May 72—MASDC 19 Feb 75 (tt. 9,286.5)—Sold to Aero Spacelines, nose, wings, lower belly pan, tail used on SGT-201-3, F-GDSG.

c/n 16656: 52-2625 KC-97G-24-BO—converted to C-97K—to MASDC 27 April 66—Sold to Aero Spacelines, parts used on SGT-201-1, F-BTGV.

c/n 17062: 53-0280 KC-97G—converted to KC-97L—to MASDC 20 Oct 65—Returned to service 03 July 70—Sold to Aero Spacelines—Sold to Airbus and stored in France for SGT-201-5 along with section 45-46 from B-377 N74603—Broken up Jan 05.

G. *Guppy Aircraft Built by Aero Spacelines and UTA (Airbus)*

c/n 15925: B-377PG N1024V—P&W R-4360—Aero Spacelines "Pregnant Guppy" ff. 16 Sep 1962—Sold to American Jet 23 Oct 74, registered as N126AJ—Registration cancelled 05 Feb 78—Broken up Van Nuys 1980—Section 45 sold to Airbus and used on SGT-201-4, F-GEAI.

c/n 15938: N1038V (N940NA)—P&W YT-34-P—Aero Spacelines "Super Guppy" converted from YC-97J 52-2693 and parts from 52-2762, ff. 31 Aug 1965—Purchased by NASA 29 Mar 79, registered N940NS—Retired AMARC flyable storage 09 July 91—Damaged right wing 1999—Preserved on display Pima Air Museum Tucson (DMA).

c/n 15937: N1037V (N422AJ/N422AU)—P&W R-4360—Aero Spacelines "Mini-Guppy" Model B-377, ff. 24 May 1967—Sold to American Jet October 1974, registered N422AJ—Sold to Aero Union July 80—re-registered N422AU Oct 81—Sold to Erickson Skycrane, Central Point, Oregon 1988—wfu 1995 flyable storage—Last flight 1997 to Tillamook, Oregon. Currently on display Tillamook Naval Air Station Museum.

c/n 001MGT: N111AS—Allison 501 turboprops, Aero Spacelines Model SGT-101-1, ff. 13 Mar 1970—Crashed on take-off at Mohave, 12 May 1970.

c/n 001SGT: N211AS—Parts from KC-97G 52-2625 and ex–Pan Am N90942 (not conclusive) including section 45—Allison 501 turboprops, Aero Spacelines Model SGT-201-1 (C-97), ff. 24 Aug 1970—Airbus Industries Nov 1971, re-registered F-BTGV—Aeromaritime then Airbus Sky-link, "1"—Retired 31 Dec 1994, total hours 14,065.—On display (EG74) Bruntingthorpe, England as of 1996.

c/n 002SGT: N212AS—Parts from KC-97G 53-0163 and 53-0121—Allison 501 turboprops. Aero Spacelines Model SGT-201-2 (C-97), ff. 26 May 1973—Airbus Industries Aug 1973, re-registered F-BPPA—Aeromaritime Aug 1973, then—Airbus Skylink, "2"—Retired 31 Dec 1994, total hours 13,482. On display (TLS) Toulouse, France.

c/n 003SGT: F-WDSG—Parts from C-97G s/n 52-2766, 53-0215 and B-377 c/n 15974—Allison 501 turboprops. UTA Industries at Le Bourget Model SGT-201(F)-3 (C-97), ff. 01 June 1982—Airbus Industries re-registered 01 July 1982 F-GDSG—Airbus Skylink, "3" F-GDSG—Retired 31 Dec 1994—Stored Airbus, Hamburg. On display (XFW) preserved.

c/n 004SGT: F-WEAI—Parts from C-97G c/n 52-0828 and B-377 (PG) c/n 15924—Allison 501 turboprops. UTA Industries at Le Bourget Model SGT-201(F)-4 (C-97), ff. 21 June 1983—Airbus Industries re-registered F-GEAI, Airbus Skylink, "4"—Purchased by NASA 17 October 1997—re-registered N941NA—Active w/NASA, Johnson Space Flight Center, Ellington Field, Houston, Texas.

c/n 005SGT: F-XXXX—C-97G 53-0280 s/n 17062, MASDC 20 Oct 1965—re-registered N49548—to UTA 20 Oct 1965 and B-377 c/n 15949, N74603 section 45-46.—Model SGT-201(F)-5 Not Built, order cancelled—C-97 on display at the Musee de l'Air at Le Bourget Airport France until 1988—Flown to Museum at Nancy—Essey which went bankrupt—Broken up Jan 2005—Tail section remained in Zruc park until 2012.

H. *B-377 and SGT Guppy Operational Flights Partial List*

The Guppy fleet of aircraft which were developed to transport cargo for NASA has operated for the agency and commercially since 1962. The list compiled from multiple logs and records although not complete provides an overall view of a very active operation.

Mission	*Cargo*	*Date*	*From/To*	*Wt*	*Acft*
62-FF	First Flight	19 Sep 62	VNY-MHV		PG
62-xx	Empty Display Air Traffic Mgrs	24 Oct 62	MHV-DEN		PG
63-xx	Maiden Flt to Cape	Jun 63	MHV-XMR		PG
63-xx	S-IV Inert Saturn I Stage	14 Jun 63	LAX-EDW	20,379	PG
63-xx	Saturn I Second Stage	13 Jul 63	XMR-		PG
63-xx	S-IV-5	20–21 Sep 63	MHR-XMR		PG
63-xx	S-IV-6	27 Sep 63	SMO-MHR		PG
63-xx	F-1 Engine Rocketdyne	29–31 Oct 63	LAX-HUA	18,416	PG
64-xx	S-IV-7	13 Feb 64	SMO-MHR		PG
64-xx	S-IV-6	21–22 Feb 64	MHR-XMR		PG
64-xx	S-IV-9	08 May 64	SMO-MHR		PG
64-xx	S-IV-7	10–12 Jun 64	NHR-XMR		PG
64-xx	S-IV-8	07 Aug 64	SMO-MHR		PG
64-xx	S-IV-9	21–22 Oct 64	MHR-XMR		PG
64-xx	S-IV-10	05 Nov 64	SMO-MHR		PG

Mission	*Cargo*	*Date*	*From/To*	*Wt*	*Acft*
65-xx	Rocket Nozzle	09 Jan 65	ONT-BQK		PG
65-xx	GT-3 Gemini-Titan 2nd Stage	20 Jan 65	BWI-XMR		PG
65-xx	GT-3 Gemini Titan Rocket	23 Jan 65	BWI-XMR		PG
65-xx	S-IV-8	23–26 Feb 65	MHR-XMR		PG
65-xx	S-IV-10	08–10 May 65	MHR-XMR		PG
65-xx	Empty Display McDonnell Acft	17 Jun 65	MHV-STL		PG
65-xx	Meteoroid Detection Sat	22 Jun 65	BWI-KSC		PG
65-FF	First Flight	31 Aug 65	VNY-MHV		SG
65-xx	S-IV-7	05 Sep 65	MHR-X-68		PG
65-xx		23 Sep 65	-IND		PG
65-xx	Instr Unit Apollo/Saturn 201	28 Oct 65	-HUA		SG
66-xx	SI-U 500FS	10 Mar 66	HUA-LGB		SG
66-xx	SI-U 500FS	11 Mar 66	LGB-MHR		SG
66-xx	S-IVB-500ST	30 Mar–01 Apr 66	MHR-HUA		SG
66-xx	S-IVB-203	04–06 Apr 66	MHR-XMR		SG
66-xx	S-1B Instr Unit Apollo/Strn 203	18 Apr 66	HUA-XMR		
66-xx	S-IVB-204	06 Apr 66	MHR-XMR		SG
66-xx	S-IVB-502	01 Jun 66	SLI-MHR		SG
66-xx	S-IVB-206	30 Jun–01 Jul 66	SLI-MHR		SG
66-xx	Surveyor II	Jul 66	LAX-XMR	6,800	SG
66-xx	S-IVB-501	12–14 Aug 66	MHR XMR		SG
66-xx	Third Stage 501-S-IVB	15 Aug 66	-XMR		SG
66-xx	Mockup TM-3	19 Aug 66	BPA-DOV-XMR		SG
66-xx	S-IVB-207	30–31 Aug 66	SLI-MHR		SG
66-xx	TM-6 and LTA-IO	27 Sep 66	BPA-DOV-XMR		PG
66-xx	S-IVB-503	11 Oct 66	SLI-MHR		SG
66-xx	S-IVB-206	02 Dec 66	SLI-MHR		SG
66-xx	S-IVB-206	13–14 Dec 66	MHR-XMR		SG
66-xx	Saturn I Instrument Unit	19 Dec 66	HUA-XMR		SG
67-xx	S-IVB-502	20–21 Feb 67	MHR-XMR		SG
67-xx	S-IVB-209	09 Mar 67	SLI-MHR		SG
67-xx	S-IVB-206	13–14 Apr 67	XMR-MHR		SG
67-FF	First Flight	24 May 67	SBA-MHV		MG
67-ASI	Ferry-empty	25 May 67	MHV-PNE		MG
67-Budd	Budd Sky Lounge	26–27 May 67	PNE-PHL-SYR-MSS-YYG-YJT-YQX-SMA-LBG		MG
67-xx	Demonstration to EADS	7–9 Jun 67	LBG-TLS		MG
67-xx	Budd Sky Lounge	10–12 Jun 67	TLS-SMA-YQX-YJT-YQM-MSS-SYR-PNE		MG
67-xx	Ferry Empty	14 Jun 67	PNE-PHL-SBA		MG
67-xx	S-IVB-504N	16 Jun 67	SLI-MHR		SG
67-xx	S-IVB-503N	24–25 Jun 67	SLI-MHR		SG
67-xx	S-IVB-505N	17 Aug 67	SLI-MHR		SG
67-ASI	Ferry-empty	24 Aug 67	SBA-MSP		MG
67-com	American Derrick	26 Aug 67	MSP-YYT-KEF-LBG-ATH-ANK-TEH	15,400	MG
67-com	Ferry-empty	02 Sep 67	TEH- MUC		MG
67-com	ELDO Blue Streak Rocket	06–12 Sep 67	MUC-KHI-CMB-SIN-CGK-DRW-UMC		MG
67-com	Ferry-empty	14 Sep 67	UMC-BNE-NAN-PPG-HNL-SBA		MG
67-xx	S-IVB-503N	27 Dec 67	MHR-MHR		SG
67-xx	S-IVB-503N	9–30 Dec 67	MHR-XMR		SG
68-xx	Skylab Orbital Workshop	01 Jan 68	LGB-HUA		SG
68-xx	Apollo 8 Instrument Unit 503	04 Jan 68	HUA-XMR		SG
68-xx	S-IVB-506N	25 Jan 68	SLI-MHR		SG
68-test	Ballast-All up weight	Feb 68	SBA-SBA		MG
68-xx	S-IVB-205	06–08 Apr 68	MHR-XMR		SG
68-xx	Apollo Module 101/Saturn 205	17 May 68	North Am CA-X68		PG
68-xx	Apollo 7 Command Module	29 May 68	LGB-LGB (rtn eng failure)		PG

Mission	*Cargo*	*Date*	*From/To*	*Wt*	*Acft*
68-xx	Apollo 7 Command Module	30 May 68	LGB-XMR		PG
68-xx	Lunar Module 3 Ascent Stage	13 Jun 68	-XMR		SG
68-xx	S-IVB-507	07 Aug 68	SLI-HMR		SG
68-xx	Cmd Module 103 Apollo VII	12 Aug 68	-XMR		SG
68-xx	S-IV-B-504N 3rd Stage Saturn	10–12 Sep 68	MHR-XMR		SG
68-xx	Instrument Unit Apollo IX	01 Oct 68	-XMR		SG
68-xx	S-IVB-211	17–18 Oct 68	SLI-MHR		SG
68-xx	S-IVB-505N	02–03 Dec 68	MHR-XMR		SG
68-xx	Third Stage Saturn V Apollo 10	05 Dec 68	-XMR		SG
68-xx	S-IVB-508	30 Dec 68	SLI-MHR		SG
69-xx	Ascent Module 5 Apollo 11	09 Jan 69	-XMR		SG
69-xx	Dcnt Lunar Module 5 Apollo 11	17 Jan 69	-XMR		
69-xx	S-IVB-506N	17–18 Jan 69	MHR-XMR		SG
69-xx	Cmd Svc Module Apollo XI	22 Jan 69	-XMR		SG
69-xx	Apollo 11 Command Module	24 Jan 69	-XMR		SG
69-xx	S-IVB-507	06–10 Mar 69	MHR-SLI		SG
69-xx	S-IVB-509	31 Mar 69	SLI-MHR		SG
69-xx	S-IVB-508	12–13 Jun 69	MHR-SLI		SG
69-xx	S-IVB-510	19 Jun 69	SLI-MHR		SG
69-xx	S-IVB-511	16 Sep 69	SLI-MHR		SG
70-ASI	First DC-10 Fuselage Section	09 Jan 70	SAN-LGB		SG
70-xx	S-IVB-509	17–20 Jan 70	MHR-XMR		SG
70-xx	S-IVB-207	01 May 70	MHR-SLI		SG
70-xx	S-IVB-510	11–12 Jun 70	MHR-XMR		SG
70-xx	S-IVB-511 3rd Stg Apollo 16	29 Jun-01 Jul 70	MHR-XMR		SG
70-xx	S-IVB-209	22 Jul 70	MHR-SLI		SG
70-xx	S-IVB-206	03 Aug 70	MHR-SLI		SG
70-xx	S-IVB-211	15 Sep 70	MHR-SLI		SG
70-xx	S-IVB-208	13 Oct 70	MHR-SLI		SG
72-xx	Goodyear Airship Europa	Jan 72	CAK- RAF Cardington, UK		MG
72-xx	Pioneer 10 Spacecraft	Feb 72	LAX-XMR		MG
72-xx	Airlock Docking Adapter	05 Oct 72	STL-XMR		SG
72-xx	Skylab Airlock Module	27 Oct 72	-XMR		SG
76-xx	X-24 and HL-10	May 76	EDW-FFO		SG
79-01	RSRA	25 Sept 79	WAL—Ames	20,245	SG
79-02	NTF Nacelle	30 Sept 79	SLC—LFI	37,811	SG
79-03	WIF Article 1&2	28 Nov 79	BED-JSC	24,002	SG
80-04	SH-2 Helicopter to NARF	14 Mar 80	NUG-NPA	18,493	SG
80-05	C-130 Fuselage 59-5133	09 Jul 80	CYS-BHM	25,830	SG
80-06	A-7D Aircraft	19 Aug 80	SJU-DFW	22,016	SG
81-07	Neutral Buoyancy Trainer	07 Jan 81	BED-JSC	22,808	SG
81-08	A-7D Aircraft	27 Feb 81	SJU-DFW	22,730	SG
81-09	A-7D Aircraft	Mar 81	SJU- DFW	22,325	SG
81-10	Tail Cone GSE from (DFRC)	27 Apr 81	DFRC-XMR	28,043	SG
81-11	SRB AFT Skirt/EMU	01 May 81	XMR-HUA-JSC	22,800	SG
81-Am Jet	Airbus/cargo SGT-201-4	Jun 81	SBA-LGB		MG
81-12	A7D Aircraft	18 Aug 81	CFB-TIK	27,930	SG
81-13	Lidar Trailer	15 Oct 81	BFK-DFRC	25,510	SG
81-14	Tail Cone GSE	25 Nov81	EDW-XMR	26,400	SG
82-Tracor	SGT-201-4 outer wings/tail	Mar 81	SBA-LBG		MG
82-15	Tail Cone GSE (White Sands)	09 Apr 82	HMN-XMR	25,800	SG
82-16	SRB AFT Skirt	29 Apr 82	LGB-XMR	32,569	SG
82-17	SRB AFT Skirt	06 May 82	HUA-HIF	26,558	SG
82-18	Tail Cone Fixture	29 Jun 82	EDW-XMR	23,744	SG
82-19	Tail Cone Containers	09 Jul 82	KSC-DFRC (EDW)	24,510	SG
82-20	Tail Cone Fixture	17 Jul 82	EDW-KSC	26,560	SG
82-21	SRB AFT Skirt	08 Aug 82	LGB-HUA	29,590	SG
82-22	OMS/RCS Pods	01 Sep 82	EDW- KSC	35,025	SG
82-23	SRB AKT Skirt	08 Sep 82	LGB-HUA	29,690	SG

Mission	*Cargo*	*Date*	*From/To*	*Wt*	*Acft*
82-24	SRB AFT Skirt	09 Sept 82	HUA-KSC	30,608	SG
82-25	F-14 Aircraft	21 Oct 82	NTD-DMA	33,559	SG
82-26	SRB AFT Skirt	27 Oct 82	LGB-HUA	29,228	SG
82-27	SRB AFT Skirt	29 Oct 82	HUA-KSC	30,258	SG
83-28	SRB SFT Skirt	04 Jan 82	LGB-HUA	29,128	SG
83-29	Atlas Centaur Fairing	08 Feb 83	SAN-KSC	20,625	SG
83-30	Tail Cone GSE	14 Apr 83	DFRF-KSC	23,181	SG
83-31	Pay Load Device	16 Apr 83	XMR-JSC	23,054	SG
83-32	RSRA	19 Apr 83	Ames-SWF	23,102	SG
83-33	OTSA Study	05 May 83	DFRF-Local	8,300	SG
83-34	PFTA/DFI	08 Jul 83	JSC-KSC	23,500	SG
83-35	Orbiter Tail Cone	15 Sep 83	XMR-PMD	23,354	SG
83-36	OMS/RCS Pods	21 Sep 83	PMD-KSC	36,615	SG
83-37	B-1 AIF	02 Nov 83	DFW- PMD	36,082	SG
84-38	SYNCOM IV (frm El Segundo)	22 Feb 84	LAX-KSC	36,517	SG
84-38A	DFI & SYNCOM CONT	01 Mar 84	KSC-JSC		SG
84-39	SYNCOM IV (frm El Segundo	29 Apr 84	LAX-KSC	37,045	SG
84-40	PFTA & SYNCOM CONT	30 Apr 84	KSC-JSC	33,588	SG
84-41	SYNCOM CONT	04 Jun 84	JSC-LAX	19,650	SG
84-41A	AFA-12 TEST ARTICLE	05 Jun 84	LAX-JSC	25,070	SG
84-42	LASAR TRAILER	06 Jul 84	KSC-DFRF (EDW)	29,941	SG
84-43	CENTAUR DEPL ADAPTER	23 Jun 84	SAN-ADW	9,150	SG
84-44	OTA/SC	30 Jul 84	HUA-SWF	36,000	SG
84-45	CENTAUR DEPL ADAPTER	17 Aug 84	ATW-SAN	8,650	SG
84-46	CENTAUR FWD ADAPTER	20 Aug 84	SAN-ATW	10,750	SG
84-47	CENTAUR FWD ADAPTER	18 Sept 84	ATW-SAN	10,950	SG
84-48A	OTA/SC	13 Oct 84	HUA-SWF	22,331	SG
84-48B	OTA/SC CONTINUE 48A	Oct 84	SWF-NUQ	35,436	SG
84-48C	OTA/SC	Nov 84	NUQ-HUA	22,331	SG
84-49	OMS PODS	18 Oct 84	EDW-KSC	22,341	SG
84-50	OMS POD/BODY FLAP	17 Nov 84	EDW-KSC	28,652	SG
84-51A	SYNCOM IV	12 Dec 84	LAX-KSC	32,492	SG
84-51B	SYNCOM IV CONT	19 Dec 84	KSC-JSC	26,893	SG
84-52	G-2 Wing	Dec 84	SAV-Bethpage	25,951	SG
85-53A	SYNCOM IV CONTAINER	24 Jan 85	JSC-TUS	19,580	SG
85-53B	F-14 Aircraft	29 Jan 85	TUS-ONT	32,755	SG
85-54	SPACE LAB TROLLEY	28 Feb 85	KSC-HUA	25,785	SG
85-55	SPACE LAB TROLLEY	15 Apr 85	HUA-KSC	25,389	SG
85-56	SYNCOM IV	26 APR 85	LAX-KSC	32,430	SG
85-57	SHUTTLE TNG EQUIP	03 May 85	PHX-JSC	8.919	SG
85-58	B-1B AIF Aircraft	22 May 85	DFW-PMD	36,082	SG
85-59	PAGOTA/ SYNCOM CONT	20 Jun 85	KSC-SAN	29,592	SG
85-60	CENTAUR CISS	28 Jun 85	SAN- KSC	23.875	SG
85-61	CENTAUR TTF	26 Jul 85	SAN-KSC	20,501	SG
85-62	F-14 Aircraft	31 Jul 85	TUC-NZY	23,857	SG
85-63	F-14 Aircraft	01 Aug 85	TUC-MEM	31,558	SG
85-64	F-14 Aircraft	10 Sep 85	TUC-SWF	30,302	SG
85-65	CENTAUR SC-1	26 Sep 85	SAN-KSC	32,314	SG
85-66	F-14 Aircraft	15 Oct 85	TUC-SWF	29,120	SG
85-67	F-14 Aircraft	23 Oct 85	TUC-SWF	29,140	SG
85-68	CENTAUR SC-2	13 Nov85	SAN-KSC	25,751	SG
85-69	CENTAUR TTF	12 Dec 85	SAN-KSC	18,043	SG
85-70	CENTAUR SC-2	21 Dec 85	SAN-KSC	31,947	SG
86-71	A-7D Aircraft	28 Jan–11 Feb 86	BLB-DFW	19,587	SG
86-72	B1-B AIF Aircraft	01 Apr 86	DFW-PMD	37,725	SG
86-72A	Guppy Prop/Tires	09 Apr 86	ELP-JSC	10,432	SG
86-73	F-14 Aircraft	20 Jun 86	SEA-SWF	34,050	SG
86-74	CENTAUR SC-2	19 Sep 86	KSC-SAN	29,137	SG
86-75	RSRA	26 Sep 86	OQU-EDW	20,722	SG

Mission	*Cargo*	*Date*	*From/To*	*Wt*	*Acft*
86-76	CENTAUR SC-1	02 Oct 86	KSC-SAN	29,784	SG
86-77	TRAILER	09 Nov 86	STL-FTW	38,652	SG
86-78	TRAILER	20 Nov 86	FTW-EDW	38,652	SG
86-79	F-14 Aircraft	06 Jan 87	SAN-SWF	25,414	SG
87-80	F-14 Aircraft	21 Jan 87	SAN-SWF	27,110	SG
87-80A	F-14 Aircraft	03 Feb 87	NAS Norfolk-SWF	29,366	SG
87-81	F-14 Aircraft	10 Feb 87	SAN-SWF	28,458	SG
87-82	F-14 Aircraft	24 Feb 87	SAN-SWF	28,112	SG
87-83	T-38 (NASA 914)	14–17 Mar 87	SLI-EFD	14,824	SG
88-84	T-38 Aircraft	24 May 89	HMN-RND	17,825	SG
89-85	A-7D Aircraft	15 May 89	TUC-LTV Dallas	22,644	SG
89-86	STA WING	19 Jun 89	Bethpage-STL	28,320	SG
89-87	HST Support	11 Sep 89	NUQ-KSC	33,320	SG
89-88	HST Support	01 Oct 89	NUQ-KSC	28,677	SG
98-747SP	747 Fuselage	04 May 98	ADM-ACT	24,450	SGT
ISSP 98-01	S1 STA Truss Test Element	22–27 Jun 98	SLI-HUA	27,341	SGT
ISSP 98-02	Node Structural Test Element	24 Aug 98	HUA-HUA	15,700	SGT
ISSP 98-03	S0-STA Truss Test Article	26–28 Oct 98	SLI-EFD	23,496	SGT
98-B	SGSF ECS Test Flight	04–05 Nov 98	EFD-EFD	16,470	SGT
ISSP 98-04	Laboratory Module	16 Nov 98	HUA-X68	36,870	SGT
ISSP 99-01	S0-STA Truss Test Element	22 Jan 99	EFD-SLI	23,496	SGT
ISSP 99-02	S1-STA Truss Flight Element	12–14 Feb 99	SLI-HUA	28,870	SGT
ISSP 99-03	Node Struct Test Element	18–19 Feb 99	HUA-EFD	24,948	SGT
ISSP 99-04	P3-STA Test Element	29–31 Mar 99	TUL-HUA	25,602	SGT
99-A	X-33 LH #2 Fuel Tank	23 Apr 99	NUQ-HUA	20,360	SGT
ISSP 99-05	P1 Truss Flight Element	19–20 May 99	SLI-HUA	27,379	SGT
ISSP 99-06	S0 Truss Flight Element	09–11 Jun 99	SLI- X-68	35,013	SGT
ISSP 99-07	S1-STA Truss Test Element	29 Jun 99	HUA-EFD	44,275	SGT
ISSP 99-08	Common Module Test Elmt	10–11 Aug 99	BFI-HUA	25,942	SGT
ISSP 99-09	S1 Structural Test Element	02 Oct 99	EFD-HUA	44,275	SGT
ISSP 99-10	S1 Truss Flight Element	06 Oct 99	HUA- X-68	33,477	SGT
99-B	X-33 RH #2 Fuel Tank	14–15 Oct 99	NUQ-PMD	20,500	SGT
ISSP 99-11	P3 Truss Flight Element	17–18 Oct 99	TUL- X-68	24,489	SGT
ISSP 00-01	S1-STA	26–27 Jan 00	HUA-SLI	35,165	SGT
ISSP 00-02	S0-STA	27 Jan-02 Feb 00	SLI-EFD	36,611	SGT
ISSP 00-03	Node-STA	02–03 Mar 00	EFD-HUA	24,927	SGT
00-A	V-131 (X-38)	11 Jul 00	EFD-EDW	31,340	SGT
00-B	T-38 (1) Aircraft	13 Jul 00	NUQ-ELP	18,660	SGT
00-C	T-38 (2) Aircraft	14–15 Jul 00	ELP-EFD	18,141	SGT
ISSP 00-04	P1-STA Flight Element	23–24 Jul 00	HUA- X-68	33,857	SGT
ISSP 00-05	Airlock Element	11 Sep 00	HUA- X-68	27,415	SGT
ISSP 00-06	S3 Truss Flight Element	07 Dec 00	TUL- X-68	24,931	SGT
ISSP 01-01	P3/P4-STA	31 Jan 01	HUA-EFD	42,000	SGT
ISSP 01-02	S5 Qual Assembly	21 Mar 01	TUL-HUA	19,118	SGT
ISSP 01-03	P3/P4-STA	25 Jun 01	EFD-HUA	42,000	SGT
ISSP 01-04	P5 Truss Flight Element	19 Jul 01	TUL- X-68	20,200	SGT
01-A	EMU	27 Jul 01	X-68 EFD	4,265	SGT
ISSP 01-05	S6 Long Spacer	16 Aug 01	TUL-EFD	24,367	SGT
ISSP 01-06	S6 IEA	07 Nov 01	HUA-EFD	25,175	SGT
ISSP 01-07	S1-STA	29 Nov-06 Dec 01	SLI-HUA	34,321	SGT
02-A	Navy SH-3 Helicopter	14–20 Jan 02	DMA-JNR	23,161	SGT
02-B	X-38/DPS	23–25 Jan 02	MHR-EFD	29,100	SGT
02-C	Navy F-14 Aircraft	05–08 Feb 02	NZY-NIP	35,396	SGT
ISSP 02-01	S5 Truss Flight Element	11–15 Mar 02	HUA- X-68	21.292	SGT
ISSP 02-02	Node-STA	07–10 Mar 02	HUA- X-68	24,985	SGT
ISSP 02-03	Common Module	21–25 Oct 02	HUA- X-68	24,247	SGT
ISSP 02-04	S6 IEA Flight Element	11–13 Nov 02	EFD- X-68	27,617	SGT
ISSP 02-05	S6 Long Spacer Flt Element	16–18 Dec 02	EFD- X-68	23,827	SGT
02-D	Gravity Probe B Test Flt	19 Dec 02	EFD-EFD	15,500	SGT

Mission	*Cargo*	*Date*	*From/To*	*Wt*	*Acft*
03-A	Gravity Probe B Test Flt	8–30 Mar 03	EFD-HSV	17,150	SGT
03-B	Gravity Probe B Test Flt	02 Apr 03	EFD-EFD	17,150	SGT
03-C	2 Navy T-38	09–11 Jun 03	GSB-RND	26,580	SGT
03-D	NASA T-38	01–02 Jul 03	X-68-EFD	16,956	SGT
03-E	V-22 Osprey w/ballast	04–05 Sep 03	AMA-PHL	26,867	SGT
04-A	2 UASF—NASA T-38 Parts	04–16 Jan 04	EFD-TIK-RND-ELP-EFD	20,925	SGT
04-B	V-22 Osprey w/ballast	01–02 Apr 04	PHL-AMA	27,604	SGT
05-A1	V-22 Osprey MV #40 w/bal	24–26 May 05	AMA-PHL	26,562	SGT
05-A2	V-22 Osprey CV #07 w/bal	27–31 May 05	PHL-AMA	27,112	SGT
05-B1	V-22 Osprey MV #39 w/bal	28–29 May 05	AMA-PHL	26,737	SGT
05-B2	V-22 Osprey MV #71 w/bal	30-Jun 01–Jul 05	PHL-AMA	26,280	SGT
05-C1	V-22 Osprey MV #38 w/bal	25–26 Jul 05	AMA-PHL	26,562	SGT
05-C2	V-22 Osprey MV #72 w/bal	27–28 Jul 05	PHL-AMA	26,262	SGT
05-D1	V-22 Osprey MV #37 w/bal	22–23 Aug 05	AMA- PHL	26,547	SGT
05-D2	V-22 Osprey MV #73 w/bal	24–25 Aug 05	PHL-AMA	26,297	SGT
05-E1	V-22 Osprey MV #36 w/bal	19 Sep 05–Oct 05	AMA-PHL	26,667	SGT
05-E2	V-22 Osprey MV #74 w/bal	05–06 Oct 05	PHL-AMA	25,210	SGT
05-F	V-22 Osprey MV #40 w/bal	14–16 Dec 05	PHL-AMA	26.150	SGT
ISSP 06-01	Common Module Test Art	08 Feb 06	X-68-HUA	24,293	SGT
06-A	V-22 Osprey MV #39 w/bal	28–30 Mar 06	PHL-AMA	26,129	SGT
06-B	NASA T-38	06 Apr 06	GPT-EFD	16,000	SGT
06-C1	V-22 Osprey MV #35 w/bal	26–27 Jun 06	AMA-PHL	26,492	SGT
06-C2	V-22 Osprey MV #38 w/bal	28–29 Jun 06	PHL-AMA	27,195	SGT
06-D1	V-22 Osprey MV #33 w/bal	25–27 Sep 06	AMA-PHL	25,685	SGT
06-D2	V-22 Osprey MV #37 w/bal	27–28 Sep 06	PHL-AMA	26,225	SGT
07-A1	V-22 Osprey MV #32 w/bal	09–11 Jan 07	AMA-PHL	25,585	SGT
07-A2	V-22 Osprey MV #36 w/bal	11–18 Jan 07	PHL-AMA	25,933	SGT
07-B1	V-22 Osprey MV #31 w/bal	10–12 Apr 07	AMA-PHL	25,525	SGT
07-B2	V-22 Osprey MV #35 w/bal	12–16 Apr 07	PHL-AMA	26,135	SGT
07-C	V-22 Osprey MV #30 w/bal	13–14 Jun 07	AMA-PHL	25,145	SGT
07-D1	V-22 Osprey MV #29 w/bal	10–11 Jul 07	AMA-PHL	25,215	SGT
07-D2	V-22 Osprey MV #33 w/bal	11–16 Jul 07	PHL-AMA	25,885	SGT
07-E1	V-22 Osprey MV #28 w/bal	10–11 Oct 07	AMA-PHL	25,631	SGT
07-E2	V-22 Osprey MV #32 w/bal	11–15 Oct 07	PHL-AMA	25,941	SGT
07-E3	V-22 Osprey MV #27 w/bal	17–23 Oct 07	AMA-PHL	25,461	SGT
ISSP07-01	SARJ P-3 STA Element	18 Dec 07	HUA-EFD	23,948	SGT
08-A1	V-22 Osprey MV #26 w/bal	21–23 Jan 08	AMA-PHL	25,111	SGT
08-A2	V-22 Osprey MV #31 w/bal	23–24 Jan 08	PHL-AMA	25,930	SGT
08-B1	V-22 Osprey MV #30 w/bal	25–26 Mar 08	PHL-AMA	25,453	SGT
09-A1	V-22 Osprey MV #27 w/bal	26–30 Jan 09	PHL-AMA	25,521	SGT
09-B1	V-22 Osprey MV #26 w/bal	14–19 May 09	PHL-AMA	25,904	SGT
ISSP 10-01	NODE-STA Element	03–06 May 10	HAS- X-68	26,253	SGT
10-A1	2 T-38s Davis-Monthan	13–14 Dec 10	DMA-HMN	25,195	SGT
10-A2	2 T-38s Davis-Monthan	15 Dec 10	DMA-HMN	25,275	SGT
11-A1	2 T-38s Davis-Monthan	08 Mar 11	DMA-HMN	25,367	SGT
11-A2	2 T-38s Davis-Monthan	09 Mar 11	DMA-HMN	25,417	SGT
11-A3	2 T-38s Davis-Monthan	11 Mar 11	DMA-HMN	17,192	SGT
11-B1	2 T-38s Davis-Monthan	22 Mar 11	DMA-HMN	25,277	SGT
11-B2	2 T-38s Davis-Monthan	23 Mar 11	DMA-HMN	25,387	SGT
11-B3	2 T-38s Davis-Monthan	25 Mar 11	DMA-HMN	25,327	SGT
11-C1	V-22 Osprey CV #164 w/bal	21–22 Apr 11	PHL-AMA	20,143	SGT
11-C1	V-22 Osprey MV #1027 w/bal	23–29 Apr 11	PHL-AMA	21,047	SGT
11-C2	V-22 Osprey CV #165 w/bal	17–18 May 11	PHL-AMA	22,318	SGT
11-C2	V-22 Osprey MV #166 w/bal	19–20 May 11	PHL-AMA	22,198	SGT
11-D1	V-22 Osprey MV #1028 w/bal	23–24 Jun 11	PHL-AMA	22,850	SGT
11-E1	V-22 Osprey CV #169 w/bal	12–13 Jul 11	PHL-AMA	22,148	SGT
11-E1	V-22 Osprey CV #170 w/bal	14–15 Jul 11	PHL-AMA	22,168	SGT
11-E2	V-22 Osprey CV #171 w/bal	19–20 Jul 11	PHL-AMA	22,135	SGT

Mission	*Cargo*	*Date*	*From/To*	*Wt*	*Acft*
11-E2	V-22 Osprey CV #1029 w/bal	26–27 Jul 11	PHL-AMA	22,850	SGT
11-F1	V-22 Osprey MV #1030 w/bal	08–09 Sep 11	PHL-AMA	22,795	SGT
11-F2	V-22 Osprey CV #175 w/bal	15–20 Sep 11	PHL-AMA	22,145	SGT
11-G1	2 NASA T-38 AC 901 & 902	11–14 Nov 11	ELP-DMA	20,725	SGT
11-H1	V-22 Osprey CV #183	13–15 Dec 11	PHL-AMA	22,816	SGT
12-A1	V-22 Osprey CV #1033	23–25 Dec 12	PHL-AMA	23,511	SGT
12-B1	NASA T-38 AC 914	12–13 Apr 12	ELP-DMA	13,061	SGT
12-C1	FFT Crew Comp	25–30 Jun 12	EFD-BFI	33,066	SGT
12-C2	FFT Aft Payload Bay	25–26 Jul 12	EFD-BFI	21,855	SGT
12-C3	FFT Fwd Payload Bay	8–09 Aug 12	EFD-BFI	20,315	SGT
12-D1	CCT-1	22 Aug 12	EFD-FFO	31,651	SGT
12-E1	Heat Shield Tool	28 Nov 12	RIV-BKF	21,096	SGT
13-A1	2 NASA T-38s 863 & 864	8–19 Mar 13	EDW-ELP	24,447	SGT
13-B1	Heat Shield & Cover	25–27 Mar 13	BKF-BED	30,469	SGT
13-C1	Heat Shield	04–05 Dec 13	MHT-KSC	35,310	SGT
14-A1	Boeing 5.5M Cryo Tank	25–27 Mar 14	BFI-HUA	36,489	SGT
14-B1	Boeing Blended Wing-Pressure box (MMB)	08–12 Dec 14	LGB-LFI	29,113	SGT
15-A1	Heat Shld Container w/STA	20–23 Apr 15	KSC-DEN	27,933	SGT
15-B1	Orion Barrel Ring	09–10 Sep 15	RIV-HAS	5,128	SGT
15-C1	NASA T-38 N923	23 Sep 15	ELP-EFD	11,341	SGT
15-D1	Orion CMA Tooling Ring	02–03 Nov 15	KSC-MFD	19,549	SGT
16-A1	Orion HS Tool & Skin	04–08 Jan 16	BKF-NUQ	23,744	SGT
16-B1	Orion HS Tool & Skin	25–26 Jan 16	NUQ-BKF	22,851	SGT
16-C1	CMTF Pressure Vessel	01 Feb 16	NGB-TTS	28,077	SGT

Glossary

Abbreviations

AAF Army Air Force
ABMA Army Ballistic Missile Agency
AC alternating current
ADI Anti-Detonate Injection
AFFTC/HO Air Force Flight Test Center History Office, Edwards AFB California
AIAA American Institute of Aeronautics and Astronautics
aka also known as
Akros Akros Dynamics Corporation
ALG Aviation Leasing Group
ALPA Airline Pilots Association
AMP Avionics Modernization Program
ANG Air National Guard
AOD Air Operations Director
AOG Aircraft On Ground
APP Auxiliary Power Plant
APU Auxiliary Power Unit
ARDC Air Research and Development Command
ASI Aero Spacelines Incorporated
ATC Air Traffic Control
ATC Approved Type Certificate
ATEL Aviation Traders Engineering Limited
ATG Air Transport Group
ATI Airbus Transport International
ATL Aviation Traders Limited
ATOG allowable takeoff gross
BAS British Air Services
BOAC British Overseas Airways Corporation
CAA Civil Aeronautics Administration
CAB Civil Aeronautics Board
CAC Continental Air Command
CAC Clark Aircraft Corporation (DC-10 and Megalifter)
CAR Civil Air Regulations
CAS Custom Air Service
CEO Chief Executive Officer
CLT cargo lift trailer (NASA)
COA certificate of airworthiness
COIL chemical oxygen-iodine laser
DAC Douglas Aircraft (Santa Monica)
DARPA Defense Advanced Research Projects Agency
DASA Daimler-Benz Aerospace Airbus
db decibels
dba doing business as
DC direct current
DFC Distinguish Flying Cross
DGAC Direction Générale de l'Aviation Civile (French Civil Aviation Authority)
DHC Dee Howard Corporation
DMA Davis Monthan, Air Force Base
DME Distance Measuring Device
DOD Department of Defense
EADS European Aeronautic Defense and Space
ECM ECM-Electronic Counter Measures
EGAT Evergreen Aviation Technologies Corporation
ELDO European Launcher Development Organisation (European Space Agency)
ESA European Space Agency
e.s.h.p. equivalent shaft horsepower
EU European Union
FAA Federal Aviation Administration
F/E Flight Engineer
FIA First International Airways
Ficon FIghter CONveyor
FOD foreign object damage
FTL Flying Tigers
GE General Electric Corporation
GETTS Glider External Tank Transport System
GIA Global International Airways
gph gallon per hour
GSA Government Services Administration
HIS Horizontal Situation Indicator
HK-1 Hughes H-4 "Spruce Goose" NX37602
IACAR Interstate Aviation Committee Aviation Register (Russian)
I.A.S. indicated air speed
IBT International Brotherhood of Teamsters
ICAO International Civil Aviation Organization
ICBM Intercontinental Ballistic Missile
IOS International Overseas Services (Fund of Funds)
IRBM Intermediate Range Ballistic Missile
ISS or ISSP International Space Station (Program)
ISSC International Space Station Program

JSC Johnson Space Center
LCF Large Cargo Freighter
LTV Ling-Temco-Vought
MAC Military Airlift Command
MAP Ministry of Aviation Industry (Soviet)
MATS Military Air Transport Service
Max ATOG maximum allowable take-off gross
Md maximum mach number
METO Maximum Except Take-Off
MG Mini Guppy
MGT Mini Guppy Turbine
MOM Ministry of General Machine-building (Soviet)
MSFC Marshall Space Flight Center
MSTS Military Sea Transport Service
n/miles nautical miles
NWA Northwest Airlines
ONAT Orvis Nelson Air Transport
ONUC Organisation des Nations unies au Congo
Pan Am Pan American World Airways
PEV pyro-electric vidicon
PG Pregnant Guppy
POW prisoner of war
psi pounds per square inch
QOR Qualitative Operational Requirement
RANSA Rutas Aereas Nacionales SA (Venezuela)
RCAF Royal Canadian Air Force
R&D research and development
RERP Reliability Enhancement and Re-Engine Program
RMI Radio Magnetic Indicator
S-I Saturn I (originally Saturn C-l) first stage
S-I-5 fifth Saturn I (SA-5) first stage
S-IC Saturn V first stage
S-II Saturn V (originally Saturn C-5) second stage
S-IU-5 Saturn SA-5 instrument unit
S-IV Saturn I (originally Saturn C-l) second stage
S-IVB Saturn V (originally Saturn C-5) third stage
S-IV Battleship non-flight S-IV stage replica for engine tests
S-V Saturn C-l third stage contemplated but dropped
SA-01 SA-1 booster's first flight qualification test
SA-1 Saturn I (originally Saturn C-l), first flight vehicle
SA-2 Saturn I (originally Saturn C-l), second flight vehicle
SAAMA San Antonio Air Materiel Area
SAC Strategic Air Command
SACTO Sacramento, California, test facility of Douglas Aircraft Co.
SA-Dl dynamic test of Saturn I (originally Saturn C-l) dummy vehicle
SA-T test booster
SA-T1 SA-1 test booster
SAT-01 first live firing of the Saturn test booster Santa Monica DAC S-IV stage fabrication facility in Santa Monica, Calif.
SAS Scandinavian Airlines System
SATIC Special Airbus Transport International Company
Saturn 1 first large space vehicle preliminary to the moon flight vehicle
Saturn IB manned earth orbital flight vehicle preceding moon flight rocket and composed of Saturn I's first stage and Saturn V's third stage
Saturn V manned moon flight launch vehicle
SCA Shuttle Carrier Aircraft
SCM Space Cargo Modified
SCTS Space Container Transport System
SEC Securities and Exchange Commission
SG Super Guppy
SGT Super Guppy Turbine 201
SGT(F) Super Guppy Turbine 201 built in France
Shp Shaft Horsepower
SILA Svensk Interkontinental Lufttrafik AB
SRB Solid Rocket Booster
SSFTWG Space Station Freedom Transportation Working Group
SST Supersonic Transport (Boeing)
STC Supplemental Type Certificate
STSA Short Term Strategic Airlifter
TAC Transmeridian Air Cargo (formerly Trans Meridian Air Cargo TMAC before 1969)
TACAN Tactical Air Navigation System
TBM Tactical Ballistic Missile
TCAS Traffic Collision Avoidance System
TEMCO Texas Engineering and Manufacturing Corporation
TLD Téléflex Lionel-Dupont (formerly Tissmétal Lionel-Dupont)
TRW Thompson, Ramo, Wooldridge
TTS Tennessee Technical Service
TWA Trans World Airlines
UCAV Unmanned Combat Air Vehicle
UN United Nations
Upgrd upgraded
USAAC United States Army Air Corps (prior to 20 June 1941)
USAAF United States Army Air Force (after 20 June 1941)
USAF United States Air Force (after 18 September 1947
USC University of Southern California
USD United States Dollars
USRA Universities Space Research Association
UTA Union de Transports A'eriens
Va design maneuvering speed
Vc design cruise speed
VE Victory in Europe
Vd design diving speed
VHF Very High Frequency
VJ Victory in Japan
VLCT Very Large Commercial Aircraft
Vmo (velocity) Maximum operating limit speed
Vne never exceed speed
VM-T Vladimir Myasishchev-Trahnsportnyy
VOR VHF Omni-directional Range
WRM War Readiness Materials
Vs 1-g stall speed
ZFW Zero Fuel Weight

IATA and FAA Airport Codes

ACT Waco, Texas
ADM Ardmore, Oklahoma
AMA Amarillo, Texas
ANK Ankara, Turkey
ATH Athens, Greece
ATW Andrews AFB, Maryland
BED Bedford, Massachusetts
BFI Boeing Field, Seattle, Washington
BFK Buckley Field, Aurora, Colorado
BKF Brooks Lake, Alaska
BLB Howard AFB, Panama
BNE Brisbane, Australia
BPA Grumman, Bethpage, New York
BQK Brunswick, Georgia
CFB Cold Lake, Canada
CGK Jakarta, Indonesia
CMB Colombo, Ceylon
DEN Denver, Colorado (Stapleton)
DFRF Dryden Research Center (EDW) California
DMA Davis-Monthan AFB, Tucson, Arizona
DOV Dover AFB, Delaware.
DRW Darwin, Australia
EDBG Berlin RAF Gatow Airport (ICAO code)
EDDT Berlin Tegal Airport (ICAO code)
EDW Edwards AFB, California
EFD Ellington AFB, Houston, Texas
EG74 Bruntingthorpe, United Kingdom (FAA Airport Code)
ELP El Paso, Texas
FFO Wright-Patterson AFB, Ohio
GPT Gulfport-Biloxi, Mississippi.
GSB Seymour-Johnson AFB, Goldsboro, North Carolina
HECA Cairo Egypt (ICAO code)
HIL Hill AFB, Utah
HMN Holloman AFB, Alamogordo, New Mexico
HNL Honolulu, Hawaii
HSV Huntsville, Alabama
HUA Redstone Field, MSFC, Huntsville, Alabama.
JSC Johnson Space Flight Center, Houston, Texas
KEF Keflavik, Iceland
KHI Karachi, Pakistan
KHUA (HUA) Redstone Airfield, MSFC—Huntsville Alabama (ICAO code)
KSC Kennedy Space Center (X68), Florida
LBG Le Bourget, Paris, France
LFBO Toulouse France—Blagnac Airport (ICAO code)
LFI Langley AFB, Virginia
LFML Marseille France—Marignane Airport (ICAO code)
LGB Long Beach, California
MFD Lahm Municipal—Mansfield, Ohio
MHR Mather AFB, Sacramento, California.
MHT Manchester, New Hampshire
MHV Mojave, California
MSS Massena, New York
MUC Munich, Germany
NAN Nadi, Fiji
NIP Jacksonville NAS, Jacksonville, Florida
NPA Naval Air Station, Pensacola, Florida
NTD Point Magu Naval Air Station, California
NUQ Moffett AFB, California
NZY North Island NAS, San Diego, California.
ONT Quonset Point, Rhode Island
PHL Philadelphia, Pennsylvania
PHX Phoenix, Arizona
PMD Palmdale, California
PNE North Philadelphia
PPG Samoa
RIV March AFB, Riverside, California
RND Randolph AFB, San Antonio, Texas
SAV Travis Field, Savannah, Georgia
SBA Santa Barbara, California
SIN Singapore
SLI Los Alamitos, California
SJU San Juan, Puerto Rico
SMA Santa Maria, Azores, Portugal
SMO Santa Monica, California
SWF Stewart Field, New York (Long Island)
SYR Syracuse, New York
TEH Tehran, Iran
TIK Tinker AFB, Oklahoma City, Oklahoma
TTS Merritt Island (KSC), Florida
TUL Tulsa, Oklahoma
UMR Woomera, Australia
WSD White Sands, New Mexico (Alamogorda HMN)
X68 NASA Shuttle Landing Field, Titusville, Florida
XFW Finkenwerder, Hamburg, Germany
XMR Skid Strip, Kennedy Space Center, Florida
VNY tephenville, Newfoundland, Canada
YQM Moncton, New Brunswick, Canada
YQX Gander, Newfoundland, Canada
YSP Ypsilanti, Michigan (Detroit)
YYG Charlottetown, Prince Edward Island, Canada
YYT St. John's, Newfoundland, Canada
0221 RAF Cardington, UK (FAA Code)

Chapter Notes

Chapter 1

1. Dennis R. Jenkins, *Magnesium Overcast* (North Branch, MN: Specialty, 2002), 9.
2. George A. Larson, "The XC-99—Convair's Very Heavy Transport," *AAHS Journal* 46, no. 3 (2001): 163–169.
3. Jenkins, *Magnesium Overcast*, 15.
4. Robert E. Bradley, *Convair Advanced Designs* (North Branch, MN: Specialty, 2010), 154–160.
5. Bradley, *Convair Advanced Designs*, 154–160.
6. *Ibid.*
7. Jenkins, *Magnesium Overcast*, 24.
8. "Convair Plans Largest," *The Eagle*, Official Publication of Consolidated Vultee Aircraft Corp, Fort Worth Texas Division, March 2, 1945, 1.
9. *Ibid.*
10. *Ibid.*
11. Don Pyeatt and Dennis R. Jenkins. *Cold War Peacemaker* (North Branch MN: Specialty, 2010), Appendix D, 201–203.
12. Jenkins, *Magnesium Overcast*, 17–18.

Chapter 2

1. Jenkins, *Magnesium Overcast*, 25.
2. "Convair Plans Largest," *The Eagle*, 28.
3. Jenkins, *Magnesium Overcast*, 28.
4. *Ibid.*, 33.

Chapter 3

1. Jenkins, *Magnesium Overcast*, 196–205.
2. "XC-99 Sets Air Cargo Record," *Aviation Week*, June 02, 1952, 18.
3. Jenkins, *Magnesium Overcast*, 31.
4. "XC-99 Sets Air Cargo Record."
5. Roy Paris, "Aviation's Amazon," *The Airman* III, no. 7 (July 1959): 10–12.
6. *Ibid.*
7. *Ibid.*
8. *Ibid.*

Chapter 4

1. NASA, *Outsized Booster Carrier Aircraft*, NASA Contractor Report R534-001, 16 March 1964.
2. Kent A. Mitchell, "Fairchild Could Have Been Aircraft Designs," *AAHS Journal* 42, no. 4 (Winter 1997): 274–284.
3. *Ibid.*, 283.
4. Jenkins, *Magnesium Overcast*, 33.

Chapter 5

1. Nicholas M. Williams, "Globemaster, the Douglas C-74," *AAHS Journal* 25, no. 2 (Summer 1980): 83–106.
2. *Ibid.*, 83.
3. Rene J. Francillon, *McDonnell Douglas Aircraft Since 1920, Volume I* (Oxford: Alden, 1988), 392.
4. Williams, "Globemaster, the Douglas C-74."
5. Douglas C-74 Globemaster (DC-7) *Press Book 1945*, 13–14.
6. Williams, "Globemaster, the Douglas C-74," 86.
7. Ronald Miller and David Sawers, *The Technical Development of Modern Aviation* (New York: Praeger, 1970).
8. *Ibid.*
9. Williams, "Globemaster, the Douglas C-74." 86.
10. F. W. Wolcott, "Preliminary Performance Results of Study of C-74 Airplane Equipped with Prop Turbine Engines," *Douglas Memorandum*, August 17, 1948.
11. Douglas C-74 *Press Book*.
12. *Ibid.*
13. *Ibid.*
14. *Ibid.*
15. *Ibid.*
16. Douglas C-74 Globemaster, www.theaviationzone.com/factsheets/c74/.asp, 2008.
17. Williams, "Globemaster, the Douglas C-74," 92.
18. *Ibid.*
19. Nicholas M. Williams, "The Bug Eyed Monster: The Douglas Model 415A, the first Globemaster and DC-7," *Air Enthusiast* 60 (November/December 1995): 40–53.
20. Earl T. Benjamin, FBI Investigation Interview Report, FBI #105-80291-31, February 09, 1960.
21. *Ibid.*
22. *Ibid.*
23. Dominick Edward Bartone, Akros Dynamics, FBI Investigation, FBI #105-80291-16, 17, 1st NR 17, 21 December 1959.
24. David Kaiser, *The Road to Dallas: The Assassination of John F. Kennedy* (Cambridge, MA: Belknap Press, 2008), 8.
25. Benjamin, FBI #105-80291-16, 17, 1st NR 17.
26. *Ibid.*
27. Frank Lloyd, Investigator Akros Dynamics Corp, *GAO Report, B-141258*, Feb. 7, 1961, 19034.
28. Chauncey Marvin Holt, *Self-Portrait of a Scoundrel* (Walterville, OR: Trine Day, 2013), 275.
29. Improper Activities in Labor, *Senate Select Committee Hearings, 86th Congress*, 25–26, 29–30 June 1959, 18976.
30. Kaiser, *The Road to Dallas*, 31.
31. Dan E. Moleda, The Hoffa Wars: *The Rise and Fall of Jimmy Hoffa* (New York: Shapolsky, 1992), 122–123.
32. Improper Activities in Labor, 18976.
33. *Ibid.*, 19103.
34. Holt, *Self-Portrait of a Scoundrel*, 203.
35. Improper Activities in Labor, 19103.
36. *Ibid.*, 19038.
37. *Ibid.*
38. *Ibid.*, 19102
39. *Ibid.*, 19088.
40. Holt, *Self-Portrait of a Scoundrel*, 277–278.
41. Improper Activities in Labor, 19089.
42. Williams, Globemaster, the Douglas C-74, 100.
43. Holt, *Self-Portrait of a Scoundrel*, 75–78, 83–85.
44. Improper Activities in Labor, 19089–19090.
45. Holt, *Self-Portrait of a Scoundrel*, 275.
46. Improper Activities in Labor, 19038.
47. *Ibid.*, 19075.
48. *Ibid.*
49. *Ibid.*
50. *Ibid.*, 19105–19106.
51. *Ibid.*
52. *Ibid.*, 19090.
53. *Ibid.*, 19091.
54. *Ibid.*, 19092.
55. "Seven Indicted in Miami in Arms Smuggling," *New York Times*, June 06, 1959.
56. Improper Activities in Labor, 19093–19095.
57. *Ibid.*, 19093.
58. Millville Ordnance Company, www.m1carbinesinc.com/carbine_millville.html, 27 May 2012.
59. Bartone, FBI #105-80291-16, 17, 1st NR 17, October 19, 1960.
60. Bartone, FBI #105-80291-48, 21.
61. *Ibid.*, 16–18.
62. *Ibid.*, 16.
63. "Guy W. Gully Sentenced," *Securities and Exchange Commission News Digest* 63 (January 1963): 1.
64. Bartone, FBI #105-80291-48, 17.
65. Pan American Bank of Miami v. Gully, Civ. A. No. 62-592, United States District Court W. D. Pennsylvania, August 1, 1963.
66. Roger D. Launius, "Orvis M. Nelson," in *The Airline Industry*, edited by William M. Leary (New York: Facts on File, 1992), 309–313.
67. *Ibid.*
68. "Flying Box Cars," *Popular Mechanics* 72, no. 1 (July 1939): 35, 114A.
69. Floyd B. Odlum, Papers 1892–1976, www.eisenhower.archives.gov/research/finding_aids/pdf/Odlum_Floyd_Papers.pdf, 08 December 2016, 2.
70. William Patrick Dean, *The ATL-98 Carvair: A Comprehensive History of the Aircraft and All 21 Airframes* (Jefferson NC: McFarland, 2008), 284–285.
71. James Herriot, *The Lord Made Them All* (New York: St. Martin's, 1981).
72. Tad Houlihan (Ted Hunt), *Flying Cowboys* (Bloomington, IN: Author House, 2004).
73. *Ibid.*, 141–142, 166.

74. *Ibid.*, 248.
75. Ted Hunt, Captain, Air Systems—Aeronaves de Panama, letter, 12 July 2010.
76. Williams, "Globemaster, the Douglas C-74," 106.
77. John Lanham, "Transocean Airlines (TOLOA)," *The "Crosswind Drift" Newsletter, EAA Chapter 1232* (October 2014): 3.
78. Williams, "Globemaster, the Douglas C-74," 106.
79. *Ibid.*
80. *Ibid.*
81. *Ibid.*
82. *Ibid.*
83. *Ibid.*
84. *Ibid.*
85. *Ibid.*
86. *Ibid.*
87. *Ibid.*
88. *Ibid.*
89. *Ibid.*
90. *Ibid.*
91. *Ibid.*

Chapter 6

1. Anthony J. Tambini, *Douglas Jumbos: The Globemaster* (Boston MA: Branden, 1999), 115.
2. Douglas C-74 (DC-7) *Press Book*, Power Plants, 19.
3. Flight Manual, USAF Series, C-124 A&C Aircraft, T.O., 1C-124A-1, 1–39.
4. *Ibid.*
5. Edward J. Bodenmiller, USAF 28th ATS, Hill Air Force Base, 1962–66.
6. Flight Manual, USAF Series, C-124 A&C Aircraft, T.O., 1C-124A-1, 1–46.
7. Collin Bakse, *Airlift Tanker: History of U.S. Airlift Tanker Forces* (Paducah, KY: Turner, 1995), 21.
8. Earl Berlin, *Douglas C-124 Globemaster II, Air Force Legends Number 206* (Simi Valley CA: Ginter, 2000), 10.
9. "South Pole Station: The First 10 Years." http://www.southpolestation.com/trivia/igy1/igy1.html, 15 September 2016.
10. Roger E. Bilstein, *Stages to Saturn* (Gainesville: University Press of Florida, 2003). 19.
11. "The Congo Airlift," http://amcmuseum.org/history/airlifts/congo_airlift.php, 12 February 2013.
12. Bakse, *Airlift Tanker*, 100.
13. Robert F. Dorr, "The Turboprop Globemaster," *Aviation News*, 13–26 October 1989, 497.
14. Berlin, *Douglas C-124 Globemaster II*, 127–129.
15. *Ibid.*, 131, 153.
16. Cal Taylor, *Remembering an Unsung Giant: The Douglas C-133 Cargomaster and Its People* (Olympia, WA: Firstfleet, 2005), 44.
17. Donald L. Elder, "Aerodynamics Project Engineer C-124 Comments," www.aero-web.org, 05 January 2011.
18. Rene J. Francillon, *McDonnell Douglas Aircraft Since 1920*, vol. I (London: Putnam, 1988), 439.
19. Dick Mulready, *Advanced Engine Development at Pratt & Whitney: The Inside Story of Eight Special Projects*, 1946–1957 (Warren PA: Society of Automotive Engineers, 2001), 20.
20. Cal Taylor, "The Douglas C-132 Logistics Transport and Air Refueler," *AAHS Journal* 50, no. 1 (Spring 2005): 20.

Chapter 7

1. Cal Taylor, "The Douglas C-132 Logistics Transport and Air Refueler," 16.
2. *Ibid.*, 19.
3. Taylor, *Unsung Giant*, 80.
4. *Ibid.*, 83.
5. *Ibid.*, 83–84.
6. Apollo "A"/Saturn C-1 Launch Vehicle Systems Report, Saturn Systems Office, 1961, Sec 18.3.3.2, 298–307.
7. Bilstein, *Stages to Saturn*, 309.
8. Taylor, *Unsung Giant*, 83, 243.
9. *Ibid.*, 244.
10. *Ibid.*, 244–246, 372.
11. *Ibid.*, 239–248.

Chapter 8

1. Williams, "Globemaster, the Douglas C-74," 86.
2. Bradley, *Convair Advanced Designs*, 154–160.
3. Dean, *The ATL-98 Carvair*, 12.
4. *Ibid.*, 6–7.
5. *Ibid.*, 12.
6. *Ibid.*, 16.
7. Air Charter Advertisement, *Flight*, 29 May 1953.
8. Peter M. Bowers, *Boeing Aircraft Since 1916* (Annapolis, MD: Naval Institute, 1993), 557–558.
9. Clay Lacy, interview, Camarillo, CA, 15 July 2015.
10. *Ibid.*
11. Brian Kerry, chief aerodynamicist, ATEL; Cliff Berrett, chief draftsman, ATEL.
12. *Ibid.*
13. *Ibid.*
14. Dean, *The ATL-98 Carvair*, 42.
15. *Ibid.*, 72.
16. *Ibid.*, 15.
17. *Ibid.*, 22.
18. *Ibid.*, 123.
19. *Ibid.*, 130–133.
20. "Flying Box Cars," *Popular Mechanics* 72, no. 1 (July 1930): 35, 114A.
21. Keith Evans, Southend personal account, email, 09 February 2016.
22. Dean, *The ATL-98 Carvair, 103–104*
23. *Ibid.*, 26–27.
24. *Ibid.*, 46
25. *Ibid.*, 28–30.
26. *Ibid.*, 96–106.
27. *Ibid.*, 107–113.
28. *Ibid.*, 114–120.
29. *Ibid.*, 121, 132–133.
30. *Ibid.*, 134, 160–164.
31. *Ibid.*, 165, 172–174.
32. *Ibid.*, 175–179.
33. Frank Moss, owner, Honduras Caribbean (Hondu Carib) Certificate, Tela, Honduras, interview, 20 July 2007.
34. Dean, *The ATL-98 Carvair*, 191, 196–199.
35. *Ibid.*, 200, 210.
36. *Ibid.*, 211–212, 224.
37. *Ibid.*, 225, 235.
38. *Ibid.*, 237, 245.
39. *Ibid.*, 247, 254–255.
40. *Ibid.*, 257, 264–267.
41. *Ibid.*, 268, 272.
42. *Ibid.*, 273, 277.
43. *Ibid.*, 284, 291–293.
44. *Ibid.*, 295, 300.
45. *Ibid.*, 301, 310.
46. *Ibid.*, 312, 332.
47. *Ibid.*, 334, 346–348.

Chapter 9

1. Bowers, *Boeing Aircraft Since 1916* (reprint 1993), 318–28.
2. "Prototype Boeing B-29 Crashes into Seattle's Frye Packing Plant on February 18, 1943," http://www.historylink.org/index.cfm?displaypage=output.cfm&file_id=2874, 03 March 2016.
3. Bowers, *Boeing Aircraft Since 1916*, 326–27.
4. *Ibid.*, 348–52.
5. *Ibid.*, 351–52.

Chapter 10

1. Bowers, *Boeing Aircraft Since 1916*, 326–27, 353–57.
2. Nicholas A. Veronico, *Stratocruiser, Boeing 377*, vol. 9 (North Branch MN: Specialty, 2001), 11.
3. "Five Airlines Reported Interested in C-97," *Aviation Week*, 28 July 1952.
4. Bill Melberg, "Good Bad and the Ugly: Remembering the Boeing Stratocruiser Airliners," *Airliners*, Winter 1960, 20–25.
5. *Ibid.*
6. CAA Accident Investigation Report 1-0080: Pan American World Airways Boeing 377, N1032U, Carolina, Brazil, 29 April 1952, 1.
7. *Ibid.*
8. Walter Schoendorf, "Clipper Good Hope," *Airways*, April 2000, 57–64.
9. *Ibid.*
10. CAA Accident Investigation Report 1-0080, 3.
11. *Ibid.*, 4.
12. Schoendorf, "Clipper Good Hope.
13. *Ibid.*, 59.
14. CAA Accident Investigation Report 1-0080, 5.
15. *Ibid.*, 6.
16. *Ibid.*, 6.
17. *Ibid.*, 6–7.
18. *Ibid.*, 7.
19. Schoendorf, "Clipper Good Hope.
20. CAA Accident Investigation Report 1-0080, 8–9.
21. *Ibid.*, 8.
22. Schoendorf, "Clipper Good Hope.
23. CAA Accident Investigation Report 1-0080, 10.
24. Schoendorf, "Clipper Good Hope.
25. CAA Accident Investigation Report 1-0080, 11–18.
26. *Ibid.*
27. *Ibid.*
28. Gregg Herken and Ken Fortenberry, "The Mystery of the Lost Clipper," *Air & Space Magazine*, September 2004.
29. "Five Airlines," *Aviation Week*, 28 July 1952.
30. KC-97G USAF Aircraft Flight Manual T.O. 1C-97 (K) G-1, 15 October 1959, 1–36.
31. Boeing Stratocruiser Pilot's Handbook, Model 377-10-34, United Airlines, 20 February 1950, 1–18.
32. Patrick Dean, observed on HC-97 of 305th in 1968.

Chapter 11

1. Roger E. Bilstein, *Stages to Saturn* (Gainesville: University Press of Florida, 1980; reprint 2003), 294.
2. *Ibid.*, 295.
3. *Ibid.*, 307–308
4. Don Stratman, comments, Santa Barbara, 07 July 2014.
5. Clay Lacy, interview, Camarillo, CA, 15 July 2015.
6. *Ibid.*
7. Robert R. Kirby and George M. Warner, *Aviation Visionary* (Upland CA: BAC, 2008), xi.
8. Clay Lacy, 15 July 2015.
9. Joseph E. Libby, "Stanley D. Weiss," in *The*

Airline Industry, edited by William M. Leary (New York: Facts on File 1992, 496.
10. Clay Lacy, 15 July 2015.
11. Roger D. Launius, "Orvis M. Nelson," in *The Airline Industry*, edited by William M. Leary (New York: Facts on File 1992), 313.
12. Clay Lacy, 15 July 2015.
13. *Ibid.*
14. *Ibid.*
15. Di Freeze, "Clay Lacy: The Planes I've Flown and the People I've Known," *Van Nuys Aviation Journal* 3, no. 9 (03 August 2003).
16. "Lloyd G Dorsett, Dorsett Industries, Inc., Norman, Oklahoma," *Sunday Oklahoman*, Oklahoma City, November 12, 1967.
17. Clay Lacy, 15 July 2015.

Chapter 12

1. Freeze, "Clay Lacy."
2. Bilstein, *Stages to Saturn*, 310.
3. Milbrey Conroy, email, 17 April 2016.
4. Clay Lacy, 15 July 2015.
5. Don Stratman, comments, 07 July 2014.
6. *Ibid.*
7. Robert S. Tripp, "Guppy," *Invention and Technology* 17, no. 4 (Spring 2002).
8. James Bugbee, FAA inspector, letter to author, 17 June 2014, 10.
9. Lloyd S. Jones, "Jack Conroy and His Flying Fish," *AAHS Journal* 157, no. 2 (2012): 82–88.
10. *Ibid.*
11. Clay Lacy, 15 July 2015.
12. Bugbee, letter, 6–8.
13. Clay Lacy, 15 July 2015.
14. Bugbee, letter, 6–8.
15. *Ibid.*
16. Clay Lacy, 15 July 2015.
17. *Ibid.*
18. *Ibid.*
19. *Ibid.*
20. Dottie Furman, "Flight Testing and Demonstration Memo," 29 October 1962.
21. Bilstein, *Stages to Saturn*, 311.
22. *Ibid.*
23. *Ibid.*
24. Furman, "Flight Testing and Demonstration Memo."
25. *Ibid.*
26. Tripp, "Guppy,"
27. Furman, "Flight Testing and Demonstration Memo."
28. Robert S. Tripp, "Glorious Guppies, Metamorphous of Monsters," unpublished paper, 1990.
29. A.M. Kaplan, chief engineer, Strato Engineering, letter to Jack Conroy, 24 October 1962.
30. *Ibid.*
31. Jack Conroy, letter to Wernher von Braun, 29 October 1962.
32. Harry H. Gorman, deputy director for administration, NASA, letter to John Conroy, 13 November 1962.
33. Conroy, letter to von Braun, 29 October 1962.
34. *Ibid.*
35. Tripp, "Guppy,"
36. Fred Weir, ASI electrician, interview, 25 January 2014.
37. Bilstein, *Stages to Saturn*, 311.
38. Don Stratman, interview, 01 June 2015.
39. Tripp, "Guppy."
40. *Ibid.*
41. NASA, "Transporting Saturn S-IV Stage," Santa Monica Shipping Report SM-44052, 11 June 1963.
42. *Ibid.*
43. Bugbee, letter, 9.
44. Contract of Air Transportation Services Contract, Aero Spacelines and NASA, 1 September 1963 through 20 June 1964, Conduct of Services, 06 September 1964.
45. NTSB LAX64A0016, incident report, N705PA.
46. United States of America v Aero Spacelines No. 20274, 361F.2d 916 (1966), United States Court of Appeals Ninth Circuit, 12 May 1966.
47. Steve Weintz, "Her Majesty's Nuclear Seaplane," https://warisboring.com/her-majesty-s-nuclear-seaplane/, 28 May 2106.
48. *Ibid.*
49. *Astronautics Year 1964: International Astronautical and Military Space/Missile Review* (London: Pergamon, 1965).
50. "Aero Spacelines Seeking to Buy Saunders Roe Flying Boats," *Aviation Week & Space Technology*, 20 January 1964, 34.
51. "A Fulbright 'Folly' of Lunar Race," *Space Daily*, 27 April 1964, 146.
52. Bob Wealthy, Saunders-Roe and the Princes Flying Boat, Presentation to the Hamburg Branch of the Royal Aeronautical Society, 03 June 2010, 194.
53. Don Stratman, ASI, telephone interview, 01 June 2015.
54. *Ibid.*
55. *Ibid.*
56. Weir, interview, 25 January 2014.
57. Bruce Stratton, ASI pilot, comments, 07 July 2014.
58. Bilstein, *Stages to Saturn*, 313.
59. "Aero Spacelines," World Airline Directory, *Flight International*, 21 March 1974, 9.
60. Clay Lacy, 15 July 2015.
61. *Ibid.*
62. Tom Roberts, Aero Spacelines, interview, 12 October 2015.
63. Aero Spacelines Inc. Invoice 6535 to Union de transports Aeriens, 05 September 1980.

Chapter 13

1. "Aero Spacelines Plans Jet Augmented Guppy," *Aviation Week & Space Technology*, 10 August 1964, 33.
2. *Ibid.*
3. Don Stratman, 01 June 2015.
4. Bilstein, *Stages to Saturn*, 314.
5. Peter C. Andrews. "Unexcelled Chemical Acquires Twin Fair Inc.," *Buffalo Courier Express*, 08 March 1962, 21.
6. Alan A. Block, *Masters of Paradise: Organized Crime and the Internal Revenue Service in the Bahamas* (New Brunswick, NJ: Transaction, 1991), 63–70.
7. Holt, *Self-Portrait of a Scoundrel.*
8. Charles Raw, Bruce Page and Godfrey Hodgson, *Do You Sincerely Want to Be Rich? The Full Story of Bernard Cornfeld and IOS* (New York: Viking, 1971), 236-237.
9. *Ibid.*, 233–234.
10. Michael F. Rizzo, *The Glory Days of Buffalo Shopping* (Charleston SC: History, 2013), 37–39.
11. "Unexcelled Chemical Files for Secondary," *Securities and Exchange Commission News Digest*, 12 February 1965, 1.
12. Richard Oulahan and William Lambert, "The Scandal in the Bahamas," *Life Magazine,* 03 February 1967, 59–74.
13. Block, *Masters of Paradise*, 63–70.
14. *Ibid.*
15. Holt, *Self-Portrait of a Scoundrel*, 278.
16. John M. Conroy vs. Unexcelled Inc., Superior Court of the State of California for Santa Barbara County, No. 82063, Judgment and Decree, 04 February 1969.
17. Robert S. Tripp, "Glorious Guppies," 11.
18. *Ibid.*, 12.
19. Lt. Col. P.G. Smith, "The Day the Super Guppy Blew Her Top," *Air Force Magazine* (April 1971).
20. *Ibid.*
21. *Ibid.*
22. Stratman, interview, 01 June 2015.
23. Smith, "The Day the Super Guppy Blew Her Top."
24. Aero Spacelines, Inc. v. United States, 530 F.2d 324 (1976), United States Court of Claims, January 28, 1976.
25. *Ibid.*, 293.
26. Milbrey Conroy, comments Santa Barbara, 07 July 2014.
27. Aero Spacelines Inc., Specializing in Outsize Cargo Airlift, sales brochure, 1966, 11.
28. Angelee Conroy, email regarding Allen Manus, 09 November 2016.
29. Oulahan and Lambert, "The Scandal In The Bahamas."
30. *Ibid.*
31. Raw, Page and Hodgson, *Do You Sincerely Want to Be Rich*, 237.
32. Bilstein, *Stages to Saturn*. 293.
33. Incentive Provisions of Saturn V Stage Contracts 79, GAO Report to Congress B-161366, p. 40 of Appendix II, 1970.
34. Marshall Space Flight Center, Saturn V Progress Office, *Saturn V Semiannual Progress Report* (July 31-December 31, 1967), 65.
35. Robert Dallek, *Flawed Giant: Lyndon Johnson and His Times, 1961–1973* (New York: Oxford University Press, 1998), 423.
36. Ritchie W. Tilson, Boeing divisional director of material, letter to Jack Conroy, 16 December 1966.
37. J.F. Kraft, Martin-Marietta, letter to Jack Conroy, 02 December 1966.
38. Aero Spacelines, Inc. v. Untied States, 530 F,2d324 (1976), Contribution to the Defense Effort.
39. Kirby, *Aviation Visionary*, xv.
40. "Business & People," *Los Angeles Times*, November 8, 1967, part III.
41. "Johnston to Head Aero Spacelines," *Aviation Week & Space Technology*, November 1967.
42. Robert L. Twiss, "Artist's Sketch Shows How 'Guppy' Aircraft Could Accommodate Boeing 747 Sections," *Seattle Times*, February 18, 1968.
43. United States General Accounting Office, Letter B-146804 to James C. Fletcher, administrator, NASA, 19 March 1973.

Chapter 14

1. John C. Goodrum, NASA logistics, letter to John Conroy, 06 February 1967.
2. Weir, interview, 25 January 2014.
3. Contract to transport Budd "People Pod" between city of Los Angeles and Aero Spacelines, May 1967.
4. John M. Conroy, letters to Clifford T. Moore, city of Los Angeles, 08 and 10 May 1967.
5. N. E. Andros, "The Strange Guppy Family," *Flight International*, February 1968, 234.
6. Holt, *Self-Portrait of a Scoundrel*, 278.
7. "Unexcelled Inc. Proposes Debenture Offering," *Securities and Exchange Commission News Digest*, No. 67–187, 28 September 1967, 2.
8. Weir, interview, 25 January 2014.
9. "Mini Guppy Stirs Alarm," *Star Free Press*, Ventura, California, 01 February 1968.

10. Raw, Page and Hodgson, *Do You Sincerely Want to Be Rich*, 242.
11. *Ibid.*
12. Robert J. Cole, "Trading in Shares Of Unexcelled, Inc., Is Halted on Amex," *New York Times*, September 04, 1970, 42.
13. Robert Harridge, Aero Spacelines, electrician, interview at Santa Barbara, 12 October 2015.
14. *Ibid.*
15. *Ibid.*
16. "Goleta's Big Flying Whales Cut Cargoes to Size," *Valley Sun*, 01 April 1970.
17. Roberts, interview, 12 October 2015.

Chapter 15

1. Weir, interview, 25 January 2014.
2. "CA Cargo Plane Crash," *The Valley News*, Van Nuys, California, 14 May 1970.
3. NTSB LAX70AL071, incident report, Crash of N111AS.
4. *Ibid.*
5. *Ibid.*
6. *Ibid.*
7. *Ibid.*
8. *Ibid.*
9. *Ibid.*
10. *Ibid.*
11. Bugbee, letter, 11–12.
12. Harridge, interview, Santa Barbara, 12 October 2015.

Chapter 16

1. Aero Spacelines Inc., sales brochure, 1966, 9.
2. Weir, interview, 25 January 2014.
3. Stratman, interview 01 June 2015.
4. *Ibid.*
5. Stratman, comments, 07 July 2014.
6. *Ibid.*
7. Weir, interview, 25 January 2014.
8. Robert Eryl Crump, "Airbus Guppies and Belugas," *Airways*, July/August 1996, 17–23.
9. Donald F. Woods, *International Logistics* (New York: McGraw Hill, 2002).
10. NASA, *Audit of Super Guppy Operations*, Johnson Space Center Report A-JS-88-008, 29 September 1989.
11. Charles J. Gillespie, NASA Flight Operations, email, 01 December 2015.
12. "Aero Spacelines Is Alive and Well," *Flight International*, 27 October 1979, 1343.
13. *Ibid.*
14. *Ibid.*
15. Allan Winn, "Super Send-Off," *Flight International*, 24 December 1997, 24–26.
16. Aero Spacelines Inc. Invoice 6148 to Union de transports Aeriens, 16 November 1979.
17. Roberts, interview, 12 October 2015.
18. A. Hurato and J. Thomas, letter from Airbus to UTA requesting a quote on additional Guppy aircraft, 06 September 1988.
19. *Ibid.*
20. "UK's Lucas Nets Tracor Aviation," *Flight International*, 12–18 June, 1991.
21. Ken Sanders, NASA, N941 Flight Operations, telephone interview, 04 April 2016.
22. *Ibid.*
23. Robert Rivers, NASA ops division instructor pilot, email, 22 December 2015.
24. Ken Sanders, interview, 04 April 2016.
25. *Ibid.*
26. NASA, *Audit of Super Guppy Operations*, Johnson Space Center Report A-JS-88-008, 29 September 1989, 3–10
27. *Ibid.*
28. *Ibid.*, 9.
29. *Ibid.*, 3–10
30. Stratman, comments, 07 July 2014.
31. Larry J. Glenn, NASA Flight Operations, emails, December 2015.
32. Eldon W. Hall and Francis C. Schwenk, "Current Trends in Large Booster Developments," *Aerospace Engineering*, May 1960, 21.
33. Kirby, *Aviation Visionary*, xv.
34. Larry Glenn, email, December 2015.
35. *Ibid.*
36. *Ibid.*
37. *Ibid.*
38. Ken Sanders, interview, 04 April 2016.
39. Larry Glenn, email, 06 January 2016.
40. *Ibid.*
41. *Ibid.*
42. *Ibid.*
43. *Ibid.*
44. *Ibid.*
45. Ken Sanders, interview, 04 April 2016.
46. *Ibid.*
47. *Ibid.*
48. Larry Glenn, email, December 2015.

Bibliography

Books

Adams, Michael R. *Ocean Station: Operations of the U.S. Coast Guard, 1940–1977*. Eastport, ME: Nor'easter, 2010.

Bakse, Collin. *Airlift Tanker: History of U.S. Airlift Tanker Forces*. Paducah, KY: Turner, 1995.

Berlin, Earl. *Douglas C-124 Globemaster II, Air Force Legends Number 206*. Simi Valley, CA: Steve Ginter, 2000.

Bickers, Richard Townshend. *Airlift, Military Air Transport: The Illustrated History*. London: Osprey, 1998.

Bilstein, Roger E. *Stages to Saturn*, Gainesville: University Press of Florida, 2003.

Bishop, John C. "Flying Tiger Line." In *The Airline Industry*, edited by William W. Leary, 181–182.

Bishop, John C. "Robert W. Prescott (Flying Tiger)." In *The Airline Industry*, edited by William W. Leary, 377–380.

Block, Alan A. *Masters of Paradise: Organized Crime and the Internal Revenue Service in the Bahamas*. New Brunswick, NJ: Transaction, 1991.

Bowers, Peter M. *Boeing Aircraft Since 1916*. Annapolis, MD: Naval Institute, 1993.

Bradley, Robert E. *Convair Advanced Designs*. North Branch, MN: Specialty, 2010.

Chin, Art. *Anything, Anytime, Anywhere: The Legacy of Flying Tiger Line, 1945–89*. Seattle: Tassels and Wings, 1993.

Dallek, Robert. *Flawed Giant: Lyndon Johnson and His Times, 1961–1973*. New York: Oxford University Press, 1998.

Dean, William Patrick. *The ATL-98 Carvair: A Comprehensive History of the Aircraft and All 21 Airframes*. Jefferson, NC: McFarland, 2008.

Door, Robert. *7th Bombardment Group/Wing 1918–1995*. Paducah, KY: Turner, 1998.

Eastwood, A. B., and J. Roach. *Piston Engine Airliner Production List*. Middlesex, England: Aviation Hobby Shop, 1991.

Francillon, Rene J. *McDonnell Douglas Aircraft Since 1920*, vol. I. London: Putnam, 1988.

Geere, Stacy T. *Lucky Me: The Life and Flights of Veteran Aviator Clay Lacy*. Virginia Beach, VA: Donning, 2010.

Gordon, Yefim. *Myasishchev M-4 & 3M*. Hinckley: Midland, 2003.

Green, William, and Gerald Pollinger. *The Aircraft of the World*. Garden City, NY: Doubleday, 1965.

Herriot, James. *The Lord Made Them All*. New York: St. Martin's, 1981.

Holt, Chauncey Marvin. *Self-Portrait of a Scoundrel*. Walterville, OR: Trine Day, 2013.

Houlihan, Tad (Ted Hunt). *Flying Cowboys*. Bloomington, IN: Author House, 2004.

Jacobsen, Meyers K., and Ray Wagner. *B-36 in Action*. Carrollton, TX: Squadron/Signal, 1980.

Jenkins, Dennis R. *Magnesium Overcast*. North Branch, MN: Specialty, 2002.

Kaiser, David. *The Road to Dallas: The Assassination of John F. Kennedy*. Cambridge, MA: Belknap Press, 2008.

Kirby, Joe. *The Bell Bomber Plant*. Charleston, SC: Arcadia, 2008.

Kirby, Robert R., and George M. Warner. *Aviation Visionary "Smilin Jack" Conroy*. Upland, CA: BAC, 2008.

Launius, Roger D. "Orvis M. Nelson." iIn *The Airline Industry*, edited by William W. Leary, 309–313.

Lawrie, Alan. *Sacramento's Moon Rockets*. Charleston, SC: Arcadia, 2015.

Menard, David W. *USAF Plus Fifteen: A Photo History, 1947–1962*: Schiffer, 1993.

Miller, Ronald E., and David Sawers. *The Technical Development of Modern Aviation*. Praeger, 1970.

Moleda, Dan E. *The Hoffa Wars: The Rise and Fall of Jimmy Hoffa*. New York: Shapolsky, 1992.

Mulready, Dick. *Advanced Engine Development at Pratt & Whitney: The Inside Story or Eight Special Projects, 1946–1957*. Warren, PA: Society of Automotive Engineers, 2001.

Myrha, David. *Broken ME 262 Jet Fighters, Part 2*. Fort Myers, FL: RCW Technology S&S, 2012.

Newton, Wesley Phillips. "Juan T. Trippe." In *The Airline Industry*, edited by William W. Leary, 464–476.

Pyeatt, Don, and Dennis R. Jenkins. *Cold War Peacemaker*. North Branch, MN: Specialty, 2010.

Raw, Charles, Bruce Page and Godfrey Hodgson. *Do You Sincerely Want to Be Rich? The Full Story of Bernard Cornfeld and IOS*. New York: Viking, 1971.

Rizzo, Michael F. *The Glory Days of Buffalo Shopping*. Charleston, SC: History, 2013.

Tambini, Anthony J. *Douglas Jumbos: The Globemaster*. Boston, MA: Branden, 1999.

Taylor, Cal. *The Douglas C-132*. Report No. *SM-18220*, Model XC-132, Design and Mock-up Data. Olympia, WA: Firstfleet, 2010.

Taylor, Cal. *Remembering an Unsung Giant: The Douglas C-133 Cargomaster and Its People*. Olympia, WA: Firstfleet, 2005.

Veronico, Nicholas A. *Boeing 377 Stratocruiser*, Airliner Tech Series Vol. 9. North Branch, MN: Specialty, 2001.

Veronico, Nicholas A., and Jim Dunn. *Giant Cargo Planes*. Osceola, WI: MBI, 1999.

Veronico, Nicholas A., and Ron Strong. *AMARG (America's Military Aircraft Boneyard)*. North Branch, MN: Specialty, 2010.

Woods, Donald F. *International Logistics*. New York: McGraw-Hill, 2002.

Articles

"Aero Spacelines." World Airline Directory. *Flight International*, 21 March 1974, 9.

"Aero Spacelines Is Alive and Well." *Flight International*, 27 October 1979, 1343.

"Aero Spacelines Plans Jet Augmented Guppy." *Aviation Week & Space Technology*, 10 August 1964, 33.

"Aero Spacelines Seeking Options to Buy Saunders Roe Flying Boats." *Aviation Week & Space Technology*, 20 January 1964, 34.

"Air Charter Limited." *Flight*, 29 May 1953.

"Airbus Develops Cargo Lifter." *Flight International*, 02 June 1999.

Andrews, Peter C. "Unexcelled Chemical Acquires Twin Fair Inc." *Buffalo Courier Express*, 08 March 1962, 21.

Andros, N.E. "The Strange Guppy Family." *Flight International*, February 1968, 233–4.

"B-36 May Tote Saturn Stage." *Huntsville Times*, 01 December 1963.

"Bahamas, Letters to the Editor." *Life Magazine*, March 17, 1967, 26.

"Battle of Kansas." *Warbird Digest* 7 (March/April 2006): 60–61.

Berry, Peter. "Boeing Stratocruiser." *AAHS Journal* 48, no. 1 (Spring 2003): 64.

Berry, Peter. "Post-War Trans-Atlantic Propliners." *Propliner* 83 (Summer 2000): 16–23.

"Big Guppies Seek a Bigger Pond." *Business Week*, September 10, 1966, 77.

"Blue, Blue Yonder." *Forbes*, April 1, 1974, 67.

Bohl, Walt. "United Airlines' Queen of the Fleet: The Boeing Stratocruiser, 1949 to 1954." *AAHS Journal* 56, no. 2 (Summer 2011): 147–149.

Bone, Andrew R. "I've Ridden in a Flying Warehouse." *Popular Science*, January 1953, 156–159.

"Breaking a Giant's Heart." *Air Force, The Official Journal of the Air Force Association* 31, no. 4 (January 1948): 16–17.

"Business & People." *Los Angeles Times*, November 8, 1967, Part III.

Chopp, Timothy A. "Flight of Deliverance." *Propliner* 76 (Autumn 1998): 34–36.

Cole, Robert J. "Trading in Shares of Unexcelled, Inc., Is Halted on Amex." *New York Times*, September 04, 1970 42.
"Conroy Charges Lockheed-Saturn Conspiracy in Engine Charter Contract." *Aviation Daily*, January 27, 1970, 14–15.
"Conroy Resigns as President of Aero Spacelines." *Santa Barbara News Press*, November 06, 1967.
"Convair Plans Largest." *The Eagle*, Official Publication of Consolidated Vultee Aircraft Corp, Fort Worth Texas Division, March 2, 1945, 1.
Crump, Robert Eryl. "Airbus Guppies & Belugas." *Airways*, July/August 1996, 17–23.
"Directory: World Airliners, An-124 Ruslan." *Flight International*, 21–27 October 2003, 58.
"The Ditching of Sovereign of the Skies." *Airways*, April 2001, 24–26.
Dorr, Robert F. " The Turboprop Globemaster." *Aviation News*, 13–26 October 1989, 497.
Esler, Dave. "The Guppies." *World's Greatest Aircraft*, 1972, 134–139.
"Explosion in Airliner." *The Morning Bulletin*, 08 December 1953, N0.29466, 1.
"FiFi Boeing B-29 Superfortress," *Warbird Digest* 7 (March/April, 2006): 50–59.
Fink, Donald E. "Hybrid Heavy-Lift Vehicle Under Study." *Aviation Week & Space Technology*, Vol. 101, No. 13, 29 July 1974. pp. 49–51.
"Five Airlines Reported Interested in C-97." *Aviation Week*, 28 July 1945.
"Flying Box Cars." *Popular Mechanics* 72, no. 1 (July 1930): 35–37, 114A.
Freeze, Di. "Clay Lacy: The Planes I've Flown and the People I've Known." *Van Nuys Aviation Journal* 3, no. 9 (03 August 2003).
"A Fulbright 'Folly' of Lunar Race." *Space Daily*, April 27, 1964, 146.
Gaskell, Keith. "European Super Guppy Operations." *Propliner* 63(Summer 1995): 26–29.
"Giant Aircraft Going to Fidel, Cuba in Market for C-74." *Arizona Daily Star*, March 21, 1959.
"Goleta's Big Flying Whales Cut Cargoes to Size." *Valley Sun*, 01 April 1970.
"Gun Smuggling Plan Smashed, 11 Seized." *Miami Herald*, May 23, 1959.
"Guppy-101 Flies." *Flight International*, 02 April 1970, 542.
"Guppy Flies Surveyor II to Cape Kennedy." *Los Angeles Times*, 19 July 1966, part 2, 2.
"Guppy Undergoes Overhaul." *Santa Barbara News Press*, November 01, 1967.
"Guy W. Gully Enters No Contest Plea." *Securities and Exchange Commission News Digest*, 15 November 1963, 2.
"Guy W. Gully Sentenced." *Securities and Exchange Commission News Digest*, No. 63-1-14, 21 January 1963, 1.
Hall, Eldon W., and Francis C. Schwenk. "Current Trends in Large Booster Developments." *Aerospace Engineering*, May 1960, 21.
Herken, Gregg, and Ken Fortenberry. "The Mystery of the Lost Clipper." *Air & Space Magazine*, September 2004.
Holcombe, Chet. "Aero Spacelines Plans Expansion." *Santa Barbara News Press*, 11 March 1968.
"Hubble Space Telescope Arrives." Eos, *Transactions American Geophysical Union* 70, no. 42 (17 October 1989): 906.
Hunter, George S. "Mini Guppy Begins Certification Testing." *Aviation Week & Space Technology*, July 03, 1967.
"Invasion of Cuba from Dominican Republic 1959." *Cuban Information Archives*, RIF 124-10294-10051, FBI record 2-1423-9TH NR 36.
"Jack Conroy." *Icarian Flying Club Newsletter, Tower Talk* 14, no. 9 (December 1962): 6.
"Johnston Quits Boeing." *Seattle Post Intelligencer*, 04 November 1967.
"Johnston to Head Aero Spacelines." *Aviation Week & Space Technology*, November 1967.
Jones, Lloyd S. "Jack Conroy and His Flying Fish." *AAHS Journal* 157, no. 2 (2012): 82–88.
Jones, Tony Merton. "The Ocean Liner of the Air." *Propliner* 72 (Autumn 1997): 13–16.
Jones, Tony Merton. "The Ocean Liner of the Air, Part Two." *Propliner* 73(Winter 1997): 32–38.
Jones, Tony Merton. "The Ocean Liner of the Air, Part Three." *Propliner* 74 (Spring 1998): 26–34.
Jones, Tony Merton. "The Ocean Liner of the Air, Part Five." *Propliner* 76 (Autumn 1998): 18–23.
Jones, Tony Merton. "The Ocean Liner of the Air, Part Six." *Propliner* 77 (Winter 1998): 18–23.
Jones, Tony Merton. " The Transglobe Blunder." *Propliner* 103 (Summer 2005): 16–23.
Kuhns, Carl. "Boeing Model 377 Stratocruiser Propeller Woes." *Aircraft Engine Historical Society*, 1–2.
Lanham, John. "Transocean Airlines (TOLOA)." *The "Crosswind Drift" Newsletter-EAA Chapter 1232*, October 2014, 3.
"Larger Guppy Aimed at S-IVB Transport." *Aviation Week and Space Technology*, 19 April 1965, 82.
Larson, George A. "The XC-99: Convair's Very Heavy Transport." *AAHS Journal* 46, no. 3 (2001): 163–169.
"Lighter Than Air Technology Broadens." *Aviation Week & Space Technology*, September 30, 1974.
"Lloyd G. Dorsett, Dorsett Industries, Inc., Norman Oklahoma," *Sunday Oklahoman*, Oklahoma City, November 12, 1967.
"Lockheed Disputes Saturn-Hercules-RB.211 Link, Admits Conroy Was Top Bidder." *Aviation Daily*, October 5, 1970, 189.
Lowther, Scott. "Virtus." *AIAA Horizons* 37, no. 5 (March/April 2012): 26–29.
"Lucas Aerospace Buys Tracor Aviation for $27 Million." *Commuter Regional Airline News*, June 17, 1991.
Marshall, John A. "Ditching of the Sovereign Seas." *Airways*, April 2001, 24–26.
Maurice, Tom. "Guppy Maker Thinking Bigger Planning New Cargo Aircraft." *Golden Coast News*, Santa Barbara, 20 July 1968.
McSurely, Alexander. "Five Airlines Reported Interested in C-97." *Aviation Week*, 28 July 1952, 47.
Mellberg, Bill. "The Good, the Bad and the Ugly: Remembering the Boeing Stratocruiser." *Airliners*, Winter 1960, 20–25.
"Military Aircraft 1954, Transports." *Flight*, 25 June 1954, 860.
"Mini Guppy Stirs Alarm." *Star Free Press*, Ventura, California, 01 February 1968.
"Missile Carrier Under Development." *Aviation Daily*, 02 April 1962, 206.
Mitchell, Kent A. "Fairchild Could Have Been Aircraft Designs." *AAHS Journal* 42, no. 4 (Winter 1997): 274–284.
"New Lift for Airships." *Time*, 29 July 1974, 80.
"New York Firm Buys Twin Fair." *Tonawanda News*, 08 March 1962, 15.
Oulahan, Richard, and William Lambert. "The Scandal in the Bahamas." *Life Magazine*, February 03, 1967, 59–74.
Paris, Roy. "Aviation's Amazon." *The Airman* III, no. 7, July 1959, 10–12.
"Plane Dismantled." *Las Vegas Review-Journal*, 27 March 2001, B2.
Platoni, Kara. "Big Idea: Megalifters Prove You're Never Too Fat to Fly." *Air and Space Magazine*, September 2008, 45–49.
"PO Studies XC-99 Use as All-Mail Plane." *Aviation Week*, December 18, 1950, 16–17.
Puntus, Boris, and Konstantin Ubalov, translated by Rauf Elyanbekov. "Myasishchev M-50 'Bounder.'" *International Air Power Review* 4 (Spring 2002): 172–181.
Redmon, Michael. "The Guppy: S.B. Airport Home to One of the Largest Airplanes in the World." *Santa Barbara Independent*, 15 March 2001.
"Saturn's Landing Rights In U.K. May Affect RB. 211 Contract with Lockheed." *Aviation Daily*, October 20, 1970, 270.
Savage, Daren. "Volumetric Air Transports: Part 1—Development of Pregnant and Super Guppies." *AAHS Journal* 53, No. 1 (Spring 2008): 21.
Schoendorf, Walter. "Clipper Good Hope." *Airways*, April 2000, 57–64.
Seideman, Jeff. "Dive to Near Disaster." *Propliner* 110 (Spring 2007): 26–27.
"Seven Indicted in Miami in Arms Smuggling." *New York Times*, June 06, 1959.
Smith, Lt. Col. P.G. "The Day the Super Guppy Blew Her Top." *Air Force Magazine*, April 1971.
"Super Planes Okayed." *San Diego Evening Tribune*, 20 March 1968.
Taylor, Cal. "The Douglas C-132 Logistics Transport and Air Refueler." *AAHS Journal* 50, no. 1 (Spring 2005): 20.
"Tenders Invited for Princesses." *Flight International*, 28 November 1963, 853.
"Think Bigger! New Planes Accommodate Outsize Cargo." *Handling & Shipping*, July 1967, 41–43.
Thomas, Ewart. "Listen, It's the Wail of the Turboprops." *Popular Mechanics*, January 1955, 105–109.
"TIFS to Sell Stock." *Securities and Exchange Commission News Digest* 69–148, 05 August 1969, 2.
"Titan Flies by Guppy." *Flight International*, February 1965.
"Tracor." Company News, *Flight International*, 28 November 1981, 1643.
Tripp, Robert S. "Glorious Guppies, Metamorphous of Monsters." Unpublished paper, 1990.
Tripp, Robert S. "Super Guppy." *Invention and Technology* 17, no. 4 (Spring 2002).
Twiss, Robert L. "Artist's Sketch Shows How 'Guppy' Aircraft Could Accommodate Boeing 747 Sections." *Seattle Times*, February 18, 1968.
"UK's Lucas Nets Tracor Aviation." *Flight International*, 12–18 June 1991.
"Unexcelled Chemical Files for Secondary." *Securities and Exchange Commission News Digest* 62-2-10, 12 February 1965, 1.

"Unexcelled Inc. Proposes Debenture Offering." *Securities and Exchange Commission News Digest* 67–187, 28 September 1967, 2.
"Up to Date N447FT at Bournemouth." *Swingtail, CL-44 Association Newsletter*, July 2015, 2–3.
"U.S. Accuses 12 of Plot to Ferry Guns by Plane." *Tucson Daily Citizen*, May 3, 1959.
"USAF Studies Modified Aircraft to Ship Large Shuttle Payloads." *Aviation Week & Space Technology*, October 15, 1984.
Vicenzi, Ugo. "Wright R 3350TC and Turbo Compound Aero Engines." *Propliner* 89 (Winter 2001): 31–37.
"Where Albatrosses Fly." *Propliner* 110 (Spring 2007): 23.
Williams, Nicholas M. "The Bug Eyed Monster: The Douglas Model 415A, the first Globemaster and DC-7." *Air Enthusiast* 60 (November/December 1995): 40–53.
Williams, Nicholas M. "Globemaster, the Douglas C-74." *AAHS Journal* 25, no. 2 (Summer 1980): 82.
Winkler, Matthias. "Deliverance Reborn." *Propliner* 87 (Summer 2001): 15–16.
Winn, Allan. "Super Send-Off." *Flight International*, 24 December 1997, 24–26.
Wolcott, F. W., chief of aerodynamics, Douglas Aircraft. "Preliminary Performance Results of Study of C-74 Airplane Equipped with Prop Turbine Engines." *Douglas Memorandum*, August 17, 1948.
"World's Largest Land Plane, Latest Flying Boxcar Take to Air; Both P&W Powered." *The Power Plant* IV, no. 10 (December 11, 1947).
"The XC-99." *Air Force, the Official Journal of the Air Force Association* 31, no. 1 (January 1948): 23.
"XC-99 Sets Air Cargo Record." *Aviation Week*, June 02, 1952, 18.

Periodicals and Newspapers

Santa Barbara News Press, 26 May 1974.
The Golden Coast News, various issues.
Los Angeles Times, 16 December 1979. Obituary of John M. "Jack" Conroy.
Los Angeles Times, 22 November 2009. Obituary of William M. Ballon.
The Trentonian, 11 August 2013. Obituary of Henry Brown Murphy.
The Valley News (Van Nuys California), 14 May1970.

Press Releases

Furman, Dottie. "The B-377PG—First Flight," 19 September 1962.
Furman, Dottie. "Conroy Aircraft Announces Plans for Super Cargo Aircraft," 18 December 1970.
Furman, Dottie. "Conroy Files Lawsuit Against Unexcelled," 16 April 1968.
Furman, Dottie. "Flight Crew—Super Guppy," 31 August 1965.
Furman, Dottie. "Flight Testing and Demonstration," 29 October 1962.
NASA. "NASA to Air Mail Saturn Rocket Parts," Release 17-63, 31 May 1963.
NASA. "Pregnant Guppy to Haul Giant Rocket," Release 21-63, 13 June 1963.

Pamphlets

Aero Spacelines Inc., Specializing in Outsize Cargo Airlift. Sales brochure, 1966.

Reports and Registers

Aero Spacelines Inc. Invoice: 6148 to Union de transports Aeriens, 16 November 1979.
Aero Spacelines Inc. Invoice: 6535 to Union de transports Aeriens, 05 September 1980.
Aero Spacelines, Inc. v. United States. 530 F.2d 324 (1976). United States Court of Claims. January 28, 1976.
Airbus Industrie, quote request, A. Hurato and J. Thomas to UTA, 02 September 1988.
Aircraft Specification No. A812, Boeing 377, 377PG, 377MG.
Akros Dynamics Corp, GAO Report, B-141258, Feb.7, 1961.
Apollo "A"/Saturn C-1 Launch Vehicle Systems Report. Saturn Systems Office, 1961.
CAA Accident Investigation Report 1–0080: Pan American World Airways Boeing 377, N1032U. Carolina Brazil, 29 April 1952.
Clark, Frank M. Megalifter U.S. Patent Summary 4,052,025, 04 October 1977.
Conroy, John M. vs. Unexcelled Inc. Superior Court of the State of California for Santa Barbara County, No. 82063 Judgment and Decree, 04 February 1969.
FBI, File 105-90291-48, Dominick Edward Bartone. 19 October 1960.
Hamilton Standard Service Bulletins 177, 193, 273, 302.
An Illustrated Chronology of NASA Marshall Center and MFSC Programs, 1960–1973, Huntsville, Alabama, May 1974.
Improper Activities in Labor, Senate Select Committee Hearings/86th Congress, 25, 26, 29, 30 June 1959.
Incentive Provisions of Saturn V Stage Contracts 79. GAO Report to Congress B-161366, p. 40 of Appendix II, 1970.
Marshall Space Flight Center, Saturn V Progress Office. Saturn V Semiannual Progress Report (July 31–Dec. 31, 1967), 65.
NASA. *Audit of Super Guppy Operations*. Johnson Space Center Report A-JS-88-008, 29 September 1989.
NASA. *Boeing 747 Aircraft with External Cargo Pod*. Vought Corporation Contractor Report 158932, July 1978.
NASA. *Boeing 747 Aircraft with Large External Pod for Transporting Outsize Cargo*. Vought Corporation Contractor Report 159067, May 1979.
NASA. *Effects of a Military Cargo Pod and Tail Fins on the Aerodynamic Characteristics of a Large Wide-Body Transport Model*. NASA Technical Memorandum 80052, February 1979.
NASA. *Outsized Booster Carrier Aircraft*. NASA Contractor Report R534-001, 16 March 1964.
NASA, *Study of an Advanced Transport Airplane Design Concept Known as FLATBED*. Lockheed-Martin Contractor Report 159337, October 1980.
NASA. *Transporting Saturn S–IV Stage*. Santa Monica Shipping Report SM-44052, 11 June 1963.
NTSB. ANC69A0049, Crash of N446T.
NTSB. LAX64A0016, Incident N705PA.
NTSB. LAX70AL071, Crash of N111AS.
NTSB. LAX70FUT04, Incident N447T.
Pan American Bank of Miami v. Gully. Civ. A. No. 62-592. United States District Court W. D. Pennsylvania. August 1, 1963.
United States General Accounting Office, letter B-146804 to James C. Fletcher, Administrator NASA, 19 March 1973.
United States of America v Aero Spacelines No 20274, 361F.2d 916 (1966), United States Court of Appeals, Ninth Circuit. May 12, 1966.
Wealthy, Bob, Saunders-Roe and the Princes Flying Boat. Presentation to the Hamburg Branch of the Royal Aeronautical Society, 03 June 2010.

Manuals

Boeing Stratocruiser Pilot's Handbook, Model 377-10-34, United Airlines Inc., 20 February 1950.
C-124 Flight Manual, USAF Series, A&C Aircraft, T.O. 1C-124A -1.
Douglas C-74 (DC-7) *Press Book*.
KC-97G USAF Aircraft Flight Manual T.O. 1C-97 (K) G-1, 15 October 1959.
Pilot's Handbook YC-97 Airplane D-7490, Boeing Aircraft Company, 15 January 1947.
XC-99 Flight Operations Instruction Model 37, Consolidated Vultee Aircraft Corporation.

Logs

Astronautics Year 1964, International Astronautical and Military Space/Missile Review. London: Pergamon Press Ltd, 1965.

Websites

Aero Spacelines Inc. v. United States. http://leagle.com/decision/1976854530F2d324_1802.xml/AERO%20SPACELINES,%20INC.%20v.%20UNITED%20STATES, 2014.
Aero Spacelines 377MG Mini Guppy. http://www.aviastar.org/air/usa/aeroguppy.php, 14 June 2012.
Airbus A300-600ST Beluga. Aerospace-technology.com. http://www.aerospace-technology.com/projects/stbeluga/, 23 December 2013.
All About Guppys. http://www.allaboutguppys.com/, 05 June 2012.
AN-124 Ruslan. http://www.volga-dnepr.com/eng/charter/fleet_development/putting_an124/, 27 Jan 2014.
The AN-225. http://gelio.livejournal.com/193025.html, 11 March 2014.
Atlant, the Special Transport Aircraft. http://englishrussia.com/2013/06/25/atlant-the-special-transport-aircraft/, 25 June 2013.
B377PG Pregnant Guppy. http://www.globalsecurity.org/military/systems/aircraft/b377pg.htm, 30 January 2014.
Big Fella XC-99. http://www.freerepublic.com/focus/f-vetscor/1152267/posts, 09 May 2013.
Boeing C-97 Stratofreighter. http://www.boeing.com/boeing/history/boeing/c97.page, 10 October 2014.
The Boeing 747. http://www.airvectors.net/avb747.html, 07 April 2014.
Boeing 747 Large Cargo Freighter Successfully

Tests Swing Tail. http://www.boeing.com/news/releases/2006/q4/061023b_pr.html, 23 October 2006.
"Boeing Superfreighter Takes Shape in Taiwan." *The Seattle Times*, 15 April 2006, http://seattletimes.nwsource.com/html/businesstechnology/2002931297_boeing15.html, 10 September 2013.
Buran. http://buran.ru/, 27 February 2014.
Buran-Energia. http://www.buran-energia.com/vmt-atlant/vmt-desc.php, 02 March 2014.
C-5C Space Cargo Modified (SCM) Galaxy. http://www.globalsecurity.org/military/systems/aircraft/c-5c.htm, 14 March 2012.
C-124 Globemaster II. http://www.kamov.net/american-aircraft/c-124-globemaster-ii/, 26 March 2012.
C-124 Globemaster II. http://www.globalsecurity.org/military/systems/aircraft/c-124.htm, 19 February 2011.
C-74 Globemaster. http://www.globalsecurity.org/military/systems/aircraft/c-74.htm, 31 December 2013.
The Congo Airlift. http://amcmuseum.org/history/airlifts/congo_airlift.php, 12 February 2013.
The Douglas A/B-26 Invader, Everything You Ever Wanted to Know About on Mark Engineering by Richard E. Fulwiler. http://napoleon130.tripod.com/id589.html, 11 June 2012.
Clay Lacy: The Planes I've Flown and People I've Known. http://www.airportjournals.com/display.cfm?varid=0308001, 17 June 2012.
Edwards Air Force Base, CA, Cargo Plane Crash, May 1970. http://www3.gendisasters.com/california/11617/edwards-air-force-base-ca-cargo-plane-crash-may-1970, 28 February 2009.
Douglas C-74 Globemaster. www.theaviationzone.com/factsheets/c74/.asp, 2008.
Dryden Flight Research Center. http://www.dfrc.nasa.gov/Gallery/Photo/Guppy/HTML/EC05-0091-47.html, April 2013.
Elder, Donald L. "Douglas Aerodynamics Project Engineer Comments on C-124." http://www.aero-web.org/specs/douglas/yc-124b.htm [ref: 35206], 05 January 2011.
Evergreen Aviation Technologies Corporation. http://en.wikipedia.org/wiki/Evergreen_Group#Evergreen_Aviation_Technologies_Corporation, 18 April 2014.
Goleta Air and Space. http://www.air-and-space.com/b-36%20variants.htm, 12 April 2012.
http://history.nasa.gov/SP-4206/notes.htm#10.
Joe Bauger Home Page. http://www.joebaugher.com/usaf_serials/1968.html, 14 October 2014.
M-52A Myasishchev Heavy Transporter Project. http://www.ussr-airspace.com/index.php?main_page=document_general_info&cPath=28_39_38_108&products_id=707, 25 February 2014.
The Making of a Flight Engineer. http://www.enginehistory.org/the_making_of_a_flight_engineer_(3).shtml, 12 April 2012.
Millville Ordnance Company. www.m1carbinesinc.com/carbine_millville.html, 27 May 2012.
The Mystery of the Lost Clipper. http://www.airspacemag.com/history-of-flight/clipper.html, 28 September 2012.
Odlum, Floyd B.: Papers, 1892–1976. www.eisenhower.archives.gov/research/finding_aids/pdf/Odlum_Floyd_Papers.pdf, 08 December 2016.
On Mark Marksman. http://en.wikipedia.org/wiki/On_Mark_Executive, 25 January 2014.
Prototype Boeing B-29 Crashes into Seattle's Frye Packing Plant on February 18, 1943. HistoryLink.org Essay 2874, http://www.historylink.org/index.cfm?displaypage=output.cfm&file_id=2874, 03 March 2016.
Russian Companies Design Space Tour Plane. http://en.ria.ru/science/20120830/175523145.html, 30 August 2012.
Saunders Roe Princes 60th Anniversary. http://www.globalaviationresource.com/reports/2012/aviation-events-saunders-roe-princess-60th/, 26 September 2014.
South Pole Station: The First 10 Years. http://www.southpolestation.com/trivia/igy1/igy1.html, 15 September 2016.
Speedbird Strats. http://www.ovi.ch/b377/articles/speedbird/, 14 March 2014.
747 Large Cargo Freighter (LCF) Dreamlifter. http://www.boeing-747.com/special_boeing_747s/747_large_cargo_freighter_l.php, 01 October 2013.
The Skytamer Archive. http://www.skytamer.com/Aero_Spacelines_B-377MG.html, 15 Dec 2013.
U.S. Air Force Fact Sheet Boeing HC-97. http://www.scribd.com/doc/60154049/HC-97G-Factsheet, 15 May 2011.
U.S. Department of Justice, FBI Investigation. NARA Record Number 124-10283-10237. www.maryferrell.org/mffweb/archive/viewer/showDoc.do?docId=79219&relPageId=16, 27 May 2012.
United Aircraft Corporation. http://www.uacrussia.ru/en/, 02 March 2014.
Weintz, Steve. Her Majesty's Nuclear Seaplane. https://warisboring.com/her-majestys-nuclear-seaplane-7043b94b09aa#.7jd5xlh9u, 28 May 2106.
World's Largest Aircraft: Antonov AN-225 Mriya. http://sometimes-interesting.com/2012/01/19/worlds-largest-aircraft-antonov-an-225-mriya/, 05 May 2014.
www.MilitaryFactory.com, 2003–2009.

Interviews

Bodenmiller, Edward J. USAF 28th ATS, Hill AFB, 1962–66.
Bourgeois, Shy. Spouse of ASI photographer Clyde Bourgeois.
Carlisle, Joel. Conroy Aviation. Telephone interview, 14 August 2015.
Conroy, Angelee. Interview, 13–15 July 2015.
Conroy, Milbrey. Widow of Jack Conroy, conversation, July 2014.
Hammond, Seth. Aero Spacelines, sheet metal specialist, conversation, July 2014.
Harridge, Robert. Aero Spacelines, electrician and flight crew member, 12 October 2015.
Jones, Lloyd S. Aero Spacelines, model builder-photographer.
Kirby, Bob. Aero Spacelines, Conroy Aircraft, conversation July 2014.
Lacy, Clay. Interview, 15 July 2015.
Lawton, Timothy. City of Santa Barbara Airport Department, interview, July 2014.
Moss, Frank. Owner, Honduras Caribbean (Hondu Carib) and N103, interviews, 2007.
Parish, Harry. Ogden-Allied, interviews, 2015.
Roberts, Tom. Aero Spacelines, fabricator and logistics, 12 October 2015.
Sanders, Ken. NASA, N941 flight crew, telephone interview, 04 April 2016.
Smothermon, Tom. Custodian, Aero Spacelines files.
Stratman, Donald. Aero Spacelines, A&P, telephone interview.
Stratton, Bruce. Aero Spacelines, captain, conversation, July 2014.

Letters and Email

Babcock, Jim. Aero Union, captain, emails, 2014–15.
Ben-Aziz, B. Aeronautics R&D, letter to Frank M. Clark, 12 December 1974.
Boeing Defense & Space Group. Letter to Clark Aircraft Company, 05 November 1990.
Brooks, R. L. Rockwell International, letter to Dr. Nigel Pridmore Brown, Clark Aviation, 16 June 1986.
Brooks, R. L. Material director, Rockwell International, letter to Frank Clark, 16 June 1986.
Bugbee, James. Letter to William Patrick Dean, June 2014.
Calhoun, R. J. Traffic manager, Fairchild Hiller, letter to Jack Conroy, 25 November 1966.
Clark, Frank M. Letter to William Patrick Dean, 14 October 2014.
Clark, Frank M. Email to William Patrick Dean, 08 April 2015.
Clark, Frank M. Letter to Lyle Eveland, Boeing Aircraft, 13 May 1997.
Clark, Frank M. Letter to Sergei K. Gromov, Energiya Scientific Industrial Complex, Moscow, 04 June 1999.
Clark, Frank M. Letter to Tom Smothermon, January 2012.
Clark, Steve. Letter to William Patrick Dean. Excerpts from Frank Clark files, 20 September 2015.
Collins, Michael. Director, National Air and Space Museum, letter to Frank Clark re: Hughes HK-1 Wing, 25 March 1975.
Conroy, John M. Aero Spacelines Inc., letter to Wernher von Braun, 29 October 1962.
Conroy, John M. Letters to Clifford T. Moore, City of Los Angeles, 08 and 10 May 1967.
Conroy, Angelee. Regarding Allen Manus, email, 09 November 2016.
Conroy, Milbrey. Widow of Jack Conroy, emails, 08 August 2015 and 17 April 2016.
Evans, Keith. Son of Geoffery Evans, Channel Air Bridge crew planning Southend, email 09 February 2016.
Fagan, Wayne. Dee Howard Foundation, emails August 2015.
Ferguson, John. Conroy Aircraft, emails, 2014.
Gillespie, Charles J. NASA Flight Operations, email, 01 December 2015.
Gillette, Walter B. Boeing senior vice president, email to Frank Clark, 23 September 2004.
Glenn, Larry J. NASA Flight Operations, emails, December 2015.
Goldwater, Senator Barry. Letter to Frank M. Clark regarding Megalifter, 23 April 1974.
Goodrum, John C. NASA logistics, letter to John Conroy, 06 February 1967.
Gorman, Harry H. Deputy director for admin-

istration, NASA, letter to John Conroy, 13 November 1962.
Hunt, Ted. Air Systems/Aeronaves de Panama, captain, emails.
Kaplan, A. M. Chief engineer, Strato Engineering, letter to Jack Conroy, 24 October 1962.
Kraft, J.F. Martin-Marietta, letter to Jack Conroy, 02 December 1966.
Reid, John G. Northrop traffic manager, letter to John M. Conroy, 23 November 1966.
Rivers, Robert. NASA ops division instructor pilot, email, 22 December 2015.
Sanders, Kenneth. NASA Flight Operations, emails, January 2016.
Sleminski, Denis. Boeing Logistics Services, letter to Frank Clark, 15 December 1997.
Sutherland, Michael. 747/767 Boeing Wichita Division, letter to Frank Clark, 08 September 1997.
Tilson, Ritchie W. Boeing divisional director of material, letter to Jack Conroy, 16 December 1966.
Weir, Fred. Aero Spacelines, electrician, emails, 2014–15.
Winch, John B., Boeing Aerospace, letter to Frank Clark, 18 February 1975.
Yardely, John F., associate administrator, Manned Space Flight, NASA, letter to Frank Clark, 23 September 1975.

Airlines, Museums, Historical Societies and Associations

Air Britain
Alabama Space and Rocket Center
American Aviation Historical Society
City of Santa Barbara Airport Department Visitors Center
Pima Air Museum
Santa Barbara Historical Society

Index